DIGITAL NETWORKS

Prentice-Hall
Series in Automatic Computation

AHO, ed., *Currents in the Theory of Computing*
AHO AND ULLMAN, *The Theory of Parsing, Translation, and Compiling,* Volume I: *Parsing*; Volume II: *Compiling*
ANDREE, *Computer Programming: Techniques, Analysis, and Mathematics*
ANSELONE, *Collectively Compact Operator Approximation Theory and Applications to Integral Equations*
BATES AND DOUGLAS, *Programming Language/One*, 2nd ed.
BLUMENTHAL, *Management Information Systems*
BRENT, *Algorithms for Minimization without Derivatives*
BRINCH HANSEN, *Operating System Principles*
BRZOZOWSKI AND YOELI, *Digital Networks*
COFFMAN AND DENNING, *Operating Systems Theory*
CRESS, et al., *FORTRAN IV with WATFOR and WATFIV*
DAHLQUIST, BJORCK, AND ANDERSON, *Numerical Methods*
DANIEL, *The Approximate Minimization of Functionals*
DEO, *Graph Theory with Applications to Engineering and Computer Science*
DESMONDE, *Computers and Their Uses*, 2nd ed.
DRUMMOND, *Evaluation and Measurement Techniques for Digital Computer Systems*
ECKHOUSE, *Minicomputer Systems: Organization and Programming (PDP-11)*
FIKE, *Computer Evaluation of Mathematical Functions*
FIKE, *PL/I for Scientific Programmers*
FORSYTHE AND MOLER, *Computer Solution of Linear Algebraic Systems*
GEAR, *Numerical Initial Value Problems in Ordinary Differential Equations*
GORDON, *System Simulation*
GRISWOLD, *String and List Processing in SNOBOL4: Techniques and Applications*
HANSEN, *A Table of Series and Products*
HARTMANIS AND STEARNS, *Algebraic Structure Theory of Sequential Machines*
JACOBY, et al., *Iterative Methods for Nonlinear Optimization Problems*
JOHNSON, *System Structure in Data, Programs, and Computers*
KIVIAT, et al., *The SIMSCRIPT II Programming Language*
LAWSON AND HANSON, *Solving Least Squares Problems*
LORIN, *Parallelism in Hardware and Software: Real and Apparent Concurrency*
LOUDEN AND LEDIN, *Programming the IBM 1130*, 2nd ed.
MARTIN, *Computer Data-Base Organization*
MARTIN, *Design of Man-Computer Dialogues*
MARTIN, *Design of Real-Time Computer Systems*
MARTIN, *Future Developments in Telecommunications*
MARTIN, *Programming Real-Time Computing Systems*
MARTIN, *Security, Accuracy, and Privacy in Computer Systems*
MARTIN, *Systems Analysis for Data Transmission*
MARTIN, *Telecommunications and the Computer*

MARTIN, *Teleprocessing Network Organization*
MARTIN AND NORMAN, *The Computerized Society*
MCKEEMAN, et al., *A Compiler Generator*
MYERS, *Time-Sharing Computation in the Social Sciences*
MINSKY, *Computation: Finite and Infinite Machines*
NIEVERGELT, et al., *Computer Approaches to Mathematical Problems*
PLANE AND MCMILLAN, *Discrete Optimization: Integer Programming and Network Analysis for Management Decisions*
POLIVKA AND PAKIN, *APL: The Language and Its Usage*
PRITSKER AND KIVIAT, *Simulation with GASP II: A FORTRAN-based Simulation Language*
PYLYSHYN, ed., *Perspectives on the Computer Revolution*
RICH, *Internal Sorting Methods Illustrated with PL/I Programs*
SACKMAN AND CITRENBAUM, eds., *On-Line Planning: Towards Creative Problem-Solving*
SALTON, ed., *The SMART Retrieval System: Experiments in Automatic Document Processing*
SAMMET, *Programming Languages: History and Fundamentals*
SCHAEFER, *A Mathematical Theory of Global Program Optimization*
SCHULTZ, *Spline Analysis*
SCHWARZ, et al., *Numerical Analysis of Symmetric Matrices*
SHAH, *Engineering Simulation Using Small Scientific Computers*
SHAW, *The Logical Design of Operating Systems*
SHERMAN, *Techniques in Computer Programming*
SIMON AND SIKLOSSY, eds., *Representation and Meaning: Experiments with Information Processing Systems*
STERBENZ, *Floating-Point Computation*
STOUTEMYER, *PL/1 Programming for Engineering and Science*
STRANG AND FIX, *An Analysis of the Finite Element Method*
STROUD, *Approximate Calculation of Multiple Integrals*
TANENBAUM, *Structured Computer Organization*
TAVISS, ed., *The Computer Impact*
UHR, *Pattern Recognition, Learning, and Thought: Computer-Programmed Models of Higher Mental Processes*
VAN TASSEL, *Computer Security Management*
VARGA, *Matrix Iterative Analysis*
WAITE, *Implementing Software for Non-Numeric Application*
WILKINSON, *Rounding Errors in Algebraic Processes*
WIRTH, *Algorithms + Data Structures = Programs*
WIRTH, *Systematic Programming: An Introduction*
YEH, ed., *Applied Computation Theory: Analysis, Design, Modeling*

DIGITAL NETWORKS

JANUSZ A. BRZOZOWSKI

University of Waterloo

MICHAEL YOELI

Technion—Israel Institute of Technology

PRENTICE-HALL, INC.

ENGLEWOOD CLIFFS, N.J.

Library of Congress Cataloging in Publication Data

BRZOZOWSKI, J A
Digital networks.

(Automatic computations)
Bibliography: p.
Includes index.
1. Digital electronics. I. Yoeli, M. (date)-
joint author. II. Title.
TK7868.D5B78 621.3815'3 75-17966
ISBN 0-13-214189-2

10 9 8 7 6 5 4 3 2

Printed in the United States of America

PRENTICE-HALL INTERNATIONAL, INC., *London*
PRENTICE-HALL OF AUSTRALIA, PTY. LTD., *Sydney*
PRENTICE-HALL OF CANADA, LTD., *Toronto*
PRENTICE-HALL OF INDIA PRIVATE LIMITED, *New Delhi*
PRENTICE-HALL OF JAPAN, INC., *Tokyo*
PRENTICE-HALL OF SOUTHEAST ASIA (PTE.) LTD., *Singapore*

To Grażyna and Nehama

CONTENTS

APPENDICES

PREFACE

Digital network engineering has developed very rapidly in recent years. Its impact is felt in many areas, especially those of digital computers, telephone switching, and digital control. The growth of digital systems engineering has been accompanied by the birth of new technologies, such as integrated circuits, large-scale integration, and MOS devices. This book is intended to be a realistic, up-to-date introduction to the theory and applications of digital networks.

For the application-oriented reader we provide a comprehensive and mathematically precise treatment of commercially available integrated-circuit modules, which have become the building blocks of modern digital systems. We introduce new mathematical models to properly explain the behavior of complex flip-flops (including master–slave and edge-sensitive types); and static MOS devices. We also make an effort toward establishing a modular design approach for large digital systems. Our presentation is at the logical design level and does not rely on electrical engineering prerequisites.

For the more theoretically inclined reader we provide mathematically nontrivial topics which are at the same time realistic and applicable. A considerable amount of the material in this book represents recent, unpublished research results. The reader will come across several problem areas that require further investigation.

Conventional switching theory originated with relay technology, but it did not keep up to date with the rapid developments of new technologies. Rather, it deals mainly with mathematical problems that presently have little relevance to actual practice in digital network design. On the other hand, a number of texts addressed to the practicing technician and engineer are not on the proper mathematical level for university undergraduates. In contrast to both these trends, we have strived for both relevance and mathematical maturity.

This book is addressed to more than one type of reader. Most of the material is suitable for an undergraduate, junior-level course for computer science and electrical engineering students. For example, the topics of Chapters 1, 2, 4 (unstarred part), 5, 6, 7, and 8 have been covered in a one-semester course for third-year computer science students at the University of Waterloo. Alternative sequences are also possible at the undergraduate level, for example Chapters 1, 2, 4 (unstarred part), 6, 7, 8, 9 (possibly with some parts omitted), and 10 (unstarred part). The first part of each chapter gives an indication of prerequisite material and points out sections that can be omitted. For a graduate course in computer science or electrical engineering, Chapters 3 through 10 constitute a suitable sequence.

Appendices A–C, together with Chapters 3 and 4, provide a good deal of general mathematical background required by computer science students.

We wish to express our gratitude for helpful comments, reactions, and criticisms of early versions of this book to the following: students at the University of Waterloo, especially G. Bouwers, T. A. Cargill, D. R. Cheriton, and M. G. Gouda; Dr. R. Knast; Professors A. Bar-Lev, I. Kidron, W. D. Little, and J. C. Majithia; and several anonymous, very helpful reviewers.

Special thanks are due to M. I. Irland, who shared the experience of teaching from our manuscripts, for constructive criticism and for contributing to the problem sections.

Both authors wish to gratefully acknowledge the assistance of both the University of Waterloo and the Technion—Israel Institute of Technology.

The research for this book was supported in part by the National Research Council of Canada under Grant A-1617, and by the Technion, under Grant 120-509.

We are greatly indebted to Teresa Miao of the University of Waterloo for her excellent typing of the various stages of the manuscript, and to Raya Anavi of the Technion, for typing extensive revisions.

J. A. BRZOZOWSKI
M. YOELI

DIGITAL NETWORKS

1 SWITCHES AND GATES

ABOUT THIS CHAPTER

This chapter introduces switches, gates, and relatively simple gate networks. The approach is largely intuitive, yet it is made quite precise through the use of the notation of propositional calculus, given in Section 1.1. For the mathematically oriented reader, we present in Appendix A additional material on propositional calculus.

In Section 1.2 we begin with switches, because their implementation and operation are extremely simple conceptually. In Section 1.3 we introduce ideal gates without attempting to explain their numerous and widely differing implementations. A mathematical model of one type of implementation, MOS gates, is presented in Appendix D; however, we consider electronic details outside the scope of our book and refer the interested reader to the literature [GAR]. Thus we restrict our attention to logical properties of gates and we consider gates as operators on binary signals.

In Section 1.4, Boolean operators are defined and some of their properties are described; many of these properties are needed in Section 1.6. In Section 1.5, some fundamental concepts related to networks of gates are discussed.

With this rather modest background we next proceed to the design examples of Section 1.6. With the aid of the propositional-calculus notation, we develop an approach to the design of some interesting networks. Word descriptions of problems are first converted to precise mathematical statements and then manipulated to obtain gate implementations. For the most part we use ideal gates; however, certain restrictions are introduced in some of the examples. For instance, we attempt to use as few gates or contacts as possible. In this connection, the reader is warned that minimal solutions are quite difficult to find for some of the examples and problems. Such problems constitute intellectually challenging puzzles, but, otherwise, the reader should not attach too much importance to absolutely minimal solutions. In general, no systematic methods are known for finding such minimal networks. The subject is treated in more detail in Chapter 5.

To the best of our knowledge, the approach of Section 1.6 is original, although countless designers have undoubtedly gone through similar steps informally.

1.1. NOTATION

Throughout this book we frequently use notation from the calculus of propositions. We now summarize this notation. For a more detailed account of the propositional calculus, the reader is referred to Appendix A.

A *proposition* is a statement that is either true or false. If p and q are propositions, the proposition $p \wedge q$ (read: p AND q) is defined to be true iff (if and only if) both p *and* q are true. The proposition $p \vee q$ (read: p OR q) is true iff either p *or* q *or* both p and q are true. The proposition $\sim p$ (read: NOT p) is true iff p is *not* true, i.e., iff p is false. We write LHS $\equiv$ RHS to state that the left-hand side and right-hand side are *equivalent*; i.e., they are either both true or are both false. For example, $x > y \equiv y < x$. Another example of an equivalence is the following:

$$p \vee q \equiv p \vee ((\sim p) \wedge q), \qquad (*)$$

where p and q denote arbitrary propositions. This and similar equivalences can be verified by means of a *truth table*, which exhausts all the possibilities for p and q, as shown in Table 1-1, where T and F stand for *true* and *false*, respectively.

Table 1-1 Truth-Table Verification of Equivalence (*)

p	q	$p \vee q$	$\sim p$	$(\sim p) \wedge q$	$p \vee ((\sim p) \wedge q)$
F	F	F	T	F	F
F	T	T	T	T	T
T	F	T	F	F	T
T	T	T	F	F	T

We refer to the symbols $\wedge$, $\vee$, and $\sim$ as *logical connectives*.

For convenience in such formulas as $((f \vee g) \vee h)$, we omit the outer parentheses and write $(f \vee g) \vee h$. Also, the $\sim$ operator has precedence over the $\vee$ and $\wedge$ operators. Thus $p \vee ((\sim p) \wedge q)$ will be written $p \vee (\sim p \wedge q)$.

1.2. SWITCHES

Simple Switches

A simple and common example of a switching device is a manually operated mechanical *switch*, e.g., a light switch or a push button. Associated with a simple switch are two terminals, t_1 and t_2, which are connected (short-circuited) when the switch is operated, and disconnected (open-circuited) when it is unoperated. A schematic diagram of a simple switch is shown in Fig. 1-1(a). At any time the switch is either operated or unoperated; consequently, its behavior can be described by a *binary*, i.e., two-valued, variable s.

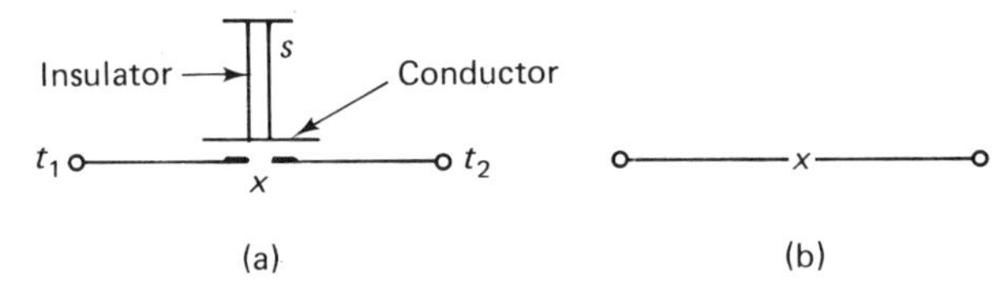

Fig. 1-1 (a) Schematic diagram of a simple switch; (b) symbolic diagram.

We use the convention that $s = 0$ when the switch is unoperated and $s = 1$ when it is operated. The choice of symbols 0 and 1 is arbitrary and the symbols have no numerical significance. Any other pair of distinct symbols would be equally acceptable. For the terminals t_1 and t_2 we introduce a binary variable x such that $x = 0$ if the path between t_1 and t_2 is open, and $x = 1$ if it is closed. Thus the performance of the switch of Fig. 1-1(a) can be specified by the statement $x = 1 \equiv s = 1$, or simply by $x = s$. To simplify the representation of switches we use the symbolic diagram of Fig. 1-1(b) rather than the schematic drawing.

A slightly more complicated switch is shown in Fig. 1-2, where two pairs

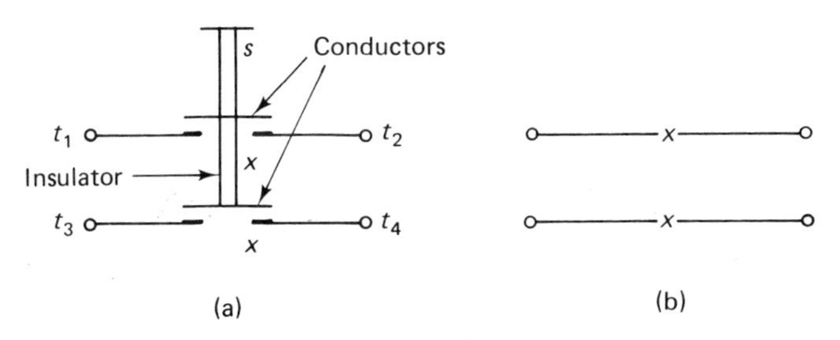

Fig. 1-2 Illustrating multiple contacts.

of terminals are closed and opened by the same motion. Since the paths (t_1, t_2) and (t_3, t_4) will be either both closed or both open, the same variable x is used to describe the state of these two paths. In general, we shall assume that any number of contacts can be operated simultaneously by one switch, and all such contacts operated by a single switch will be labeled with the same variable.

A different switch with multiple contacts is shown in Fig. 1-3. As before, both paths (t_1, t_2) and (t_3, t_4) are affected by the single motion of the switch. However, when the switch is unoperated ($s = 0$), the top path is open ($x = 0$) and the bottom path is closed ($y = 1$). When the switch is operated ($s = 1$), $x = 1$ and $y = 0$. The performance of this switch can be specified by

$$x = 1 \equiv s = 1$$

$$y = 1 \equiv s = 0 \equiv \sim(s = 1).$$

We use the notation $\triangleq$ to mean "is defined as."

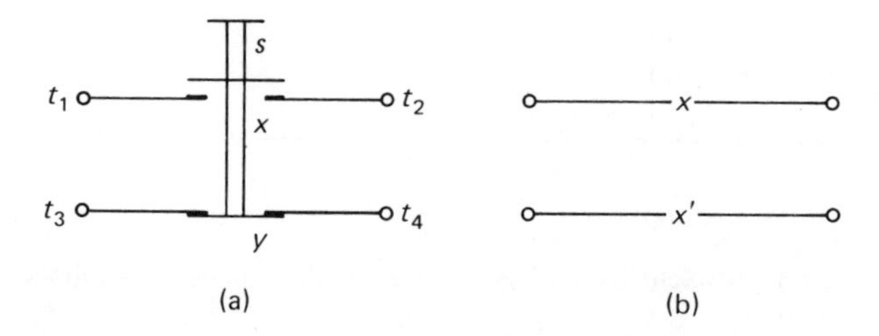

Fig. 1-3 Illustrating complementary contacts.

Let x' denote the *complement* of x, where $0' \triangleq 1$ and $1' \triangleq 0$. Then the performance of the switch in Fig. 1-3 can also be described by the equations $s = x$ and $s' = y$, or simply by $y = x'$. This relationship is represented by the symbolic diagram of Fig. 1-3(b). Contacts that are open (closed) when the switch is unoperated are called *normally open* or *n.o.* (*normally closed* or *n.c.*) contacts.

Note that, if a switch resembling the schematic diagram of Fig. 1-3(a) were constructed, it would be possible to depress the vertical rod partially in such a way that both paths (t_1, t_2) and (t_3, t_4) would be open. In our idealized model, such situations are not permitted. We assume that a single switch can have any number of n.o. and n.c. contacts. The switch is either unoperated (all n.o. contacts are open and all n.c. contacts are closed) or it is operated (all n.o. contacts are closed and all n.c. contacts are open).

A switch that is operated by an electromagnet is called a *relay*.

Contact Networks

More complicated behavior can be obtained by connecting contacts together to form contact networks. Figure 1-4 shows two connections of two

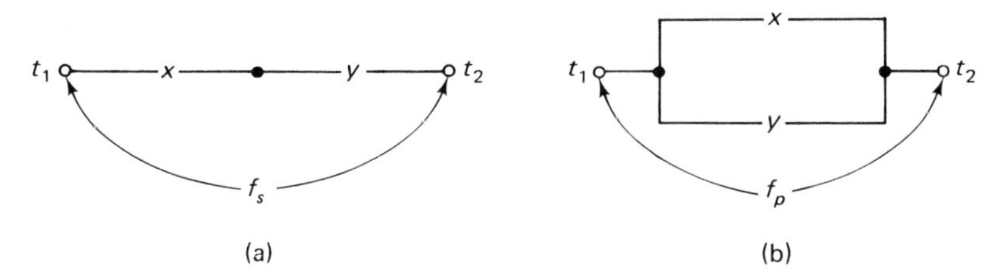

Fig. 1-4 Series and parallel connections of contacts.

contacts x and y operated by independent switches. We assign a binary variable f_s to represent the state of the path (t_1, t_2) in the *series* connection, and a binary variable f_p in the *parallel* connection. The variables x and y are the *independent* or *input* variables, whereas the variables f_s and f_p are *dependent* or *output* variables.

The behavior of the networks of Fig. 1-4 can be described in words as follows: The series circuit is closed iff both x *and* y are closed, and the parallel circuit is closed iff either x *or* y *or* both are closed. Applying the notation introduced earlier, we have

$$f_s = 1 \equiv x = 1 \wedge y = 1$$
$$f_p = 1 \equiv x = 1 \vee y = 1.$$

In many applications we are not interested in the internal structure of a network but only in its input–output behavior. Different structures may realize the same behavior. This is illustrated in Fig. 1-5. In Fig. 1-5(a)

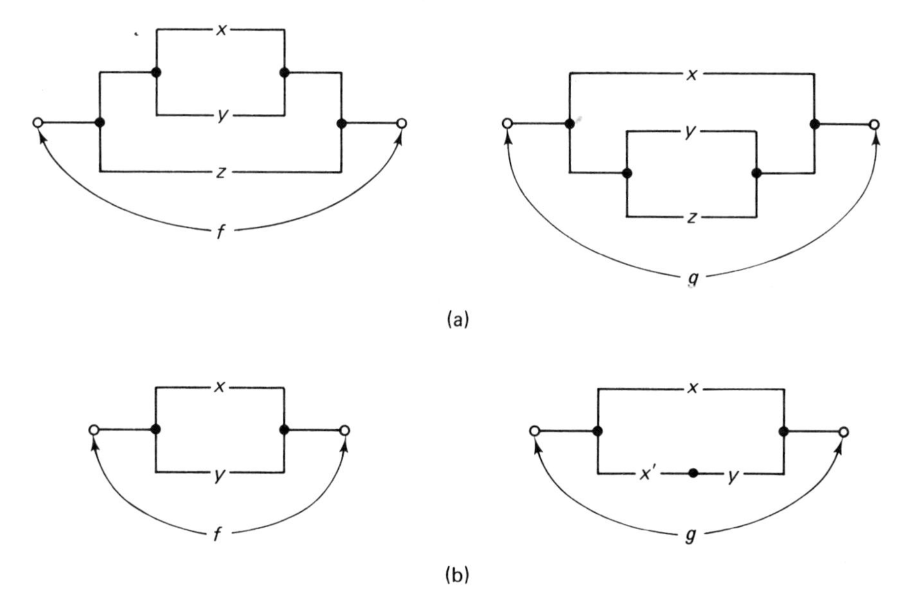

Fig. 1-5 Illustrating equivalent contact networks.

we have

$$f = 1 \equiv (x = 1 \lor y = 1) \lor z = 1$$
$$g = 1 \equiv x = 1 \lor (y = 1 \lor z = 1).$$

One easily verifies that the two right-hand sides are equivalent, and consequently $f = 1 \equiv g = 1$; i.e., both structures have the same behavior. Similarly, we have, for Fig. 1-5(b),

$$f = 1 \equiv x = 1 \lor y = 1$$
$$g = 1 \equiv x = 1 \lor (x = 0 \land y = 1) \equiv x = 1 \lor (\sim(x = 1) \land y = 1).$$

Now $p \lor q \equiv p \lor (\sim p \land q)$ for all propositions p and q, as we have verified earlier, by means of truth tables. Again it follows that $f = 1 \equiv g = 1$.

We point out that we have used simple switches only as a means of introducing switching networks. In practice, electronic gates are used in place of mechanical switches. The basic concepts of such gates are discussed next.

1.3. GATES

Electronic computers and other digital systems use electronic switching devices called *gates*. A general symbol for a 2-input, 1-output gate is shown in Fig. 1-6. Each of the two inputs x and y, as well as the output z, is capable

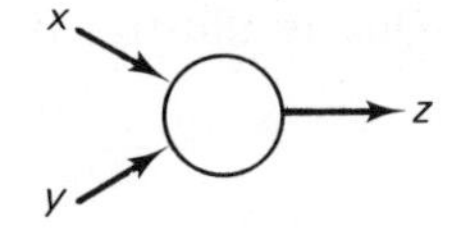

Fig. 1-6 2-input gate symbol.

of being in either one of two different states. Most frequently these states are two well-separated voltage ranges. By convention, we associate the symbols 0 and 1 with the lower and higher voltage ranges, respectively. This convention is referred to as *positive logic*, whereas the opposite convention is called *negative logic*. There are many physical realizations of such gates, using transistors as the basic device [GAR]. However, the discussion of the electronic circuits is beyond our scope. We treat gates as abstract mathematical models in which the inputs and outputs are capable of assuming the values 0 and 1. A 2-input AND gate, for example, may be specified by the equivalence

$$z = 1 \equiv x = 1 \land y = 1.$$

A list of basic gate types together with their characteristic equivalences is given in Table 1-2. The table also gives some frequently used alternative symbols for these gate types.

Gates may be composed to form gate networks as shown in Fig. 1-7. Gate

Table 1-2 Basic Gate Types

Gate Type	Gate Symbol	Alternate Symbol	Characteristic Equivalence
INVERTER	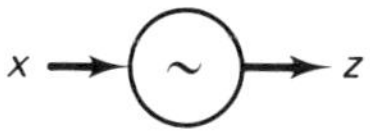	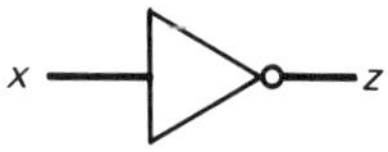	$z = 1 \equiv \sim (x = 1)$
AND	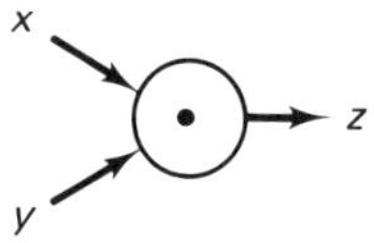	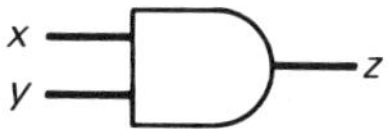	$z = 1 \equiv x = 1 \wedge y = 1$
OR	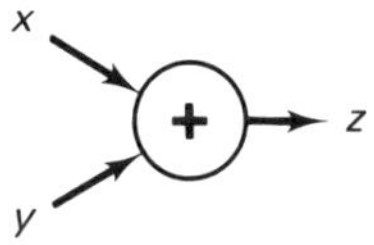	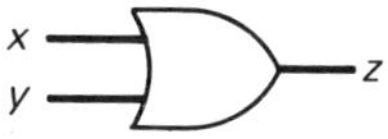	$z = 1 \equiv x = 1 \vee y = 1$
XOR	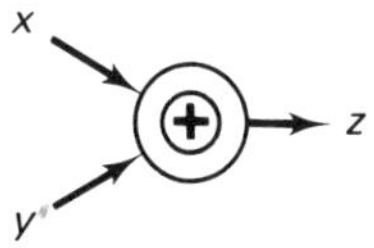	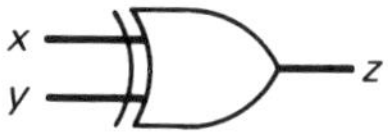	$z = 1 \equiv x \neq y$
EQUIV	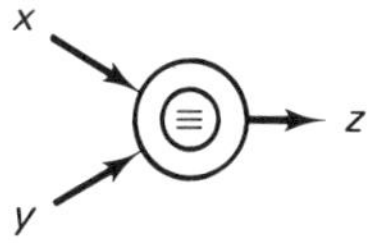	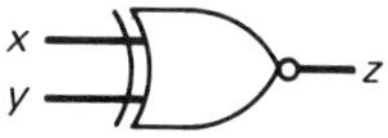	$z = 1 \equiv x = y$
NAND	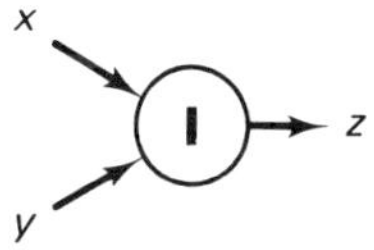	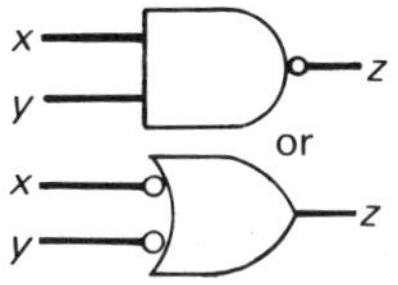 or	$z = 1 \equiv \sim (x = 1 \wedge y = 1)$
NOR	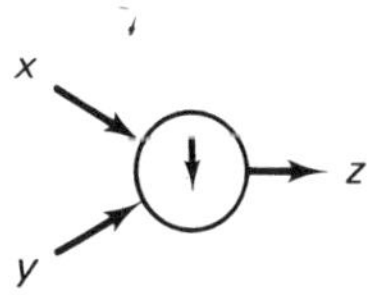	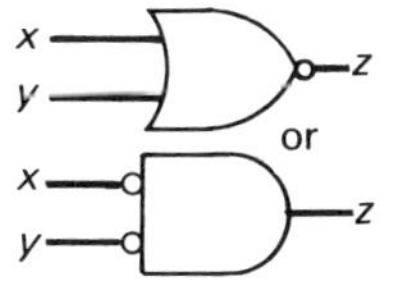 or	$z = 1 \equiv \sim (x = 1 \vee y = 1)$

networks with the same input–output behavior are *equivalent*. One easily verifies that the two gate networks in Fig. 1-7(a) are equivalent, as are the gate networks in Fig. 1-7(b). Indeed, Fig. 1-7 corresponds in an obvious way to Fig. 1-5, and the same arguments apply.

In the next section we introduce a more conventional algebraic approach to the study of switching networks.

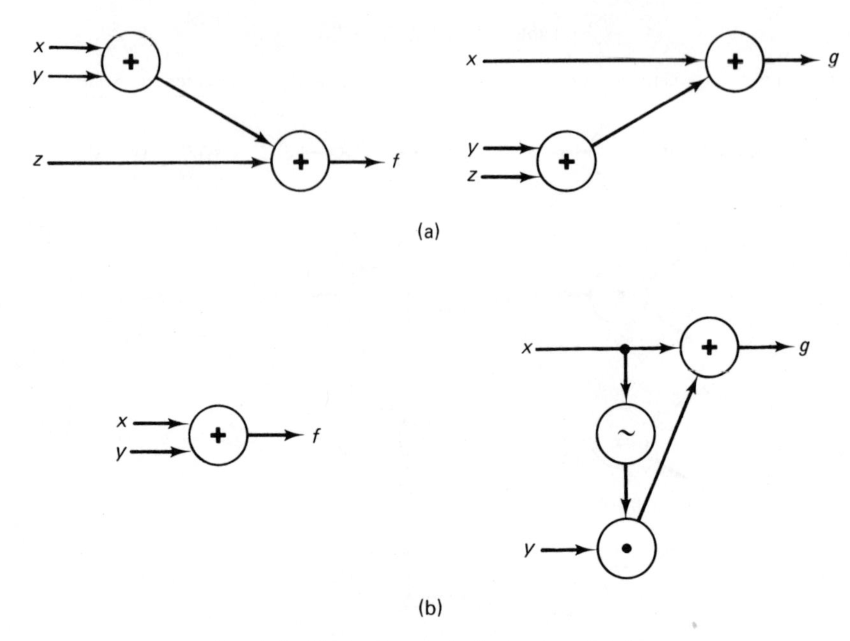

Fig. 1-7 Illustrating equivalent gate networks.

1.4. BOOLEAN OPERATORS

It is possible to proceed with the development of the theory of switching networks by using propositional calculus. However, it is more conventional to develop an algebraic approach. Such an approach is notationally more convenient; moreover, it leads to some nontrivial applications of Boolean algebra to the theory of switching networks, as will be seen later.

Consider the AND gate introduced in the previous section. As is customary, we have chosen the values of the binary inputs x and y and of the output z to be 0 and 1. If we form a table to show the dependence of z on x and y, we obtain:

x	y	z
0	0	0
0	1	0
1	0	0
1	1	1

We can take advantage of the fact that the operation performed by the AND gate coincides with ordinary multiplication $(\cdot)$ restricted to 0 and 1. Thus the AND gate can be described simply by the equation $z = x \cdot y$ rather than by the equivalence

$$z = 1 \equiv x = 1 \wedge y = 1.$$

Similarly, the notation for the operation performed by the OR gate is simplified by writing $z = x + y$. The $(+)$ is called *Boolean addition* and is defined by the following table:

x	y	z
0	0	0
0	1	1
1	0	1
1	1	1

Unfortunately, Boolean addition corresponds to ordinary addition restricted to 0 and 1 *only in three of the four cases.* Nevertheless, the symbol $+$ is widely adopted for the OR operation.

Table 1-3 Table Defining Boolean Operators

x	x'
0	1
1	0

x	y	$x \cdot y$	$x + y$	$x \oplus y$	$x \circleddash y$	$x \mid y$	$x \downarrow y$
0	0	0	0	0	1	1	1
0	1	0	1	1	0	1	0
1	0	0	1	1	0	1	0
1	1	1	1	0	1	0	0

In Table 1-3 we define several common Boolean operators. These operators are closely related to the logical connectives. More precisely, we have the equivalences listed in Table 1-4.

Table 1-4 Relating Boolean Operators to Logical Connectives

1. $x' = 1 \equiv \sim(x = 1)$
2. $x \cdot y = 1 \equiv x = 1 \wedge y = 1$
3. $x + y = 1 \equiv x = 1 \vee y = 1$
4. $x \oplus y = 1 \equiv (x = 1 \wedge y = 0) \vee (x = 0 \wedge y = 1)$
5. $x \circleddash y = 1 \equiv (x = 1 \wedge y = 1) \vee (x = 0 \wedge y = 0)$
6. $x \mid y = 1 \equiv \sim(x = 1 \wedge y = 1)$
7. $x \downarrow y = 1 \equiv \sim(x = 1 \vee y = 1)$

Various properties of the Boolean operators $+$, $\cdot$, and $'$ are listed in Table 1-5.

Table 1-5 Properties of Boolean Operators $\cdot$, $+$, and $'$

B1. $x + x = x$	B1$'$. $x \cdot x = x$	(idempotent laws)
B2. $x + y = y + x$	B2$'$. $x \cdot y = y \cdot x$	(commutative laws)
B3. $x + (y + z) = (x + y) + z$	B3$'$. $x \cdot (y \cdot z) = (x \cdot y) \cdot z$	(associative laws)
B4. $x + (x \cdot y) = x$	B4$'$. $x \cdot (x + y) = x$	(absorption laws)
B5. $x + 0 = x$	B5$'$. $x \cdot 1 = x$	(laws for 0 and 1)
B6. $x + 1 = 1$	B6$'$. $x \cdot 0 = 0$	(laws for 0 and 1)
B7. $x + x' = 1$	B7$'$. $x \cdot x' = 0$	(laws for complementation)
B8. $(x')' = x$		(laws for complementation)
B9. $x + (y \cdot z) = (x + y) \cdot (x + z)$	B9$'$. $x \cdot (y + z) = (x \cdot y) + (x \cdot z)$	(distributive laws)
B10. $(x + y)' = x' \cdot y'$	B10$'$. $(x \cdot y)' = x' + y'$	(De Morgan's laws)

For all binary variables x, y, and z.

The properties listed in Table 1-5 can be proved directly by *tables of combinations*, which are similar to truth tables. For example, the proof of B10 is given below:

x	y	$x + y$	$(x + y)'$	x'	y'	$x' \cdot y'$	**B10**
0	0	0	1	1	1	1	T
0	1	1	0	1	0	0	T
1	0	1	0	0	1	0	T
1	1	1	0	0	0	0	T

In view of the associativity of $\cdot$ and $+$, we write $x \cdot y \cdot z$ for $(x \cdot y) \cdot z = x \cdot (y \cdot z)$, as usual. Similarly, expressions such as $x_1 \cdot x_2 \cdot \ldots \cdot x_n$, $x + y + z$, and $x_1 + x_2 + \cdots + x_n$ are unambiguous. These remarks also apply to the logical connectives $\wedge$ and $\vee$ (see Appendix A, Table A-2).

We are now in a position to provide alternative definitions of the gate types listed in Table 1-2, by means of Boolean operators. These alternative definitions are summarized in Table 1-6. This table also indicates various conversion rules (which correspond in an evident way to rules in Table A-3, Appendix A).

Table 1-6 Input–Output Equations for Basic Gate Types

Gate Type	Input–Output Equation
INVERTER	$z = x'$
AND	$z = x \cdot y$
OR	$z = x + y$
XOR	$z = x \oplus y = (x \cdot y') + (x' \cdot y) = (x + y) \cdot (x' + y')$
EQUIV	$z = x \circledequal y = (x \cdot y) + (x' \cdot y') = (x + y') \cdot (x' + y)$
NAND	$z = x \mid y = (x \cdot y)' = x' + y'$
NOR	$z = x \downarrow y = (x + y)' = x' \cdot y'$

Note: x and y are inputs; z is an output.

So far we have only discussed 1-input and 2-input gates. However, various multi-input gates are in extensive use as primitive switching devices. In Table 1-7 we extend the definition of some basic 2-input gate types to multi-input gates.

We have already mentioned the unambiguity of the expressions $x_1 \cdot x_2 \cdot \ldots \cdot x_n$ and $x_1 + x_2 + \cdots + x_n$. Similarly, the expression $x_1 \oplus x_2 \oplus \cdots \oplus x_n$ is unambiguous, since the $\oplus$ operation is associative (verify this!). On the other hand, neither the NAND nor the NOR operator is associative (verify this by suitable counterexamples!), and the 3-input

Table 1-7 Multi-input Gate Types

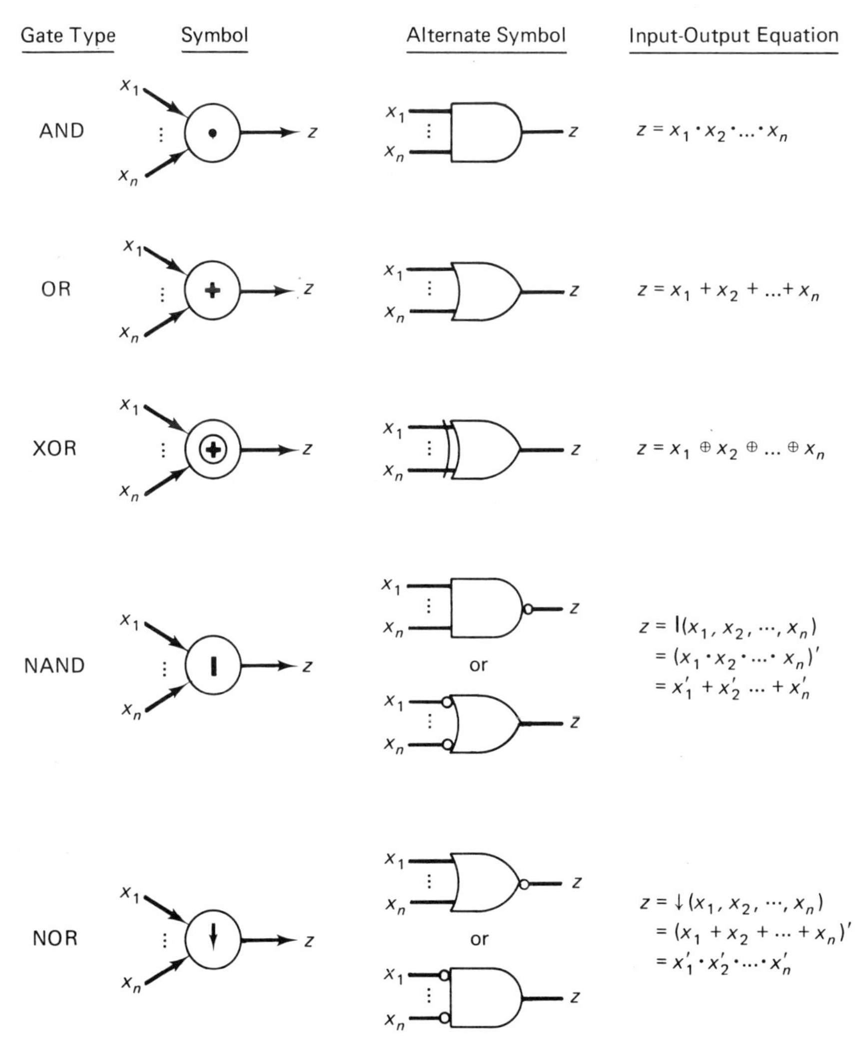

Gate Type	Symbol	Alternate Symbol	Input-Output Equation
AND			$z = x_1 \cdot x_2 \cdot \ldots \cdot x_n$
OR			$z = x_1 + x_2 + \ldots + x_n$
XOR			$z = x_1 \oplus x_2 \oplus \ldots \oplus x_n$
NAND		or	$z = \mid(x_1, x_2, \cdots, x_n)$ $= (x_1 \cdot x_2 \cdot \ldots \cdot x_n)'$ $= x_1' + x_2' \ldots + x_n'$
NOR		or	$z = \downarrow(x_1, x_2, \cdots, x_n)$ $= (x_1 + x_2 + \ldots + x_n)'$ $= x_1' \cdot x_2' \cdot \ldots \cdot x_n'$

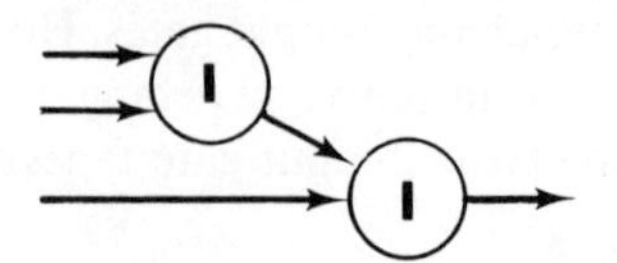

Fig. 1-8 A 3-input gate network *not* equivalent to a 3-input NAND gate.

NAND gate, for example, is *not* equivalent to the gate network shown in Fig. 1-8.

In Section 1.6 we show how logical connectives and Boolean operators can be used for the design of contact and gate networks. In this connection, the equivalences listed in Table 1-8, as well as the properties of the $\oplus$ operator

Table 1-8 Useful Equivalences

1. $x = 0 \equiv x' = 1$
2. $x \oplus y = 1 \equiv x \neq y$
3. $x ⊜ y = 1 \equiv x = y$
4. $x \cdot y' = 1 \equiv x > y$
5. $x' + y = 1 \equiv x \leq y$

listed in Table 1-9, will prove useful. (In Table 1-8 the symbols $>$ and $\leq$ are used in the natural way; i.e., the symbols 0 and 1 are considered to denote the integers zero and one.)

All the equivalences of Table 1-8, as well as the identities of Table 1-9, can be immediately verified by means of tables of combinations. $\sum x_i$ in Table 1-9 denotes the *arithmetic* sum of $x_1, \ldots, x_n$. To prove the equivalences (8) in Table 1-9, mathematical induction on n can be used (give a detailed proof!). Alternatively, since the $\oplus$ operation coincides with modulo-2 addition, elementary modular arithmetic provides an immediate proof of (8).

Table 1-9 Properties of XOR Operator

1. $(x \oplus y) \oplus z = x \oplus (y \oplus z)$
2. $x \oplus y = y \oplus x$
3. $x \oplus 0 = x$
4. $x \oplus 1 = x'$
5. $x \oplus x = 0$
6. $x \cdot (y \oplus z) = x \cdot y \oplus x \cdot z$†
7. $x \oplus y \oplus 1 = x' \oplus y = x \oplus y' = x ⊜ y$†
8. $x_1 \oplus x_2 \oplus \cdots \oplus x_n = 0 \equiv \sum x_i$ is even
   $x_1 \oplus x_2 \oplus \cdots \oplus x_n = 1 \equiv \sum x_i$ is odd

†We assume that the operators $\cdot$ and $'$ have precedence over $\oplus$.

1.5. GATE NETWORKS

Gate networks differ from contact networks in several respects. There is a direction associated with the gate from input to output. Binary signals are applied at the inputs $x_1, \ldots, x_n$, and the resulting output is determined by the gate type. The binary values do not represent closed and open paths, but rather the voltage levels of the signals, say high and low.

An input to a gate network may be a constant (i.e., 0 or 1) or a binary variable, x_i. The construction of a switching network begins with these inputs, which are combined by means of gates. Connecting the outputs of several gates as inputs to a new gate G and considering the output of G corresponds to performing a Boolean operation on the inputs of G.

We now describe some properties of gate networks and introduce some terminology.

1. The number of inputs to a gate is the *fan-in* of the gate. Physical gates have several inputs, but large fan-in is generally not practical.
2. A gate output may be connected to several gate inputs; their number is called the *fan-out* of the gate. The *maximal fan-out* of a physical gate circuit depends on various factors and will be discussed further in Chapter 2.
3. The input of a gate may be connected to an external input or to an output of a gate.
4. The output of a gate may be connected to an external output or to the input of a gate but never to an output of another gate. Such a connection would lead to an ill-defined function when the outputs disagree. Note that we are dealing with mathematical gate models, *not* with their physical implementations. Particular gate implementations provide hardware facilities called *wired-OR* and *wired-AND*. However, in our mathematical model these hardware facilities correspond to separate gates. More will be said about this later.
5. Any point in a gate network can be an external output.

Notice that the above allows the existence of loops in a gate network, for example, when a gate output is connected to one of its own inputs.

Definition 1

A gate network is *loop-free* iff, starting at any point in the network and proceeding via connections through gates in the direction input to output, it is not possible to reach the same point twice.

In a loop-free network, we can define another property:

6. A gate whose inputs are all external is said to be a *level 1* gate. Inductively, if the inputs to a gate G are either external or they are outputs of gates of level $< k$ and at least one input is an output of a level $(k - 1)$ gate, then G is a *level* k gate. A loop-free network is a *level* k network iff k is the maximal level of all its gates.

One verifies that in a loop-free network the notion of level is well defined and the gates can be classified into a finite number of levels. The analysis of a loop-free network can then proceed starting with level 1 gates, whose outputs are determined solely by the external inputs. The outputs of level k gates ($k > 1$) are determined by the outputs of all the lower-level gates and the external inputs.

Obviously, every loop-free gate network satisfies the condition that the outputs of all gates are uniquely determined by the external inputs. Such a network is called *combinational*. This concept will be made more precise in Chapter 2.

In practice, it takes some time to switch a gate output from one value to another, when its inputs change. Thus there is a delay associated with the propagation of a signal through a gate (called *propagation delay*), and it is therefore desirable to reduce the number of levels in a network if speed of operation is essential. However, gate delay can be ignored for the present. We shall return to this topic later.

For now we restrict our attention to loop-free networks. The case of networks with loops will be discussed in Chapter 6.

1.6. DESIGN EXAMPLES

The material developed so far has immediate applications to the design of various switching networks. In the design of gate networks we first assume that all gate types of Tables 1-2 and 1-7 are available, and we consider a solution with fewer gates preferable to one with more gates. In the design of contact networks, one usually assumes the total number of contacts to be the decisive cost factor.

Example 1

By a *binary word* $x \triangleq x_1, x_2, \ldots, x_n$ we mean an ordered set of n binary symbols. Let $x = x_1, x_2, x_3$ and $y = y_1, y_2, y_3$ be binary words. Construct a gate network that indicates whether $x = y$.

Solution

Let z be the output. Then $z = 1 \equiv x = y$. Now $x = y$ iff each coordinate x_i is equal to the corresponding coordinate y_i. This is represented by the equivalence†

$$(x = y) \equiv (x_1 = y_1) \wedge (x_2 = y_2) \wedge (x_3 = y_3). \qquad (1)$$

Next, from Table 1-8(3), we have $(x = y) \equiv (x \circledequiv y = 1)$. Hence (1) can be rewritten as

$$(x = y) \equiv (x_1 \circledequiv y_1 = 1) \wedge (x_2 \circledequiv y_2 = 1) \wedge (x_3 \circledequiv y_3 = 1). \qquad (2)$$

†The parentheses in statements such as (1) could be omitted without leading to ambiguity. However, they often improve the readability of the statements.

Since [by Table 1-4(2)], $(x = 1) \wedge (y = 1) \equiv (x \cdot y = 1)$, we have

$$(x = y) \equiv [(x_1 \circledequiv y_1) \cdot (x_2 \circledequiv y_2) \cdot (x_3 \circledequiv y_3) = 1]. \quad (3)$$

Thus $z = 1$ is equivalent to the right-hand side of (3), or

$$z = (x_1 \circledequiv y_1) \cdot (x_2 \circledequiv y_2) \cdot (x_3 \circledequiv y_3). \quad (4)$$

The network can be drawn immediately (Fig. 1-9) from the input–output equation (4).

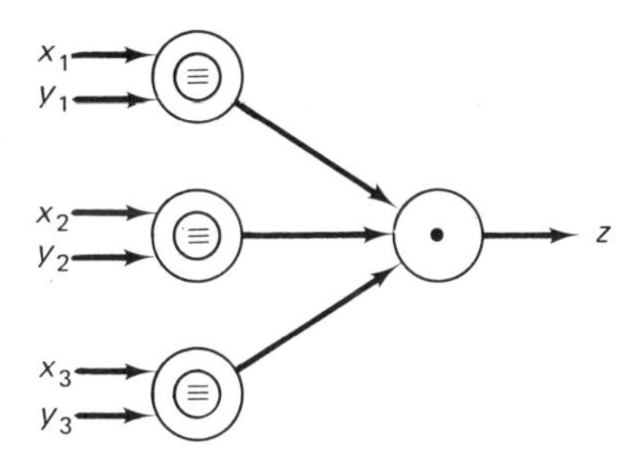

Fig. 1-9 Gate network for Example 1.

Note that the example generalizes to binary words of any length n. The number of gates used in the extended solution is $n + 1$.

Example 2

Design a contact network with inputs $x_1, \ldots, x_5$ which indicates whether the condition

$$x_1 \geq x_2 \geq x_3 \geq x_4 \geq x_5$$

is satisfied.

Solution

Let z be the output. Clearly, the given condition can be restated as

$$(z = 1) \equiv (x_1 \geq x_2) \wedge (x_2 \geq x_3) \wedge (x_3 \geq x_4) \wedge (x_4 \geq x_5). \quad (5)$$

Since $x \geq y$ is equivalent to $x + y' = 1$ [Table 1-8(5)], (5) becomes

$$(z = 1) \equiv [(x_1 + x_2') = 1] \wedge \cdots \wedge [(x_4 + x_5') = 1]$$

and hence

$$z = (x_1 + x_2') \cdot (x_2 + x_3') \cdot (x_3 + x_4') \cdot (x_4 + x_5').$$

The contact network is shown in Fig. 1-10. In Section 4.7 we discuss

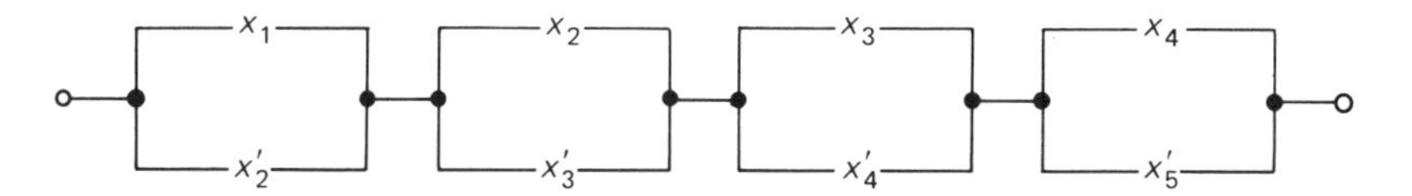

Fig. 1-10 Contact network for Example 2.

methods by which the minimality of this network can be proved. Again the method generalizes to n inputs.

Example 3

Design a 5-input gate network that satisfies the following conditions:

1. If all inputs are 0, the output is 0.
2. Whenever a single input is changed, the output changes.

Solution

Let the inputs be $x_1, \ldots, x_5$ and the output z. We denote by $\sum x_i$ the arithmetic sum of $x_1, \ldots, x_5$, i.e., the number of x_i's equal to 1. If $\sum x_i = 0$, then $z = 0$, by condition 1. If a single input is set to 1, i.e., $\sum x_i$ is changed from 0 to 1, z will change from 0 to 1. If $\sum x_i$ is now changed from 1 to 2, z will change from 1 to 0. Continuing this reasoning, we see that $z = 1$ iff $\sum x_i$ is odd. Hence

$$\begin{aligned}(z = 1) &\equiv (\textstyle\sum x_i \text{ is odd}) \\ &\equiv (x_1 \oplus x_2 \oplus x_3 \oplus x_4 \oplus x_5 = 1). \qquad \text{[Table 1-9(8)]}\end{aligned}$$

Thus

$$z = x_1 \oplus x_2 \oplus x_3 \oplus x_4 \oplus x_5.$$

The network consists of a single 5-input XOR gate. The solution to the extended problem of n inputs requires one n-input gate.

Example 4

Design a 6-input gate network which indicates whether two 3-bit words differ in exactly one coordinate.

Solution

Let the two 3-bit words be $x = x_1, x_2, x_3$ and $y = y_1, y_2, y_3$ and denote the output by z. It is convenient to introduce auxiliary variables w_1, w_2, and w_3, where

$$\begin{aligned}w_i = 1 &\equiv x_i \neq y_i \\ &\equiv (x_i \oplus y_i = 1). \qquad \text{[Table 1-8(2)]}\end{aligned}$$

Thus

$$w_i = x_i \oplus y_i.$$

It follows that

$$z = 1 \equiv \textstyle\sum w_i = 1.$$

Note now that

$$(\textstyle\sum w_i = 1) \equiv (\sum w_i \text{ is odd}) \wedge (\sum w_i \neq 3).$$

Furthermore,

$$(\textstyle\sum w_i \text{ is odd}) \equiv (w_1 \oplus w_2 \oplus w_3 = 1) \qquad \text{[Table 1-9(8)]}$$

and

$$\begin{aligned}(\textstyle\sum w_i \neq 3) &\equiv [\sim(\textstyle\sum w_i = 3)] \\ &\equiv [\sim(w_1 = 1 \wedge w_2 = 1 \wedge w_3 = 1)] \\ &\equiv [(w_1 \cdot w_2 \cdot w_3)' = 1]. \quad \text{[Table 1-4(1, 2)]}\end{aligned}$$

Summarizing, we obtain

$$(z = 1) \equiv (w_1 \oplus w_2 \oplus w_3 = 1) \wedge [(w_1 \cdot w_2 \cdot w_3)' = 1].$$

Thus

$$z = (w_1 \oplus w_2 \oplus w_3) \cdot (w_1 \cdot w_2 \cdot w_3)' = (w_1 \oplus w_2 \oplus w_3) \cdot [|(w_1, w_2, w_3)].$$

The corresponding gate network contains six gates. A network with fewer gates can be obtained by observing that

$$\begin{aligned}(\textstyle\sum w_i = 1) \equiv [(\textstyle\sum w_i \text{ is odd}) \wedge \sim(\textstyle\sum w_i = 3)] \\ \vee [\sim(\textstyle\sum w_i \text{ is odd}) \wedge (\textstyle\sum w_i = 3)]\end{aligned} \tag{6}$$

since the situation $\sum w_i$ even and $\sum w_i = 3$ is impossible. It follows that (6) can be rewritten as

$$\begin{aligned}(z = 1) \equiv [w_1 \oplus w_2 \oplus w_3 = 1 \wedge (w_1 \cdot w_2 \cdot w_3)' = 1] \\ \vee [(w_1 \oplus w_2 \oplus w_3)' = 1 \wedge (w_1 \cdot w_2 \cdot w_3) = 1].\end{aligned}$$

Thus

$$\begin{aligned}z &= (w_1 \oplus w_2 \oplus w_3) \cdot (w_1 \cdot w_2 \cdot w_3)' + (w_1 \oplus w_2 \oplus w_3)' \cdot (w_1 \cdot w_2 \cdot w_3) \\ &= (w_1 \oplus w_2 \oplus w_3) \oplus (w_1 \cdot w_2 \cdot w_3). \qquad \text{(Table 1-6)}\end{aligned}$$

The resulting 5-gate network is shown in a convenient, *detached* form in Fig. 1-11, where the label w_i on two different input lines means that both are to be connected to the w_i output.

Note that this problem extends naturally to two n-bit words. However, the particular solution given here cannot be directly extended.

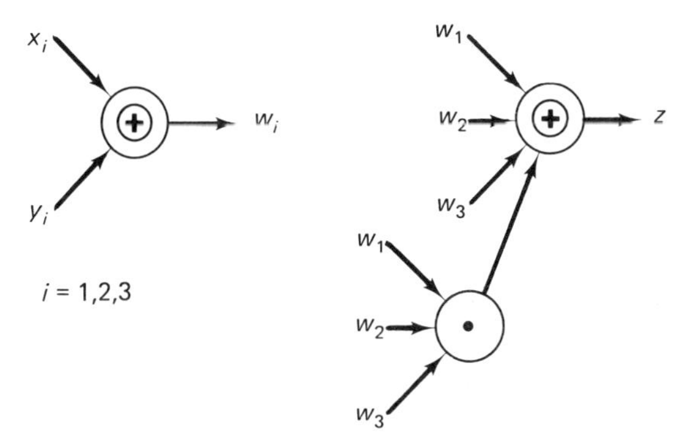

Fig. 1-11 Gate network for Example 4.

Frequently, the designer of gate networks is restricted in his choice of gate types. In the following two examples we assume that only (multi-input) AND and OR gates are available. On the other hand, we assume further that together with any input x_i, its complement x_i' is also available. This is, indeed, often the case in an actual design environment.

We also introduce the following notation. Let $x = x_1, x_2, \ldots, x_n$ be a binary word. Then $\perp x$ will denote the (nonnegative) integer whose base-2 representation is x; i.e.,

$$\perp x \triangleq \sum_{i=1}^{n} x_i \cdot 2^{n-i}.$$

For example, $\perp 1101 = 13$, $\perp 001110 = 14$.

Example 5

Let x be a 4-bit word. Design a gate network with output z such that $z = 1 \equiv \perp x > 9$.

Solution

First note that $9 = \perp 1001$. For $\perp x > 9$ to hold, we must have $x_1 = 1$. However, this condition is not sufficient; i.e., we must exclude the cases 1000 and 1001. This can be done as follows:

$$(z = 1) \equiv (\perp x > 9) \equiv (x_1 = 1) \wedge [\sim(x_2 = 0 \wedge x_3 = 0)].$$

Thus

$$z = x_1 \cdot (x_2' \cdot x_3')' = x_1 \cdot [(x_2')' + (x_3')'] \qquad \text{[Table 1-5(B10')]}$$

$$= x_1 \cdot (x_2 + x_3). \qquad \text{[Table 1-5(B8)]}$$

The gate network is shown in Fig. 1-12.

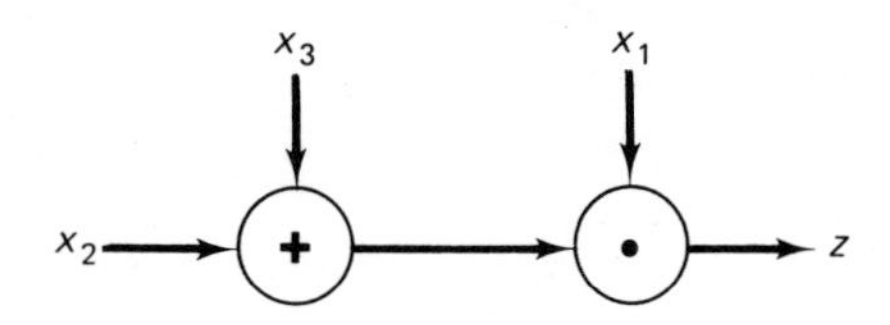

Fig. 1-12 Gate network for Example 5.

Example 6

Let x and y be 2-bit words. Design a gate network with output z such that $z = 1 \equiv \perp x > \perp y$.

Solution

For $\perp x > \perp y$ to hold, one of the following two exclusive conditions must be satisfied: either (a) $x_1 > y_1$ or (b) $(x_1 = y_1) \wedge (x_2 > y_2)$. However, one frequently arrives at a more economical network if inclusive rather than exclusive conditions are considered. In the example under discussion, one easily verifies that $\perp x > \perp y$ holds iff at least one of the following two conditions is met:

1. $x_1 > y_1$.
2. $(x_1 \geq y_1) \wedge (x_2 > y_2)$.

Thus

$$\begin{aligned}(z = 1) &\equiv (x_1 > y_1) \vee [(x_1 \geq y_1) \wedge (x_2 > y_2)] \\ &\equiv (x_1 \cdot y_1' = 1) \vee [(x_1 + y_1' = 1) \wedge (x_2 \cdot y_2' = 1)] && \text{(Table 1-8)} \\ &\equiv [(x_1 \cdot y_1') + ((x_1 + y_1') \cdot x_2 \cdot y_2') = 1]. && \text{(Table 1-4)}\end{aligned}$$

Thus

$$z = x_1 \cdot y_1' + (x_1 + y_1') \cdot x_2 \cdot y_2'.$$

In writing the last line we have made the customary assumption that the AND operator $(\cdot)$ has precedence over the OR operator $(+)$. The corresponding network is shown in Fig. 1-13.

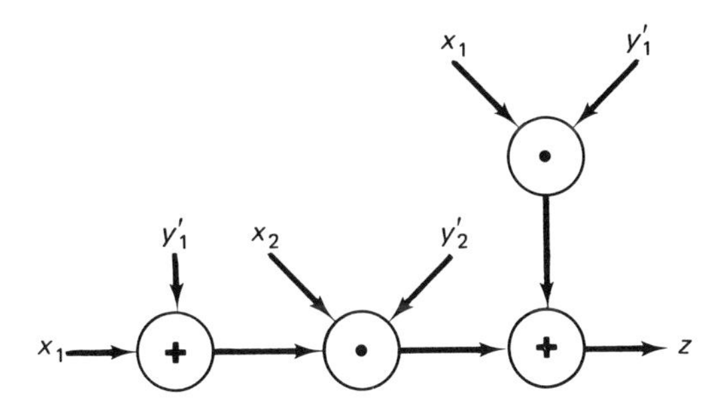

Fig. 1-13 Gate network for Example 6.

Example 7

Design a gate network with input word $x = x_1, \ldots, x_5$ and output word $z = z_0, z_1, \ldots, z_5$ such that $\perp z = 1$ PLUS $\perp x$, where PLUS denotes arithmetic addition of integers. Assume that all the basic gate types listed in Table 1-2 are available.

Solution

We start with an example of adding 1, using base-2 arithmetic. Let $x = 10111$. Then z may be obtained by the following base-2 addition:

$$\begin{array}{rr l} & 0\ 0\ 1\ 1\ 1 & \longleftarrow \text{carry bits} \\ x = & 1\ 0\ 1\ 1\ 1 & \multirow{2}{*}{\Big\} \text{ADD}} \\ & 1 & \\ \hline z = & 0\ 1\ 1\ 0\ 0\ 0 & \longleftarrow \text{result} \end{array}$$

This base-2 addition is comparable to the following base-10 addition:

$$\begin{array}{rl} 0\ 0\ 1\ 1\ 1 & \leftarrow \text{carry bits} \\ 9\ 0\ 9\ 9\ 9 & \\ 1 & \Big\}\ \text{ADD} \\ \hline 0\ 9\ 1\ 0\ 0\ 0 & \leftarrow \text{result} \end{array}$$

We now introduce the following notation for the base-2 addition scheme:

$$\begin{array}{rrl} & c_0\ c_1\ c_2\ c_3\ c_4 & \leftarrow \text{carry bits} \\ x = & x_1\ x_2\ x_3\ x_4\ x_5 & \\ & 1 & \Big\}\ \text{ADD} \\ \hline z = & z_0\ z_1\ z_2\ z_3\ z_4\ z_5 & \leftarrow \text{result} \end{array}$$

The following equations hold:

$$z_5 = x_5 \oplus 1 = x'_5$$

$$z_i = x_i \oplus c_i \qquad i \in \{1, 2, 3, 4\}$$

$$z_0 = c_0$$

$$c_4 = x_5$$

$$c_{i-1} = x_i \cdot c_i \qquad i \in \{1, 2, 3, 4\}.$$

These equations may be implemented as shown in Fig. 1-14. The gate network of Fig. 1-14 may also be implemented by using a (commercially available) 2-input, 2-output building block, called a *half-adder* (HA), as shown in Fig. 1-15.

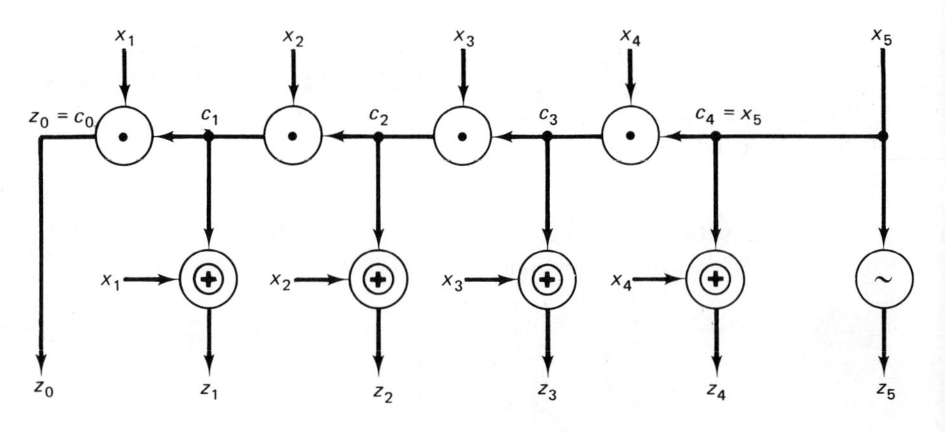

Fig. 1-14 Gate network for Example 7.

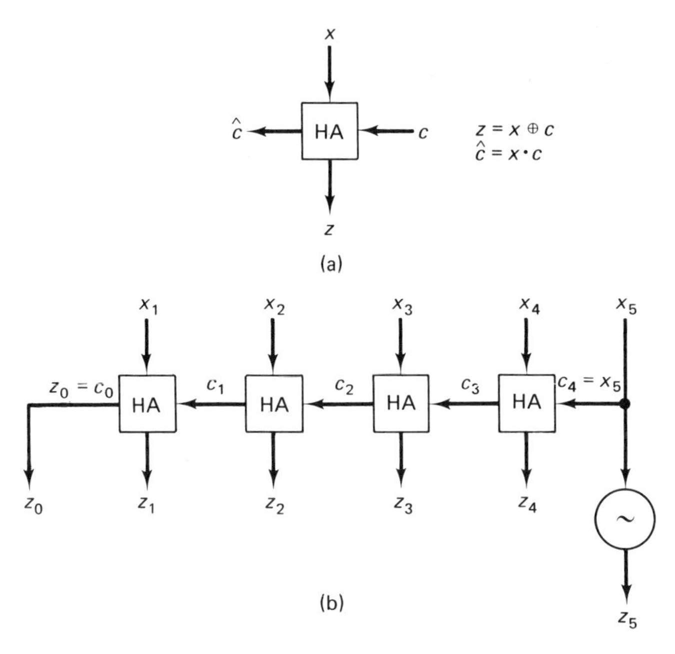

Fig. 1-15 (a) Symbol and definition of half-adder (HA); (b) implementation of gate network of Fig. 1-14, using half-adders.

PROBLEMS

1. (a) Find expressions for f and g in the two contact networks shown in Fig. P1-1.

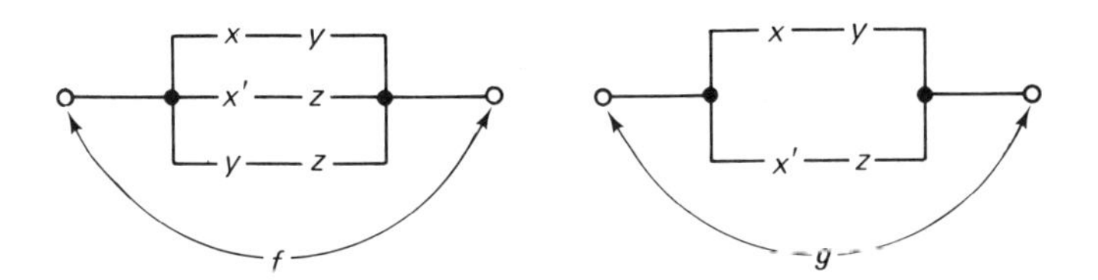

Figure P1-1

(b) Show that the networks are equivalent by evaluating the tables of combinations for f and g.

(c) Repeat part (b) by using the identities of Table 1-5.

(d) Assume that the x and x' contacts are operated by a switch as shown in Fig. 1-3. If we take into account the partially operated (*transient*) state of the switch where both x and x' are open, are the two networks still equivalent?

2. Determine whether the two gate networks in Fig. P1-2 are equivalent. Prove your answer.

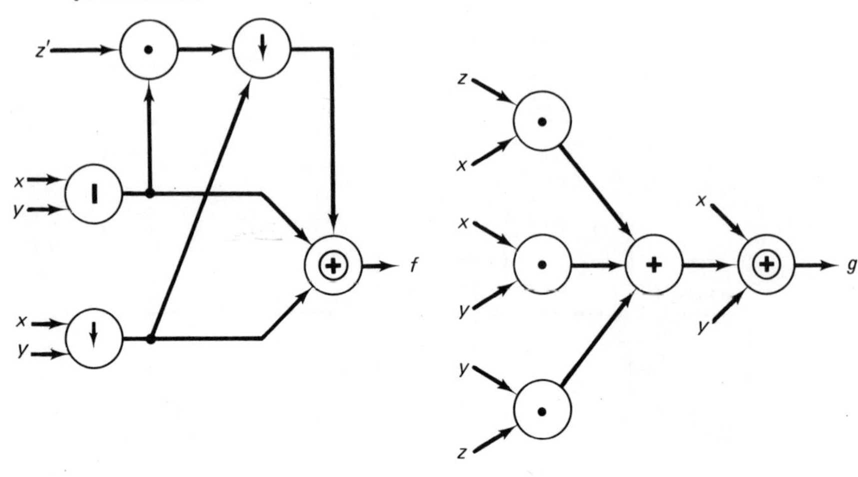

Figure P1-2

3. Repeat Problem 2 for the gate networks of Fig. P1-3.

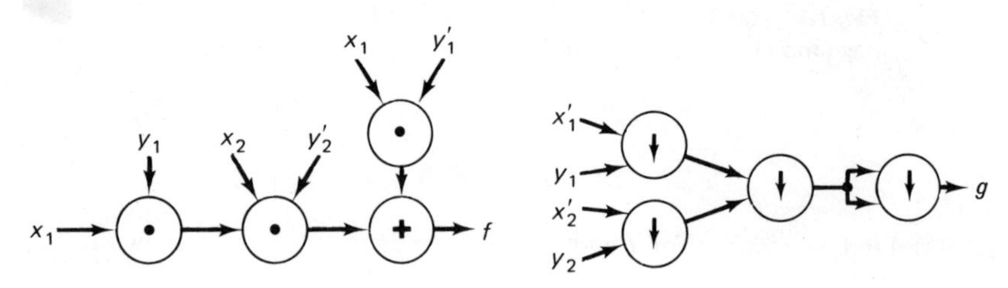

Figure P1-3

4. (a) Prove the following:
   (1) $x \mid x = x'$
   (2) $(x \mid y)' = x \cdot y$
   (3) $x' \mid y' = x + y$
   (b) Derive a network, composed of 2-input NAND gates only, which is equivalent to a 2-input XOR gate.

5. Prove or give a counterexample:
   (a) $x + (y \oplus z) = (x + y) \oplus (x + z)$
   (b) $x + (y \circleddash z) = (x + y) \circleddash (x + z)$
   (c) $x \oplus y = z \equiv x \oplus z = y$
   (d) $x + y = x \oplus y \oplus x \cdot y$

6. Assume that all the gate types of Table 1-7† are available. Find networks to satisfy the conditions specified below. Use as few gates as possible. The inputs

†You may assume that inverters are also available in all the problems that use Table 1-7.

are always $x_1, \ldots, x_n$ and the output is z.

(a) $z = 1$ iff exactly two of x_1, x_2, x_3 are 1 (two gates).
(b) $z = 1$ iff exactly two of x_1, x_2, x_3, x_4 are 1 (three gates).
(c) $z = 1$ iff three or more of $x_1, x_2, \ldots, x_5$ are 1. This function is called a *majority function*. (Solutions with eight gates are easy to find. There is also a seven-gate solution.)
(d) $z = 1$ iff exactly one of x_1, x_2, x_3 is 1 and exactly one of x_3, x_4, x_5 is 1.

7. Let x be a 4-bit word. Design a gate network with a 3-bit output z such that $\perp z = \sum x_i$, where $\sum$ denotes arithmetic addition. Use only gates of Table 1-2. (*Hint:* Consider using the half-adders of Example 7. The problem can be solved with nine gates.)

8. Let x and y be 4-bit words. Design a gate network with output z such that $z = 1 \equiv \perp x > \perp y$. Assume that you have available INVERTERS, 2-input AND gates, 2-input OR gates, and also 2-bit comparators, i.e., networks as shown in Fig. 1-13. (There is a solution with three 2-bit comparators and three gates.)

9. (a) Design a gate network, as shown in Fig. P1-4, specified as follows: $z_j = 1$ iff $x_j = 1$ and $x_i = 0$ for every $i < j$. All cells except the extreme ones are identical. Use no more than $3n - 4$ gates of Table 1-2.

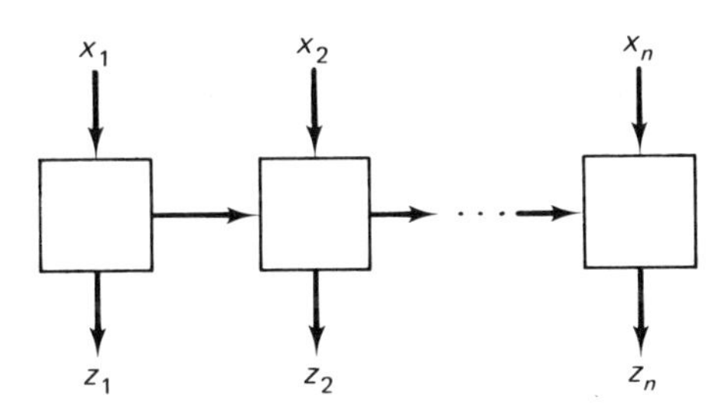

Figure P1-4

(b) Draw the corresponding contact network. Use no more than $2n - 1$ contacts.

10. Let x and y be 4-bit words. Design a gate network with a 4-bit output word z such that: if $\perp x \geq \perp y$, then $z = x$; otherwise, $z = y$. Use the gates of Table 1-7.

11. Design gate networks (with input word $x = x_1, \ldots, x_n$) satisfying the conditions given below. Use the gates of Table 1-7.

(a) $z = 1$ iff there are two consecutive bits in x that are both 1.
(b) $z = 1$ iff there are three consecutive bits in x that are all 0.
(c) $z = 1$ iff there are two consecutive bits in x that are equal.
(d) $z = 1$ iff x contains the substring 101.

12. Design a gate network with a 4-bit input x and a 4-bit output z such that

$$\text{for } \perp x \leq 9, \perp z = 3 \text{ PLUS } \perp x$$
$$\text{for } \perp x > 9, \text{ the output is irrelevant.}$$

Use the gates of Table 1-7. (The representation of a decimal digit d, $0 \leq d \leq 9$, by the binary code for 3 PLUS d is called the *excess-three* code.)

13. (a) Design a "selector" network with inputs x_1, x_0, and s and output z such that

$$z = x_0 \quad \text{if } s = 0 \qquad \text{and} \qquad z = x_1 \quad \text{if } s = 1.$$

(b) Generalize the network of part (a) to one with four "data" inputs, x_3, x_2, x_1, and x_0; two "control" inputs, s_1 and s_0; and one output, z. If the control inputs represent the integer i, $0 \leq i \leq 3$, the output z is to "select" x_i. In other words, $z = x_i$ if $\perp(s_1, s_0) = i$.

14. Design a "scaler" network with a 4-bit data word x, a 2-bit control word s, and a 4-bit output z. If $\perp s = i$, the output z is to be the word x shifted to the left by i positions. The most significant i bits are lost and the least significant i bits are to be replaced by 0's. For example, if $\perp s = i = 2$, z should be of the form $x_3, x_4, 0, 0$. Use the gates of Table 1-7.

15. Design a gate network with a 4-bit input word, $x = x_1, \ldots, x_4$; a 4-bit output word, $y = y_1, \ldots, y_4$; and a control input, c. If $c = 1$, the output word should be the input word shifted by one position to the right "cyclically"; i.e., $y_2 = x_1$, $y_3 = x_2$, $y_4 = x_3$, and $y_1 = x_4$. Similarly, if $c = 0$, y should be x shifted by one position to the left cyclically. You are allowed to use only multi-input AND and OR gates and INVERTERS. Use as few gates as possible.

16. Design a 6-input gate network which indicates whether a 3×2 binary matrix contains two rows that are equal. Use the gates of Table 1-7.

17. Sometimes it is necessary to cross two wires. If we are constrained to a plane, we require some circuitry to perform this task. Show how to cross two wires on a plane using only XOR gates (Fig. P1-5). List the properties of the $\oplus$ operation that make your circuit work.

Figure P1-5

2 COMBINATIONAL INTEGRATED CIRCUIT MODULES

ABOUT THIS CHAPTER

The reader has some familiarity now with idealized gates of Chapter 1. We next turn to a more realistic treatment of gate networks by introducing some modern building blocks, called *integrated circuit* (IC) packages. It has become more economical to produce a number of gates in a single "package" or "module," and therefore the design of networks using single gates as building blocks is obsolete.

From experience, manufacturers have realized that certain functional units frequently appear as subnetworks of larger digital networks. Many such functional units have therefore become commercially available as IC packages. We analyze the logic diagrams of several such packages and derive their input–output behavior. Some applications of these packages are given in the Problems section. The design of more complex digital networks using these and other packages will be discussed in Chapters 8 and 9.

For some of the packages we describe only the external characteristics and postpone the logical design until Chapter 5. We suggest to the reader that he also may wish to postpone certain details of this chapter until he reads Chapters 4 and 5. For example, the arithmetic-logic unit of Section 2.9 may be a good candidate for this. Also, it may be a good idea to return to Chapter 2 after reading Chapter 5.

The reader may find it difficult to understand at this point why certain details of the design are what they are. For example, some units are enabled by a 1-signal, others by a 0-signal. We could easily make the presentation more "elegant" in spots by avoiding such differences. However, we retain them in the interest of realism.

A word about "buzzwords." We use such terms as SSI, TTL, MOS, etc. The nonexpert reader should not get discouraged, for we do not expect extensive knowledge of these topics on his part. We comment on these concepts in sufficient detail for the nonexpert.

Our sources for this material are manufacturers' catalogs [INT, MOT3, RCA, SIG, TI1, TI2]. We have omitted all circuit details because they are not within the scope of our book and are not needed for our purposes. Also, we have applied the notation introduced in Chapter 1 to provide precise descriptions of the logical behavior of the packages.

The material in this chapter is restricted to combinational (see Section 2.1) modules. Other modules will be discussed in Chapters 7 and 8.

2.1. COMBINATIONAL NETWORKS

By a *combinational single-output network* we mean a network that has the following properties. It has a finite number n of binary inputs $x_1, x_2, \ldots, x_n$ and a binary output z, as shown in Fig. 2-1. Furthermore, at any time t the value of the output z depends only on the values of the inputs at time t; i.e., "the present output depends only on the present input." Such a system has no memory, since there is no dependence of the output on the past input values. Since the output value depends only on the present input combination, we have the name *combinational network*. This, however, is an idealized model. In practice, the output responds to a change of the inputs after a delay Δ. This delay Δ is called the *propagation delay* of the network.

In general, a *combinational multi-output network* has m outputs $z_1, z_2,$

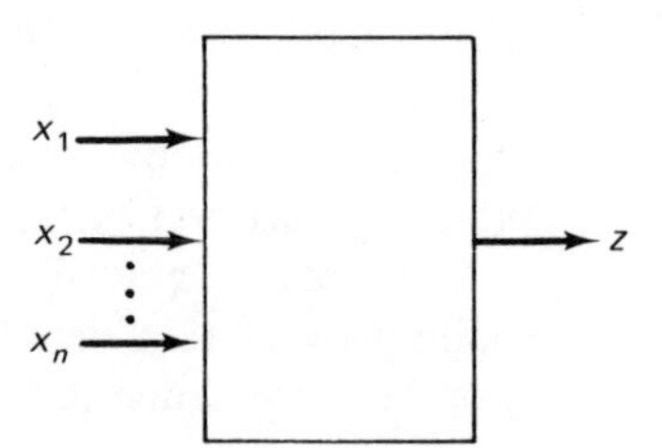

Figure 2-1

$\ldots, z_m$, each of which has the properties of the output z above. To analyze a multi-output network, we may perform the analysis of a single-output network m times, for each z_i in turn. For this reason, it is sufficient to consider single-output networks below.

The performance of a single-output combinational network is conveniently discussed by means of the concept of a *function*. Any mapping f of a set X into a set Y is called a *function from X into Y*, written $f: X \longrightarrow Y$. For every element x in X there exists a unique element y in Y such that $f(x) = y$. We shall sometimes write $f: x \mapsto y$ to state that the function f *assigns* the *value* y to the *argument* x. More information about functions is given in Appendix C.

Let V_n denote the set of all binary n-tuples. Thus $V_1 = \{0, 1\}$. Then the performance of the (single-output) combinational network of Fig. 2-1 can be specified by a function $f: V_n \longrightarrow \{0, 1\}$ such that $f(x_1, x_2, \ldots, x_n) = z$, where $(x_1, \ldots, x_n)$ is an arbitrary input combination and z is the corresponding output. Such functions are called *Boolean*. The network function f can be described by a table of combinations listing all 2^n possible input combinations $(x_1, \ldots, x_n)$ together with the corresponding output values $f(x_1, \ldots, x_n) = z$. We say that the network *implements* or *realizes* the function f.

In this chapter we assume that the binary variables x_i and z_j are realized by voltage levels. Furthermore, we use positive logic terminology; i.e., 0 will be represented by a low voltage (e.g., 0 volts, written 0 V) and 1 by a high voltage (e.g., 3.5 V). In actual networks the transitions from 0 to 1 or from 1 to 0 are not instantaneous, but the corresponding voltage signals have waveforms as shown in Fig. 2-2. Figure 2-2 also indicates a frequently

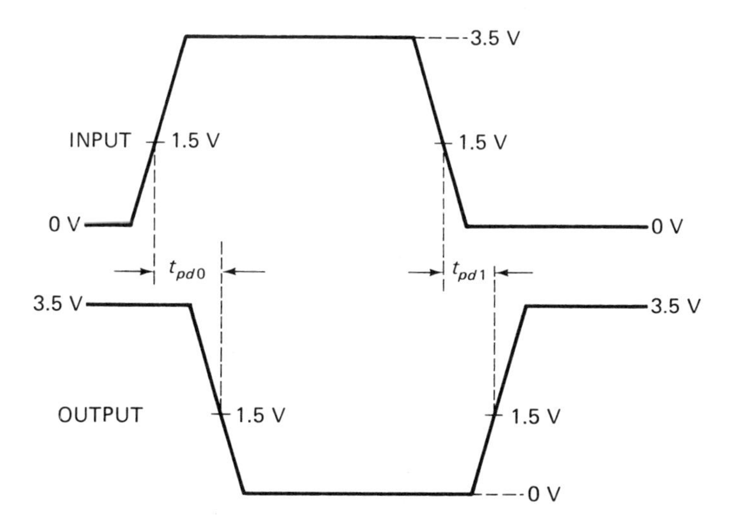

Fig. 2-2 More realistic voltage waveforms.

used method of defining propagation delays precisely. The times t_{pd0} and t_{pd1} are referred to as *propagation delay time to logical 0* and *propagation delay time to logical 1*, respectively.

2.2. INTEGRATED CIRCUITS

Integrated circuit (IC) technologies are presently widely used for the implementation of gate networks. A large collection of order-by-catalog IC modules, called *IC packages*, is now available to the designer. The total number of gates inside a single IC package may vary from a few gates (SSI or small-scale integration) up to many hundreds of gates (LSI or large-scale integration). The intermediate range of 12 to 100 gates per IC package is referred to as MSI (medium-scale integration). The size of an IC package is, however, largely determined by the number of its outside connectors (pins) rather than the number of its gates. This and other hardware con-

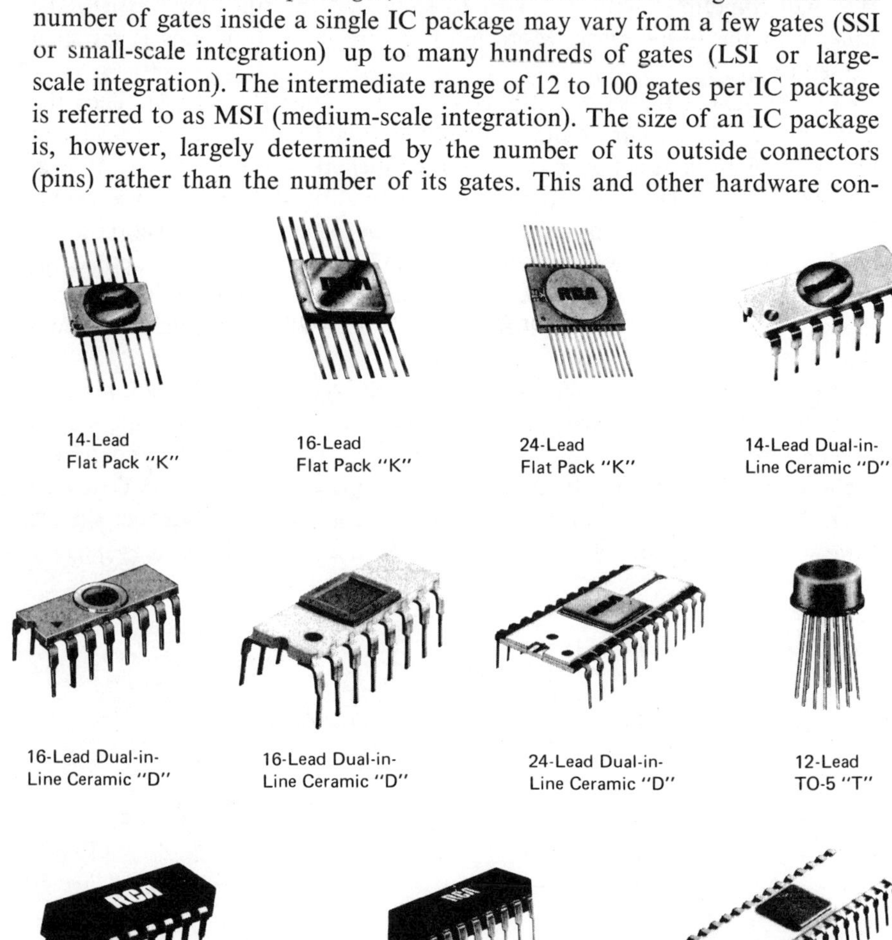

Fig. 2-3 Various types of IC packages. (*Courtesy* RCA Solid State Division.)

siderations make the use of IC packages with high gate-to-pin ratios especially attractive. The number of pins of SSI and MSI packages varies from 10 to 24, and a typical 24-pin ("flat") package has the following dimensions (in inches): $0.6 \times 1.2 \times 0.1$. Various types of IC packages are shown in Fig. 2-3.

IC packages are available in various *families*, each family being characterized by a particular combination of hardware features, such as speed, power consumption, maximal fan-out, temperature range, and noise immunity. Presently, a widely used family, available from various manufacturers, is the TTL (transistor–transistor–logic) Series 74 family. Some of its typical gate characteristics are as follows:

propagation delay: 10 ns (1 ns = 1 nanosecond = 10^{-9} second)

power dissipation (*per gate*): 10 mW (1 mW = 1 milliwatt = 10^{-3} watt)

maximal fan-out (when connected to other gates of this family): 10

temperature range: 0 to 70°C (Celsius)

The TTL Series 54 family is similar to the TTL Series 74 family except for its wider temperature range (−55 to 125°C). Other frequently used families offer either higher-speed or lower-power consumption, as illustrated in the following examples:

Family	Typical Gate Propagation Delay	Typical Power Dissipation per Gate
TTL Series 54/74	10 ns	10 mW
TTL Series 54H/74H	6 ns	22 mW
TTL Series 54L/74L	33 ns	1 mW
ECL 10,000	2 ns	25 mW
COS/MOS	50 ns	Less than 1 μW† (1 μW = 10^{-6} W)

†Under the condition of no input changes.

Here ECL refers to *E*mitter-*C*oupled *L*ogic, presently providing higher speed than TTL. Both TTL and ECL are representatives of the *bipolar technologies*. MOS (*M*etal *O*xide *S*emiconductor) technology differs considerably from bipolar technologies. From the user's viewpoint with which we are concerned, the main advantages of MOS over bipolar technology are higher circuit density, enabling a higher gate-to-pin ratio (for packages of equal size) and lower power dissipation. COS/MOS, a particular MOS family, stands for *CO*mplementary-*S*ymmetry/MOS. This family is of particular interest in view of its very low power dissipation.

Modules within the same family can be interconnected without any interface problems; i.e., they are *compatible*. As to compatibility among different families, the reader is referred to the manufacturers' catalogs for further information.

Wired logic refers to the capability of tying together the outputs of gates to realize either the AND ("implied AND") or the OR ("implied OR") function without additional hardware or components. For example, the outputs of ECL NOR gates can be tied together, under suitable circumstances, to form their OR function.

See Fig. 2-4 for wired-logic symbols used commercially. In this book we shall not use these symbols, using instead the appropriate gate symbol.

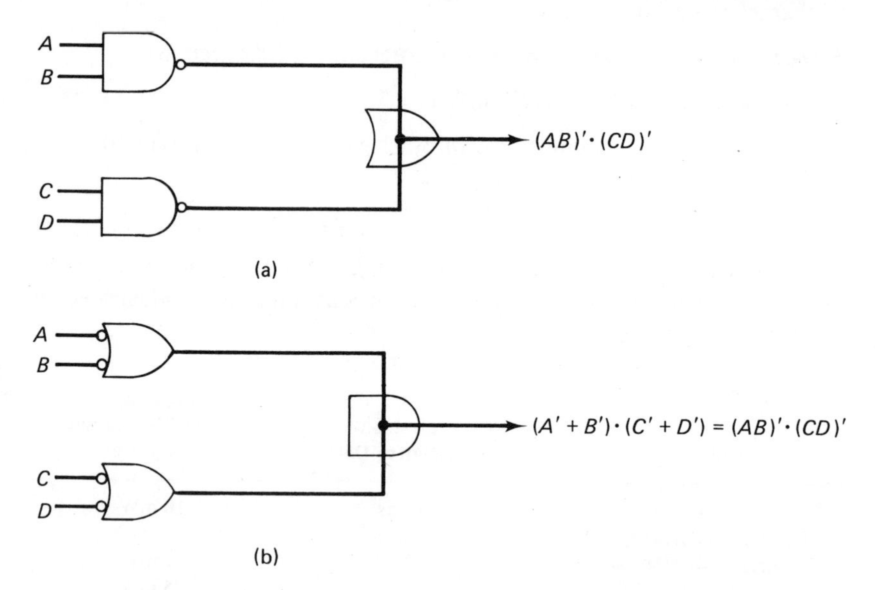

Fig. 2-4 Commonly used wired logic symbols. (a) Incorrect symbol—the proper symbol is AND; (b) correct symbol.

2.3. SMALL-SCALE INTEGRATION

Examples of SSI Packages

In many of the presently available IC families, NAND gate modules play an important role. This is partly due to circuit considerations (such as circuit simplicity and output signal amplification), the details of which are beyond the scope of this book. As for the logic design viewpoint, it can be shown that NAND gates are sufficient to implement any combinational network. We show in Chapter 4 that any output variable can be represented as a sum of products of the input variables and their complements. Later in this section we show how any such sum of products may be implemented by means of NAND gates.

To illustrate the availability of NAND gate packages, we provide in

Table 2-1 a list of such packages of the TTL Series 74 family. Although INVERTERS can be made out of 2-input NAND gates, the use of the special six-INVERTER package 7404 (see Table 2-1) is frequently preferable. We also mention the availability of the four-XOR package 7486, the four-AND package 7408, and the four-NOR package 7402 (see Table 2-1). In each of these packages two pins are used for "ground" (0 V) and supply voltage (5 V). The remaining pins provide the required input and output terminals.

Table 2-1 Some TTL Series 74 SSI Packages

Package Code (see Note)	Number of Pins	Type of Gates	Number of Gates (per Package)	Number of Inputs (per Gate)	Maximal Fan-out	Maximal Propagation Delay (ns)†
7400	14	NAND	4	2	10	22
7410	14	NAND	3	3	10	22
7420	14	NAND	2	4	10	22
7430	14	NAND	1	8	10	22
7404	14	INVERTER	6	1	10	22
7486‡	14	XOR	4	2	10	30
7408	14	AND	4	2	10	27
7402	14	NOR	4	2	10	22

Note: These and other SSI packages are available from Texas Instruments [TI1, TI2] (code prefix SN), Signetics [SIG] (code prefix N), and others.

†This is the propagation delay of each gate in the package.

‡An MSI package included here for convenience.

Applications of SSI Packages

Suppose that an output variable z is specified as a sum of products of the input variables and their complements; e.g.,

$$z = x_1x'_2x_3 + x'_1x'_3x_4x_5x_6 + x_2x'_3x_4x'_5x'_6.\dagger$$

To obtain a NAND-gate representation (see Table 1-7), we simply replace each product P_i of z by $(P'_i)'$. Thus

$$z = (P'_1)' + (P'_2)' + (P'_3)'.$$

Recall that $|(x_1, \ldots, x_n) \triangleq (x_1x_2 \cdots x_n)' = x'_1 + x'_2 + \cdots + x'_n$. Then

$$\begin{aligned} z &= |(P'_1, P'_2, P'_3) \\ &= |[|(x_1, x'_2, x_3), |(x'_1, x'_3, x_4, x_5, x_6), |(x_2, x'_3, x_4, x'_5, x'_6)]. \end{aligned}$$

If we assume that only the uncomplemented inputs $x_1, \ldots, x_6$ are available, this implementation requires five INVERTERS, two 3-input NAND gates, and two 5-input NAND gates. If only SSI packages of Table 2-1 are to be

†This particular expression is of no significance; it was chosen arbitrarily, as an example. Also, we frequently omit the dot indicating multiplication, writing e.g. $x_1x'_2x_3$ for $x_1 \cdot x'_2 \cdot x_3$.

used, one 7404, one 7410, and two 7430 packages are required. The corresponding gate network is shown in Fig. 2-5. Note that unused NAND-gate inputs are to be connected either to a constant 1 or to other connected inputs.

In Chapter 5 we discuss methods of simplifying Boolean expressions. Such methods can be used to minimize the total number of SSI packages required to implement a given output function. However, even after simplification, NAND gates of more than 8 inputs may be required. In this

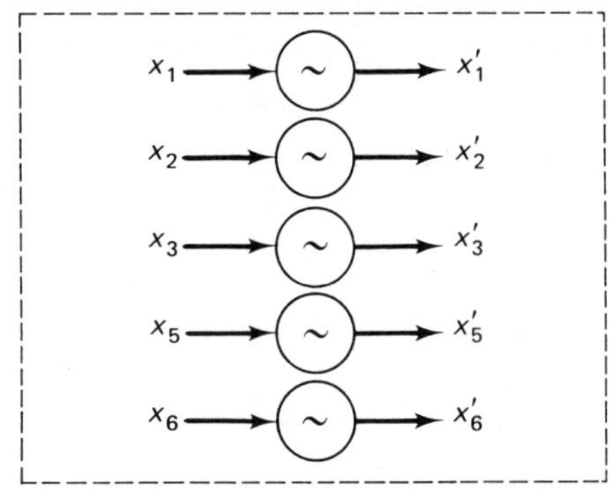

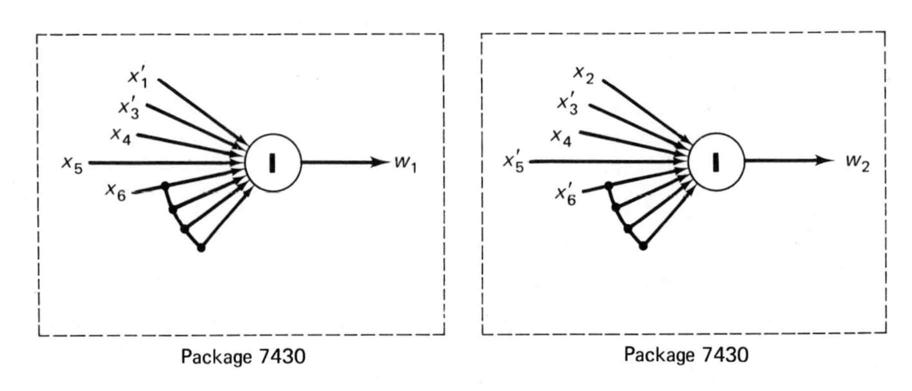

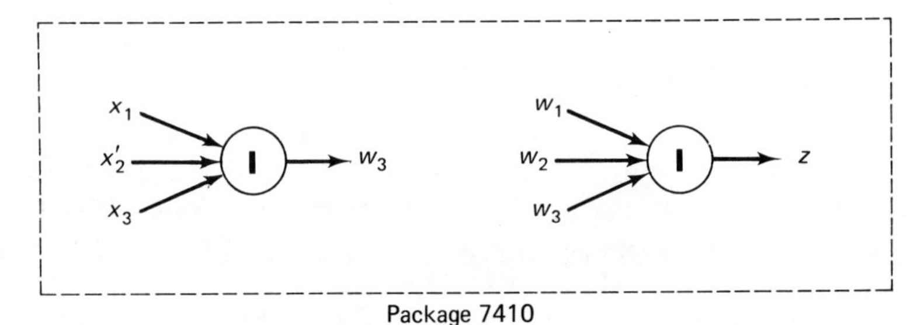

Fig. 2-5 Example of SSI implementation of a combinational network.

case, $m + 1$ 8-input NAND gates ($m \leqq 8$) and m INVERTERS can be combined to form the equivalent of an $8m$-input NAND gate. To illustrate this idea, Fig. 2-6 shows how four 3-input NAND gates together with three inverters may be connected to form the equivalent of a single 9-input NAND gate.

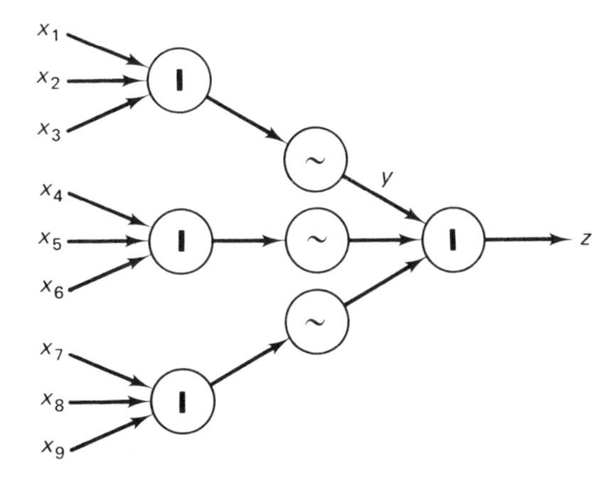

Fig. 2-6 Formation of a 9-input NAND gate by means of 3-input NAND gates.

Indeed, we have

$$y = [|(x_1, x_2, x_3)]' = x_1 x_2 x_3$$

and

$$\begin{aligned} z &= |(x_1 x_2 x_3, x_4 x_5 x_6, x_7 x_8 x_9) = (x_1 x_2 \cdots x_9)' \\ &= |(x_1, x_2, \ldots, x_9). \end{aligned}$$

Similar remarks apply to NOR gates and product of sums expressions. For example,

$$\begin{aligned} &(x_1 + x_2')\cdot(x_3 + x_1')\cdot(x_3 + x_4) \\ &\quad = [(x_1 + x_2')']'\cdot[(x_3 + x_1')']'\cdot[(x_3 + x_4)']' \\ &\quad = [(x_1 + x_2')' + (x_3 + x_1')' + (x_3 + x_4)']' \\ &\quad = \downarrow[\downarrow(x_1, x_2'), \downarrow(x_3, x_1'), \downarrow(x_3, x_4)]. \end{aligned}$$

2.4. COMPARATORS

As the first example of an MSI package available in the TTL Series 74, we discuss in this section the 4-bit comparator. An *n-bit comparator* is a gate network that compares two n-bit words x and y and determines whether† $\perp x > \perp y$, or $\perp x = \perp y$, or $\perp x < \perp y$. To simplify our notation, we associate with any proposition p the binary variable $[p]$, where $[p] = 1$

†Recall that $\perp x$ is the integer denoted by the binary word x.

iff p is true. Furthermore, let $X \triangleq \perp x$ and $Y \triangleq \perp y$. Then the outputs of a comparator can be denoted by $[X > Y]$, $[X = Y]$, and $[X < Y]$.

A $4m$-bit comparator, $m \geq 1$, can be obtained as the cascade connection of m identical MSI packages known as *4-bit magnitude comparators*. To specify their performance, let $n > 4$ and let

$$X = \perp x = \perp(x_1, \ldots, x_n)$$
$$Y = \perp y = \perp(y_1, \ldots, y_n)$$
$$\hat{X} \triangleq \perp(x_5, \ldots, x_n)$$
$$\hat{Y} \triangleq \perp(y_5, \ldots, y_n).$$

A 4-bit magnitude comparator has 11 inputs, namely $x_1, \ldots, x_4, y_1, \ldots, y_4$, $[\hat{X} > \hat{Y}]$, $[\hat{X} = \hat{Y}]$, $[\hat{X} < \hat{Y}]$, and produces as outputs $[X > Y]$, $[X = Y]$, and $[X < Y]$.

The gate networks realizing a 4-bit comparator can be derived by the design methods of Section 1.6. We assume that the gate types available are INVERTER and multi-input NAND. We first derive expressions using $'$, $+$, and $\cdot$ and then convert them into INVERTER-NAND networks, using the technique of Section 2.3. Let $w_i \triangleq x_i \circleddash y_i$ $(i = 1, 2, 3, 4)$. The $[X = Y]$-output equation is easily obtained (see Section 1.6. Example 1):

$$X = Y \equiv x_1 = y_1 \wedge x_2 = y_2 \wedge x_3 = y_3 \wedge x_4 = y_4 \wedge \hat{X} = \hat{Y}.$$

Thus

$$[X = Y] = w_1 \cdot w_2 \cdot w_3 \cdot w_4 \cdot [\hat{X} = \hat{Y}].$$

To derive w_i, we use Table 1-6,

$$w_i = x_i \circleddash y_i = x_i' \cdot y_i' + x_i \cdot y_i.$$

The $[X > Y]$-output equation is derived as follows (see also Section 1.6, Example 6):

$$\begin{aligned} X > Y \equiv x_1 > y_1 &\vee (x_1 = y_1 \wedge x_2 > y_2) \vee (x_1 = y_1 \wedge x_2 = y_2 \wedge x_3 > y_3) \\ &\vee (x_1 = y_1 \wedge x_2 = y_2 \wedge x_3 = y_3 \wedge x_4 > y_4) \\ &\vee (x_1 = y_1 \wedge \cdots \wedge x_4 = y_4 \wedge \hat{X} > \hat{Y}). \end{aligned}$$

Thus

$$\begin{aligned} [X > Y] = x_1 \cdot y_1' &+ w_1 \cdot x_2 \cdot y_2' + w_1 \cdot w_2 \cdot x_3 \cdot y_3' + w_1 \cdot w_2 \cdot w_3 \cdot x_4 \cdot y_4' \\ &+ w_1 \cdot w_2 \cdot w_3 \cdot w_4 \cdot [\hat{X} > \hat{Y}]. \end{aligned}$$

The $[X < Y]$-output equation is obtained similarly:

$$\begin{aligned} [X < Y] = x_1' \cdot y_1 &+ w_1 \cdot x_2' \cdot y_2 + w_1 \cdot w_2 \cdot x_3' \cdot y_3 + w_1 \cdot w_2 \cdot w_3 \cdot x_4' \cdot y_4 \\ &+ w_1 \cdot w_2 \cdot w_3 \cdot w_4 \cdot [\hat{X} < \hat{Y}]. \end{aligned}$$

The gate networks for the $[X = Y]$- and $[X > Y]$-outputs, using NAND gates and INVERTERS, are shown in Fig. 2-7. Rather similar networks are used,

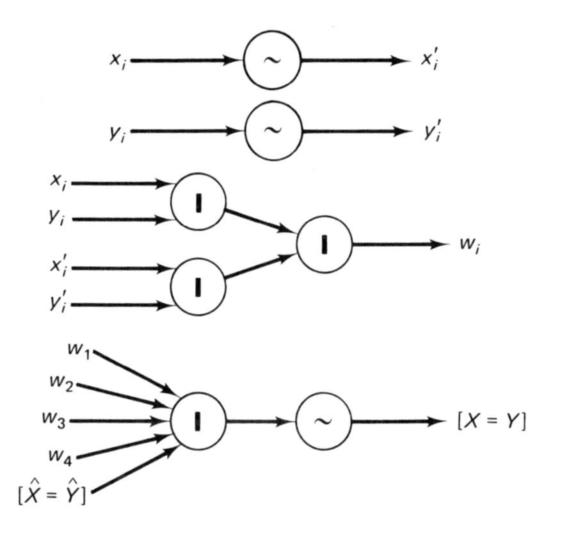

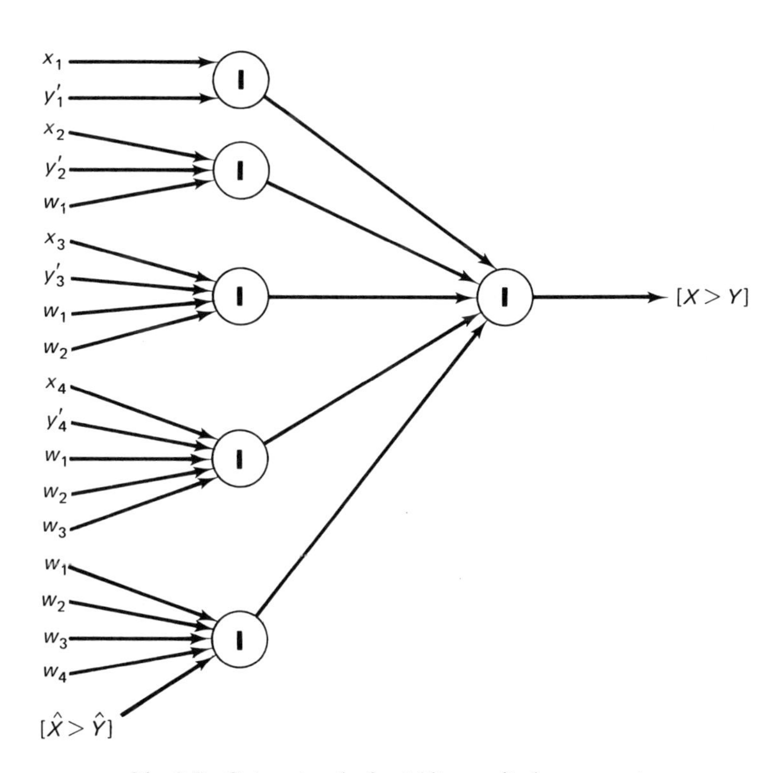

Fig. 2-7 Gate networks for 4-bit magnitude comparator.

in various commercially available IC packages, e.g., TTL 7485. A 12-bit comparator composed of three 4-bit comparators is shown in Fig. 2-8.

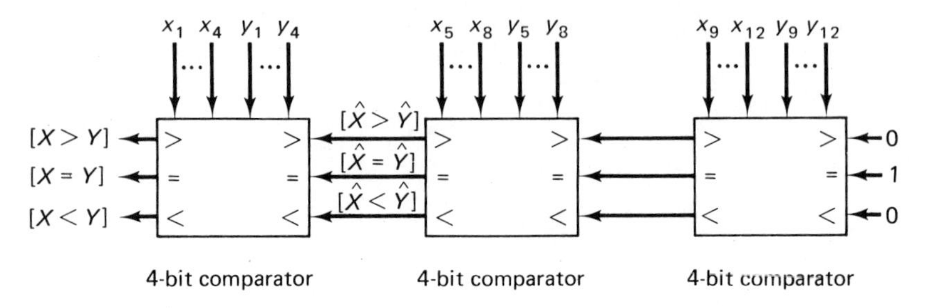

Fig. 2-8 12-bit comparator.

2.5. SELECTORS AND DECODERS

A *data selector* (or *multiplexer*) is a combinational gate network, as shown in Fig. 2-9. The inputs of a data selector are a binary *data word* $d = d_0, \ldots, d_{n-1}$; a binary *select word* $s = s_0, \ldots, s_{m-1}$, where $2^m = n$; and an *enable bit* or *strobe* e. Its output is a single bit z specified as follows:

$$z = e' \cdot d_{\perp s}.$$

Thus (when $e' = 1$), if $\perp s$ represents the decimal number i, z is equal to the ith bit of the data word.

A data selector can be used to implement an arbitrary Boolean function f on m variables. For this application, s is the input, e is set to 0, and $d_{\perp x}$ is set to $f(x)$ for every binary m-tuple x. One of the applications

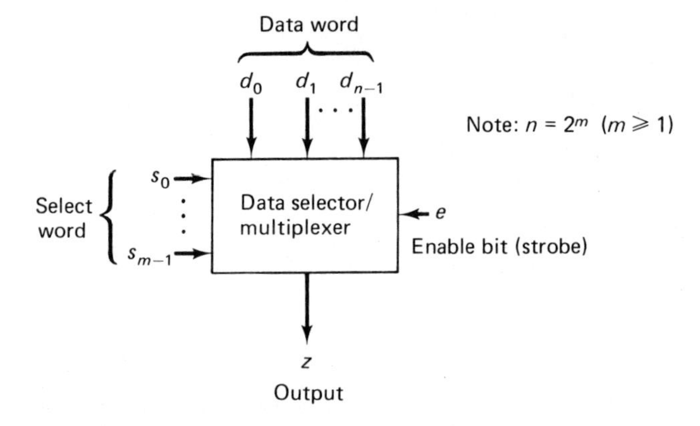

Fig. 2-9 Data selector/multiplexer.

of data selectors (multiplexers) is *time division multiplexing*, i.e., the high-speed sampling of n low-speed data sources in cyclical order, and the transmission of the sampled data over a single high-speed data channel. See Problem 1 in Chapter 9.

Data selectors/multiplexers for $m = 4$ ("16-line to 1-line") and $m = 3$ ("8-line to 1-line") are available as, e.g., TTL packages 74150 and 74151, respectively. The 74151 package provides, together with the true output z, its complement z'. The 74150 selector provides the complementary output z' only. The TTL package 74153 contains two "4-line to 1-line" (i.e., $m = 2$) data selectors/multiplexers.

A *decoder/demultiplexer* (see Fig. 2-10) connects a single-bit data input d to one out of $n = 2^m$ $(m \geq 1)$ binary outputs $z_0, \ldots, z_{n-1}$, selected by means of the m-bit address s, provided the *enable bit* (or *strobe*) e is set to 0. All the other outputs are set to 1. Precisely, the performance of a decoder/demultiplexer is specified by

$$z_i = 0 \equiv (e = 0) \wedge (d = 0) \wedge (i = \perp s).$$

In other words, if $d = e = 0$, then $z_{\perp s} = 0$ and all the other outputs are 1, and if $d + e = 1$, then all the outputs are 1.

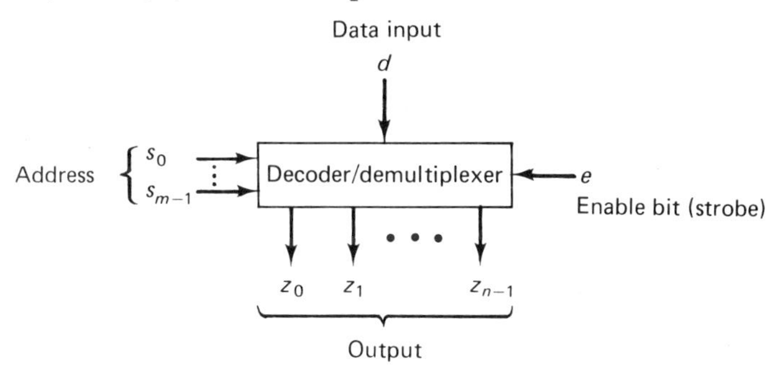

Fig. 2-10 Decoder/demultiplexer.

A decoder/demultiplexer for $m = 4$ ("4-line to 16-line") is available, e.g., as TTL IC package 74154. The package 74155 contains two decoder/demultiplexers for $m = 2$ ("2-line to 4-line").

The terminology may be somewhat confusing. The name "n-line to 1-line" multiplexer refers to the fact that there are n data bits that are channeled to one line. This is done with the aid of the m select bits. It would be natural to refer to the demultiplexer of Fig. 2-10 as a "1-line to n-line" demultiplexer. However, when the unit is used as a decoder, d is set to 0, and the m-bit address is decoded on n lines. For this reason the unit is called an "m-line to n-line" decoder.

2.6. PRIORITY ENCODERS

Consider the following problem. There are n lines $p_0, \ldots, p_{n-1}$ with line p_i having priority over line p_j iff $i < j$. The condition $p_i = 1$ indicates that a request on line p_i is present; $p_i = 0$ indicates the absence of such a request. The function of a *priority encoder* is to detect the line with highest priority among all the lines requesting service (see Fig. 2-11). The encoder

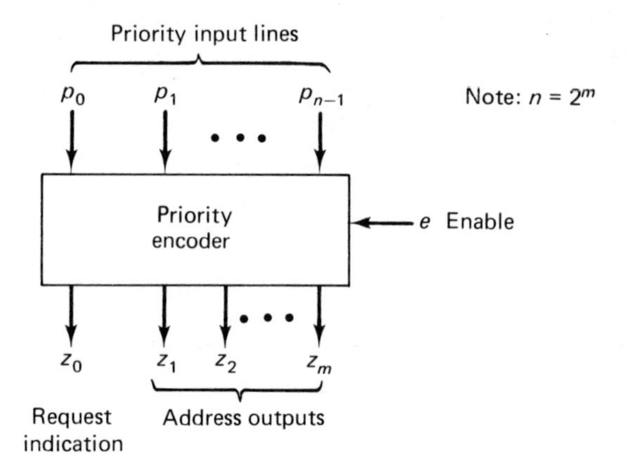

Fig. 2-11 Priority encoder.

has $n = 2^m$ *priority input* lines $p_0, \ldots, p_{n-1}$; an *enable* input e; and $m + 1$ output lines $z_0, z_1, \ldots, z_m$. When $e = 0$, the encoder is disabled and all the outputs are 0. When $e = 1$, the output z_0 tells whether any requests are present:

$$z_0 = 0 \qquad \text{if there are no requests, i.e., } p_0 + \cdots + p_{n-1} = 0$$

$$z_0 = 1 \qquad \text{if there are one or more requests, i.e., } p_0 + \cdots + p_{n-1} = 1.$$

When $e = 1$, the remaining outputs produce the code word (the "address") corresponding to the highest priority level present as follows: If i is the least value such that $p_i = 1$, then $\perp(z_1, \ldots, z_m) = i$. When $e = 1$ and there are no requests, the outputs $z_1, \ldots, z_m$ are irrelevant.

A high-speed (5 ns) 8-line priority encoder similar to the network defined above is available, e.g., in the ECL-10,000 family as package MC10165 [MOT3]. A low-speed ($\approx 1\ \mu s = 10^{-6}$ s) 8-line priority encoder is available as MOS/LSI package TMS 2801 [TI1]. Recently TTL packages implementing 10-line (TTL 74147) and 8-line (TTL 74148) priority encoders have also become available [TI2].

The design of a priority encoder using SSI packages will be described in Chapter 5.

2.7. CODE CONVERTERS

Binary-to-BCD Converters

BCD-to-binary and binary-to-BCD conversions are extensively used in computer systems. By a *binary representation* of an integer $N \geq 0$, we mean a binary word x such that $\perp x = N$. A *BCD representation* of N is obtained from its decimal representation by replacing each digit by its 4-bit binary representation. Leading 0's may be omitted. Thus 001100100110 and 1100100110 are both BCD representations of 326.

IC implementations of binary-to-BCD converters can be obtained by means of a sufficient number of copies of a single module, as we now proceed to show. Given $x = x_1, \ldots, x_n$, let $N_j \triangleq \perp(x_1, \ldots, x_j)$. Clearly $\perp x = N_n$ and

$$N_j = 2N_{j-1} + x_j, \tag{1}$$

where + is used in the conventional arithmetic sense. The iterative application of (1), starting (for example) from $j = 4$, will produce the sequence $N_4, \ldots, N_n = \perp x$. To illustrate, let $x = 1110110$. Then

$$\begin{aligned} N_3 &= 7 \\ N_4 &= 2 \times 7 + 0 = 14 \\ N_5 &= 2 \times 14 + 1 = 29 \\ N_6 &= 2 \times 29 + 1 = 59 \\ N_7 &= 2 \times 59 + 0 = 118 = \perp x. \end{aligned} \tag{2}$$

A cellular network implementing (2) is shown symbolically in Fig. 2-12(a). The performance of each cell [see Fig. 2-12(b)] is specified by

$$2d + c = 10\bar{c} + \bar{d}, \tag{3}$$

where d and $\bar{d}$ are decimal digits, whereas c and $\bar{c}$ are binary. Each row of cells in Fig. 2-12(a) performs the operation (1) for the corresponding value of j. Note that two cells in Fig. 2-12(a) are shown without $\bar{c}$-outputs. One verifies that these $\bar{c}$-outputs will be 0 for every 7-bit input word x. Thus the cellular network of Fig. 2-12(a) will perform correctly for every 7-bit input x.

Consider now replacing each decimal digit in Fig. 2-12(a) by its BCD representation. Correspondingly, we replace each vertical decimal rail by four binary rails, and the typical cell of Fig. 2-12(b) by the cell of Fig. 2-13(a), where y and $\bar{y}$ are the BCD representations of d and $\bar{d}$, respectively. In view of (3) the performance of the cell of Fig. 2-13(a) is specified by

$$2 \times \perp y + c = 10\bar{c} + \perp\bar{y} \qquad \text{where } \perp y \leq 9,\ \perp\bar{y} \leq 9 \tag{4}$$

or

$$16y_1 + 8y_2 + 4y_3 + 2y_4 + c = 10\bar{c} + (8\bar{y}_1 + 4\bar{y}_2 + 2\bar{y}_3 + \bar{y}_4).$$

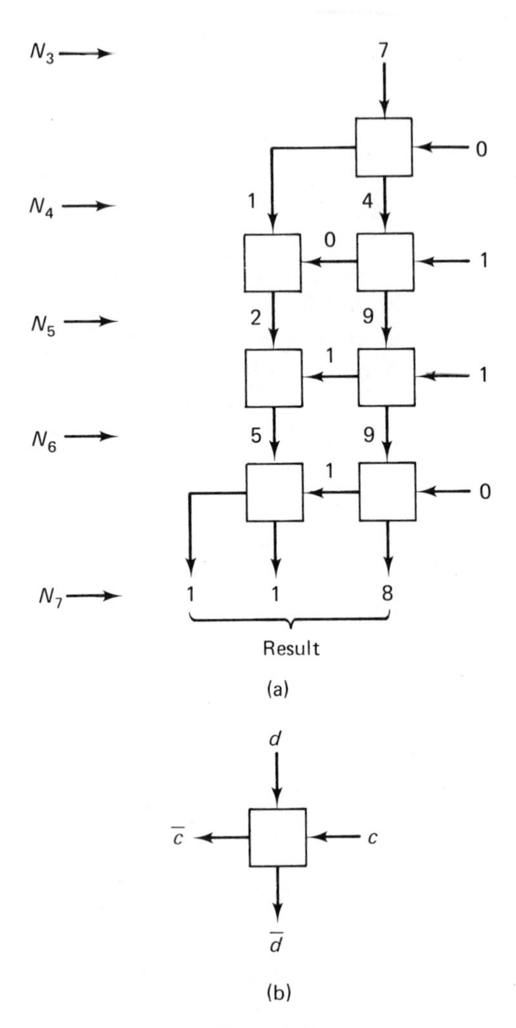

Figure 2-12

Note that $\bar{y}_4 = c$. Hence the cell of Fig. 2-13(a) can be replaced by the arrangement of Fig. 2-13(b), where BDM ("binary-to-decimal module") is the basic module by means of which arbitrary binary-to-BCD converters can be constructed.

If the y-input to a BDM is known to represent a decimal digit less than 5 (i.e., $\bar{c} = 0$), the BDM evidently degenerates into the simple through-connection of Fig. 2-13(c). Summarizing our observations we obtain for $n = 7$ the binary-to-BCD converter shown in Fig. 2-14, where D_j denotes the BCD representation of N_j. The particular binary values shown in Fig. 2-14 correspond to the example of Fig. 2-12(a).

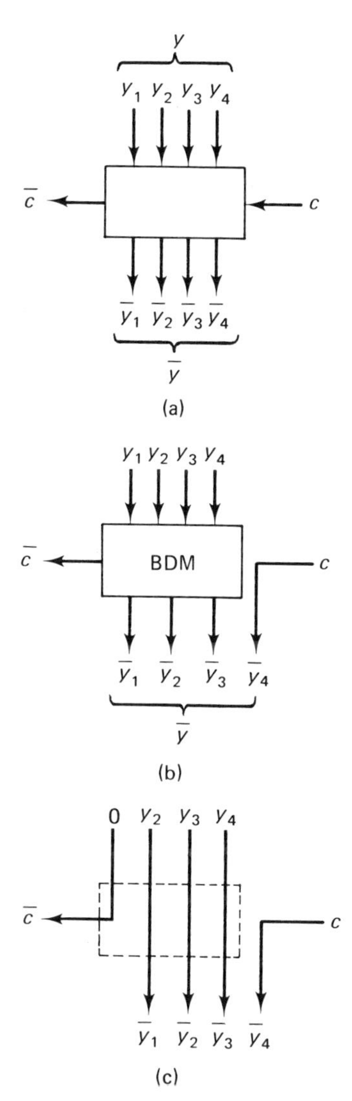

Figure 2-13

BCD-to-Binary Converters

A network performing BCD-to-binary conversion for integers up to 99 is shown in Fig. 2-15. The output is $z = z_1, \ldots, z_7$ and $N_j \triangleq \perp(z_1, \ldots, z_j)$. D_j denotes again the BCD representation of N_j. The particular binary values shown in Fig. 2-15 illustrate the conversion of $D_7 = 01011001$ (i.e., $N_7 = 59$) into $z = 0111011$. The basic step performed by each row of the network corresponds to a division of N_j by 2 to provide the quotient N_{j-1} and the remainder z_j. The basic module in Fig. 2-15 is the DBM, specified by

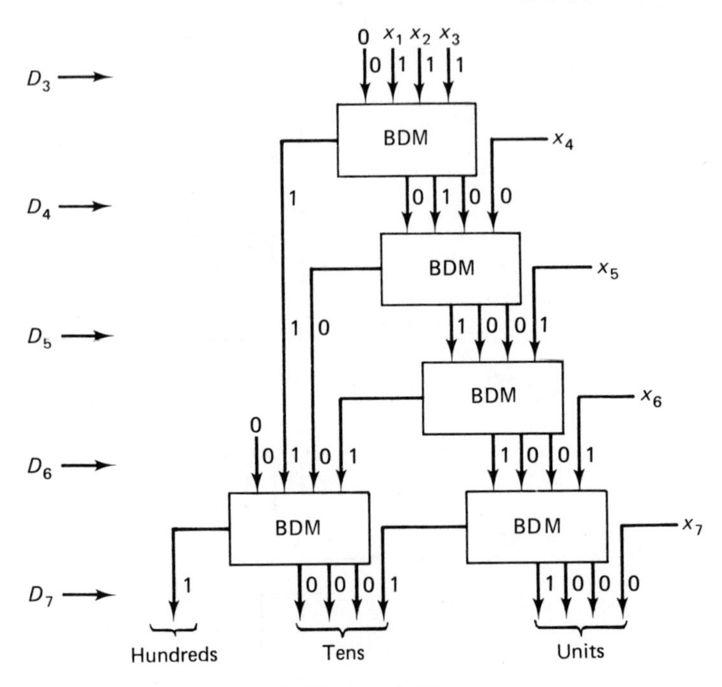

Fig. 2-14 Binary-to-BCD converter.

the converse function of a BDM. Referring to the notation of Fig. 2-15(b), we have

$$10x_1 + \perp(x_2, x_3, x_4, 0) = 2 \times \perp y$$

or

$$5x_1 + \perp(x_2, x_3, x_4) = \perp y.$$

The reader will verify that the network of Fig. 2-15(a) performs as required for all integers from 00 up to 99. The method is easily extended to more than 2 digits.

A TTL IC package MC4001 [MOT1] containing both a BDM and a DBM is available commercially. It has 4 inputs for the word to be converted and 8 outputs forming two output words, one for the BDM and one for the DBM. More complex modules for these types of conversion are also available, e.g., SN 74184 and SN 74185A [TI2].

For further reading on BCD-to-binary conversion, see [BEO].

Other Converters

Frequently, the decimal digits 0, 1, . . . , 9 are represented by a binary 1-out-of-10 code, the digit j being represented by the 10-bit word $\underbrace{0\ldots0}_{j\text{ zeros}}10\ldots0$. Combinational networks with 4 inputs and 10 outputs which convert from the BCD code to this 1-out-of-10 code are widely used and

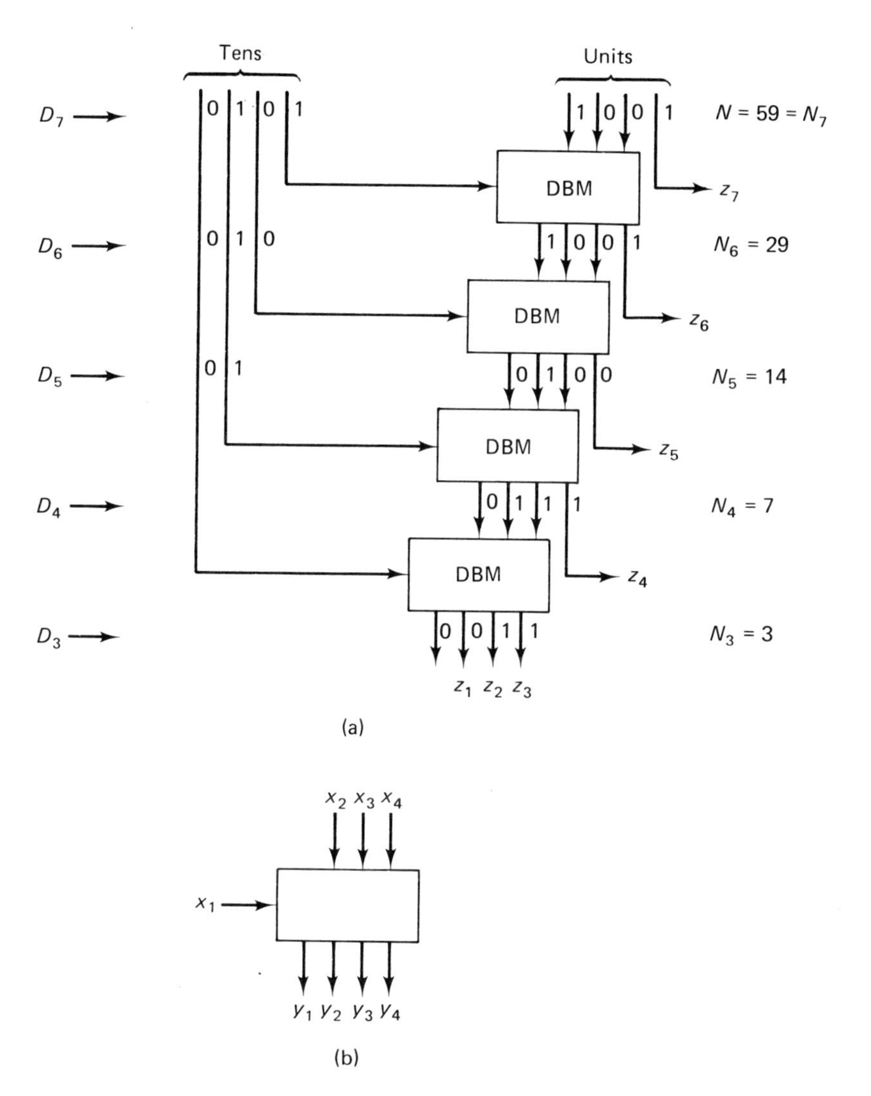

Fig. 2-15 BCD-to-binary converter.

are available in a variety of IC packages. They are usually referred to in the manufacturers' catalogs as "BCD-to-decimal decoder." One such IC package is the TTL 7442, which, as is frequently the case in this technology, provides the inverted, rather than the true, outputs. For any input word x, which is not a BCD code (i.e., $\perp x > 9$), all the outputs of the 7442 package are 1.

For various applications, 4-bit codes other than the BCD code are preferable for the representation of decimal digits. We show two such codes, together with the BCD code, in Table 2-2.

Table 2-2 BCD, EXCESS 3, and EXCESS 3 GRAY Codes

Decimal Digit	BCD	EXCESS 3	EXCESS 3 GRAY
0	0000	0011	0010
1	0001	0100	0110
2	0010	0101	0111
3	0011	0110	0101
4	0100	0111	0100
5	0101	1000	1100
6	0110	1001	1101
7	0111	1010	1111
8	1000	1011	1110
9	1001	1100	1010

The EXCESS 3 code for j, $0 \leq j \leq 9$, is the binary code for $j + 3$. This code is preferable to the BCD code in connection with the design of base-10 adders and subtractors. Note that in the EXCESS 3 code the digits j and $9 - j$ are always represented by complementary code words. (A code with this property is sometimes called *self-complementing*.)

GRAY codes have the property that any two consecutive code words differ in exactly one coordinate. In this connection the first code word is considered as consecutive to the last one. The EXCESS 3 GRAY code of Table 2-2 is an example of such a code. A 16-word GRAY code is shown in Table 2-3. Note that the EXCESS 3 GRAY code for j (Table 2-2) is the same as the GRAY code for $j + 3$ in Table 2-3.

GRAY codes have applications in connection with analog-to-digital conversions. Assume, for example, that an analog signal changes con-

Table 2-3 Sixteen-Word GRAY Code

Integer	GRAY Code
0	0000
1	0001
2	0011
3	0010
4	0110
5	0111
6	0101
7	0100
8	1100
9	1101
10	1111
11	1110
12	1010
13	1011
14	1001
15	1000

tinuously from the value 7 to the value 8. If BCD encoding is used, the outputs would change from 0111 to 1000. In reality the four outputs might not change simultaneously and a bit combination such as 0101 might occur during the transition. Such transitional outputs would provide erroneous information as to the state of the input signal. A GRAY code, on the other hand, prevents the occurrence of such erroneous transitional output combinations.

IC packages are available that convert essentially from the EXCESS 3 and EXCESS 3 GRAY codes (Table 2-2) to the 1-out-of-10 code mentioned earlier. Examples are the TTL packages 7443 (EXCESS 3) and 7444 (EXCESS 3 GRAY).

For a variety of other code converters, the reader is referred to the various manufacturers' catalogs.

2.8. STATIC READ-ONLY MEMORIES

One often deals with constant data (such as trigonometric constants) which need not be modified but are looked up periodically. If we store such data in a digital memory, we do not require the ability to write into the memory after the initial loading. This leads to more economical "read-only" memories (ROM's).

Normally, memories do not belong to combinational networks, and will be discussed in Chapter 8, in connection with sequential networks. However, the ROM described is a combinational network. Memory here corresponds to the fact that certain constants are provided to the network at manufacturing time.

A static ROM consists of a *logic array* and a decoder†. For our purposes a logic array may be considered as a *cellular* network, as shown in Fig. 2-16(a). Each cell in this network performs as indicated in Fig. 2-16(b). Thus one verifies that

$$z_j = a_1^j x_1 + a_2^j x_2 + \cdots + a_n^j x_n \qquad \text{for all } j \in \{1, \ldots, k\}.$$

The inputs to such a logic array are the x_i's and its outputs are the z_j's. The a_i^j-inputs to each cell are binary constants usually fixed during the manufacturing process. We shall refer to the $n \times k$ matrix $A \triangleq ||a_i^j||$ as the *memory matrix* of the logic array, where the subscript specifies the row and the superscript the column of A.

A static ROM may be visualized as shown in Fig. 2-17, where $n = 2^m$. If A is the memory matrix of the logic array and we denote the ith row of A by the (pseudo-APL) notation $A[i;]$, the performance of the ROM of Fig. 2-17 is given by

$$z = e \cdot A[1 \text{ PLUS } \perp s;],$$

†This decoder differs from that of Fig. 2-10 in that the outputs are inverted. Here, to "select" x_i means to make $x_i = 1$.

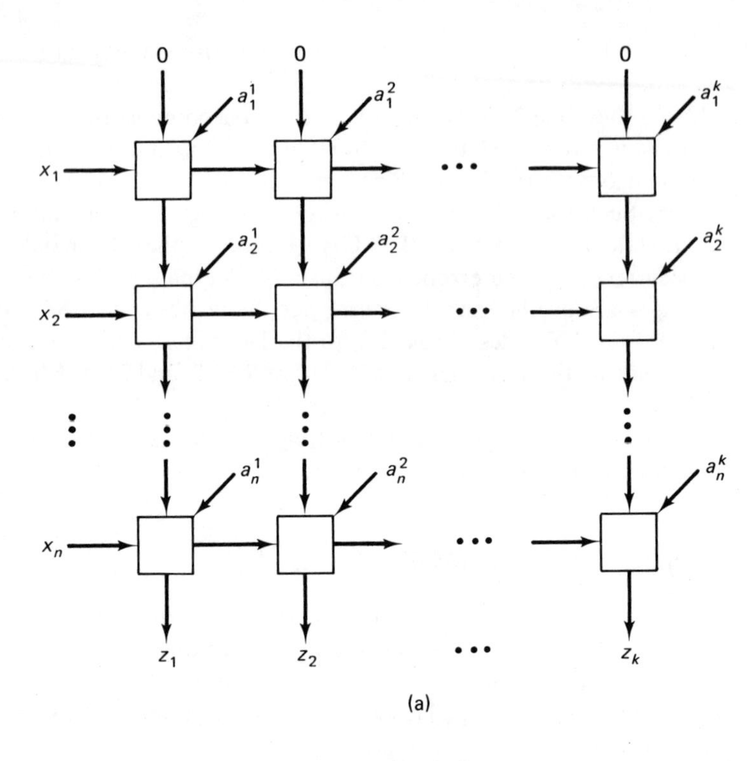

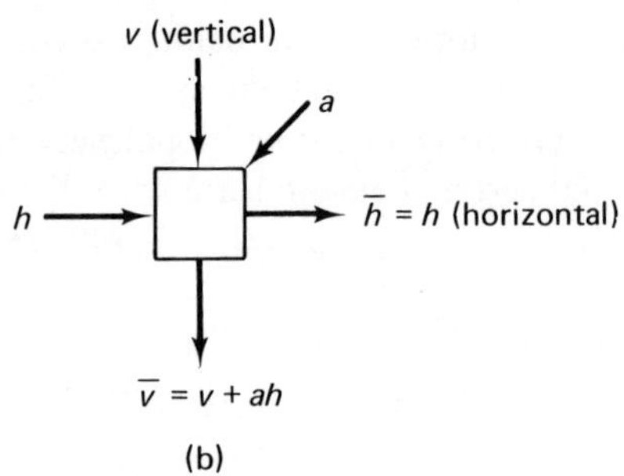

Fig. 2-16 Logic array as cellular network.

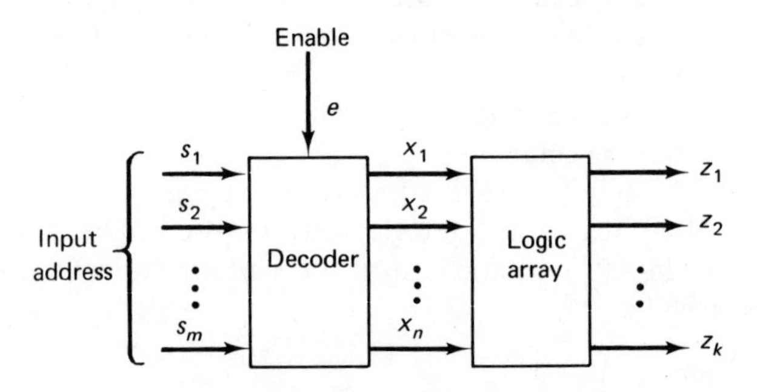

Fig. 2-17 Model of static ROM.

where $e \cdot (z_1, \ldots, z_k) \triangleq e \cdot z_1, \ldots, e \cdot z_k$, and PLUS refers to arithmetic addition. This can be checked as follows. If $\perp s = i - 1$, say, the decoder selects x_i (assuming that $e = 1$). Thus $x_i = 1$ and $x_j = 0$ for all $j \neq i$. Now

$$z_j = a_1^j x_1 + \cdots + a_n^j x_n = a_i^j.$$

Thus $z = a_i^1, a_i^2, \ldots, a_i^k$ is the ith row of A.

Note that the ROM of Fig. 2-17 could be used to realize k different Boolean functions f_j of m variables. For this application, the address corresponds to the m variables, the $2^m = n$ values of function f_j correspond to $a_1^j, a_2^j, \ldots, a_n^j$, and z_j gives the value of $f_j(s_1, s_2, \ldots, s_m)$. More will be said about this in Chapter 8.

Various ROM's are commercially available, as will be discussed in more detail in Chapter 8.

2.9. ARITHMETIC-LOGIC UNIT

Parallel Adders

We start this section by briefly reviewing some design aspects of parallel adders, including speed-up techniques. Fast adders are extensively used in modern high-speed digital computers. We then discuss an example of an MSI-implemented ALU (arithmetic-logic unit) that incorporates fast adding facilities.

An *n-bit parallel adder* has as its inputs two n-bit words $x = x_1, \ldots, x_n$ and $y = y_1, \ldots, y_n$ as well as an *input carry* bit c_n. Its outputs are an n-bit *sum* word $s = s_1, \ldots, s_n$ and an *output carry* bit c_0. The performance of the adder is characterized by the equation

$$\perp x \text{ PLUS } \perp y \text{ PLUS } c_n = \perp(c_0, s),$$

where PLUS denotes arithmetic addition of integers, and c_0, s is the $(n + 1)$-bit word $c_0, s_1, \ldots, s_n$. A 1-bit adder is usually referred to as a (single-bit) *full-adder* (in contrast to a half-adder, which adds two bits without a carry). For $n = 1$ the equation above simplifies to

$$x_1 \text{ PLUS } y_1 \text{ PLUS } c_1 = \perp(c_0, s_1). \tag{5}$$

Evidently, an n-bit parallel adder can be obtained by the cascade connection of n single-bit full-adders (n identical "cells"), as shown in Fig. 2-18. (The reader will verify that this implementation corresponds to the usual algorithm for addition of two binary numbers by hand.) This type of parallel adder is known as a *ripple-carry adder* (because the carry propagates or "ripples through" the individual cells). To obtain the output functions of a full-adder, observe that its characteristic equation (5) implies the following:

$$s_1 = 1 \equiv (x_1 \text{ PLUS } y_1 \text{ PLUS } c_1) \in \{1, 3\}$$
$$c_0 = 1 \equiv (x_1 \text{ PLUS } y_1 \text{ PLUS } c_1) \geq 2.$$

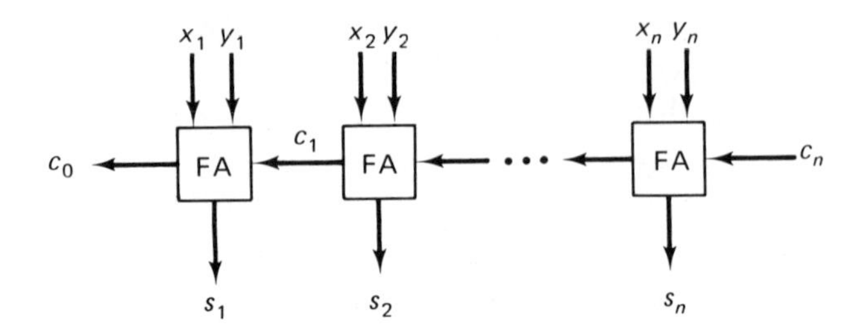

Fig. 2-18 *n*-bit parallel ripple-carry adder (FA represents a single-bit full adder).

Thus

$$s_1 = x_1 \oplus y_1 \oplus c_1 \qquad \text{[by Table 1-9(8), p. 12]}$$

and

$$c_0 = x_1 y_1 + x_1 c_1 + y_1 c_1.$$

The overall propagation delay of a ripple-carry adder (in the worst case) is n times the delay time required by a single-bit full-adder to produce the correct carry output. If this cell delay is, say, 20 ns and $n = 40$, the overall propagation delay of 800 ns may exceed the desired add time of the computer. Hence a variety of schemes have been introduced to speed up the operation of parallel adders. One such scheme, known as *0-level carry look-ahead,* is based on the following computations.

Consider a 4-bit parallel adder and assume the notation of Fig. 2-18. We have

$$c_3 = x_4 y_4 + x_4 c_4 + y_4 c_4 = g_4 + p_4 c_4, \tag{6}$$

where $g_i \triangleq x_i y_i$ and $p_i \triangleq x_i + y_i$. The g_i's and p_i's are called *1-carry generate* and *1-carry propagate* variables, respectively. Note that, if $g_i = 1$, it is known that the carry c_{i-1} out of cell i must be 1 regardless of the value of the carry c_i into cell i. Hence the name "1-carry generate". Also, if $p_i = 1$, either there is a carry generated in cell i, if $x_i y_i = 1$, or the carry c_i into cell i will be "propagated" through cell i, in the sense that if $x_i \oplus y_i = 1$, then $c_{i-1} = c_i$. Some authors use the name "carry propagate" for the function $x_i \oplus y_i$; here $p_i = x_i + y_i = (x_i \oplus y_i) \oplus x_i y_i$. Similarly,

$$c_2 = g_3 + p_3 c_3 = g_3 + p_3 g_4 + p_3 p_4 c_4 \tag{7}$$

$$c_1 = g_2 + p_2 c_2 = g_2 + p_2 g_3 + p_2 p_3 g_4 + p_2 p_3 p_4 c_4 \tag{8}$$

$$c_0 = g_1 + p_1 c_1 = g_1 + p_1 g_2 + p_1 p_2 g_3 + p_1 p_2 p_3 g_4 + p_1 p_2 p_3 p_4 c_4. \tag{9}$$

If we assume a propagation delay of 10 ns per gate, then a direct implementation of equation (9) would require only 30 ns to produce the correct output carry c_0, whereas 80 ns is needed in a 4-bit ripple-carry adder. Alternatively, note that

$$c_3 = 0 \equiv (x_4 \text{ PLUS } y_4 \text{ PLUS } c_4) < 2 \equiv (x'_4 \text{ PLUS } y'_4 \text{ PLUS } c'_4) \geq 2.$$

Hence

$$c'_3 = x'_4 y'_4 + x'_4 c'_4 + y'_4 c'_4 = {}^0g_4 + {}^0p_4 c'_4,$$

where ${}^0g_i \triangleq x'_i y'_i = p'_i$ and ${}^0p_i \triangleq x'_i + y'_i = g'_i$ are the *0-carry generate* and *0-carry propagate* variables, respectively. It follows that the *1-carry look-ahead* equations (6)–(9) remain correct if one replaces c_i by c'_i, g_i by ${}^0g_i$, and p_i by ${}^0p_i$. We shall refer to the resulting equations as *0-carry look-ahead* equations.

In the following discussion of an ALU, we illustrate applications of these look-ahead equations.

Arithmetic-Logic Unit

A 4-bit ALU is available [SIG, TI1] in the TTL Series 74 as a single 24-pin MSI package (74181). The unit can be set to perform various arithmetic operations on two 4-bit words, as well as all possible bit-by-bit Boolean operations. We shall provide a complete analysis of this unit, leading to a detailed description of its performance. To be consistent with this book, we modify somewhat the notation used by the manufacturer. See Fig. 2-19 for a block diagram of the unit.

By means of its *mode control* input m, the unit can be set to operate either in the LOGIC mode ($m = 1$) or the ARITHMETIC mode ($m = 0$).

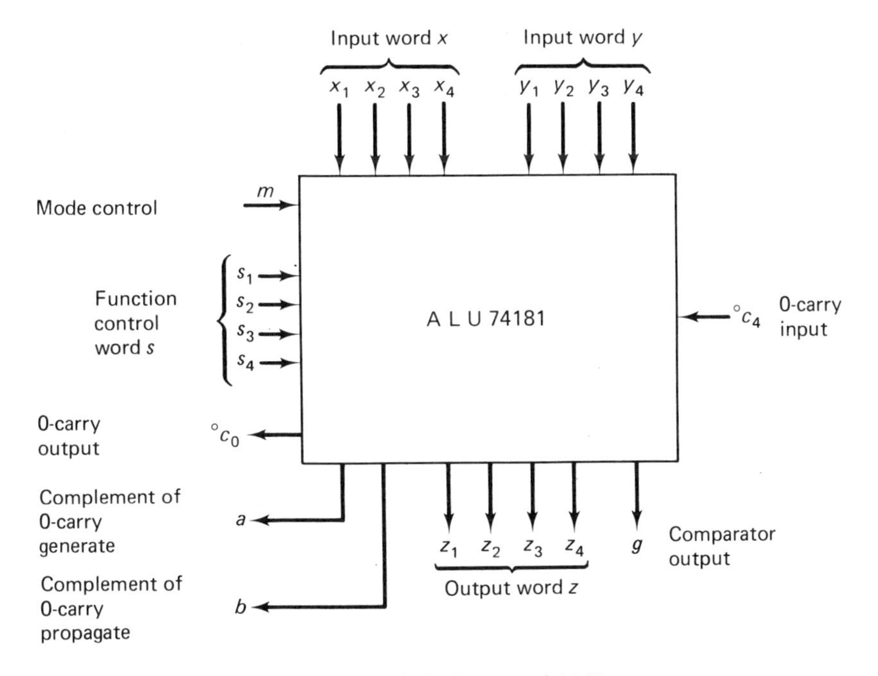

Fig. 2-19 Block diagram of ALU.

The unit has two 4-bit *data* inputs, $x = x_1, x_2, x_3, x_4$ and $y = y_1, y_2, y_3, y_4$; a 4-bit *function control* input, $s = s_1, s_2, s_3, s_4$; and a *0-carry* input, ${}^0c_4$. Its main output is a 4-bit word, $z = z_1, z_2, z_3, z_4$. If the unit is operated in its LOGIC mode, the relevant gate networks are shown in Fig. 2-20(a) and (b). Every setting of s determines a Boolean function $f_{\perp s}: V_2 \rightarrow \{0, 1\}$. This function is applied to every pair (x_i, y_i) to produce $z_i = f_{\perp s}(x_i, y_i)$, $i \in \{1, 2, 3, 4\}$. The correspondence $s \mapsto f_{\perp s}$ is easily derived from Fig. 2-20(a) and (b). For example, let $s = 1001$; i.e., $\perp s = 9$. Recalling that $\downarrow(x_1, \ldots, x_n) \triangleq (x_1 + \cdots + x_n)'$, we have

$$\begin{aligned} u_i &= \downarrow(x_i y_i, 0) = (x_i y_i)' \\ v_i &= \downarrow(0, y_i, x_i) = (x_i + y_i)' \\ w_i &= u_i \oplus v_i \\ z_i &= (1 + t_i) \oplus w_i = w_i', \end{aligned}$$

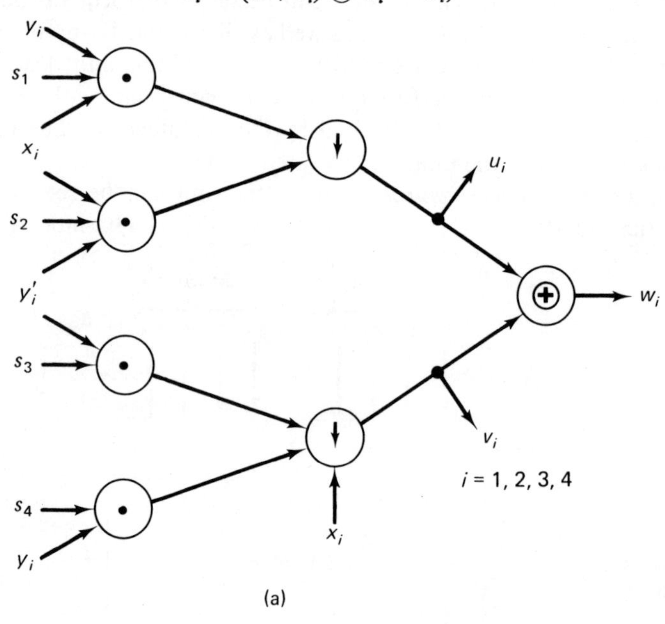

(a)

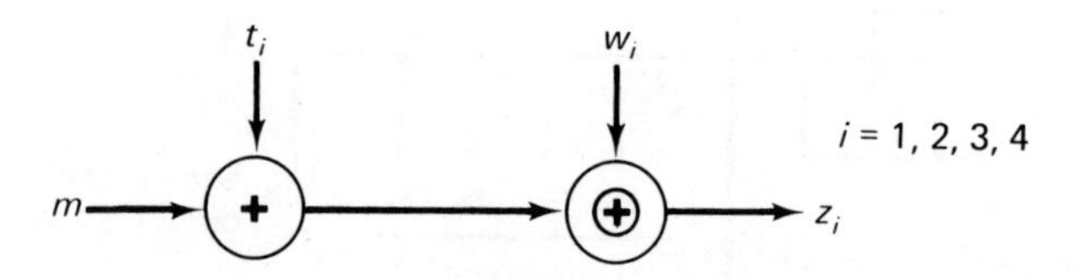

(b)

Fig. 2-20 Gate networks for ALU.

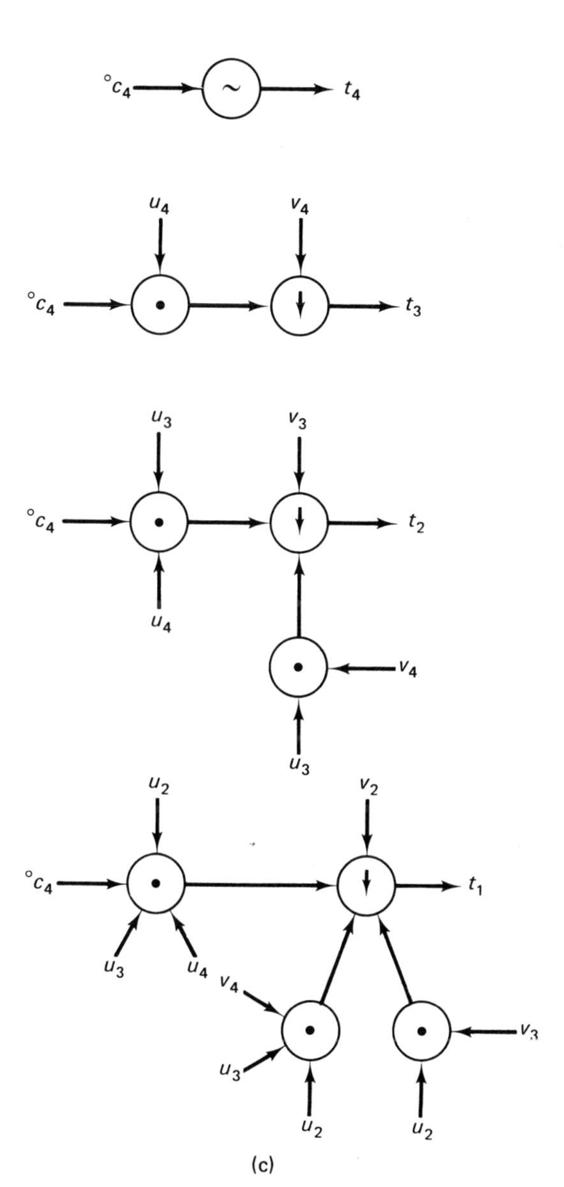

(c)

Fig. 2-20 cont'd

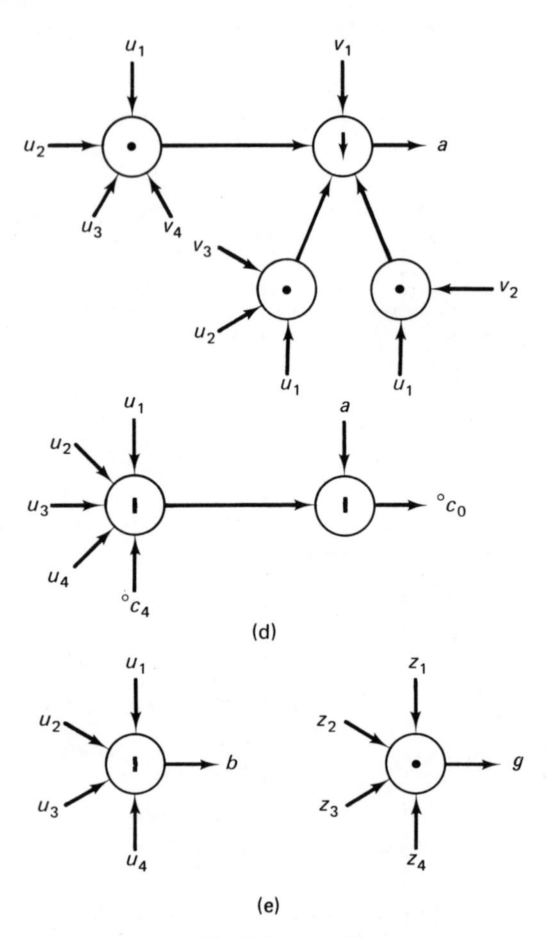

Fig. 2-20 cont'd

since $m = 1$ in the LOGIC mode. (The variable t_i is described later; clearly in the LOGIC mode its value is irrelevant.) Hence

$$f_9(x_i, y_i) = [(x_i y_i)' \oplus (x_i + y_i)']' = x_i \circledcirc y_i.$$

Evaluating all the functions $f_0, \ldots, f_{15}$, we obtain Table 2-4.

Assume now that the unit is operating in its ARITHMETIC mode ($m = 0$) and that s is again set to 1001. Referring again to Fig. 2-20 (a) and (b), we have

$$u_i = (x_i y_i)' = {}^0p_i \tag{10}$$

$$v_i = (x_i + y_i)' = {}^0g_i \tag{11}$$

$$w_i = u_i \oplus v_i = x_i \oplus y_i$$

$$z_i = w_i \oplus t_i = x_i \oplus y_i \oplus t_i.$$

Table 2-4 Logic Functions Generated by ALU Package 74181

$\perp s$	$f_{\perp s}(x_i, y_i) =$
0	x_i'
1	$x_i \downarrow y_i$
2	$x_i' y_i$
3	0
4	$x_i \mid y_i$
5	y_i'
6	$x_i \oplus y_i$
7	$x_i y_i'$
8	$x_i' + y_i$
9	$x_i \ominus y_i$
10	y_i
11	$x_i y_i$
12	1
13	$x_i + y_i'$
14	$x_i + y_i$
15	x_i

The networks producing the t_i's are shown in Fig. 2-20(c). One easily verifies that they correspond to the 0-carry look-ahead equations derived in the preceding subsection. Indeed, we have

$$t_4 = ({}^0c_4)' = c_4$$

$$t_3 = (v_4 + u_4{}^0c_4)' = (c_3')' = c_3$$

$$t_2 = (v_3 + u_3v_4 + u_3u_4{}^0c_4)' = (c_2')' = c_2$$

$$t_1 = (v_2 + u_2v_3 + u_2u_3v_4 + u_2u_3u_4{}^0c_4)' = (c_1')' = c_1.$$

It follows that the unit now performs as a 0-level carry look-ahead adder on x and y with $c_4 = ({}^0c_4)'$ as input carry bit and z as sum word. The unit also produces an inverted output carry ${}^0c_0 = c_0'$. The corresponding network, which implements the relevant 0-carry look-ahead equation, is shown in Fig. 2-20(d).

The performance of the ALU in its ARITHMETIC mode for other settings of s is easily derived from our preceding observations. For example, for $s = 1100$, we have

$$u_i = \downarrow(x_iy_i, x_iy_i') = x_i' = (x_ix_i)'$$

$$v_i = \downarrow(0, 0, x_i) = x_i' = (x_i + x_i)'.$$

Since these expressions for u_i and v_i are obtained from (10) and (11) by replacing y_i by x_i, it follows that the ALU now performs the addition $\perp x$ PLUS $\perp x$ PLUS c_4. The result is a left shift of x; i.e.,

$$c_0, z_1, z_2, z_3, z_4 = x_1, x_2, x_3, x_4, c_4.$$

For $s = 0110$, we obtain

$$u_i = \downarrow(0, x_i y_i') = (x_i y_i')'$$
$$v_i = \downarrow(y_i', 0, x_i) = (x_i + y_i')'.$$

Thus the ALU performs the addition

$$\perp x \text{ PLUS } \perp y' \text{ PLUS } c_4,$$

where $y' = y_1', y_2', y_3', y_4'$. For $c_4 = 1$, this corresponds to *2's-complement subtraction*, since $\perp x$ PLUS $\perp y'$ PLUS $1 = \perp x$ MINUS $\perp y \triangleq \perp x$ PLUS $(2^4 - \perp y)$. Note that for $s = 0110$ and $c_4 = 0$, the sum z will be 1111 iff $x = y$. Since $g = z_1 z_2 z_3 z_4$ [see Fig. 2-20(e)], the output g indicates whether $x = y$. Alternatively, we also have $g = 1 \equiv x = y$ for the setting $m = 1$ and $\perp s = 9$. Table 2-5 lists some of the arithmetic operations performed by the ALU for various settings of s. In this table we use the notation $x \square y \triangleq x_1 \square y_1, \ldots, x_4 \square y_4$ for any Boolean operation $\square$.

Table 2-5 Arithmetic Operations of the ALU Package 74181

$\perp s$	$\perp(c_0, z_1, z_2, z_3, z_4)$
0	$\perp x$ PLUS c_4
1	$\perp(x + y)$ PLUS c_4
6	$\perp x$ MINUS ($\perp y$ PLUS c_4')
8	$\perp x$ PLUS $\perp(x \cdot y)$ PLUS c_4
9	$\perp x$ PLUS $\perp y$ PLUS c_4
11	$\perp(x \cdot y)$ MINUS c_4'
12	$\perp x$ PLUS $\perp x$ PLUS c_4
13	$\perp(x + y)$ PLUS $\perp x$ PLUS c_4
15	$\perp x$ MINUS c_4'

A $4m$-bit ALU can be obtained by the cascade connection of m 4-bit ALUs. The adder operation of such a $4m$-bit ALU can be speeded up by introducing *higher-level carry look-ahead*. The relevant arrangement for $m = 4$ is shown in Fig. 2-21.

Referring to Fig. 2-20(d) and (e), we have

$$b = (u_1 u_2 u_3 u_4)'$$
$${}^0c_0 = |(a, |(u_1, u_2, u_3, u_4, {}^0c_4)).$$

Thus

$${}^0c_0 = a' + b' \cdot {}^0c_4.$$

This equation for ${}^0c_0$ has the same form as (6).

It follows that the output equations for the (1-level) look-ahead carry generator of Fig. 2-21 can be obtained from (6)–(8), by suitably renaming the variables. A corresponding MSI package (74182) is available.

The diagrams of Fig. 2-20 are functionally correct but do not reflect the details of actual hardware implementation. The computation of overall

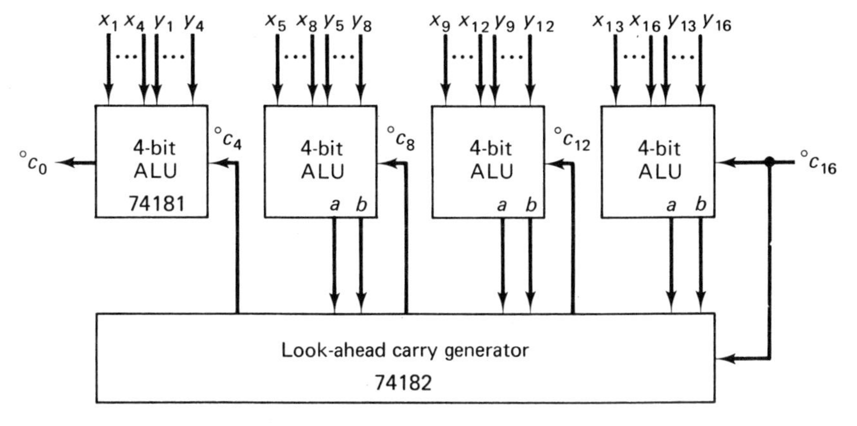

Fig. 2-21 16-bit ALU with carry look-ahead.

propagation delays using only such diagrams is not realistic. For the reader's information we quote in Table 2-6 typical figures given in [TI1].

Table 2-6 Typical ALU Addition Times (from [TI1])

Number of Bits	Package Count 74181	Package Count 74182	Total Addition Time (ns)
4	1	—	24
16	4	—	60
16	4	1	36
32	8	—	120
32	8	3	60

PROBLEMS

1. Let $x = x_1, \ldots, x_8$ and $y = y_1, \ldots, y_8$ be binary words. Design a network with output z such that $z = 1$ iff $x = y$ as follows:
   (a) Use any packages from Table 2-1, p. 31.
   (b) Use only INVERTER and NAND packages from Table 2-1.
   In each case, try to minimize the total number of packages.
2. Let $x = x_1, x_2$ and $y = y_1, y_2$ be binary words. Design a network with $z = 1$ iff $x = y$ using packages from Table 2-1, where the network propagation delay is to be ≤ 44 ns. Assume that complemented inputs are available.
3. Design a data selector for $m = 2$ ("4-line to 1-line") using packages of Table 2-1. Complemented inputs other than e' are not available. Do not use more than three packages.
4. Design an 8-line to 1-line data selector using two 4-line to 1-line data selectors. You may use a few auxiliary gates (AND, OR, and INVERTER types).

5. Design a 2-line to 4-line ($m = 2$) decoder/demultiplexer using the packages of Table 2-1. Complemented inputs are not available. Do not use more than three packages.

6. Design a 3-line to 8-line ($m = 3$) decoder/demultiplexer using two 2-line to 4-line demultiplexers. A few auxiliary gates are allowed. Complemented inputs are not available.

7. Design a 4-line to 16-line ($m = 4$) decoder/demultiplexer using two BCD-to-decimal converters TTL 7442 (see Section 2.7). You may use a few auxiliary gates. The enable input for the decoder may be omitted.

8. Design a network N to operate as follows. There are three data inputs x_1, x_2, x_3, a control input c, and an enable input E. If $E = 1$, the output w should be 0. When $E = 0$ and $c = 0$, w should be 1 iff the majority of the data inputs are 1. When $E = 0$ and $c = 1$, w should be 1 iff the number $\perp x = \perp(x_1, x_2, x_3)$ is even. Design the network using an 8-line to 1-line data selector/multiplexer. Assume that constant inputs 0 and 1 and complemented inputs are available.

9. Show how an arbitrary Boolean function of four variables can be implemented using an 8-line to 1-line data selector/multiplexer only. Assume that constant inputs 0 and 1 and complemented inputs are available.

10. (a) Show how an arbitrary Boolean function of six variables can be implemented using data selectors/multiplexers for $m = 2$.
(b) Same as part (a), but for $m = 3$.
(c) Same as part (a), but for $m = 4$.

11. Design a 16-line priority encoder using the following modules: 4-line priority encoders, 4-line to 1-line data selectors, and a few gates, if necessary.

12. Design a 4-bit BCD adder. The unit has two 4-bit *data* inputs, $x = x_1, x_2, x_3, x_4$ and $y = y_1, y_2, y_3, y_4$, and a *carry* input c_{IN}. Its outputs are a 4-bit *data* word $z = z_1, z_2, z_3, z_4$ and a carry output c_{OUT}. The data inputs are restricted to BCD representations, i.e., $0 \leq \perp x \leq 9$ and $0 \leq \perp y \leq 9$. The unit performs BCD addition; i.e., $0 \leq \perp z \leq 9$ and

$$10c_{OUT} \text{ PLUS } \perp z = \perp x \text{ PLUS } \perp y \text{ PLUS } c_{IN}.$$

You may use any of the packages described in this chapter, but no more than two packages, and a few auxiliary gates.

13. (a) Design a sorting unit for two 4-bit words. The unit has two 4-bit *data* inputs w and x, and two 4-bit *data* outputs y and z. If $\perp w > \perp x$, then $y = w$ and $z = x$; otherwise, $y = x$ and $z = w$.
You may use any of the packages described in this chapter, except the ALU. Eight packages are sufficient.
(b) Use five 2-word sorting units from part (a) to design a 4-word sorting network. The network has four 4-bit *data* inputs. Its outputs are the four data inputs, rearranged by descending magnitude. Thus, if w, x, y, z are the data outputs, then $\perp w \geq \perp x \geq \perp y \geq \perp z$ and the set $\{w, x, y, z\}$ coincides with the set of data inputs.

3 BOOLEAN ALGEBRAS

ABOUT THIS CHAPTER

In Chapter 1 we introduced the algebra applicable to switches and gates (which turns out to be a *Boolean algebra*) in a rather informal way. The purpose of this chapter is to present Boolean algebra along with some more general algebraic systems from the mathematical point of view. This chapter establishes a more rigorous foundation for the treatment of Boolean functions and expressions in Chapter 4 and of the minimization algorithms in Chapter 5. The topics covered here are, we believe, a useful background for several other areas of theoretical computer science and related subjects.

The application-oriented reader may find this treatment somewhat too detailed for his purposes. For this reason, the book is written in such a way that most of this chapter can be omitted without affecting the readability of other chapters. The nonmathematical reader should read only Section 3.2,† and then proceed to the "unstarred" sections of Chapter 4. The "starred" sections, marked *, contain topics that can be omitted without affecting the understanding of unstarred sections.

†The reader unfamiliar with the terminology and notation of set theory is referred to Section B.1 of Appendix B before proceeding with Section 3.2.

*3.1. INTRODUCTION

As general background material we provide an introduction to set theory in Appendix B and a brief summary of the properties of functions and relations in Appendix C.

An axiomatic treatment of Boolean algebra is given in Section 3.2. It is rigorous but does not provide much insight into the structure of Boolean algebras. For instance, the 19 axioms for Boolean algebras are not clearly motivated. To remedy this, we give in Sections 3.3 through 3.6 a more systematic presentation of Boolean algebras through posets, semilattices, and lattices. These algebraic systems are more general and serve to place Boolean algebras in proper perspective. Moreover, these systems have many other applications and appear rather frequently. For example, the subsets of a set form a semilattice under intersection, and the integers form a semilattice under the maximum operation. Similarly, lattices have applications, for example, to networks where transient conditions are taken into account, and to Boolean functions that involve only the OR and AND operations.

We also relate semilattices, lattices, and Boolean algebras to posets, thus gaining a new convenient representation. Finally, in Section 3.7 we show that every finite Boolean algebra is isomorphic to the algebra of all the subsets of a finite set. Many difficult problems become transparent when the set analogy is used.

In Chapter 4 we study Boolean algebras of Boolean functions. Section 4.5 provides a brief summary of the relevant algebraic concepts and puts the algebra of switches and gates in proper perspective.

3.2. BOOLEAN ALGEBRAS

The *Cartesian product* $X \times Y$ of two sets X and Y is the set of all ordered pairs (x, y), where $x \in X$ and $y \in Y$. For example, if $X = \{1, 2\}$, $Y = \{a, b, c\}$, then

$$X \times Y = \{(1, a), (1, b), (1, c), (2, a), (2, b), (2, c)\}.$$

A *unary operation* on a set X is a function from X into X. For example, for the set $\mathbf{Z}$ of integers, the function $f: z \mapsto z + 1$, $z \in \mathbf{Z}$, is a unary operation on $\mathbf{Z}$. A *binary operation* on X is a function from $X \times X$ into X. For example, addition $(+)$ of integers is a binary operation on $\mathbf{Z}$. Instead of using the functional notation "$+((2, 3)) = 5$," we usually write "$2 + 3 = 5$."

An *algebraic system* is a set X together with one or more operations that satisfy specified laws. We now define an algebraic system called a Boolean algebra.

Definition 1

A *Boolean algebra* is an algebraic system $B = \langle X, +, \cdot, ', 0, 1 \rangle$, where X is a set, $+$ and $\cdot$ are binary operations on X, $'$ is a unary operation on X, and 0 and 1 are two distinct elements of X (i.e. $0 \neq 1$) such that all the laws of Table 3-1 are satisfied. The elements 0 and 1 are called the *universal bounds* of B.

Table 3-1 Laws of Boolean Algebra

L1. $x + x = x$	L1′. $x \cdot x = x$	(idempotent laws)
L2. $x + y = y + x$	L2′. $x \cdot y = y \cdot x$	(commutative laws)
L3. $x + (y + z) = (x + y) + z$	L3′. $x \cdot (y \cdot z) = (x \cdot y) \cdot z$	(associative laws)
L4. $x + (x \cdot y) = x$	L4′. $x \cdot (x + y) = x$	(absorption laws)
L5. $x + 0 = x$	L5′. $x \cdot 1 = x$	(laws for 0 and 1)
L6. $x + 1 = 1$	L6′. $x \cdot 0 = 0$	
L7. $x + x' = 1$	L7′. $x \cdot x' = 0$	(laws for complements)
L8. $(x')' = x$		
L9. $x + (y \cdot z) = (x + y) \cdot (x + z)$	L9′. $x \cdot (y + z) = (x \cdot y) + (x \cdot z)$	(distributive laws)
L10. $(x + y)' = x' \cdot y'$	L10′. $(x \cdot y)' = x' + y'$	(De Morgan's laws)

For all x, y, and z in X.

The simplest Boolean algebra is the algebra $B_0 = \langle \{0, 1\}, +, \cdot, ', 0, 1 \rangle$, where the underlying set has only 2 elements, $+$ and $\cdot$ are the OR and AND operations, and $'$ is the complement operation as defined in Table 1-3. The fact that, by Table 1-5, 0 and 1 satisfy all the laws of Table 3-1 constitutes a proof that B_0 is indeed a Boolean algebra. (See pp. 9 and 10.)

Our second example is provided by $P(S)$, the set of all subsets of a set S. The terminology is summarized in Appendix B, Section B.1. Let $S \triangleq \{a, b, c\}$ be any set of 3 elements. We are interested in $P(S)$, where $P(S) = \{\varnothing, \{a\}, \{b\}, \{c\}, \{a, b\}, \{b, c\}, \{a, c\}, \{a, b, c\}\}$. The algebraic system $B(S) = \langle P(S), \cup, \cap, \overline{}, \varnothing, S \rangle$ can be shown to be a Boolean algebra, where the following correspondence is used:

General Concept	Corresponding Concept in Our Example
X	$P(S)$
$+$	$\cup$
$\cdot$	$\cap$
$'$	$\overline{}$
0	$\varnothing$
1	S

We claim that $B(S)$ satisfies all the laws of Table 3-1, after the proper substitution is made. We illustrate law L9 translated to $B(S)$. For example,

$$\{a, c\} \cup (\{a, b\} \cap \{b, c\}) = (\{a, c\} \cup \{a, b\}) \cap (\{a, c\} \cup \{b, c\}),$$

as the reader can easily verify. This law holds for all subsets X, Y, and Z of S; i.e.,

$$X \cup (Y \cap Z) = (X \cup Y) \cap (X \cup Z).$$

The interested reader will find more information in Appendix B. In particular, Table B-1 corresponds to Table 3-1 in this chapter, and this correspondence can be used to show that the system $B(S)$ defined for any nonempty set S is a Boolean algebra.

As a third example of a Boolean algebra, consider the set Y of positive integers,

$$Y \triangleq \{1, 2, 3, 5, 6, 10, 15, 30\}.$$

Let $\mathrm{lcm}(x, y)$ and $\gcd(x, y)$ be the least common multiple of x and y and the greatest common divisor of x and y, respectively. We establish the following correspondence:

General Concept	Corresponding Concept in Our Example
X	Y
$x + y$	$\mathrm{lcm}(x, y)$
$x \cdot y$	$\gcd(x, y)$
x'	$30/x$
0_B	1
1_B	30

Here we designate the special elements of B by 0_B and 1_B, to avoid confusion with the integers. (The integer 1 in the set Y acts as the zero element of the Boolean algebra, for example.)

One can verify that the system above is a Boolean algebra, i.e., that the laws of Table 3-1 hold. We illustrate L9 in the new terminology:

$$\mathrm{lcm}(6, \gcd(10, 15)) = \gcd(\mathrm{lcm}(6, 10), \mathrm{lcm}(6, 15)).$$

Examples of L7 and L7′ are

$$\mathrm{lcm}(3, 30/3) = 30 \quad \text{and} \quad \gcd(5, 30/5) = 1.$$

The reader of only unstarred material may now proceed to Chapter 4.

*3.3. POSETS

Definition and Representation

(For a further discussion of relations, the reader is referred to Appendix C.)

A *binary relation R on X* is any subset of $X \times X$. For instance, if

$X = \{1, 2, 3\}$, an example of a binary relation on X is

$$R \triangleq \{(1, 2), (1, 3), (2, 3)\}.$$

The elements 1 and 2 of X are related by R; we denote this by $(1, 2) \in R$ or $1R2$. It is easily seen that this relation represents the familiar "less than" relation for numbers and, rather than using the general symbol R, we can also write

$$1 < 2, \quad 1 < 3, \quad \text{and} \quad 2 < 3.$$

A binary relation R on X is called a *partial order* (relation) iff it satisfies the following three conditions:

1. *Reflexive Law:* xRx, for all $x \in X$.
2. *Antisymmetric Law:* xRy and yRx implies $x = y$, for all $x,y \in X$.
3. *Transitive Law:* xRy and yRz implies xRz, for all $x,y,z \in X$.

For example, the relation $<$ is not a partial order because it fails to satisfy the reflexive law: $1 < 1$ is false, or $1 \nless 1$. On the other hand, the relation $\leq$ for integers is a partial order, as one can easily verify.

For $X = \{1, 2, 3, 4\}$, the relation $\leq$ is formally represented by

$$\{(1, 1), (1, 2), (1, 3), (1, 4), (2, 2), (2, 3), (2, 4), (3, 3), (3, 4), (4, 4)\}.$$

This notation is convenient for theoretical arguments but is awkward for practical purposes. A better representation is the "poset diagram" ("poset" for *p*artially *o*rdered *set*), which is illustrated in Fig. 3-1, for the example above. For this example, the diagram is intuitively clear; this notion will be made more precise later.

Figure 3-1

Another example of a partial order is $\subseteq$ for sets.

The *converse* of a binary relation R on X is the binary relation $R^{-1} \triangleq \{(y, x) \mid (x, y) \in R\}$. For example, the converse of $\leq$ on $\mathbf{Z}$ is $\geq$ and the converse of $\subseteq$ is $\supseteq$ for sets.

It is customary to use the symbol $\leq$ to denote any partial order on any set, and the symbol $\geq$ will denote the converse of $\leq$. If $\leq$ is a partial order, then so is $\geq$. This is sometimes called the *duality principle*. Furthermore, we write $x < y$ or $y > x$ for $x \leq y \wedge x \neq y$.

Definition 2

A *poset* is an ordered pair $\langle X, \leq \rangle$, where X is a set and $\leq$ is a partial order on X.

It follows from the duality principle that if $\langle X, \leq \rangle$ is a poset, then so is $\langle X, \geq \rangle$. $\langle X, \geq \rangle$ is called the *dual* of $\langle X, \leq \rangle$.

Examples of infinite posets are $\langle \mathbf{Z}, \leq \rangle$ and $\langle \mathbf{Z}, \geq \rangle$. To give an example of a finite poset, we set $X \triangleq \{1, 2, 3, 4, 5\}$ and

$$\leq \triangleq \{(1, 1), (2, 2), (3, 3), (4, 4), (5, 5), (2, 1), (3, 1), (4, 1), (5, 1), (4, 2), (5, 2)\}.$$

Then $\langle X, \leq \rangle$ is a poset represented in Fig. 3-2(a). In this representation, called a *relation diagram*, the points (*nodes*, *vertices*) correspond to the elements of X, and an arrow from node i to node j indicates that $i \leq j$.

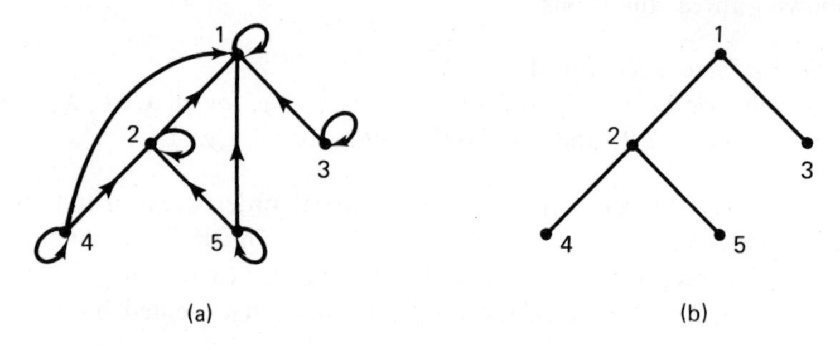

Fig. 3-2 Example of a poset: (a) relation diagram; (b) poset diagram.

A simplified diagram called the *poset diagram*, which represents the same poset, is shown in Fig. 3-2(b). The relation diagram can be derived from the poset diagram by applying the following rules:

1. Convert every line into an arrow pointing upward.
2. Add a self-loop ($\circlearrowleft$) at each node.
3. Add all the arrows required to make the relation represented by the diagram a transitive relation.

Covering Relations

The preceding ideas on poset representations can be formulated more precisely by means of the following notion of "covering."

Definition 3

Let $\langle X, \leq \rangle$ be a poset. We say that x *covers* y (written $x \gtrdot y$ or $y \lessdot x$) if $x > y$ and there exists no $z \in X$ such that $x > z > y$.

The poset diagram for a finite poset is obtained as follows. Each element is represented by a node. In general, if $x > y$, then x is placed above y. A line joins x and y iff $x \gtrdot y$, or $y \gtrdot x$.

The rules in the preceding subsection for converting a poset diagram into the corresponding relation diagram are based on the following proposition, which is easily verified (see Appendix C for notation).

Proposition 1

Let $\langle X, \leq \rangle$ be a finite poset and $\lessdot$ the corresponding covering relation. Then $\leq\, = (\lessdot)^*$.

In other words, the partial ordering of a finite poset is the reflexive and transitive closure of its covering relation. Rule 2 above produces the reflexive closure of $\lessdot$ and rule 3 the reflexive and transitive closure $(\lessdot)^* = \leq$.

Bounds

Definition 4

If $\langle X, \leq \rangle$ is a poset, $S \subseteq X$ and $x \in X$, then x is a *lower bound* of S iff $x \leq s$ for all $s \in S$. A lower bound x of S is called *greatest lower bound* of S, written glb(S), iff $y \leq x$ for every lower bound y of S. The concepts of *upper bound* and *least upper bound* of S, written lub(S), are defined dually, i.e., by interchanging $\leq$ with $\geq$.

Clearly, if glb(S) exists, it is unique, and the same holds for lub(S). If $S = \{x, y\}$, we write glb(x, y) and lub(x, y) rather than glb($\{x, y\}$) and lub($\{x, y\}$).

Referring to the 5-element poset of Fig. 3-2, we observe that 1 and 2 are both upper bounds of 4 and 5, and $2 =$ lub(4, 5). Also, $1 =$ lub(2, 3) and $2 =$ lub(2, 4). Indeed, any set of elements in Fig. 3-2 has an lub. On the other hand, no lower bounds exist for {2, 3}, {4, 5}, {3, 4}, and {3, 5}.

In the 8-element poset $\langle X, \leq \rangle$ whose poset diagram is given in Fig. 3-3,

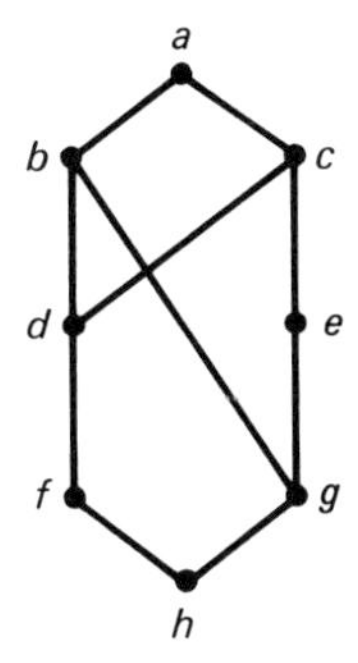

Fig. 3-3 Example of an 8-element poset.

every set of elements has at least one upper bound and at least one lower bound. For example, the upper bounds of $\{f, g\}$ are b, c and a, but none of them is an lub of f and g. Similarly, b and c have no glb. On the other hand, $a =$ lub(b, c), $c =$ lub(e, f) $=$ lub(c, f), $h =$ glb(e, f), $f =$ glb(c, f), $h =$ glb(X), $a =$ lub(X), etc.

Poset Isomorphism

The concept of isomorphism plays an important role in abstract algebra. For posets, we define isomorphism as follows. We write $f\colon X \cong Y$ to state that $f\colon X \longrightarrow Y$ is bijective (one-to-one and onto). (See Appendix C.)

Definition 5

Let $P = \langle X, \leq \rangle$ and $Q = \langle Y, \sqsubseteq \rangle$ be posets. A function $f\colon X \longrightarrow Y$ is an *isomorphism* of P to Q, written $f\colon P \cong Q$, iff $f\colon X \cong Y$ and, furthermore, $x \leq y \equiv f(x) \sqsubseteq f(y)$, for all $x,y \in X$. The function $f\colon X \longrightarrow Y$ is a *dual isomorphism* of P to Q iff $f\colon \langle X, \leq \rangle \cong \langle Y, \sqsupseteq \rangle$.

For example, the poset shown in Fig. 3-4, with set inclusion $\subseteq$ as partial order, is dually isomorphic to the poset of Fig. 3-2. A corresponding dual isomorphism f is given by $f\colon 1 \mapsto \{1\}$, $2 \mapsto \{1, 2\}$, $3 \mapsto \{1, 3\}$, $4 \mapsto \{1, 2, 5\}$, $5 \mapsto \{1, 2, 4\}$.

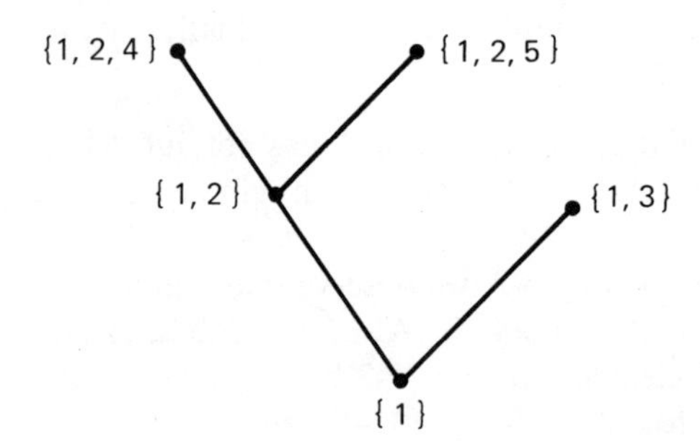

Fig. 3-4 Example of a poset with $\subseteq$ as partial order.

*3.4. SEMILATTICES

Definition 6

A *semilattice* is an algebraic system $\langle X, \square \rangle$, where X is a set and $\square$ is a binary operation on X which satisfies the following laws, for all $x,y,z \in X$:

SL1. $x \square x = x$ (idempotent law)

SL2. $x \square y = y \square x$ (commutative law)

SL3. $x \square (y \square z) = (x \square y) \square z$ (associative law)

The intersection $\cap$ of sets is an example of a semilattice operation, in view of S1′–S3′ (Table B-1, Appendix B), and so is $\cup$, in view of S1–S3 (Table B-1). More precisely, we have the following proposition.

Proposition 2

Let X be a set. Then $\langle P(X), \cap \rangle$ and $\langle P(X), \cup \rangle$ are semilattices.

Proof

By means of Proposition 1 of Appendix B, one easily shows that $\cap$ and $\cup$ are binary operations on $P(X) \triangleq \{S \mid S \subseteq X\}$; i.e., $S,T \in P(X)$ implies that $S \cap T \in P(X)$ and $S \cup T \in P(X)$. By Table B-1, S1–S3 and S1′–S3′, both operations satisfy the semilattice laws SL1–SL3. ▯

The maximum and minimum operations also provide examples of semilattices. For any two integers x and y, let $x \lceil y \triangleq \max(x, y)$ and $x \lfloor y \triangleq \min(x, y)$. Then $\langle \mathbf{Z}, \lceil \rangle$ and $\langle \mathbf{Z}, \lfloor \rangle$ are semilattices, and more generally, for any subset $X \subseteq \mathbf{Z}$, $\langle X, \lceil \rangle$ and $\langle X, \lfloor \rangle$ are semilattices.

A finite semilattice $\langle X, \square \rangle$ can, of course, be defined by means of a table, specifying X and $\square$. See Table 3-2 for an example. (The reader should verify that the operation $\square$ in Table 3-2 is indeed a semilattice operation!)

Table 3-2 Example of 5-Element Semilattice

$X = \{1, 2, 3, 4, 5\}$

$\square$	**1**	**2**	**3**	**4**	**5**
1	1	1	1	1	1
2	1	2	1	2	2
3	1	1	3	1	1
4	1	2	1	4	2
5	1	2	1	2	5

With every semilattice $\langle X, \square \rangle$ one may associate a partial order on X, as shown next.

Proposition 3

Let $\langle X, \square \rangle$ be a semilattice. Then the relation $\leq$ on X defined by

$$x \leq y \equiv x \square y = x \qquad \text{for every } x,y \in X$$

is a partial order on X.

Proof

By the definition of a partial order we have to show that $\leq$ is reflexive, antisymmetric, and transitive. SL1 implies the reflexive law $x \leq x$, for every $x \in X$. Next, assume that $x \leq y$ and $y \leq x$. By the definition of $\leq$, we have $x \square y = x$ and $y \square x = y$. By SL2, $x \square y = y \square x$. Hence $x = y$. Thus the relation $\leq$ satisfies the antisymmetric law. To prove transitivity, let $x \leq y$ and $y \leq z$, i.e., $x \square y = x$ and $y \square z = y$. Now,

$$\begin{aligned} x \square z &= (x \square y) \square z \\ &= x \square (y \square z) \qquad \text{(by SL3)} \\ &= x \square y \\ &= x. \end{aligned}$$

Hence $x \leq z$. Thus $\leq$ is indeed transitive. ▯

By Proposition 1(j) of Appendix B, the partial orders associated with the semilattices $\langle P(X), \cap\rangle$ and $\langle P(X), \cup\rangle$ are $\subseteq$ and $\supseteq$, respectively. The partial order associated with $\langle \mathbf{Z}, \lfloor\rangle$ is the usual "less than or equal" relation $\leq$ defined on integers. For the semilattice of Table 3-2, the partial order is the converse of that given by Fig. 3-2. (Verify this!)

Proposition 1(d), (e), and (f) of Appendix B turn out to be special cases of the following general result on semilattices.

Proposition 4

Let $\langle X, \square\rangle$ be a semilattice and $\leq$ its associated partial order. Then $x \square y = \text{glb}(x, y)$; i.e., for every $x,y,z \in X$:

(a) $x \square y \leq x$.
(b) $x \square y \leq y$.
(c) If $z \leq x$ and $z \leq y$, then $z \leq x \square y$.

Proof

(a) $(x \square y) \square x$
$= (x \square x) \square y$ (by SL2 and SL3)
$= x \square y$. (by SL1)
Thus $x \square y \leq x$.

(b) $x \square y = y \square x$ (by SL2)
$y \square x \leq y$. [by (a)]
Thus $x \square y \leq y$.

(c) Let $z \leq x$ and $z \leq y$. Then $z \square x = z$ and $z \square y = z$.
Hence $z \square (x \square y) = (z \square x) \square y = z \square y = z$. Thus $z \leq x \square y$. ▯

*3.5. LATTICES

Basic Properties

We now turn to lattices, i.e., to algebraic systems with two semilattice operations, suitably interrelated.

Definition 7

A *lattice* L is an algebraic system $L = \langle X, +, \cdot\rangle$, where X is a set and $+$ and $\cdot$ are binary operations on X which obey, for all $x,y,z \in X$, the following laws:

L1. $x + x = x$	L1′. $x \cdot x = x$	(idempotent laws)
L2. $x + y = y + x$	L2′. $x \cdot y = y \cdot x$	(commutative laws)
L3. $x + (y + z)$ $= (x + y) + z$	L3′. $x \cdot (y \cdot z) = (x \cdot y) \cdot z$	(associative laws)
L4. $x + (x \cdot y) = x$	L4′. $x \cdot (x + y) = x$.	(absorption laws)

It is customary, especially in switching theory, to denote the lattice operations by the symbols $+$ and $\cdot$ and to refer to them as "lattice addition" and "lattice multiplication." However, the reader must clearly distinguish between these lattice operations and the corresponding arithmetic operations. Evidently, arithmetic addition and multiplication are *not* idempotent, nor do these operations obey the absorption laws! Furthermore, the general lattice operation symbols $+$ and $\cdot$ are not to be confused with the operations $+$ and $\cdot$ defined on the particular set $\{0, 1\}$ in Section 1.4. It is true, however, that under the definitions of Section 1.4, the system $\langle\{0, 1\}, +, \cdot\rangle$ forms a lattice.

Definition 7 is equivalent to the definition of a lattice as an algebraic system $L = \langle X, +, \cdot\rangle$ such that $\langle X, +\rangle$ and $\langle X, \cdot\rangle$ are semilattices and the absorption laws L4 and L4′ are satisfied.

Examples of lattices are $\langle P(X), \cup, \cap\rangle$, where X is an arbitrary set (in view of Table 1, S1–S4 and S1′–S4′, of Appendix B), as well as $\langle Y, \lceil, \lfloor\rangle$, where Y is an arbitrary set of integers. Furthermore, if $L = \langle X, +, \cdot\rangle$ is a lattice, then so is $L^D = \langle X, \cdot, +\rangle$, the *dual* lattice of L. This is due to the fact that the set of laws in Definition 7 remains the same if the symbols $+$ and $\cdot$ are interchanged.

Now let $L = \langle X, +, \cdot\rangle$ be a lattice. We denote by $\leq$ the partial order associated with the semilattice $\langle X, \cdot\rangle$. Thus

$$(x \leq y) \equiv (x \cdot y = x) \qquad \text{for all } x, y \in X. \tag{1}$$

The following proposition shows that $\geq$, the converse of $\leq$, is the partial order associated with the semilattice $\langle X, +\rangle$.

Proposition 5

Let $L = \langle X, +, \cdot\rangle$ be a lattice. Then

$$(x + y = x) \equiv (x \geq y) \qquad \text{for every } x, y \in X.$$

Proof

Assume that $x + y = x$. Then

$$\begin{aligned} y \cdot x &= y \cdot (x + y) \\ &= y \cdot (y + x) && \text{(by L2)} \\ &= y. && \text{(by L4}'\text{)} \end{aligned}$$

Thus $y \leq x$; i.e., $x \geq y$.

Assume now that $x \geq y$. Then $y \leq x$; i.e., $y \cdot x = y$. Hence

$$\begin{aligned} x + y &= x + (y \cdot x) \\ &= x + (x \cdot y) && \text{(by L2}'\text{)} \\ &= x. && \text{(by L4)} \end{aligned}$$

Thus $x + y = x$. □

By combining Propositions 4 and 5, we immediately obtain the following result.

Proposition 6

Let $L = \langle X, +, \cdot \rangle$ be a lattice and $\leq$ the partial order defined by (1). Then, for every $x,y,z \in X$:

(a) $x \cdot y \leq x$.
(b) $x \cdot y \leq y$.
(c) If $z \leq x$ and $z \leq y$, then $z \leq x \cdot y$.
(d) $x + y \geq x$.
(e) $x + y \geq y$.
(f) If $z \geq x$ and $z \geq y$, then $z \geq x + y$.

Lattices as Posets

Lattices can alternatively be characterized as posets in which every set $\{x, y\}$ of two elements has a greatest lower bound, denoted by glb(x, y) and a least upper bound denoted by lub(x, y). Precisely we have:

Theorem 1

Let $L = \langle X, +, \cdot \rangle$ be a lattice and let $\leq$ be the partial order on X defined by (1). Then in the poset $\langle X, \leq \rangle$, $x \cdot y = \text{glb}(x, y)$ and $x + y = \text{lub}(x, y)$, for every $x,y \in X$.

Conversely, let $\langle X, \leq \rangle$ be a poset, in which glb(x, y) and lub(x, y) exist for every $x,y \in X$. Then $\langle X, +, \cdot \rangle$ is a lattice, where $x + y \triangleq \text{lub}(x, y)$ and $x \cdot y \triangleq \text{glb}(x, y)$.

Proof

The first part of this theorem is an immediate consequence of Proposition 6.

Now let $\langle X, \leq \rangle$ be a poset in which $x \cdot y \triangleq \text{glb}(x, y)$ and $x + y \triangleq \text{lub}(x, y)$ exist for every $x,y \in X$. Clearly, $+$ and $\cdot$ are binary operations on X. The verification of L1–L4 and L1′–L4′ is routine. For example, to verify L3, let $u = x + (y + z)$ and $v = (x + y) + z$. Then $u \geq x$ and $u \geq y + z$, which in turn implies that $u \geq y$ and $u \geq z$. Thus $u \geq x + y$, since u is an upper bound of $\{x, y\}$ and $x + y$ is the least upper bound of $\{x, y\}$. Similarly, $u \geq (x + y) + z = v$. By a similar argument, $v \geq u$ and $u = v$ follows. ◻

Universal Bounds, Complements, and Distributivity

Definition 8

A *lattice with universal bounds* is an algebraic system $L = \langle X, +, \cdot, 0, 1 \rangle$, where $\langle X, +, \cdot \rangle$ is a lattice and 0 and 1 are distinct elements of X (i.e.

$0 \neq 1$) that satisfy the condition

$$0 \leq x \leq 1 \qquad \text{for every } x \in X.$$

(The reader should be aware of the fact that the symbols 0 and 1 do not have their usual arithmetic meaning.)

One easily verifies that $L(X) \triangleq \langle P(X), \cup, \cap, \varnothing, X\rangle$, where X is a nonempty set, is a lattice with universal bounds. It is also easily seen that every finite lattice has universal bounds. On the other hand, the lattice $\langle \mathbf{Z}, \lceil, \lfloor \rangle$ has no universal bounds.

Proposition 7

Let $L = \langle X, +, \cdot, 0, 1\rangle$ be a lattice with universal bounds. Then the following laws hold, for every $x \in X$:

$$\text{L5. } x + 0 = x \qquad \text{L5}'.\ x \cdot 1 = x$$

$$\text{L6. } x + 1 = 1 \qquad \text{L6}'.\ x \cdot 0 = 0.$$

Proof

These are immediate consequences of the definition of $\leq$ in lattices, and of Proposition 5. □

In a lattice L with universal bounds we can define the relation of complementarity. Let $L = \langle X, +, \cdot, 0, 1\rangle$. Then define a binary relation C on X by

$$xCy \equiv (x + y = 1) \wedge (x \cdot y = 0) \qquad \text{for all } x, y \in X.$$

Clearly xCy implies yCx, so the relation C is symmetric and we say that x is a complement of y, and vice versa.

A lattice with universal bounds is *complemented* iff every element has a complement. The lattice $L(X)$ (mentioned above) is complemented, since every subset S of X has as complement the set $\bar{S} \triangleq X - S$. Indeed, one easily verifies that $S \cap (X - S) = \varnothing$ and $S \cup (X - S) = X$. As will be seen later, the complements in $L(X)$ are unique. An example of a complemented lattice in which complements are not unique is given in Table 3-3 and its poset diagram is given in Fig. 3-5. Note that aCb, aCc, and bCc. An example of a lattice with universal bounds that is not complemented is $\langle\{0, 1, 2\}, \lceil, \lfloor, 0, 2\rangle$ (see Fig. 3-6). In this lattice the integer 1 has no complement.

Table 3-3 Example of Complemented Lattice

$X = \{0, a, b, c, 1\}$

+	**0**	***a***	***b***	***c***	**1**
0	0	a	b	c	1
a	a	a	1	1	1
b	b	1	b	1	1
c	c	1	1	c	1
1	1	1	1	1	1

·	**0**	***a***	***b***	***c***	**1**
0	0	0	0	0	0
a	0	a	0	0	a
b	0	0	b	0	b
c	0	0	0	c	c
1	0	a	b	c	1

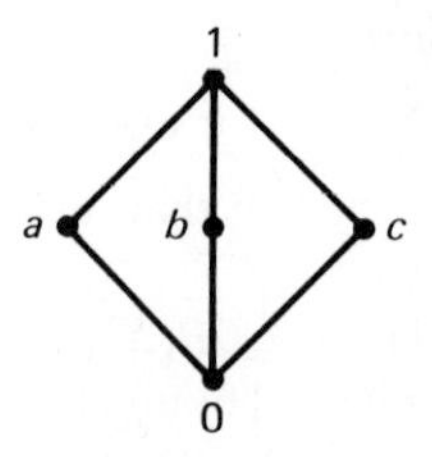

Fig. 3-5 The lattice of Table 3-3.

Fig. 3-6 A lattice which is not complemented.

We will be interested in lattices where complements are unique. For such lattices we denote by x' the unique complement of x. If $L = \langle X, +, \cdot, 0, 1\rangle$ is a lattice with universal bounds and unique complementation, the following laws are obeyed by every $x \in X$:

$$\text{L7. } x + x' = 1 \qquad \text{L7}'. \ x \cdot x' = 0.$$

Furthermore, since xCx' implies $x'Cx$ and complementation is unique, we have:

$$\text{L8. } (x')' = x.$$

Distributive lattices, to be discussed next, are a family of lattices with unique complementation.

Definition 9

A lattice $L = \langle X, +, \cdot\rangle$ is *distributive* iff it satisfies, for all $x,y,z \in X$,

$$\text{L9. } x + (y \cdot z) = (x + y)\cdot(x + z) \qquad \text{L9}'. \ x\cdot(y + z) = (x\cdot y) + (x\cdot z).$$

The conditions L9 and L9′ are not independent, as shown by the following theorem.

Theorem 2

In any lattice $L = \langle X, +, \cdot\rangle$, the following conditions are equivalent:

$$\text{L9. } x + (y\cdot z) = (x + y)\cdot(x + z) \qquad \text{for all } x,y,z \in X$$

$$\text{L9}'. \ x\cdot(y + z) = (x\cdot y) + (x\cdot z) \qquad \text{for all } x,y,z \in X.$$

Proof

Assume that L9′ holds. Then

$$(x + y)\cdot(x + z)$$

$$= ((x + y)\cdot x) + ((x + y)\cdot z) \qquad \text{(by L9}')$$

$$= x + (x \cdot y) + (x \cdot z) + (y \cdot z) \qquad \text{(by L1}', \text{L2}', \text{and L9}')$$
$$= x + (y \cdot z). \qquad \text{(by L4)}$$

Thus L9′ implies L9.

To show that L9 implies L9′, consider $L^D = \langle X, \cdot, + \rangle$, the dual lattice of L. With respect to L^D, the roles of L9 and L9′ are interchanged. It therefore follows from the first part of this proof that L9 implies L9′. ▯

Theorem 3

In a distributive lattice, complements (where they exist) are unique.

Proof

Let $L = \langle X, +, \cdot, 0, 1 \rangle$ and let $x, y, z \in X$. If xCy and xCz, then

$$x \cdot y = 0 \qquad x + y = 1$$
$$x \cdot z = 0 \qquad x + z = 1.$$

Now $1 = x + y$ implies $z = (x + y) \cdot z = (x \cdot z) + (y \cdot z)$. Since $x \cdot z = 0$, we have $z = y \cdot z$. Similarly, $1 = x + z$, $y = (x + z) \cdot y = (x \cdot y) + (z \cdot y) = z \cdot y = y \cdot z$. Thus $y = z$. ▯

An Application of Lattice Theory: Transient Phenomena in Gates [YO-RI]

So far our model of a combinational switching network has been based on various simplifying assumptions. Two of these were:

1. Signal changes (from 0 to 1 or from 1 to 0) occur instantaneously; i.e., signal transient times are zero.
2. Gate input changes affect the gate output immediately; i.e., gate delays are zero.

As was pointed out in Chapter 2, in actual physical realizations of switching networks neither of these two assumptions is met, and it depends on the application we have in mind whether signal transient times and gate delays are negligible or not.

We now introduce a ternary (3-valued) mathematical model which will enable us to treat signal transient phenomena by mathematically precise methods. It will also become clear under which circumstances our earlier binary analysis is not satisfactory and is to be replaced by the ternary approach.

To illustrate the ternary approach, consider the inverter gate in Fig. 3-7(a). Figure 3-7(b) indicates the transient phenomenon associated with a change of the input signal x from 0 to 1. The period from t_1 until t_2 is the x-signal rise time. If signals 0 and 1 are represented by voltage levels v_0 and v_1, respectively,

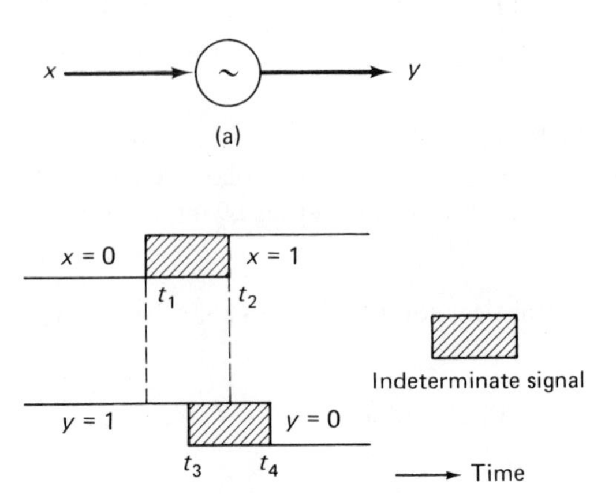

Fig. 3-7 Transient analysis of inverter gate.

and $v_0 < v_1$, then the time between t_1 and t_2 is required for the input voltage to rise continuously from v_0 to v_1. During this transition period the signal is *indeterminate*, i.e., cannot be definitely recognized as either 0 or 1. We now introduce a third symbol, $\frac{1}{2}$, to denote such an indeterminate (transient) signal. Referring to Fig. 3-7(b), we have $x(t) = \frac{1}{2}$ for $t_1 < t < t_2$.

Owing to the physically unavoidable gate delays, the start of the x-signal transition does not affect the y signal immediately. Hence, in Fig. 3-7(b), $t_3 > t_1$. Similarly, the y signal reaches its steady state only some time after the x signal (i.e., $t_4 > t_2$). Thus $y(t) = \frac{1}{2}$ for $t_3 < t < t_4$. However, in the mathematical model we are about to set up, we assume that $t_1 = t_3$ and $t_2 = t_4$. This assumption yields, in fact, a worst-case analysis, as far as difficulties caused by transient conditions ("combinational hazards") are concerned. This should be intuitively clear, since the assumption may be interpreted as extending the transition period of the x signal up to t_4, and starting the y-signal transition at time t_1 rather than t_3.

If we let $t_3 = t_1$ and $t_2 = t_4$, we may consider the inverter gate as a device that instantaneously transforms a ternary input signal $x \in \{0, \frac{1}{2}, 1\}$ into $y = 1 - x$. To develop this viewpoint further, we introduce the appropriate mathematical tool, *ternary switching algebra*.

Let $L_0 \triangleq \{0, \frac{1}{2}, 1\}$. Consider the poset $\langle L_0, \leq \rangle$, where $\leq$ denotes the usual less than or equal relation. The corresponding poset diagram is shown in Fig. 3-8. The poset $\langle L_0, \leq \rangle$ evidently forms a lattice. The lattice operations $+$ and $\cdot$ coincide with the operations $\lceil$ and $\lfloor$, respectively; i.e., for $x, y \in L_0$,

$$x + y = x \lceil y$$

$$x \cdot y = x \lfloor y.$$

1
$\frac{1}{2}$
0 **Fig. 3-8** Poset diagram for L_0.

The corresponding tables are shown in Table 3-4.

Table 3-4 Lattice Operations in L_0

$+$	**0**	$\frac{1}{2}$	**1**
0	0	$\frac{1}{2}$	1
$\frac{1}{2}$	$\frac{1}{2}$	$\frac{1}{2}$	1
1	1	1	1

$\cdot$	**0**	$\frac{1}{2}$	**1**
0	0	0	0
$\frac{1}{2}$	0	$\frac{1}{2}$	$\frac{1}{2}$
1	0	$\frac{1}{2}$	1

One easily verifies that the lattice L_0 is distributive. Thus all the laws L1–L6, L9, L1′–L6′, L9′ are valid. We now define the unary operation * in L_0 by means of Table 3-5.

Table 3-5 Definition of *

x	x^*
0	1
$\frac{1}{2}$	$\frac{1}{2}$
1	0

Proposition 8

The system $\langle L_0, +, \cdot, * \rangle$ satisfies the following laws:
(a) $(x + y)^* = x^* \cdot y^*$.
(b) $(x \cdot y)^* = x^* + y^*$.
(c) $(x^*)^* = x$.

Proof

(a) We have as immediate consequence of the definition of *:

$$x \leq y \text{ iff } y^* \leq x^*.$$

One verifies that $(x \ulcorner y)^* = x^* \llcorner y^*$ and the claim follows.
(b) The proof is similar to that of Part (a).
(c) Obvious. □

On the other hand, $\frac{1}{2} + (\frac{1}{2})^* = \frac{1}{2} \cdot (\frac{1}{2})^* = \frac{1}{2}$. Thus the system $\langle L_0, +, \cdot, * \rangle$ is clearly not a Boolean algebra because, for example, the element $\frac{1}{2} \in L_0$ has no complement in the lattice $\langle L_0, +, \cdot \rangle$.

We shall return to this model in Chapters 4 and 10.

*3.6. BOOLEAN LATTICES

Boolean Lattices and Boolean Algebras

Definition 10

A *Boolean lattice* is a lattice with universal bounds, $L = \langle X, +, \cdot, 0, 1 \rangle$, which is complemented and distributive.

By Theorem 3, every element x of L has a unique complement x'. In view of Table 1, S1–S9 and S1′–S9′, of Appendix B, the lattice $L(X) \triangleq \langle P(X), \cup, \cap, \varnothing, X \rangle$ is a Boolean lattice for every set X. Laws S10 and S10′ of Table 1 of Appendix B also generalize to arbitrary Boolean lattices:

Proposition 9

Let $L = \langle X, +, \cdot, 0, 1 \rangle$ be a Boolean lattice. Then for all $x, y \in X$:

L10. $(x + y)' = x' \cdot y'$ L10′. $(x \cdot y)' = x' + y'$. (De Morgan's laws)

Proof

$$(x + y) \cdot (x' \cdot y') = (x \cdot x' \cdot y') + (y \cdot x' \cdot y') = 0.$$

Also,

$$\begin{aligned}(x + y) + (x' \cdot y') &= (x + y + x') \cdot (x + y + y') \\ &= (y + 1) \cdot (x + 1) = 1 \cdot 1 = 1.\end{aligned}$$

Hence $(x + y) C (x' \cdot y')$. Since complements are unique in Boolean lattices (Theorem 3), we have $(x + y)' = x' \cdot y'$. Thus L10 holds; L10′ follows similarly. □

In a Boolean lattice the function $x \mapsto x'$ defines a unary operation on its elements. The corresponding algebraic system which includes this operation is called a Boolean algebra. Precisely, we have the following summarizing definition.

Definition 11

A *Boolean algebra* is an algebraic system $B = \langle X, +, \cdot, ', 0, 1 \rangle$, where X is a set, $+$ and $\cdot$ are binary operations on X, $'$ is a unary operation on X, and 0 and 1 $(0 \neq 1)$ are elements of X such that all the laws of Table 3-1 are satisfied; i.e., $\langle X, +, \cdot, 0, 1 \rangle$ is a Boolean lattice, and $'$ is its unary operation of complementation.

As we have done for semilattices, we can associate with each Boolean algebra $B = \langle X, +, \cdot, ', 0, 1 \rangle$ a poset $\langle X, \leq \rangle$, where for any x and y in B, $x \leq y \equiv x \cdot y = x$. For completeness we also summarize in Table 3-6 the properties of the partial order $\leq$ in Boolean algebras. Compare this with Proposition 1 of Appendix B.

Table 3-6 Properties of $\leq$ in Boolean Algebras

(a) $x \leq x$ (reflexivity)
(b) $x \leq y$ and $y \leq x$ implies $x = y$ (antisymmetry)
(c) $x \leq y$ and $y \leq z$ implies $x \leq z$ (transitivity)
(d) $x \cdot y \leq x$
(e) $x \cdot y \leq y$
(f) $z \leq x$ and $z \leq y$ implies $z \leq x \cdot y$
(g) $x + y \geq x$
(h) $x + y \geq y$
(i) $z \geq x$ and $z \geq y$ implies $z \geq x + y$
(j) The following statements are all equivalent:
(1) $x \leq y$ (2) $x \cdot y = x$ (3) $x + y = y$ (4) $x \cdot y' = 0$ (5) $y' \leq x'$

For all x, y, and z in a Boolean algebra.

Isomorphism and Duality

For Boolean algebras we define isomorphism as follows:

Definition 12

Let $B = \langle X, +, \cdot, ', 0, 1 \rangle$ and $A = \langle X_A, \sqcup, \sqcap, \overline{}, 0_A, 1_A \rangle$ be Boolean algebras. A function $f: X \longrightarrow X_A$ is an *isomorphism* of B to A iff $f: X \cong X_A$, $f(0) = 0_A$, and $f(1) = 1_A$, and furthermore, for every $x, y \in X$, $f(x + y) = f(x) \sqcup f(y)$, $f(x \cdot y) = f(x) \sqcap f(y)$, and $f(x') = \overline{f(x)}$.

To illustrate Definition 12, we consider $B(\{0, 1\})$ and the Boolean algebra A, given by Table 3-7. One easily verifies that the following function f is an

Table 3-7 Example of 4-Element Boolean Algebra A

$$A = \langle X_A, +, \cdot, ', 0, 1 \rangle$$
$$X_A = \{0, a, b, 1\}$$

+	0	a	b	1
0	0	a	b	1
a	a	a	1	1
b	b	1	b	1
1	1	1	1	1

·	0	a	b	1
0	0	0	0	0
a	0	a	0	a
b	0	0	b	b
1	0	a	b	1

x	0	a	b	1
x′	1	b	a	0

isomorphism of $B(\{0, 1\})$ to A:

$$f: \varnothing \mapsto 0, \qquad \{0\} \mapsto a, \qquad \{1\} \mapsto b, \qquad \{0, 1\} \mapsto 1.$$

The poset diagrams of the algebras A and $B(\{0, 1\})$ are shown in Fig. 3-9.

An isomorphism between two Boolean algebras A and B is related to an isomorphism between the corresponding posets as follows.

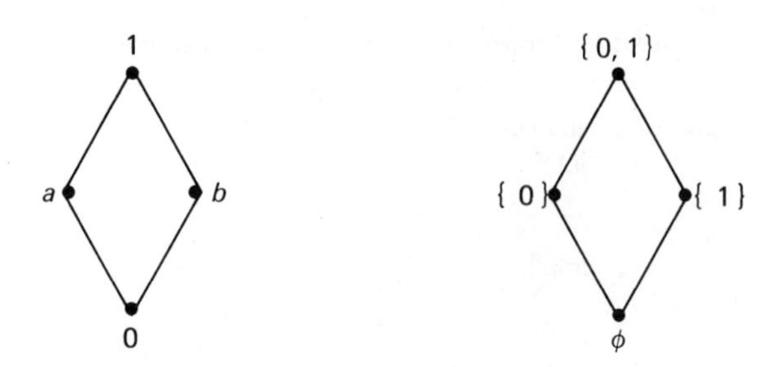

Fig. 3-9 Illustrating isomorphic Boolean algebras.

Proposition 10

Let $B = \langle X, +, \cdot, ', 0, 1\rangle$ and $A = \langle X_A, \sqcup, \sqcap, {}^{-}, 0_A, 1_A\rangle$ be two Boolean algebras and let $P = \langle X, \leq\rangle$ and $Q = \langle X_A, \sqsubseteq\rangle$ be the corresponding posets, where $(x \leq y) \equiv (x \cdot y = x)$ in B and $(x \sqsubseteq y) \equiv (x \sqcap y = x)$ in A. Then a function $f\colon X \longrightarrow X_A$ is an isomorphism of B to A iff f is a poset isomorphism of P to Q.

Proof

The verification of this is routine and is left to the reader. □

In Section 3.5 we defined the dual lattice L^D of a given lattice L. We now extend the concept of duality to Boolean algebras.

Theorem 4

Let $B = \langle X, +, \cdot, ', 0, 1\rangle$ be a Boolean algebra. Then the system $B^D \triangleq \langle X, \cdot, +, ', 1, 0\rangle$ is also a Boolean algebra, and the function $x \mapsto x'$, $x \in X$ is an isomorphism of B to B^D.

Proof

First note that the set of laws in Table 3-1 remains the same if we interchange $+$ with $\cdot$ and 0 with 1. Thus B^D is a Boolean algebra. The function $'\colon X \longrightarrow X$ is injective, since $x' = y'$ implies $(x')' = (y')'$; i.e., $x = y$. The function is also surjective, since for every $x \in X$, $(x')' = x$. Thus $'$ is a bijection. Now $0' = 1$, $1' = 0$, and for every $x, y \in X$, $(x + y)' = x' \cdot y'$, by L10, and $(x \cdot y)' = x' + y'$, by L10′. Finally, the requirement $f(x') = \overline{f(x)}$ of Definition 12 is trivially met. □

*3.7. CHARACTERIZATIONS OF BOOLEAN ALGEBRAS

Sets of Axioms for Boolean Algebras

In defining Boolean algebra we have used the 19 laws of Table 3-1. This set of axioms was reached in a rather natural way, since we approached

Boolean algebras through semilattices, lattices, lattices with 0 and 1, distributed lattices, and complemented lattices. The axioms of Table 3-1 are not, however, independent. For example, it is a rather easy exercise to verify the following result.

Theorem 5

Let $B = \langle X, +, \cdot, ', 0, 1 \rangle$ be an algebraic system that satisfies the following laws of Table 3-1: Ln and Ln' for $n \in \{2, 3, 5, 6, 7, 9\}$. Then B is a Boolean algebra.

It is possible to define Boolean algebras using considerably fewer axioms than in Theorem 5. An example of this is given below [HUN].

Theorem 6

Let $A = \langle X, +, \cdot, ' \rangle$ be an algebraic system satisfying, for all $x,y,z \in X$,

$$x + y = y + x$$

$$x + (y + z) = (x + y) + z$$

$$(x' + y')' + (x' + y)' = x$$

$$x \cdot y = (x' + y')'.$$

Then there exist elements 0 and 1 in X such that $B = \langle X, +, \cdot, ', 0, 1 \rangle$ is a Boolean algebra.

Proof

The proof of this theorem is *not* straightforward. The interested reader is referred to Problem 19, where hints are suggested for constructing a proof.

□

Several other axiom systems can be used to define Boolean algebras; however, further discussion of this topic is beyond the scope of this book. For completeness we briefly discuss the characterization of Boolean algebras by Boolean rings.

Definition 13

A *Boolean ring* is an algebraic system $A = \langle X, \oplus, \cdot, 0, 1 \rangle$ where X is a set; $\oplus$ and $\cdot$ are binary operations on X; 0 and 1 are distinct elements of X, and the following laws are satisfied for all $x,y,z \in X$:

BR1. $(x \oplus y) \oplus z = x \oplus (y \oplus z)$ BR1′. $(x \cdot y) \cdot z = x \cdot (y \cdot z)$

BR2. $x \oplus y = y \oplus x$ BR2′. $x \cdot y = y \cdot x$

BR3. $x \oplus 0 = x$ BR3′. $x \cdot 1 = x$

BR4. $x \oplus x = 0$ BR4′. $x \cdot x = x$

BR5. $x \cdot (y \oplus z) = (x \cdot y) \oplus (x \cdot z)$.

(The reader will note that, in view of BR1, BR3, and BR4, the system $\langle X, \oplus \rangle$ is a group. It is Abelian in view of BR2. A Boolean ring is, of course, a particular case of a ring.)

Let X be a nonempty set. Then the system $\langle P(X), \Delta, \cap, \varnothing, X \rangle$ is a Boolean ring, in view of Proposition 2 and Table 1 of Appendix B. Similarly, Tables 5 and 9 of Chapter 1 show that $\langle \{0, 1\}, \oplus, \cdot, 0, 1 \rangle$ is a Boolean ring.

The following theorem interrelates Boolean algebras and Boolean rings.

Theorem 7

Let $A = \langle X, \oplus, \cdot, 0, 1 \rangle$ be a Boolean ring. Define operations $+$ and $'$ on X by

$$x + y \triangleq x \oplus y \oplus x \cdot y$$

$$x' \triangleq x \oplus 1.$$

Then $B(A) = \langle X, +, \cdot, ', 0, 1 \rangle$ is a Boolean algebra.

Conversely, Let $B = \langle X, +, \cdot, ', 0, 1 \rangle$ be a Boolean algebra. Define the operation $\oplus$ (XOR) on X by

$$x \oplus y = (x \cdot y') + (x' \cdot y).$$

Then $A(B) = \langle X, \oplus, \cdot, 0, 1 \rangle$ is a Boolean ring. Furthermore, $B(A(B)) = B$ and $A(B(A)) = A$.

Proof

The proof of this theorem is routine and is left to the reader. □

Representation of Finite Boolean Algebras

We now proceed to a structural characterization of finite Boolean algebras [HAR].

Definition 14

An element $a > 0$ of a Boolean algebra $B = \langle X, +, \cdot, ', 0, 1 \rangle$ is an *atom* iff for all $x \in X$, either $a \cdot x = a$ or $a \cdot x = 0$.

Proposition 11

(a) a is an atom iff $a \gtrdot 0$.
(b) If a_1, a_2 are atoms, then $a_1 \neq a_2$ iff $a_1 \cdot a_2 = 0$.
(c) For any atom a and $x \in X$, $a \leq x \equiv \sim(a \leq x')$.

Proof

(a) If a is an atom but $a \gtrdot 0$ is false, there exists an $x \in X$ such that $a > x > 0$. Then $a \geq x$ and $a \cdot x = x$. But $a \cdot x$ is either 0 or a; hence either $x = 0$ or $x = a$, a contradiction. Conversely, if $a \gtrdot 0$ and there exists an $x \in X$ such that $a \cdot x \neq 0$ and $a \cdot x \neq a$, then $a > a \cdot x > 0$, again a contradiction.

(b) Assume that $a_1 \neq a_2$ and $a_1 \cdot a_2 \neq 0$. Then $a_1 \cdot a_2 = a_1$, because a_1 is an atom and $a_1 \cdot a_2 = a_2$ because a_2 is an atom. Thus $a_1 = a_2$, a contradiction. Conversely, $a_1 \cdot a_2 = 0$ implies $a_1 \neq a_2$, for if $a_1 = a_2$, then $a_1 \cdot a_2 = a_1 > 0$.

(c) If $a \leq x$ and $a \leq x'$, then $a \cdot x = a$ and $a \cdot x' = a$. But then $0 = (a \cdot x) \cdot (a \cdot x') = a \cdot a = a$, a contradiction. If $a \not\leq x$ and $a \not\leq x'$, then $a \cdot x \neq a$ and $a \cdot x' \neq a$. This means that $a \cdot x = 0 = a \cdot x'$ since a is an atom. Hence $0 = a \cdot x + a \cdot x' = a \cdot (x + x') = a$, a contradiction. ▯

Proposition 12

Let $x > 0$ be an element of a finite Boolean algebra B. Then there exists an atom $a \in X$ such that $x \geq a$.

Proof

A sequence of elements $x_1, \ldots, x_k$ of X such that $x_1 > x_2 > \cdots > x_k$ is called a *strictly descending chain* of length k. Clearly all elements of such a chain are different and since X is finite, there exists only a finite number of such chains. $x > 0$ is a strictly descending chain starting with x and ending in 0. Let

$$x = x_1 > x_2 > \cdots > x_{k-1} > x_k = 0$$

be a longest such chain. Then $x_{k-1} > 0$, and by Proposition 11(a), x_{k-1} is an atom. Since $x \geq x_{k-1}$, the proposition is proved. ▯

It will be useful to extend Boolean multiplication to sets of elements of a Boolean algebra. For $S_1, S_2 \subseteq X$, define

$$S_1 \cdot S_2 = \{z \mid z = x \cdot y,\ x \in S_1,\ y \in S_2\}.$$

It is easy to verify that this multiplication of subsets of X is commutative and associative.

Lemma 1

Let x be an element of a finite Boolean algebra B. Let A be the set of all atoms of B and let

$$y = \sum_{a_i \in A \cdot \{x\}} a_i,$$

where $\sum$ represents Boolean addition. Then

$$A \cdot \{x\} - \{0\} = \{a \in A \mid a \leq x\} \qquad \text{and} \qquad x = y.$$

Proof

The reader can verify that $A \cdot \{x\} - \{0\} = \{a \mid a \leq x \text{ and } a \text{ is an atom}\}$. Then since $x \geq a_i$ for all $a_i \in A \cdot \{x\}$, it follows that $x \geq y$. Now, $y \not\geq x$ is equivalent to $x \cdot y' \neq 0$, by Table 3-6(j). By Proposition 12, there exists an atom a such that $x \cdot y' \geq a$. Thus $x \geq a$, $a \in A \cdot \{x\}$, and therefore $y \geq a$. But $x \cdot y' \geq a$ also implies $y' \geq a$. Thus $0 = y \cdot y' \geq a$, a contradiction. Hence $x \cdot y' = 0$, $y \geq x$, and $y = x$ follows. ▯

Theorem 8 (Stone)

Every finite Boolean algebra $B = \langle X, +, \cdot, ', 0, 1\rangle$ is isomorphic to $B(A)$, where A is the set of atoms in B.

Proof

By Proposition 12, since X has at least one element greater than 0, it has at least one atom. Define the relation $\tau \subseteq X \times P(A)$ as follows: $x \,\tau\, (A \cdot \{x\} - \{0\})$ for any $x \in X$. τ is a function from X to $P(A)$ because:

(a) For each x in X there exists $Y = A \cdot \{x\} - \{0\}$ such that $x \,\tau\, Y$.

(b) $x \,\tau\, Y_1$ and $x \,\tau\, Y_2$ implies $Y_1 = Y_2$, since each $x \in X$ is related by τ to only one $Y \in P(A)$.

Hence, if $x \,\tau\, Y$, we can write $Y = \tau(x)$.

Next τ is surjective (onto) because $\varnothing$ is the image of 0, and each nonempty subset Y of A is the image of $x = \sum_{a_i \in Y} a_i$ in B, since one verifies that $A \cdot (\sum_{a_i \in Y} a_i) - \{0\} = Y$. By Lemma 1,

$$A \cdot \{x_1\} - \{0\} = A \cdot \{x_2\} - \{0\}$$

implies

$$x_1 = \sum_{a_i \in A \cdot \{x_1\}} a_i = \sum_{a_i \in A \cdot \{x_2\}} a_i = x_2.$$

Thus, for all x_1, x_2 in X, $x_1 \neq x_2$ implies $\tau(x_1) \neq \tau(x_2)$. Therefore, τ is also injective (one-to-one). Altogether, τ is a bijection from X to $P(A)$; i.e., τ establishes a one-to-one correspondence between the elements of X and $P(A)$ or, in symbols, $\tau : X \cong P(A)$.

To complete the proof we show that

(a) $\tau(x + y) = \tau(x) \cup \tau(y)$

(b) $\tau(x \cdot y) = \tau(x) \cap \tau(y)$

(c) $\tau(x') = A - \tau(x) = \overline{\tau(x)}$

For part (a), $\tau(x + y) = A \cdot \{x + y\} - \{0\} \subseteq A$. Thus

$$\begin{aligned} a \in \tau(x + y) &\equiv (a \leq x + y) \wedge (a \text{ is an atom}) \\ &\equiv (a \cdot (x + y) = a) \wedge (a \in A) \\ &\equiv (a \cdot x + a \cdot y = a) \wedge (a \in A). \qquad (*) \end{aligned}$$

Now $a \cdot x$ can be either a or 0, since a is an atom and this is also true of $a \cdot y$. However, we cannot have $a \cdot x = 0$ and $a \cdot y = 0$ since $a \cdot x + a \cdot y = a > 0$. Hence, (*) implies

$$((a \cdot x = a) \vee (a \cdot y = a)) \wedge (a \in A). \qquad (**)$$

But clearly (**) implies (*) and we have

$$\begin{aligned} (a \cdot x + a \cdot y = a) \wedge (a \in A) &\equiv ((a \cdot x = a) \vee (a \cdot y = a)) \wedge (a \in A) \\ &\equiv ((a \cdot x = a) \wedge (a \in A)) \vee ((a \cdot y = a) \wedge (a \in A)) \end{aligned}$$

$$\begin{aligned}
&\equiv ((a \leq x) \wedge (a \in A)) \vee ((a \leq y) \wedge (a \in A)) \\
&\equiv (a \in (A \cdot \{x\} - \{0\})) \vee (a \in (A \cdot \{y\} - \{0\})) \\
&\equiv (a \in \tau(x)) \vee (a \in \tau(y)) \\
&\equiv a \in (\tau(x) \cup \tau(y)).
\end{aligned}$$

Thus $a \in \tau(x + y) \equiv a \in (\tau(x) \cup \tau(y))$ and $\tau(x + y) = \tau(x) \cup \tau(y)$.

Part (b) follows by a similar argument. For part (c),

$$\begin{aligned}
a \in \tau(x') &\equiv a \in (A \cdot \{x'\} - \{0\}) \\
&\equiv (a \leq x') \wedge (a \in A) \\
&\equiv (a \not\leq x) \wedge (a \in A) \qquad \text{[by Proposition 11(c)]} \\
&\equiv (a \notin (A \cdot \{x\} - \{0\})) \wedge (a \in A) \\
&\equiv (a \in A) \wedge (a \notin \tau(x)) \\
&\equiv a \in (A - \tau(x)) \\
&\equiv a \in \overline{\tau(x)},
\end{aligned}$$

and (c) follows. ▯

Corollary 1

Every finite Boolean algebra B has 2^n elements for some $n \geq 1$.

Proof

If B has n atoms, there are 2^n subsets of A. ▯

Corollary 2

For each n, the Boolean algebra of 2^n elements is unique up to isomorphism.

Proof

All Boolean algebras with 2^n elements are isomorphic to the Boolean algebra of all subsets of a set of n elements. ▯

Example 1

Let B have 4 elements. Two of these elements correspond to the universal bounds; call these 0 and 1 and let the third element be x. One verifies that the remaining element must be x'. Thus the poset of B has the form shown in Fig. 3-10(a). Clearly the atoms of B are x and x'. Representing each element of B as a sum of atoms, we have

$$0 = 0, \quad x = x, \quad x' = x', \quad \text{and} \quad 1 = x + x'.$$

Let $\{a, b\}$ be a set of two elements. The algebra of subsets of $\{a, b\}$ isomorphic to B is shown in Fig. 3-10(b). From the poset diagram of B we can

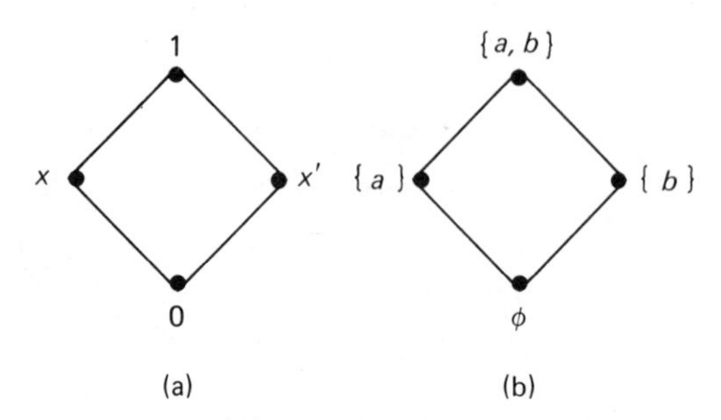

Fig. 3-10 A Boolean algebra with 4 elements.

construct the addition, multiplication, and complement tables as shown in Table 3-8.

Table 3-8 Operations in B

$+$	0	x	x'	1
$\mathbf{0}$	0	x	x'	1
$\mathbf{x}$	x	x	1	1
$\mathbf{x'}$	x'	1	x'	1
$\mathbf{1}$	1	1	1	1

$\cdot$	0	x	x'	1
$\mathbf{0}$	0	0	0	0
$\mathbf{x}$	0	x	0	x
$\mathbf{x'}$	0	0	x'	x'
$\mathbf{1}$	0	x	x'	1

y	y'
$\mathbf{0}$	1
$\mathbf{x}$	x'
$\mathbf{x'}$	x
$\mathbf{1}$	0

PROBLEMS

1. Prove that laws L1, L1′, L4, and L4′ of Table 3-1 may be derived from the remaining laws of Table 3-1 (p. 59).

2. Prove that in any Boolean algebra, $x + y = 1$ and $x \cdot y = 0$ always implies $y = x'$. Use only laws of Table 3-1 to prove this.

3. (a) Prove that the following law holds in any Boolean algebra:

$$x \cdot y + x' \cdot z = x \cdot y + x' \cdot z + y \cdot z.$$

Use only laws of Table 3-1 to prove this.

*(b) Show that the corresponding law does not hold in L_0 (p. 72).

4. (a) Let $Y \triangleq \{1, 2, 3, 4, 6, 12\}$ and define for $x, y \in Y$:

$$x + y = \text{lcm}(x, y)$$
$$x \cdot y = \text{gcd}(x, y)$$
$$x' = 12/x$$
$$0_B = 1$$
$$1_B = 12$$

Show that $B = \langle Y, +, \cdot, ', 0_B, 1_B \rangle$ is not a Boolean algebra.

*(b) Prove that the system $L = \langle Y, +, \cdot, 0_B, 1_B \rangle$ is a distributive lattice with universal bounds.

5. Let B be an arbitrary Boolean algebra. Define the operation $\oplus$ in B by

$$x \oplus y \triangleq (x \cdot y') + (x' \cdot y).$$

Prove that the operation $\oplus$ satisfies the following laws:

$$x \oplus y = y \oplus x$$

$$x \oplus 0 = x$$

$$x \oplus x = 0$$

$$(x \oplus y) \oplus z = x \oplus (y \oplus z)$$

$$x \cdot (y \oplus z) = (x \cdot y) \oplus (x \cdot z)$$

For your proof you may use only the laws of Table 3-1.

6. The NOR operation $\downarrow$ is defined in a Boolean algebra by

$$x \downarrow y = (x + y)' = x' \cdot y'.$$

It is possible to describe a finite Boolean algebra B uniquely by giving its "NOR table," i.e., by specifying $x \downarrow y$ for each x and y in B. Such a NOR table for $B = \{a, b, c, d\}$ is shown in Fig. P3-1.

$\downarrow$	a	b	c	d
a	c	d	d	c
b	d	d	d	d
c	d	d	a	a
d	c	d	a	b

Figure P3-1

(a) Find the 0 and 1 elements of B.
(b) Find the complement of each element in B.
(c) Construct the multiplication table for B.

7. Let the operation $*$ be defined in a Boolean algebra by $a * b = a'b$. Is this operation (a) idempotent? (b) associative? (c) commutative? (d) complete? That is, can xy, $x + y$, and x' be expressed in terms of the $*$ operation (and perhaps the constants 0 and 1) alone? Prove your answers.

***8.** *Note:* Refer to Appendix C for terminology. Let $X = \{a, b, c, d\}$ and let $P = \{(a, b), (b, b), (b, c), (b, d)\}$.
(a) Find the reflexive-and-transitive closure of P.
(b) Find a relation $R \supseteq P$ such that R is a total order on X, i.e., a partial order such that for all $x, y \in X$, either xRy or yRx.

***9.** Let X be the set of positive integers. For $x, y \in X$, define xRy iff x is a multiple of y. Show that $\langle X, R \rangle$ is a poset. Show that every pair of elements in X has a glb and an lub. Does X have universal bounds?

***10.** Find all the nonisomorphic posets of the form $\langle X, \leq \rangle$, where X has 3 elements. Define $x \cdot y = \text{glb}(x, y)$ and $x + y = \text{lub}(x, y)$. For each poset $\langle X, \leq \rangle$, is $\langle X, \cdot \rangle$ a semilattice? Is $\langle X, +, \cdot \rangle$ a lattice?

***11.** Find all the nonisomorphic lattices $\langle X, +, \cdot \rangle$, where X has 5 elements. (*Hint:* Apply the result of Problem 10.) Are any of these lattices (a) distributive? (b) complemented? (c) Boolean?

***12.** Construct the "addition" (lub) and "multiplication" (glb) tables for the lattice whose poset diagram is shown in Fig. P3-2.

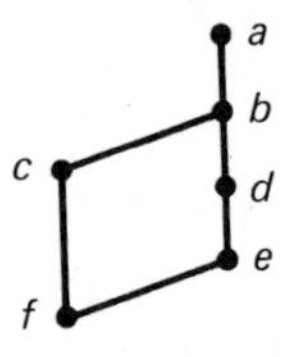

Figure P3-2

***13.** The table of Fig. P3-3 is a multiplication table for an 8-element semilattice S. Draw the poset diagram of S. Show that complementation and addition can be defined in such a way that S forms a Boolean algebra.

$\cdot$	a	b	c	d	e	f	g	h
a	a	g	a	a	g	g	g	a
b	g	b	b	b	g	b	g	g
c	a	b	c	c	g	b	g	a
d	a	b	c	d	e	f	g	h
e	g	g	g	e	e	e	g	e
f	g	b	b	f	e	f	g	e
g	g	g	g	g	g	g	g	g
h	a	g	a	h	e	e	g	h

Figure P3-3

***14.** Consider the following conjecture: "In any finite lattice, $a + c = b + c$ and $ac = bc$ imply $a = b$."

(a) Prove that the conjecture is false by finding a counterexample.

(b) Prove that the conjecture is true under the additional assumption that the lattice is distributive.

***15.** In any lattice $L = \langle X, +, \cdot, 0, 1 \rangle$, let $a \in X$ and let $X_a = \{x \mid x \in X \wedge x \leq a\}$ be the set of all elements $\leq a$. It is easy to verify that $L_a = \langle X_a, +, \cdot, 0, a \rangle$ is also a lattice, with universal bounds 0 and a. Consider the following conjecture: "If L is a finite complemented lattice, then so is L_a."

(a) Prove that this conjecture is false by finding a counterexample.

(b) Prove that the conjecture is true under the additional assumption that the lattice L is distributive.

***16.** Let B be a finite Boolean algebra, and let $S = \{c \in B \mid c < 1$ and, for every element $x \in B$, $c + x = 1$ or $c + x = c\}$. Prove:

(a) If $c,d \in S$, then $c + d = 1$ iff $c \neq d$.

(b) If an element c is in S and its complement c' is also in S, then the set S contains exactly two elements, c and c'.

***17.** Find the atoms of the Boolean algebra of Problem 13 and express each element >0 as a sum of the atoms.

***18.** Let B be a Boolean algebra. Prove that multiplication of subsets (Section 3.7) in B is commutative and associative. Is it idempotent?

***19.** Using the definition of A in Theorem 6, prove the following in the order suggested: For all $x,y \in X$,

(a) $x + x' = x' + x''$

(b) $x'' = x$

(c) $x + x' = y + y'$

In view of part (c), we can define the element $1 = x + x'$ in X.

(d) $1 + 1 = 1$

(e) $x + 1' = x$

(f) $x + x = x$

4 BOOLEAN FUNCTIONS AND EXPRESSIONS

ABOUT THIS CHAPTER

In Chapters 1 and 2 we discussed combinational networks in a rather informal way. The concepts introduced there are made more precise and are extended in this chapter. The reader of the unstarred material should read Sections 4.1 through 4.4 and the summary in Section 4.6. He can then proceed to Chapter 5, which is more application-oriented.

We strongly recommend that the reader of the starred material read all of Section 4.6. Section 4.5 is a continuation of the material in Section 3.5 on transient phenomena and will be required again only in Section 10.3. Section 4.7 describes classes of Boolean functions of particular interest, but no subsequent material relies on it.

4.1. INTRODUCTION

Consider the gate network of Fig. 4-1, with input variables x_1, x_2, and x_3 and output variable z. Using the ideas of Chapter 1, we can describe the network by the following *expression* for z:

$$(x_1 + x_2')\cdot x_3 + x_1\cdot x_2$$

In this case there is an obvious correspondence between the form of the expression and the *structure* of the network.

On the other hand, one is often interested only in the input–output *behavior* of a network. It then suffices to give, for each input 3-tuple (x_1, x_2,

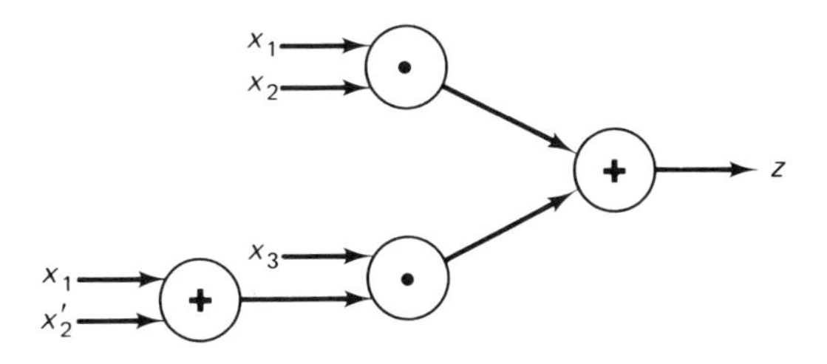

Fig. 4-1 A gate network.

x_3) of 0's and 1's, the corresponding binary value of z. In other words, we are interested in the *function* mapping the set of all binary 3-tuples into the set $\{0, 1\}$. Table 4-1 shows the function corresponding to the network of Fig. 4-1. These notions will now be treated more precisely.

Table 4-1 Function Corresponding to Fig. 4-1

x_1	x_2	x_3	z
0	0	0	0
0	0	1	1
0	1	0	0
0	1	1	0
1	0	0	0
1	0	1	1
1	1	0	1
1	1	1	1

4.2. BOOLEAN FUNCTIONS

As was pointed out before, the simplest Boolean algebra is the algebra $B_0 = \langle\{0, 1\}, +, \cdot, ', 0, 1\rangle$, consisting of two elements 0 and 1, as shown in Table 4-2.

Table 4-2 Operations in B_0

+	0	1
0	0	1
1	1	1

$\cdot$	0	1
0	0	0
1	0	1

x	x'
0	1
1	0

This algebra is of great importance for applications in switching networks. We will now define an algebra of functions in which the operations are derived from B_0. For a set S, S^n is the cartesian product of n copies of S, for $n > 0$; i.e.,

$$S^n \triangleq \underbrace{S \times \cdots \times S}_{n \text{ times}} = \{(x_1, x_2, \ldots, x_n) \mid x_i \in S \text{ for } 1 \leq i \leq n\}.$$

Thus $\{0, 1\}^n$ denotes the set of all binary n-tuples. This set was called V_n in Chapter 2.

Definition 1

A *Boolean function* of n variables is any function f from $\{0, 1\}^n$ into $\{0, 1\}$, for $n > 0$.

To each n-tuple $x = (x_1, x_2, \ldots, x_n) \in \{0, 1\}^n$, the function f assigns a unique value $f(x) \in \{0, 1\}$. We can represent any Boolean function by a *table of combinations* in which all the n-tuples and the corresponding function values are listed. The table of combinations has 2^n rows since there are 2^n distinct n-tuples. Table 4-3(a) illustrates the tables of combinations for two functions f and g, both mapping $\{0, 1\}^2$ to $\{0, 1\}$.

Table 4-3 Examples of Boolean Functions

(x_1, x_2)	$f(x)$	$g(x)$	$(f+g)(x)$	$(f \cdot g)(x)$	$f'(x)$
0 0	0	0	0	0	1
0 1	1	1	1	1	0
1 0	1	0	1	0	0
1 1	0	1	1	0	1
	(a)			(b)	

In forming Boolean functions, in each row of the table of combinations there are two choices, 0 or 1. Thus there are $2 \times 2 \times \cdots \times 2$ (2^n times) possible different tables of combinations. It follows that there are 2^{2^n} distinct Boolean functions of n variables. We now define three operations on Boolean functions.

Definition 2

Given Boolean functions $f\colon \{0, 1\}^n \longrightarrow \{0, 1\}$ and $g\colon \{0, 1\}^n \longrightarrow \{0, 1\}$, define, for each n-tuple $x = (x_1, \ldots, x_n) \in \{0, 1\}^n$,

$$(f + g)(x) = f(x) + g(x)$$
$$(f \cdot g)(x) = f(x) \cdot g(x)$$
$$f'(x) = (f(x))',$$

where on the right-hand side of each equation the addition, multiplication, and complementation are the operations of B_0 in Table 4-2. This is well defined since, for each x, the values $f(x)$ and $g(x)$ are elements of $\{0, 1\}$.

In the table of combinations for $(f + g)$, $(f \cdot g)$, and f' we are constructing these new functions from f and g, row by row, using the operations of B_0. Table 4-3(b) illustrates these operations on functions.

Define two special functions $|0|$ and $|1|$ as follows: For all $x \in \{0, 1\}^n$,

$$|0|(x) = 0 \qquad |1|(x) = 1.$$

These two functions map all the n-tuples into 0 and 1, respectively.

Theorem 1

Let B_n be the set of all Boolean functions of n variables. Then B_n is a Boolean algebra under the operations of Definition 2 with $|0|$ and $|1|$ as universal bounds.†

Proof

One must simply verify that the operations on functions as given in Definition 2 satisfy the laws for Boolean algebras. For example, to verify commutativity of $+$, we must show that

$$(f + g) = (g + f) \tag{1}$$

Now, to prove that two functions are equal, we must verify that they have the same table of combinations. In other words, we must show that for each n-tuple $x \in \{0, 1\}^n$,

$$(f + g)(x) = (g + f)(x). \tag{2}$$

But, by definition, $(f + g)(x) = f(x) + g(x)$ and $(g + f)(x) = g(x) + f(x)$. Since addition in B_0 is commutative, it follows that (2) holds in each row. Hence (1) also holds. The remaining laws follow by a similar argument. $\square$

We now define a binary relation $\leq$ on B_n, where a binary relation on B_n is any subset of $B_n \times B_n$. The reader of the starred material is already familiar with binary relations. The reader of the unstarred material may view this concept simply as follows. Given any pair $(f, g) \in B_n \times B_n$, either f is related to g by the relation $\leq$ (to be defined) and we write $f \leq g$, or f is not related to g, $f \nleq g$. Note that $\leq$ is used here as an abstract symbol and $f \nleq g$ does not necessarily imply $f > g$. In fact, f and g may not be related by $\leq$. It is true, however, that if $f \leq g$, then $g \geq f$, by definition of $\geq$. We write $f < g$ or $g > f$ as a short form for $f \leq g$ and $f \neq g$.

For $f, g \in B_n$, we now define $f \leq g$ iff $(f \cdot g) = f$; i.e., for each $x \in \{0, 1\}^n$, $(f \cdot g)(x) = f(x)$. In the tables of combinations, $f \leq g$ iff g has a 1, whenever f has a 1. Note that this notion is simply an extension of the natural order $\leq$ for $\{0, 1\}$, i.e., $0 \leq 0$, $0 \leq 1$, $1 \leq 1$, to columns of the tables of combinations or, equivalently, to 2^n-tuples.

It is easy to verify that for all $f, g, h \in B_n$, the following laws hold:

1. Reflexive: $f \leq f$.
2. Antisymmetric: $f \leq g$ and $g \leq f$ implies $f = g$.
3. Transitive: $f \leq g$ and $g \leq h$ implies $f \leq h$.

†The elements 0 and 1 of a Boolean algebra $B = \langle X, +, \cdot, ', 0, 1 \rangle$ are called *universal bounds*.

For example, to verify the transitive law, we must check that for all $x \in \{0, 1\}^n$, $f(x) \leq g(x)$, and $g(x) \leq h(x)$ implies $f(x) \leq h(x)$. This, of course, is true, since $f(x)$ and $g(x)$ are elements of $\{0, 1\}$ and $\leq$ is transitive on the set $\{0, 1\}$, as can be verified by checking all the possibilities.

A binary relation that is reflexive, antisymmetric, and transitive is called a *partial order*. Thus $\leq$ is a partial order on B_n. One also verifies that for any f in B_n,

$$|0| \leq f \leq |1|.$$

An *atom function* is a function $m \in B_n$ that has precisely one 1 in its table of combinations. Since there are 2^n rows in the table, there are exactly 2^n different atom functions in B_n.

Theorem 2

Every Boolean function $f \in B_n$ is equal to the sum of all the atom functions m such that $m \leq f$.

Proof

If $f = |0|$, then it is the sum of no atom functions. (This is a degenerate case.) Otherwise, suppose that f has k 1's in its table of combinations. Then there are exactly k atom functions m_i, $i = 1, \ldots, k$, such that $m_i \leq f$, where m_i is the atom function that is 1 in the row x of the table corresponding to the ith 1-entry of f. (We are numbering the 1's of f in the order of appearance from top to bottom, say.) Every other atom function m will have a 1 where f has a 0, so that $m \leq f$ is false. Now form the function $g = m_1 + m_2 + \cdots + m_k$ according to the rule for addition of functions given in Definition 2. Clearly we get $g = f$. ☐

(This result also follows from Lemma 1, Chapter 3, since one verifies that the atom functions are indeed the atoms of the Boolean algebra B_n.)

For example, the function $f \in B_2$ shown in Table 4-4 is the sum $m_1 + m_2 + m_3$.

Table 4-4 Atom Functions for f

x		f	m_1	m_2	m_3
0	0	0	0	0	0
0	1	1	1	0	0
1	0	1	0	1	0
1	1	1	0	0	1

$$\begin{matrix}0\\1\\0\\0\end{matrix} + \begin{matrix}0\\0\\1\\0\end{matrix} + \begin{matrix}0\\0\\0\\1\end{matrix} = \begin{matrix}0\\1\\1\\1\end{matrix}$$

The representation of a function as a sum of atom functions is unique up to commutativity.

4.3. BOOLEAN EXPRESSIONS

Basic Definitions

We now introduce the concept of Boolean expression in order to develop a notation for Boolean functions [HAR].

Definition 3

Let $0, 1, x_1, \ldots, x_n$ be distinct symbols. *Boolean expression* over $x_1, \ldots, x_n$ is defined inductively.

1. $0, 1, x_1, x_2, \ldots, x_n$ are Boolean expressions.
2. If E and F are Boolean expressions, then so are $(E + F)$, (EF), and E'.
3. Nothing else is a Boolean expression unless its being so follows from a finite number of applications of Rules 1 and 2.

At this point, Boolean expressions represent an infinite set of well-formed strings of symbols. We relate them to Boolean functions next.

Definition 4

Let $\mathcal{E}_n$ be the set of all Boolean expressions over n variables. Define a mapping $|\ |: \mathcal{E}_n \longrightarrow B_n$ as follows.

1. The expressions $0, 1, x_1, \ldots, x_n$ are mapped to the functions $|0|, |1|, |x_1|, \ldots, |x_n|$, respectively, where $|x_i|$ is defined by

$$|x_i|(a_1, \ldots, a_n) = a_i \text{ for all } (a_1, \ldots, a_n) \in \{0, 1\}^n.$$

   Thus the column for $|x_i|$ in the table of combinations is identical to the ith input column.

2.
$$|(E + F)| = (|E| + |F|)$$
$$|(EF)| = (|E| \cdot |F|)$$
$$|E'| = |E|'.$$

The mapping $|\ |$ assigns to each expression $E \in \mathcal{E}_n$ a unique Boolean function $|E| \in B_n$. For example, let $E = ((x_1 x_2') + x_2)$. Then

$$|E| = (|(x_1 x_2')| + |x_2|) = ((|x_1| \cdot |x_2'|) + |x_2|)$$
$$= ((|x_1| \cdot |x_2|') + |x_2|).$$

Using the laws of Boolean algebra B_n we can verify that

$$((|x_1| \cdot |x_2|') + |x_2|) = (|x_1| + |x_2|) \cdot (|x_2'| + |x_2|)$$
$$= (|x_1| + |x_2|) \cdot (|1|) = (|x_1| + |x_2|).$$

We have now related expressions to functions in the sense that each expression defines a unique function. Note, however, that there is an infinite number of expressions denoting a given function. For example,

$$|x_1| = |(x_1x_1)| = |((x_1x_1)x_1)| = \text{etc.}$$

Define an equivalence relation $\equiv$ on $\mathcal{E}_n$ as follows: For $E,F \in \mathcal{E}_n$,

$$E \equiv F \qquad \text{iff } |E| = |F|.$$

(An equivalence relation on a set X is a relation R that is reflexive, transitive, and symmetric, where the last property is defined by: For all $x, y \in X$, xRy implies yRx. See Section 2 of Appendix C.) Thus two expressions are equivalent iff they denote the same function. We shall manipulate expressions to obtain more desirable equivalent forms. For example, $x_1x_1 \equiv x_1$. Note that we omit parentheses whenever this does not introduce ambiguity. Also, as usual, multiplication has precedence over addition.

Canonical Expressions

Definition 5

Expressions of the form x_i or x_i' are called *literals*. Expressions in $\mathcal{E}_n$ of the form $\tilde{x}_{i_1}\tilde{x}_{i_2} \cdots \tilde{x}_{i_k}$, where the $\tilde{x}_{i_j}$ are literals, are called *products*. Products of the form $\tilde{x}_1\tilde{x}_2 \cdots \tilde{x}_n$ are called *minterms*. (Thus a minterm is a product of n literals in which *each variable* x_i *must appear* either as x_i or x_i' *in the* ith *place* of the minterm.)

One verifies that if $\tilde{x}_1 \cdots \tilde{x}_n$ is a minterm, then $|\tilde{x}_1 \cdots \tilde{x}_n|$ is an atom function and that each atom function can be represented by a minterm. For suppose that $\tilde{x}_1 \cdots \tilde{x}_n$ is a minterm. Then $f \triangleq |\tilde{x}_1 \cdots \tilde{x}_n| = |\tilde{x}_1| \cdot \ldots \cdot |\tilde{x}_n|$, by Definition 4. Now, $f = 1$ iff $|\tilde{x}_i| = 1$, for $i = 1, 2, \ldots, n$. If $\tilde{x}_i = x_i$, then x_i must be 1 in order to make $|\tilde{x}_i| = 1$, and if $\tilde{x}_i = x_i'$, then x_i must be 0 to make $|\tilde{x}_i| = 1$. Hence there is a unique n-tuple $(a_1, \ldots, a_n)$ such that $|\tilde{x}_1 \cdots \tilde{x}_n|\,(a_1, \ldots, a_n) = 1$. Therefore, $|\tilde{x}_1 \cdots \tilde{x}_n|$ is an atom function. Conversely, suppose that m is an atom function and its (single) 1 entry corresponds to the n-tuple $(a_1, \ldots, a_n)$. Form the expression $\tilde{x}_1 \cdots \tilde{x}_n$ as follows. If $a_i = 0$, set $\tilde{x}_i = x_i'$, and if $a_i = 1$, set $\tilde{x}_i = x_i$. One then verifies that $\tilde{x}_1 \cdots \tilde{x}_n$ is the minterm representing m. It follows from Theorem 2 that *every Boolean function f has a representation as a sum of minterms, which is unique except for the order of the minterms.* Such an expression is called the *canonical sum* for f.

In Table 4-5, we show all the atom functions for B_2. In the last column we list all the minterms: A minterm is listed in the row in which the corresponding atom function has value 1.

We illustrate the canonical sum for the function f of Table 4-4:

$$F_{CS} \triangleq x_1'x_2 + x_1x_2' + x_1x_2.$$

Table 4-5 Atom Functions of B_2

(x_1, x_2)	$(\lvert x_1\rvert' \cdot \lvert x_2\rvert')$	$(\lvert x_1\rvert' \cdot \lvert x_2\rvert)$	$(\lvert x_1\rvert \cdot \lvert x_2\rvert')$	$(\lvert x_1\rvert \cdot \lvert x_2\rvert)$	**Minterms**
0 0	1	0	0	0	$x_1'x_2'$
0 1	0	1	0	0	$x_1'x_2$
1 0	0	0	1	0	x_1x_2'
1 1	0	0	0	1	x_1x_2

In an analogous way we can show that every Boolean function f has a canonical representation as a product of special expressions called *maxterms*, which are expressions of the form $\tilde{x}_1 + \tilde{x}_2 + \cdots + \tilde{x}_n$, where each $\tilde{x}_i$ is a literal. Such an expression is called the *canonical product* for f.

To obtain the canonical product for f, first find the canonical sum for f'. Then use the following generalized forms of De Morgan's laws to obtain $f = (f')'$:

$$(E_1 + E_2 + \cdots + E_m)' = E_1'E_2' \cdots E_m' \tag{3}$$

$$(E_1E_2 \cdots E_m)' = E_1' + E_2' + \cdots + E_m'. \tag{4}$$

The proof of these laws follows by induction on m and is left to the reader.

An example will make this clear. Let $f' = \lvert x_1x_2' + x_1'x_2\rvert$, where $f \in B_2$. The canonical sum for f' is $G = x_1x_2' + x_1'x_2$. By (3) we have $G' = (x_1x_2' + x_1'x_2)' = (x_1x_2')'(x_1'x_2)'$, and by (4), $G' = (x_1' + x_2)(x_1 + x_2')$. Therefore, the canonical product for f is

$$F_{CP} \triangleq (x_1' + x_2)(x_1 + x_2').$$

(The canonical product can also be obtained by applying Lemma 1 of Chapter 3 to B_n^D, the dual Boolean algebra of B_n.)

Each maxterm defines a *maxterm function*, which is a function with exactly one 0 in its table of combinations. If the 0 occurs for the n-tuple $a_1, a_2, \ldots, a_n$, then the maxterm function is described by the expression $\hat{x}_1 + \hat{x}_2 + \cdots + \hat{x}_n$, where $\hat{x}_i = x_i$ if $a_i = 0$ and $\hat{x}_i = x_i'$ if $a_i = 1$.

These ideas are illustrated in Table 4-6, where we list each maxterm in the row in which the corresponding maxterm function has a 0. One can verify that the canonical product of a function f is obtained by multiplying all

Table 4-6 Maxterms and Minterms for B_3

$x_1\ x_2\ x_3$	**Minterms**	**Maxterms**	f
0 0 0	$x_1'x_2'x_3'$	$x_1 + x_2 + x_3$	0
0 0 1	$x_1'x_2'x_3$	$x_1 + x_2 + x_3'$	0
0 1 0	$x_1'x_2x_3'$	$x_1 + x_2' + x_3$	1
0 1 1	$x_1'x_2x_3$	$x_1 + x_2' + x_3'$	0
1 0 0	$x_1x_2'x_3'$	$x_1' + x_2 + x_3$	0
1 0 1	$x_1x_2'x_3$	$x_1' + x_2 + x_3'$	1
1 1 0	$x_1x_2x_3'$	$x_1' + x_2' + x_3$	0
1 1 1	$x_1x_2x_3$	$x_1' + x_2' + x_3'$	1

the maxterms corresponding to 0's of f. The canonical sum and the canonical product for the function f of Table 4-6 are

$$F_{CS} = x_1'x_2x_3' + x_1x_2'x_3 + x_1x_2x_3$$
$$F_{CP} = (x_1 + x_2 + x_3)(x_1 + x_2 + x_3')(x_1 + x_2' + x_3')$$
$$(x_1' + x_2 + x_3)(x_1' + x_2' + x_3).$$

It is seen that if the table of a function f is available, the canonical forms can be written by inspection. On the other hand, if only an expression for f is available, it is rather tedious to construct the table for f by the application of the $|\ |$ mapping. It is sometimes convenient to manipulate the given expression and convert it to a sum-of-products form. This can always be done as follows:

1. Use De Morgan's laws (3) and (4) and the law $(E')' = E$ until all complements are removed except those that appear on single variables. The expression now involves only sums and products.
2. "Multiply out" the expression obtained in step 1, getting a sum-of-products expression.

Example

$$f = ((xy + x'z) + y)' = (xy + x'z)'y' = (xy)'(x'z)'y'$$
$$= (x' + y')(x + z')y'. \qquad \text{Here step 1 is completed.}$$
$$f = (x'x + xy' + x'z' + y'z')y' = xy' + x'z'y' + y'z'.$$

From a sum of products it is relatively easy to construct the table of combinations. First, obtain the function corresponding to each product. This is easily done. For example, if $E = xy'z'$, then we put a 1 for all n-tuples, where $x = 1$, $y = 0$, and $z = 0$. Finally, add the functions corresponding to the products. From the table of combinations it is then straightforward to find the canonical forms.

An alternative method of obtaining the canonical sum from any sum-of-products expression is to use the following, where P is any product expression:

3. (a) $1 = x + x'$; (b) $P(x + x') = Px + Px'$; (c) $P + P = P$.

For example, if f is a function of variables w, x, y, and z and $P = w'x$ is a product in a sum-of-products expression for f, then $w'x = w'x(y + y') = w'xy + w'xy' = w'xy(z + z') + w'xy'(z + z') = w'xyz + w'xyz' + w'xy'z + w'xy'z'$. Since the same minterm may be generated from two different products, Rule 3(c) is used to eliminate repetitions. The process terminates when every product is a minterm (has n literals), for then the sum is the canonical sum.

Of course, dual rules can be used for manipulating products of sums in this manner. Thus

$$w' + x = w' + x + yy' = (w' + x + y)(w' + x + y')$$
$$= (w' + x + y + zz')(w' + x + y' + zz'), \text{ etc.}$$

Functions of One or Two Variables

The set of all functions of one variable is shown in Table 4-7. Two of the functions are the constant functions $|0|$ and $|1|$ and the remaining two are the

Table 4-7 B_1

x	$\|0\|$	$\|1\|$	$\|x\|$	$\|x'\|$
0	0	1	0	1
1	0	1	1	0

identity function $|x|$ and the *complement function* $|x'|$. Next B_2 has 16 functions and is of considerable interest. The functions $|0|, |1|, |x|, |x'|, |y|$, and $|y'|$ are degenerate in that the first two are constant functions and the next four depend only on one variable. The 10 nondegenerate functions are shown in Table 4-8.

Table 4-8 Nondegenerate Functions of B_2

$x\ y$	$\|xy\|$	$\|xy'\|$	$\|x'y\|$	$\|x'y'\|$	$\|xy' + x'y\|$	$\|xy + x'y'\|$	$\|x + y\|$	$\|x + y'\|$	$\|x' + y\|$	$\|x' + y'\|$
0 0	0	0	0	1	0	1	0	1	1	1
0 1	0	0	1	0	1	0	1	0	1	1
1 0	0	1	0	0	1	0	1	1	0	1
1 1	1	0	0	0	0	1	1	1	1	0

The reader will note that some of these functions have already been introduced in Table 1-3, (Chapter 1) using special symbols ⊜, ⊕, |, and ↓ rather than only +, ·, and ′. It is customary to extend the notion of Boolean expression to include these symbols. For example, $x \oplus y = xy' + x'y$, etc.

As a further simplification, we shall not make a notational distinction between Boolean functions and Boolean expressions, unless this is necessary. Usually the meaning is clear from the context. For example, we freely write $f = xy' + x'y = x \oplus y$, as we have done in Chapters 1 and 2 for Boolean operators.

4.4. ANALYSIS OF COMBINATIONAL NETWORKS

Combinational Networks Revisited

Recall that a *combinational network* (Section 2.1) has the following properties. It has a finite number n of binary inputs $x_1, x_2, \ldots, x_n$ and a finite number m of binary outputs $z_1, z_2, \ldots, z_m$, as shown in Fig. 4-2. Furthermore, at any time t the value of each output $z_i(t)$ depends only on the values of the inputs at time t; i.e., $z_i(t) = f_i(x_1(t), x_2(t), \ldots, x_n(t))$ for $i = 1, 2, \ldots, m$, where $f_i \in B_n$.

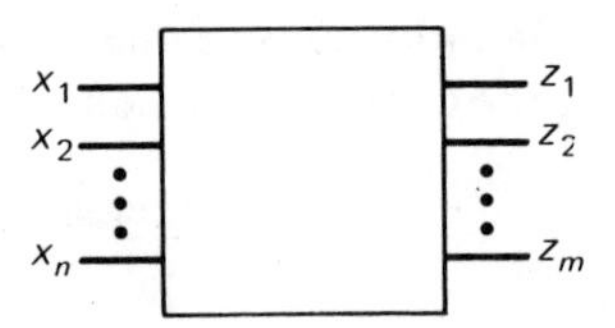

Figure 4-2

In this model, the input–output behavior of the combinational network can be described as a mapping from $\{0, 1\}^n$ into $\{0, 1\}^m$. If we restrict our attention to a single output z_j, that part of the network behavior can be described by a Boolean function f_j in B_n:

$$f_j: \{0, 1\}^n \longrightarrow \{0, 1\}.$$

If the function f_j can be found, then we can predict the value of z_j for any given n-tuple $(a_1, a_2, \ldots, a_n)$ of input values. The process can be repeated for the remaining outputs; thus the behavior of any combinational circuit can be characterized by m Boolean functions of n variables. For convenience, we will temporarily restrict our attention to networks with a single output z. The first problem that we consider is that of *analysis* of combinational networks: Given a particular network, what does it do? In view of the previous remarks, this is equivalent to finding the Boolean function f for the output z.

Series–Parallel Contact Networks

We begin with a simple class of two-terminal contact networks.

Definition 6

A *series–parallel* (SP) contact network is defined inductively:

1. The following are SP contact networks:
   (a) o—   —o (an open circuit)
   (b) o———o (a closed circuit)
   and for $i = 1, 2, \ldots, n$,
   (c) o—x_i—o (a single n.o. contact x_i)
   (d) o—x_i'—o (a single n.c. contact x_i').

2. If N_1 and N_2 are SP contact networks, then so are

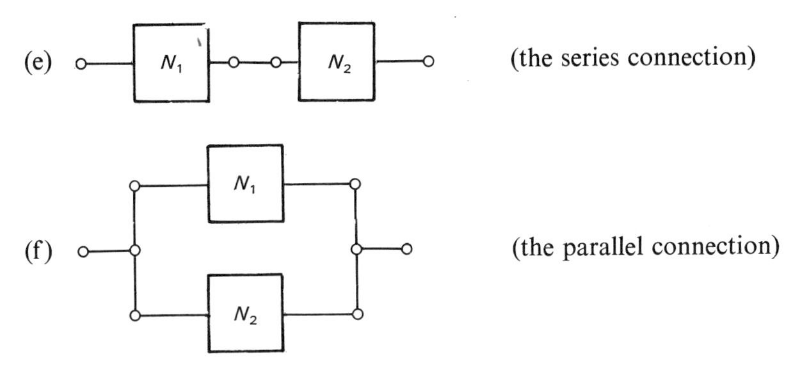

The analysis of SP networks is particularly simple. The Boolean expressions 0 and 1 correspond to the open circuit and closed circuit, respectively. Networks of types (c) and (d) are assigned the Boolean expressions x_i and x_i', respectively. Then we proceed by induction. If N_i is represented by F_i, for $i = 1, 2$, then the series connection of N_1 and N_2 is represented by F_1F_2 and the parallel connection by $F_1 + F_2$, where the addition and multiplication are those of Boolean expressions over n variables. Now if a network N is represented by expression F, the output function of N is $|F|$.

The inputs of an SP network are the independent contact variables $x_1, x_2, \ldots, x_n$. The output z is the variable representing the state of the two external terminals of the network. Note also that the operation of complementation is available only for single variables and is obtained by using n.c. (normally closed) contacts. The operation of multiplication (addition) is performed by the series (parallel) connection.

As an example, consider the network of Fig. 4-3. A Boolean expression

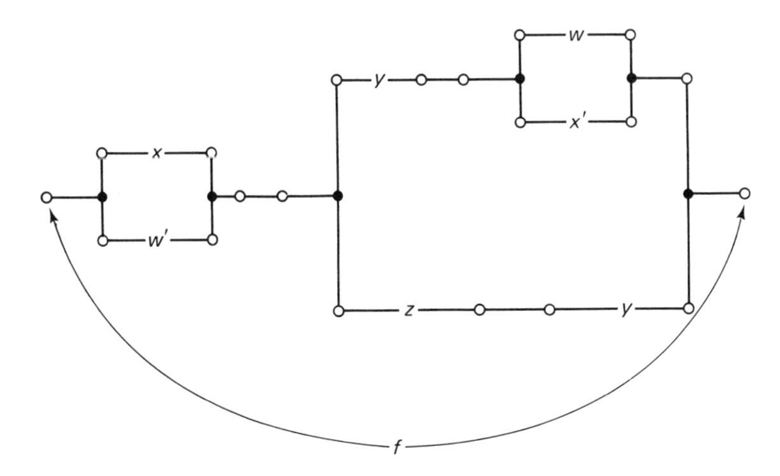

Figure 4-3

for the network is obtained by inspection:

$$f = (x + w')(y(w + x') + zy).$$

This expression can be converted to a sum-of-products form, as explained above. Thus

$$f = (x + w')(yw + yx' + zy)$$
$$= wxy + xyz + w'x'y + w'yz.$$

We remark that, since addition and multiplication of Boolean functions is associative, we can extend the networks in 2(e) and 2(f) of Definition 6 to series and parallel connections of several networks.

General Contact Networks

Many useful contact networks are not of the SP type. In general, any finite collection of contacts connected in any arbitrary manner defines a two-terminal contact network as long as two terminals are singled out to represent the output. Of course, in such a network we may have useless contacts that do not affect the output behavior, for example, contacts not connected to anything else. In practical networks such contacts will not be included, but their presence does not affect the analysis method described below.

If the output terminals t_1 and t_2 are connected, there must exist at least one path p between t_1 and t_2. Suppose that the contacts encountered in path p are $\tilde{x}_{p1}, \tilde{x}_{p2}, \ldots, \tilde{x}_{pk}$, where the literals are not necessarily distinct. Then, if path p is closed, we must have $\tilde{x}_{p1} = \tilde{x}_{p2} = \cdots = \tilde{x}_{pk} = 1$. Equivalently, the path product $\tilde{x}_{p1}\tilde{x}_{p2} \cdots \tilde{x}_{pk}$ must equal 1. In general, there may be an infinite number of paths between t_1 and t_2, since some paths may contain closed loops. However, such paths need not be considered for the following reason. Suppose that $p = (\tilde{x}_{p1}, \tilde{x}_{p2}, \ldots, \tilde{x}_{pi}, \ldots, \tilde{x}_{pj}, \ldots, \tilde{x}_{pk})$ is a path in which the terminal reached by the path $\tilde{x}_{p1}, \ldots, \tilde{x}_{pi}$ is the same as the terminal reached by $\tilde{x}_{p1}, \ldots, \tilde{x}_{pi}, \ldots, \tilde{x}_{pj}$. Then there is also a shorter path, $p_1 = (\tilde{x}_{p1}, \ldots, \tilde{x}_{pi}, \tilde{x}_{p(j+1)}, \ldots, \tilde{x}_{pk})$. Now if p causes t_1 and t_2 to be connected for a given input n-tuple, then all the literals appearing in p must be 1 for that n-tuple. But the literals of p_1 are a subset of the literals of p; hence all the literals of p_1 are also 1 for this input n-tuple, and p_1 itself causes t_1 and t_2 to be connected. Thus, if we include the path p_1 in our analysis, the loop in path p need not be considered. Clearly, it is sufficient to include only paths without loops. We are using here the fact that the product p corresponding to path p is of the form $p = q_1q_2q_3$ and that there also exists another path $p_1 = q_1q_3$. Now p and p_1 will be included in the output function f of the network in the form $f = p + p_1 + \cdots$, since f will be closed iff one or more paths are closed. However,

$$p + p_1 = q_1q_3(q_2 + 1) = q_1q_3 = p_1.$$

Hence p need not be included.

In a finite contact network there is obviously a finite number of paths without loops. The output function can therefore be described by the sum of all such path products.

Example 1

In the network of Fig. 4-4, there are 8 paths without loops. By inspection, the function f can be described by the expression

$$f = x'y'z' + x'y'yyz + x'yz + x'yyyz' + xyy'yz + xyz' + xyyy'z' + xyz.$$

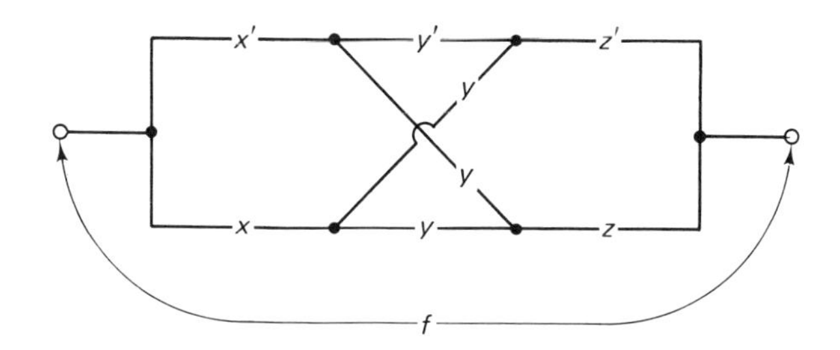

Figure 4-4

Simplifying, we have

$$\begin{aligned} f &= x'y'z' + x'yz + x'yz' + xyz' + xyz \\ &= x'y'z' + x'y(z + z') + xy(z + z') \\ &= x'y'z' + x'y + xy = x'y'z' + y \\ &= x'z' + y. \end{aligned}$$

This simpler expression shows that the network of Fig. 4-5 that has only 3 contacts is equivalent to that of Fig. 4-4. We shall return to the problem of simplifying networks later.

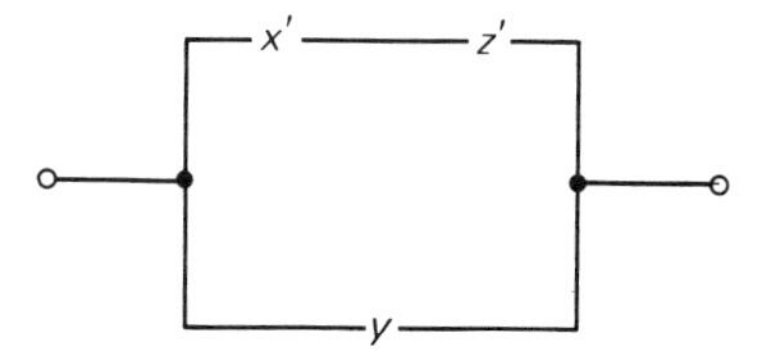

Figure 4-5

Notice that, if a path contains both a variable and its complement, the corresponding path product is always 0. This means that there is no input n-tuple that will cause the path to be closed, and such paths can be omitted.

We also remark that although the expression obtained by path enumeration for the network of Fig. 4-4 properly describes the function f, i.e., the input–output behavior of the network, the expression does not represent the structure of the network. In fact, there is no expression that is in one-to-one correspondence with the structure of the network. In the special case of SP networks, such a correspondence does exist because the series connection corresponds to multiplication and the parallel connection to addition.

In summary, to analyze a two-terminal contact network we enumerate all loop-free paths, and the output function f can be described by the sum of all such path products. A table of combinations for f can then be constructed. From the table we can predict by inspection the value of the output for any given input n-tuple, and so the analysis is complete.

If we remove the restriction that there are only two output terminals, a contact network with r terminals can realize $r(r-1)/2$ Boolean functions, one for each pair of terminals. Such a network can be analyzed by path enumeration by repeating the process for f_{ij}, $i, j = 1, 2, \ldots, r$, $i < j$, where f_{ij} is the function defined by the pair (i, j) of terminals. A more systematic approach uses Boolean matrices for this analysis. We refer the reader to the literature [HAR] for further details.

Loop-Free Gate Networks

We now relate the analysis of gate networks as introduced in Chapter 1 to Boolean expressions. The discussion is limited to loop-free networks. Networks with loops are treated in Chapter 6.

It is convenient here to include rather trivial gates; a lead with a 0, 1 or x_i input and a 0, 1 or x_i output, respectively, falls into our definition of gate. Connecting the outputs of several gates as inputs to a new gate G and considering the output of G corresponds to performing Boolean operations on several Boolean expressions and obtaining a new Boolean expression. Thus it is seen that the algebra of Boolean expressions of n variables provides a natural model for the algebra of networks of gates. If E is an expression representing a single-output gate network N, then $|E|$ is the output function realized by N.

Consider, as an example, the network of Fig. 4-6 with 4 inputs and 3 outputs. Begin at the top left part of Fig. 4-6. The inputs x_1 and x_2 are first thought of as degenerate gates represented by the expressions x_1 and x_2. Now the outputs of these degenerate gates are combined by means of an OR gate to produce the signal y_1. Clearly this corresponds to forming a new

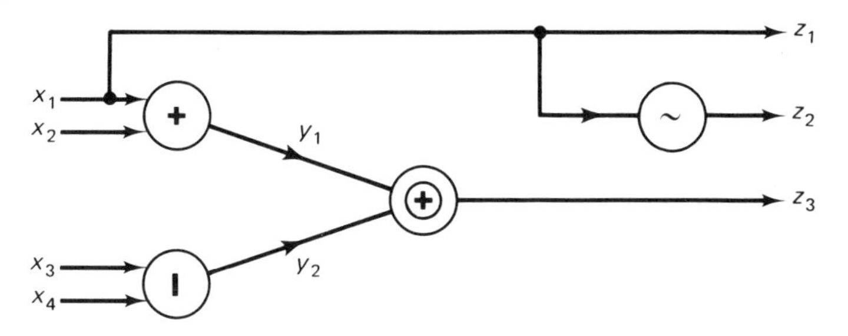

Fig. 4-6 Gate network to be analyzed.

Boolean expression y_1, where $y_1 = x_1 + x_2$. Continuing in this manner we obtain $y_2 = (x_3 \mid x_4)$, and finally $z_1 = x_1, z_2 = x_1'$ and $z_3 = (x_1 + x_2) \oplus (x_3 \mid x_4)$. Thus by using the algebra of Boolean expressions, we are able to obtain the expressions and hence the functions performed by each output.

If the fan-out of all the gates in a network is 1, then the Boolean expression (generalized to allow all Boolean operators, i.e., $+, \cdot, ', \oplus, \equiv, \mid, \downarrow$) corresponds in a natural way to the network structure. However, the example of Fig. 4-7 shows that such a simple correspondence no longer exists if the fan-out is > 1. An expression describing the network behavior is

$$F = ((xy) \mid z) + ((xy) \downarrow z).$$

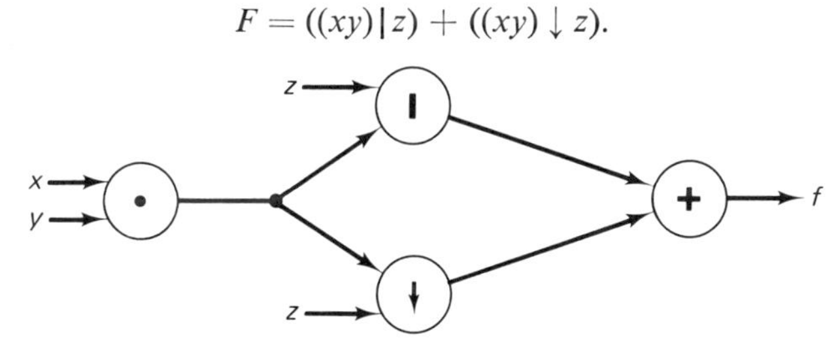

Figure 4-7

The term (xy) is repeated twice in the expression, but the corresponding gate appears only once in the network. This is similar to the case of general contact networks.

*4.5. TERNARY FUNCTIONS: MORE ABOUT TRANSIENT PHENOMENA

We can now continue to discuss the transient phenomena in combinational networks. The reader may wish to refer to Section 3.5 to recall some of the concepts introduced there.

Recall that $L_0 = \{0, \frac{1}{2}, 1\}$ and that the system $\langle L_0, +, \cdot, ^* \rangle$ is a distributive lattice. The operations defined in L_0 may be extended to functions from L_0^n to L_0 in the usual way [YO-RI].

Definition 7

(a) For functions $f: L_0^n \longrightarrow L_0$ and $g: L_0^n \longrightarrow L_0$, define $f + g$, $f \cdot g$, and f^* as follows. For every $x \in L_0^n$,

$$(f + g)(x) = f(x) + g(x)$$
$$(f \cdot g)(x) = f(x) \cdot g(x)$$
$$f^*(x) = (f(x))^*$$

(b) Define the functions $||0||: L_0^n \longrightarrow L_0$ and $||1||: L_0^n \longrightarrow L_0$ as follows:

$$||0||(x) = 0, \; ||1||(x) = 1 \qquad \text{for every } x \in L_0^n.$$

Theorem 3

Let $L_n \triangleq \{f \mid f: L_0^n \longrightarrow L_0\}$. Then the system $\langle L_n, +, \cdot, *, ||0||, ||1|| \rangle$ is a distributive lattice with universal bounds. Furthermore, it obeys the laws of Proposition 8, Chapter 3.

Proof

Use arguments similar to those applied to prove Theorem 1. □

We now wish to associate with every Boolean expression over n variables a function in L_n.

Definition 8

Let $\mathcal{E}_n$ be the set of all Boolean expressions over n variables. We define a mapping $|| \; ||: \mathcal{E}_n \longrightarrow L_n$ as follows.

1. The expressions 0 and 1 are mapped to the functions $||0||$ and $||1||$, respectively.
2. $||x_i||(x) = x_i$ for every $x = (x_1, \ldots, x_n) \in L_0^n$.
3. $||(E + F)|| = ||E|| + ||F||$
   $||(EF)|| = ||E|| \cdot ||F||$
   $||E'|| = (||E||)^*$.

By means of Definition 8, we associate with every Boolean expression $E \in \mathcal{E}_n$ a function $||E||: L_0^n \longrightarrow L_0$. This function is an extension of the function $|E|: B_0^n \longrightarrow B_0$ defined earlier (see Definition 4). Namely, we have:

Proposition 1

Let $E \in \mathcal{E}_n$. Then $|E|$ is the restriction of the function $||E||$ to B_0^n; i.e., for every $x \in B_0^n$, $||E||(x) = |E|(x)$.

Proposition 1 is easily verified.

It is important to point out that $|E| = |F|$ does not imply $||E|| = ||F||$. To illustrate, let $E = x_1x_2 + x_1'x_3$ and $F = x_1x_2 + x_1'x_3 + x_2x_3$. Here $|E| = |F|$, but $||E||\,(\frac{1}{2}, 1, 1) = \frac{1}{2}$, whereas $||F||\,(\frac{1}{2}, 1, 1) = 1$. It follows that the corresponding gate networks (see Fig. 4-8) have the same steady-state behavior but different transient behavior.

Note that $||E||\,(0, 1, 1) = 1$, $||E||\,(1, 1, 1) = 1$, but $||E||\,(\frac{1}{2}, 1, 1) = \frac{1}{2}$. This indicates that the output of the E-realization of Fig. 4-8(a) is 1 for both the input combinations (0, 1, 1) and (1, 1, 1) but becomes indeterminate during a transition between these input combinations. There exists, therefore, the possibility that during such a transition, the circuit in question [Fig. 4-8(a)] will produce a spurious 0-pulse on its output lead. Such a false pulse might become harmful, e.g., by changing the state of a memory device to which the output is connected. This difficulty might be overcome by using a slow-acting memory device that would be insensitive to a short spurious pulse. However, this remedy is unacceptable if fast-acting networks are required. In this case, the network of Fig. 4-8(a) might be replaced by the network of Fig. 4-8(b), since $|F| = |E|$ but $||F||\,(\frac{1}{2}, 1, 1) = 1$. Thus the circuit of Fig. 4-8(b) cannot produce a spurious output pulse during a transition between the input combinations (0, 1, 1) and (1, 1, 1).

We shall return to the discussion of transients in Chapter 10.

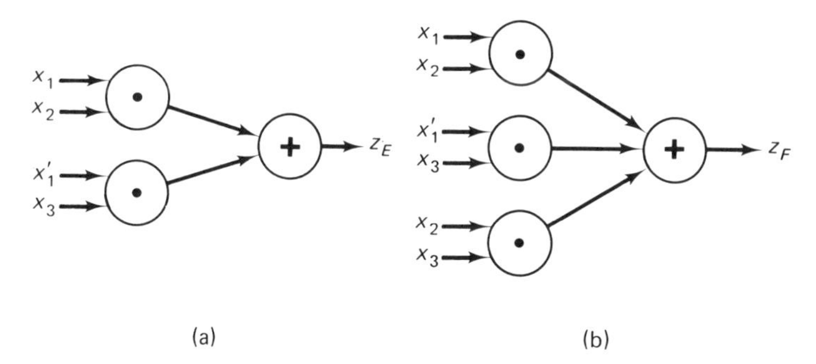

Fig. 4-8 Two gate networks with same steady-state but different transient behavior. (a) Gate network realizing $E = x_1x_2 + x'_1x_3$. (b) Gate network realizing $F = x_1x_2 + x'_1x_3 + x_2x_3$.

4.6. SUMMARY OF ALGEBRAIC CONCEPTS

Facts for the Reader of Unstarred Material

In Chapter 3 the reader of starred material learned that for each $k \geq 1$ there exists a Boolean algebra with 2^k elements. Thus there is the familiar Boolean algebra B_0 of $\{0, 1\}$ (introduced in Chapter 1) with 2 elements and there are Boolean algebras with 4, 8, 16, 32, . . . elements. Furthermore, if a Boolean algebra is finite, we showed that the number of elements in it must be some power of 2. All Boolean algebras with 2^k elements are isomorphic; in this sense, there are no other finite Boolean algebras but those with 2, 4, 8, 16, . . . elements! Now the algebras B_n of functions of n variables have 2^{2^n} elements and are rather special (they are called *free*). Thus, for switching network applications, we are dealing with Boolean algebras $B_0, B_1, B_2, B_3, \ldots$, with 2, 4, 16, 256, ... elements.

The mathematical approach in Chapter 3 allowed us to obtain these results for the general case when we established an isomorphism between any finite Boolean algebra B and a Boolean algebra of all subsets of some set. The reader of the unstarred material should try to prove, using only the definition of Section 3.2, that there are no Boolean algebras with 3, 5, 6, 7, 9, ... elements. This should convince him of the power of the starred approach.

*Summary of Starred Material

A number of algebraic concepts have been introduced. It is the purpose of this section to show clearly how these concepts are interrelated. First, we point out that the algebra of equivalence classes of Boolean expressions over n variables $x_1, \ldots, x_n$ is often called the *free Boolean algebra generated by n elements*. This is viewed as follows. Starting with a given set $\{x_1, \ldots, x_n\}$ of independent elements (called *free generators*), we construct the smallest Boolean algebra that contains these elements. Thus we must adjoin such elements as x_1', $x_1 + x_2$, $x_1 \cdot x_2$, $x_1 \cdot x_1' \triangleq 0$, etc. Since this expanded set must obey all the laws of Boolean algebra, we must recognize that $x_1 \cdot x_1$ is the same element as x_1, etc.

In general, two Boolean expressions represent the same element of the free Boolean algebra if one can be obtained from the other by applying valid laws of Boolean algebra. Clearly two such expressions denote the same Boolean function. Conversely, one can show that if two expressions denote the same Boolean function, then one can be obtained from the other by using only valid laws of Boolean algebra. (Both expressions can be converted to the same canonical sum.) It follows now that two expressions represent the same element of the free Boolean algebra iff they denote the same function, i.e., are equivalent, as in Section 4.3. Thus the algebra of Boolean functions of n variables is isomorphic to the free Boolean algebra generated by n elements.

Another example of an algebra isomorphic to a free Boolean algebra can be derived from Definition 1 of Appendix A. Namely, consider all formulas in $p_1, \ldots, p_n$, which contain only the connectives $\sim$, $\wedge$, and $\vee$. The equivalence classes of these formulas (with respect to the equivalence relation $\equiv$) form a Boolean algebra (in the evident sense), which is isomorphic to the free Boolean algebra generated by n elements.

It follows that the free Boolean algebra generated by n elements has 2^{2^n} elements. Clearly, not every finite Boolean algebra is free. For each $m \geq 2$, there exists a Boolean algebra with 2^m elements; this algebra is isomorphic to a free algebra iff $m = 2^n$ for some $n \geq 1$.

As another illustration of a free algebraic system generated by n elements, we mention the *free distributive lattice with 0 and 1* generated by n elements. This is the lattice of equivalence classes of Boolean expressions without complements, where two expressions are considered equivalent if one can be transformed into the other by using only L1–L6, L9 and L1′–L6′, L9′. This lattice is isomorphic to the lattice of Boolean functions which can be denoted by Boolean expressions without using complementation. For example, the free distributive lattice with 0 and 1 generated by two elements x_1 and x_2 is $L_2 = \{0, x_1, x_2, x_1x_2, x_1 + x_2, 1\}$. In other words, every expression without complementation over x_1 and x_2 can be reduced to one of the 6 expressions by using only the axioms of the distributive lattice with 0 and 1. In general, the number of elements in the free distributive lattice with n generators is not known.

We now summarize the algebraic systems related to Boolean algebra in Fig. 4-9. This is a "poset diagram," where the partial order is inclusion in the sense that lattices are semilattices satisfying additional requirements,

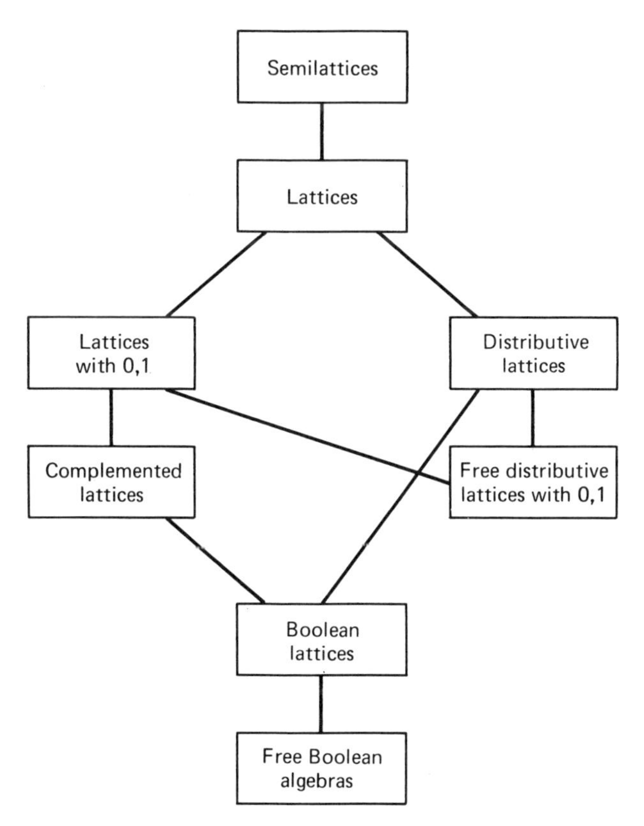

Fig. 4-9 A "poset diagram" for algebraic systems considered.

etc. The lattices L_n of Section 4.5 are distributive lattices with 0 and 1 but are not complemented and hence not Boolean.

Concentrating now on Boolean lattices or algebras, we present a summarizing diagram in Fig. 4-10, where β_m denotes the Boolean algebra with 2^m elements and B_n is the Boolean algebra of functions of n variables, isomorphic to the free Boolean algebra on n generators. Thus $B_n \cong \beta_{2^n}$. In the center column of Fig. 4-10 we show various finite Boolean algebras. In the left column are shown Boolean algebras $B(X)$, as defined in Section 3.2 and in the right column the Boolean algebras of functions.

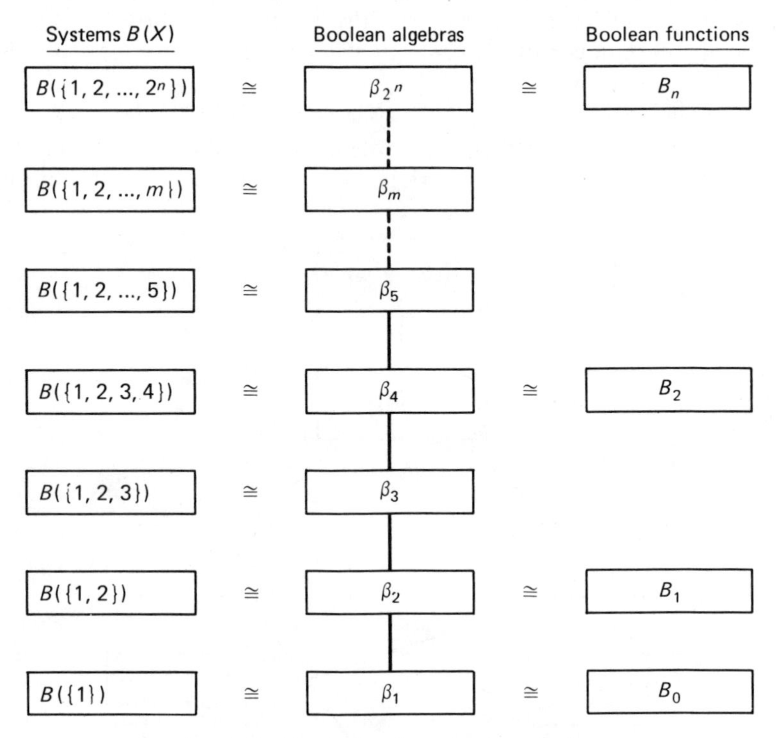

Fig. 4-10 Illustrating finite Boolean algebras.

Finally, note that the algebra of Boolean operators $+$, $\cdot$, and $'$ introduced in Section 1.4 is B_2.

*4.7. SPECIAL PROPERTIES OF BOOLEAN FUNCTIONS

We close this chapter by a brief discussion of Boolean functions with special properties [HAR].

Equivalence Classes of Boolean Functions

Consider the functions $f = zx + y'$ and $g = xy + z'$. They are distinct as elements of B_3 but are related because both can be realized by the same network by changing only the input variables, as shown in Fig. 4-11.

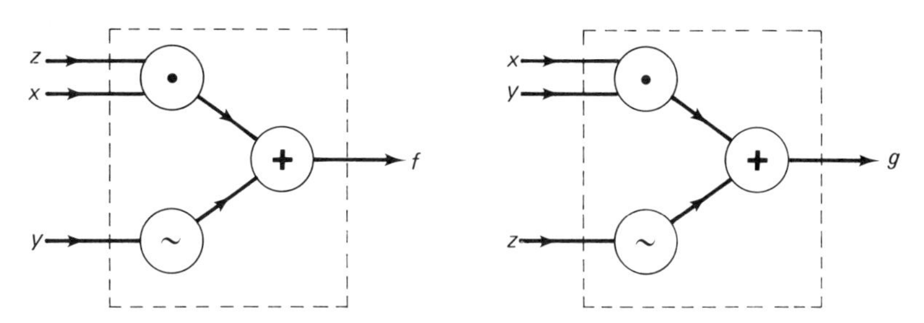

Fig. 4-11 Illustrating Π-equivalent functions.

Under the assumption that the inputs are available in both complemented and uncomplemented forms, the function $h = z'x + y$ can be realized by the same network as f, as shown in Fig. 4-12.

These observations suggest the following definitions.

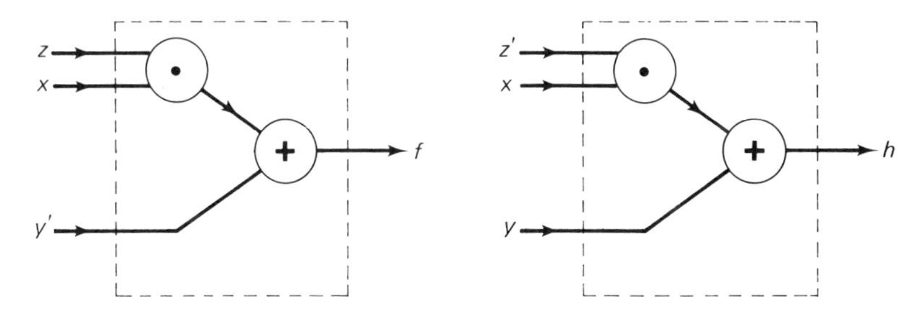

Fig. 4-12 Illustrating Γ-equivalent functions.

Definition 9

(a) Let $\pi = (i_1, \ldots, i_n)$ be a permutation of the ordered set $(1, \ldots, n)$. If f is a Boolean function in B_n, define f_π as follows:

For each $x = (x_1, \ldots, x_n) \in \{0, 1\}^n$, $f_\pi(x_1, \ldots, x_n) = f(x_{i_1}, \ldots, x_{i_n})$.

(b) Let $\gamma = (j_1, \ldots, j_n) \in \{0, 1\}^n$. Define f_γ: For each $x \in \{0, 1\}^n$, $f_\gamma(x_1, \ldots, x_n) = f(x_1^{j_1}, \ldots, x_n^{j_n})$. γ is called a *complementation of* $(x_1, \ldots, x_n)$, where $x^i = x$ if $i = 1$ and $x^i = x'$ if $i = 0$.

Definition 10

Let $f,g \in B_n$. Define the relations:

(a) $f \sim_\Pi g$ (f is *permutation-equivalent* or *Π-equivalent* to g) iff $g = f_\pi$ for some permutation π.

(b) $f \sim_{\Gamma} g$ (f is *complementation-equivalent* or *Γ-equivalent* to g) iff $g = f_\gamma$ for some complementation γ.

(c) $f \sim_{\Pi\Gamma} g$ (f is *$\Pi\Gamma$-equivalent* to g) iff $g = (f_\pi)_\gamma$ for some π and γ.

One easily verifies that each relation above is an equivalence relation on B_n. Intuitively, $f \sim_{\Pi\Gamma} g$ iff g can be obtained from f by a permutation and/or complementation of the input variables. Under the assumption that both complemented and uncomplemented inputs are available, $f \sim_{\Pi\Gamma} g$ implies that f and g can be realized by the same gate network. It is also clear that $f \sim_{\Pi\Gamma} g$ implies that, if N_f is a contact network realizing f with m contacts, then there is an m-contact network N_g realizing g. In other words, f and g have the same cost if contacts are used as a cost measure, provided that the cost of a normally open contact is the same as that of a normally closed contact.

We leave it to the reader to verify that if $g = (f_\pi)_\gamma$, then there exists γ_0 such that $g = (f_{\gamma_0})_\pi$. Thus changing the order of π and γ in Definition 10(c) results in the same equivalence relation. For example, consider $f = x_1 + x_2x_3$, $\pi = (3, 2, 1)$, and $\gamma = (1, 1, 0)$. Then $f_\pi = x_3 + x_2x_1$ and $(f_\pi)_\gamma = x_3' + x_2x_1$. In general, $(f_\pi)_\gamma \neq (f_\gamma)_\pi$. Here $f_\gamma = x_1 + x_2x_3'$ and $(f_\gamma)_\pi = x_3 + x_2x_1'$. However, let $\gamma_0 = (0, 1, 1)$. Then $f_{\gamma_0} = x_1' + x_2x_3$ and $(f_{\gamma_0})_\pi = x_3' + x_2x_1 = (f_\pi)_\gamma$.

For a detailed discussion of permutation and complementation operators, we refer the reader to [HAR].

The equivalence relations above can be used to reduce the problem of finding minimal networks in the following way. In principle, having found a network with m contacts or gates or gate inputs for a given function f, one can enumerate all networks with less than m contacts, etc. From these one can find the minimal-cost network. This is not a practical approach because of the very large numbers of networks and functions to be considered. The number of Boolean functions can be reduced by examining only one representative for each equivalence class $[f]_{\Pi\Gamma}$ of all the functions $\Pi\Gamma$-equivalent to f.

Let $N = 2^{2^n}$ be the number of functions of n variables and N_Π, N_Γ, and $N_{\Pi\Gamma}$ be the number of equivalence classes of $\sim_\Pi$, $\sim_\Gamma$, and $\sim_{\Pi\Gamma}$, respectively. We list these numbers in Table 4-9.

Table 4-9 Number of Equivalence Classes

n	N	N_Π	N_Γ	$N_{\Pi\Gamma}$
1	4	4	3	3
2	16	12	7	6
3	256	80	46	22
4	65,536	3,984	4,336	402
5	2^{32}	37,333,248	134,281,216	1,228,158

It is seen that although there are reductions, the numbers increase very rapidly with n. It is reasonable to tabulate the minimal networks for the 402 equivalence classes of $\sim_{\Pi\Gamma}$ for 4 variables, but for larger n this becomes prohibitive. Even for 4 variables a given function chosen from the 65,536 possible functions must first be converted to some canonical form in order to determine its equivalence class. This in itself is a rather lengthy process. One must conclude therefore that enumeration can be ruled out as a practical method for finding minimal networks.

Symmetric Functions

Functions that are invariant under permutation of the input variables constitute an interesting subset of B_n.

Definition 11

$f \in B_n$ is *symmetric* iff for each permutation π of $(1, \ldots, n)$, $f_\pi = f$. Let S_n be the set of all symmetric functions in B_n.

For $x \in \{0, 1\}^n$, denote by $w(x)$ the number of 1's in x (the *weight* of x). For $x,y \in \{0, 1\}^n$, define the equivalence relation $\leftrightarrow$ by $x \leftrightarrow y$ iff $w(x) = w(y)$. Let w_k denote the equivalence class consisting of all n-tuples with weight k, $w_k \triangleq \{x \in \{0, 1\}^n \mid w(x) = k\}$. Let s_k be the function defined by $s_k(x) = 1$ iff $x \in w_k$. For $k = 0, 1, \ldots, n$, each s_k is called an *elementary symmetric function*. The fact that each s_k is symmetric is obvious from the canonical sum for s_k:

$$
\begin{aligned}
s_0 &= x_1'x_2' \cdots x_n' \\
s_1 &= x_1x_2' \cdots x_n' + x_1'x_2x_3' \cdots x_n' + \cdots + x_1'x_2' \cdots x_{n-1}'x_n \\
&\vdots \\
s_n &= x_1x_2 \cdots x_n.
\end{aligned}
$$

If f is any symmetric function, $\pi = (i_1, \ldots, i_n)$ and $f(x_1, \ldots, x_n) = 1$, then also $f(x_{i_1}, \ldots, x_{i_n}) = 1$. Obviously, a permutation does not affect the weight of $(x_1, \ldots, x_n)$. Furthermore, if $x = (x_1, \ldots, x_n)$ and $x \leftrightarrow y$, then there exists a permutation $\pi = (i_1, \ldots, i_n)$ of $(1, \ldots, n)$ such that $y = (x_{i_1}, \ldots, x_{i_n})$. It follows that if $f(x) = 1$ for some n-tuple $x \in \{0, 1\}^n$ such that $w(x) = k$, then also $f(y) = 1$ for every y with $w(y) = k$. Clearly, if $w(x) = k$ and $f(x) = 1$, then $f \geq s_k$.

Theorem 4

A Boolean function is symmetric iff it can be expressed as a sum of elementary symmetric functions.

Proof

Clearly any sum of symmetric functions is again symmetric. Conversely, suppose that f is a symmetric function and is represented by $f = m_1 + \cdots + m_r$ where $m_1, \ldots, m_r$ are the atom functions of f. If $w(m_1) = k_1$, then by our previous arguments, $f \geq s_{k_1} \geq m_1$ and

$$f = s_{k_1} + m_1 + m_2 + \cdots + m_r$$
$$= s_{k_1} + m_2 + \cdots + m_r.$$

Continuing this argument we have

$$f = s_{k_1} + \cdots + s_{k_r}.$$ □

There are $n + 1$ elementary functions and hence 2^{n+1} symmetric functions, including 0 and 1. Introduce the notation $s_{\{k_1, \ldots, k_j\}} = s_{k_1} + \cdots + s_{k_j}$. Then each function in S_n is designated by s_K, where $K \subseteq \{0, 1, \ldots, n\}$. One verifies that

$$s_K + s_J = s_{K \cup J}$$
$$s_K \cdot s_J = s_{K \cap J}$$
$$s'_K = s_{\{0, 1, \ldots, n\} - K}.$$

Thus S_n is itself a Boolean algebra, a subalgebra of B_n. The atoms of S_n are the elementary symmetric functions. For $n = 2$, S_n is shown in Fig. 4-13. Because of the special properties of S_n, symmetric functions can be realized by more regular structures than arbitrary Boolean functions, as we will show.

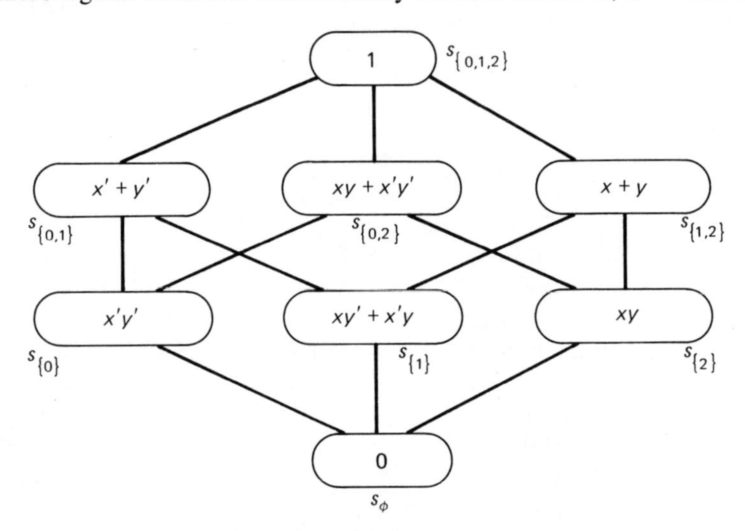

Fig. 4-13 Poset diagram of S_2.

Figure 4-14(a) shows a contact network that can realize every symmetric function of three variables. Each elementary function s_i is realized by the terminal pair (t, s_i). To obtain, for example, $s_{\{0,2\}}$, join the terminals s_0 and s_2. The new terminal together with t realizes $s_{\{0,2\}}$. All other symmetric functions are realizable in a similar way, and that network generalizes in an obvious fashion to any number of variables. Note that if a particular function is realized by the canonical network of Fig. 4-14(a), certain contacts are not used and can be eliminated. One can verify that the canonical network uses $n^2 + n$ contacts. It can be shown that for any symmetric function, at least n contacts are superfluous. *Therefore, every symmetric function can be realized by a network with n^2 or fewer contacts.* In our example for $s_{\{0,2\}}$, the superfluous contacts are marked ⊛. A simplified network is shown in Fig. 4-14(b). Note also that the networks obtained by this construction are in general not series–parallel.

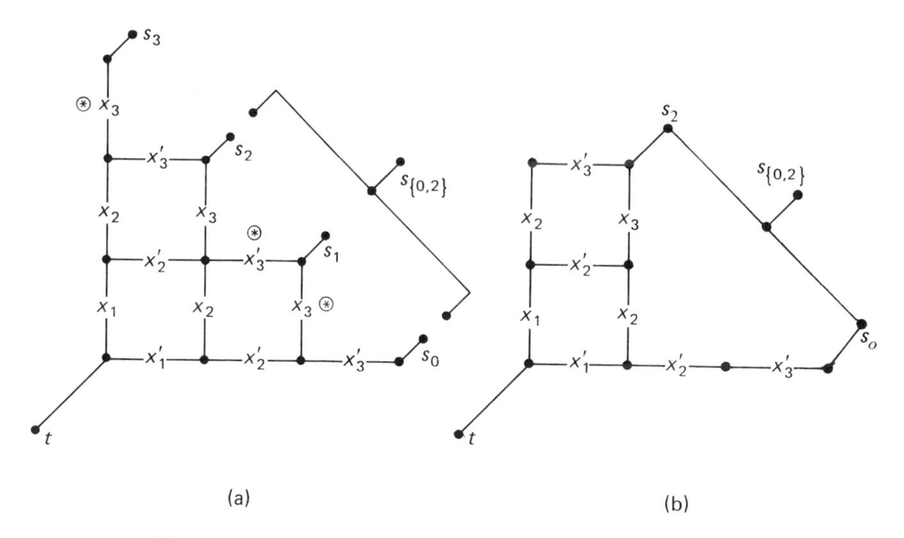

Fig. 4-14 (a) Canonical realization of symmetric functions for $n = 3$. (b) Simplified network for $s_{\{0,2\}}$.

Another approach to the realization of symmetric functions is provided by "iterative" networks. To illustrate the idea, consider the function s_2 of 5 variables, which is 1 when exactly two of the five inputs are 1. Figure 4-15 shows the iterative structure consisting of 5 identical cells (except for the first cell which has no "carry" inputs). The typical cell i receives carry information from the cell to its left whether the number of 1's to the left of input x_i is 0, 1, 2, or (3 or more). It then adds the information provided by x_i and passes the updated carries to cell $i + 1$. The logic of the typical cell is shown

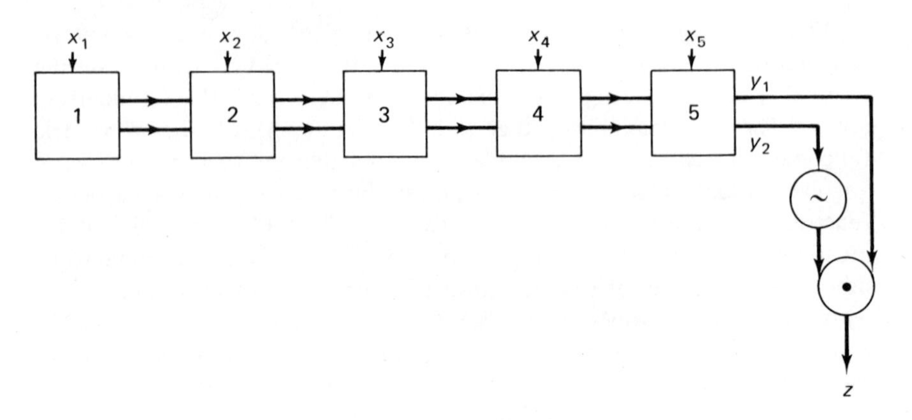

Fig. 4-15 An iterative realization of s_2.

in Table 4-10. The output z is obtained from the last cell. We have $z = 1$ iff the carry-out of the last cell is 2; i.e., $y_1 = 1, y_2 = 0$. Note that one advantage of this type of design is that the network is extended to more input variables by simply adding more cells.

The notion of symmetry can be generalized to subsets of the set of all variables. For example, the function $f = vwx + y'z'$ is symmetric in the variable sets $\{v, w, x\}$ and $\{y, z\}$ in the sense that it is unchanged if the variables in either set are permuted. If one takes into account such symmetries, specialized realizations can be obtained [HAR].

Table 4-10 Behavior of Typical Cell

Carry-in	Coded Carry-in	x_i	Coded Carry-out
0	00	0	00
	00	1	01
1	01	0	01
	01	1	10
2	10	0	10
	10	1	11
3 or more	11	0	11
	11	1	11

Essential Variables and Literals

Let F be a Boolean expression over the variables $x_1, \ldots, x_n$. Denote by $F_{x'_i}$ and F_{x_i} the expressions obtained from F by replacing all appearances of x_i by 0 and 1, respectively. Thus, if $F = x + yz'$, then $F_{x'} = 0 + yz'$ and $F_y = x + 1z'$. Extend this notation to any number of variables; thus $F_{x'y} = 0 + 1z'$. In this way we define F_P for any product expression P.

Each F_P defines a Boolean function $f_P = |F_P|$, and it is easily verified that $|E| = |F|$ implies $|E_P| = |F_P|$. Thus the definition of f_P is independent of the choice of expression used to denote f. We say that each product P defines a *valuation* on the set $V(P)$ of variables of P; if $\tilde{x}_i$ is a literal appearing in P, then $\tilde{x}_i$ is assigned the value 1. We call f_P a *residue* of f or the *P-residue* of f.

A physical interpretation of the residues is obtained as follows. If a network realizes f, then the same network will realize f_P if each variable $x_i \in V(P)$ is replaced by the appropriate constant input (0 or 1).

A function f *depends* on a variable x_i iff $f_{x'_i} \neq f_{x_i}$. Clearly if f depends on x_i then either x_i or x'_i must appear in every valid expression F for f, for if both x_i and x'_i are absent then $f_{x'_i} = f_{x_i}$. On the other hand, if f does not depend on x_i, one can always find an expression for f that involves neither x_i nor x'_i. This follows since for any f we can write

$$f = x_i f_{x_i} + x'_i f_{x'_i}.$$

If f does not depend on x_i, then $f_{x_i} = f_{x'_i}$ and $f = f_{x_i}$.

If f depends on all its variables, it is called *nondegenerate;* otherwise, it is *degenerate.*

The notion of dependence can be refined as follows:

Definition 12

A function f is *positive in* x_i iff $f_{x'_i} < f_{x_i}$; f is *negative in* x_i iff $f_{x'_i} > f_{x_i}$. If f is either positive or negative in x_i, then it is *unate in* x_i.

Correspondingly, f is *positive* (*negative*, *unate*) iff for each x_i it is positive in x_i (negative in x_i, unate in x_i). An expression F is positive in x_i if x_i appears in F but not x'_i. Similarly, define expressions negative in x_i, unate in x_i, positive, negative, and unate.

Definition 13

A literal $\tilde{x}_i$ is *essential* for a function f iff it appears in every sum of products for f.

Theorem 5

Let f be a Boolean function and x_i a variable of f.

(a) Neither x_i nor x'_i is essential iff $f_{x'_i} = f_{x_i}$.
(b) x'_i but not x_i is essential for f iff $f_{x'_i} > f_{x_i}$ (i.e., f is negative in x_i).
(c) x_i but not x'_i is essential for f iff $f_{x'_i} < f_{x_i}$ (i.e., f is positive in x_i).
(d) Both x_i and x'_i are essential for f iff $f_{x'_i}$ and f_{x_i} are incomparable (i.e., neither $f_{x'_i} \leq f_{x_i}$ nor $f_{x_i} \leq f_{x'_i}$).

Proof

Part (a) was already proved and is stated here only for completeness. For (b) suppose that x'_i is essential for f but x_i is not. Then a sum of products for f can be written in the form $F = x_i' G + H$, where G and H do not involve x_i or x_i'. Now $f_{x'_i} = G + H$ and $f_{x_i} = H$. If $|G + H| = |H|$, then

$$f = x_i'G + H = x_i'G + x_i'H + x_iH = x_i'(G + H) + x_iH = x_i'H + x_iH = H,$$

contradicting the fact that x_i' is essential. Hence $f_{x'_i} \neq f_{x_i}$ and since clearly $f_{x'_i} \geq f_{x_i}$, we have $f_{x'_i} > f_{x_i}$. Conversely, suppose that $f_{x'_i} > f_{x_i}$. Then $f_{x'_i} = f_{x'_i} + f_{x_i}$. Now

$$f = x_i'f_{x'_i} + x_if_{x_i} = x_i'(f_{x'_i} + f_{x_i}) + x_if_{x_i} = x_i'f_{x'_i} + f_{x_i}$$

and x_i is not essential for f. If x_i' is also not essential, then f does not depend on x_i and $f_{x'_i} = f_{x_i}$. This is a contradiction. Part (c) follows by a similar argument. For (d), let $f = x_i'f_{x'_i} + x_if_{x_i}$ be such that both x_i' and x_i are essential. If $f_{x'_i} \geq f_{x_i}$, then $f_{x'_i} = f_{x_i} + f_{x'_i}$ and $f = x_i'(f_{x_i} + f_{x'_i}) + x_if_{x_i} = x_i'f_{x'_i} + f_{x_i}$, contradicting the fact that x_i is essential. Similarly, $f_{x'_i} \leq f_{x_i}$ leads to a contradiction. Finally, if $f_{x'_i}$ and f_{x_i} are comparable, then x_i and x_i' cannot both be essential. ☐

The notion of essential literals can be useful in establishing the minimality of a contact network. For example, in $f = w(x + yz)$ all the literals are essential, so at least 4 contacts are required. Since the series–parallel network corresponding to $w(x + yz)$ uses 4 contacts, it is minimal. Similarly, in $f = xy + x'y'z$ all literals are essential and the corresponding network with 5 contacts is minimal. (This method can be used to prove the minimality of the network in Fig. 1-10, p. 15.)

Positive Functions

The set P_n of positive functions of n variables is isomorphic to the free distributive lattice with 0 and 1 on n generators. We now give another characterization of positive functions.

Definition 14

$f \in B_n$ is *monotone increasing* iff for all $x, y \in \{0, 1\}^n$, $x \leq y$ implies $f(x) \leq f(y)$, where $x \leq y$ iff $x_1 \leq y_1, \ldots, x_n \leq y_n$.

Theorem 6

f is positive iff it is monotone increasing.

Proof

One verifies that $0, 1, x_1, \ldots, x_n$ are all monotone increasing and that if f and g are monotone increasing, then so are $f + g$ and fg. Thus all positive functions are monotone increasing. Conversely, suppose that f is monotone increasing. If $f = 0$, then it is positive. Otherwise, let $a = x_{i_1} \cdots x_{i_k}x'_{i_{k+1}} \cdots x'_{i_n}$ be an atom of f, where $k \geq 0$ and the variables have been rearranged for convenience so that the first k variables are uncomplemented. Since f is monotone increasing, all the atoms of the form $b = x_{i_1} \cdots x_{i_k}\tilde{x}_{i_{k+1}} \cdots \tilde{x}_{i_n}$ must be in f. Thus $f \geq x_{i_1} \cdots x_{i_k} \triangleq P$. Therefore, each atom of the type above is "covered" by a positive product P. If $x'_1 \cdots x'_n$ is an atom, then f is identically 1. In all cases f is positive. ☐

Threshold Functions

In recent years, considerable attention has been devoted to certain types of functions realized by devices called *threshold gates*. However, threshold gates are not widely used at present, except for some special cases. For this reason we discuss them only briefly. On the other hand, threshold functions are of some interest in connection with pattern classification [NIL].

An n-input threshold gate is a device with n binary inputs $x_1, \ldots, x_n$ and one binary output z. Associated with each input is a real number w_i called the *weight* of x_i and with the output another real number T called the *threshold* of the gate. The logical behavior of the threshold gate is defined by

$$z = 0 \qquad \text{if } \sum_{i=1}^{n} w_i x_i < T$$

$$z = 1 \qquad \text{if } \sum_{i=1}^{n} w_i x_i \geq T.$$

Note that in this context the logical values 0 and 1 of the x_i are interpreted as real values; i.e., the product $w_i x_i$ and the summation is performed as in ordinary arithmetic.

Definition 15

A Boolean function $f \in B_n$ is a *threshold* function iff there exist $n + 1$ real numbers $w_1, \ldots, w_n$ and T such that

$$f(x_1, \ldots, x_n) = 1 \text{ iff } \sum_{i=1}^{n} w_i x_i \geq T. \qquad (*)$$

In geometrical terms, the equation $\sum_{i=1}^{n} w_i x_i = T$ defines a hyperplane in n-dimensional space. A function is a threshold function if there exists such a plane which separates the points of f (i.e. points x such that $f(x) = 1$) from those of f' in the sense that all the points of f are either on the plane or above it. For this reason, threshold functions are also called *linearly separable* functions.

For example, consider Fig. 4-16 for $n = 2$. The function $x_1 + x_2$ is linearly separable, for we can find a suitable hyperplane (in the case of two variables, this is just a straight line). For example, let $w_1 = 1$, $w_2 = 1$, and $T = \frac{1}{2}$. Then all points of $f = x_1 + x_2$ lie above the line

$$w_1 x_1 + w_2 x_2 = T; \qquad \text{i.e., } x_1 + x_2 = \tfrac{1}{2}.$$

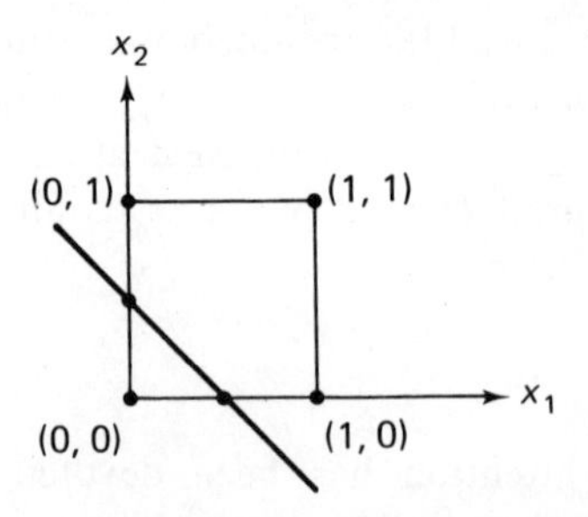

Fig. 4-16 Illustrating linear separability of $f = x_1 + x_2$.

On the other hand, one can verify that no such straight line exists for $f = x_1 \oplus x_2$. Hence XOR is not a threshold function.

The $(n + 1)$-tuple $w = (w_1, w_2, \ldots, w_n; T)$ of real numbers is called a *realization* of a threshold function if it satisfies

$$f(x_1, \ldots, x_n) = 1 \qquad \text{iff } \sum_{i=1}^{n} w_i x_i \geq T.$$

Table 4-11 shows some common threshold functions and their realizations.

Table 4-11 Common Threshold Functions

Function	Expression	Realization
NOT	x'	$(-1; 0)$
OR	$x + y$	$(1, 1; 1)$
AND	xy	$(1, 1; 2)$
NAND	$(xy)'$	$(-1, -1; -1)$
NOR	$(x + y)'$	$(-1, -1; 0)$
MAJORITY	$(xy + yz + xz)$	$(1, 1, 1; 2)$

From this it follows that every Boolean function can be realized by a network of threshold gates, since AND, OR, and NOT are threshold functions.

From the definition of threshold function, it follows that one can test whether a function is a threshold function by testing whether the corresponding set of 2^n inequalities [one for each $(x_1, \ldots, x_n) \in \{0, 1\}^n$],

$$\sum_{i=1}^{n} w_i x_i \geq T \qquad \text{when } f = 1$$

$$\sum_{i=1}^{n} w_i x_i < T \qquad \text{when } f = 0$$

is consistent, i.e., possesses a solution $(w_1, \ldots, w_n; T)$.

Consider, for example $x_1 + x_2 x_3'$ and the analysis in Table 4-12.

Table 4-12

x_1 x_2 x_3	$\sum w_i x_i$	f
0 0 0	0	0
0 0 1	w_3	0
0 1 0	w_2	1
0 1 1	$w_2 + w_3$	0
1 0 0	w_1	1
1 0 1	$w_1 + w_3$	1
1 1 0	$w_1 + w_2$	1
1 1 1	$w_1 + w_2 + w_3$	1

We find the inequalitities:

$$
\begin{array}{ll}
0 < T & w_2 \geq T \\
w_3 < T & w_1 \geq T \\
w_2 + w_3 < T & w_1 + w_3 \geq T \\
 & w_1 + w_2 \geq T \\
 & w_1 + w_2 + w_3 \geq T
\end{array}
$$

Such systems of inequalities are of interest for many problems other than threshold functions and have been studied extensively. One can use, for example, techniques of linear programming to test whether such a system is consistent [MUR].

If a function f has a realization $(w_1, \ldots, w_n; T)$, where $w_1, \ldots, w_n, T$ are real numbers, then one can always approximate each w_i and T by a rational number with arbitrary accuracy and obtain a rational realization. If all the weights and T are next multiplied by their least common denominator, the resulting integers also satisfy the inequalities. Hence every threshold function has an integral realization.

Second, the choice of the numerical values 0 and 1 for the logical values 0 and 1 of the x_i is arbitrary. However, if one associated any two real values a and b, $a \neq b$, with the logical values 0 and 1, respectively, in the definition of threshold functions, i.e., $v(x_i) \triangleq a$ when $x_i = 0$, $v(x_i) \triangleq b$ otherwise, the new definition with (*) of Definition 15 replaced by

$$f(x_1, \ldots, x_n) = 1 \qquad \text{iff } \sum_{i=1}^{n} v(x_i) w_i \geq T$$

is equivalent to Definition 15 in the sense that f is a threshold function iff it is a threshold function under this new definition.

Finally, if f is a threshold function, then so is its complement and any function that is $\Pi\Gamma$-equivalent to f.

PROBLEMS

1. For the network of Fig. P4-1, find (a) a sum-of-products expression (as simple as possible); (b) the canonical sum; and (c) the canonical product.

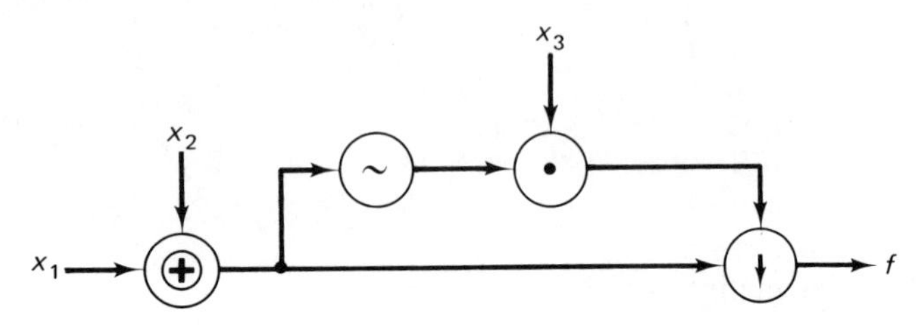

Figure P4-1

2. Find a sum-of-products expression (as simple as possible) for the network of Fig. P4-2.

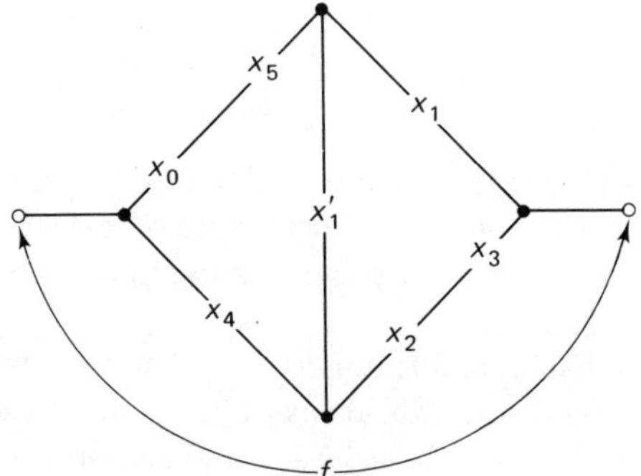

Figure P4-2

3. For each function given below, construct the table of combinations and find the canonical sum and the canonical product.

(a) $f = (y'(x + z))'$ (b) $g = (x + y) | (x + z)$

(c) $wx' + xyz$ (d) $[(x | y) | (y' | z)] | [(x | y) | (y' | z)]$

4. For $f = wx' + xy'z + xyz'$, find the canonical sum by algebraic manipulation of expressions (see Section 4.3).

5. For the expressions found in Problems 1(a) and 2:

(a) Find a series–parallel contact network.

(b) Find a two-level network (AND gates feeding an OR gate). Assume that complemented inputs are available.

(c) Find a two-level network using only NAND gates (complemented inputs are available).

(d) For the function of Problem 1, find a two-level network using only NOR gates (complemented inputs are available).

6. Each Boolean expression below denotes a Boolean function in $B_3 = \{f \mid f: \{0, 1\}^3 \longrightarrow \{0, 1\}\}$:

$$E_1 = x, \quad E_2 = x + x'y' + xz, \quad E_3 = y'z, \quad E_4 = x + y',$$
$$E_5 = (x' + y + z')', \quad E_6 = xy', \quad E_7 = xy'z.$$

(a) How many distinct functions are denoted by the seven expressions? List this set S of functions.

*(b) Draw the poset diagram for $\langle S, \leq \rangle$. Note that $\langle S, \leq \rangle$ is a lattice and you need not prove this.

*(c) Is S a complemented lattice? Justify your answer.

*(d) Is S a distributive lattice? If so, give a proof; if not, give a counter-example.

***7.** Test whether the following functions f and g are (1) Π-equivalent; (2) Γ-equivalent; (3) $\Pi\Gamma$-equivalent.

(a) $f = x_1x_3 + x_1'x_2'x_3'$
$g = x_1'x_2x_3' + x_2'x_3$

(b) $f = x_1'x_2'x_3' + x_1x_2'x_3 + x_1x_2x_3$
$g = x_1'x_2'x_3 + x_1'x_2x_3' + x_1x_2'x_3$

(c) $f = x_1x_2' + x_1'x_2 + x_1'x_3' + x_1'x_4' + x_3'x_4'$
$g = x_2x_3 + x_2'x_3' + x_1'x_4 + x_1'x_3' + x_3'x_4'$

***8.** For each function f given below, find a contact network with N contacts and prove that it is minimal. Some of these problems are easy, others quite hard.

(a) $f_1 = x_1x_3' + x_1x_4 + x_1x_2 + x_2x_3'x_4$; $N = 5$.
(b) $f_2 = x_1x_3 + x_1x_2x_5 + x_2x_3x_4 + x_4x_5$; $N = 5$.
(c) $f_3 = x_1'x_2'x_3' + x_1'x_3x_4' + x_2x_3'x_4$; $N = 7$.
(d) $f_4 = x_1x_2'x_3' + x_1x_3x_4 + x_1x_2'x_3$; $N = 4$.
(e) $f_5 = x_1x_5x_6 + x_2x_3x_4 + x_2x_3x_5x_6$; $N = 7$.

***9.** Which of the following functions are symmetric?

(a) $f_1 = x_1x_2 + x_1x_3 + x_2x_3 + x_1x_2'x_3$.
(b) $f_2 = x'(y \oplus z) + (x \oplus y \oplus 1)z + xy'z'$.
(c) $f_3 = x_1'x_2x_3x_4' + x_1x_2'x_3'x_4 + x_1'x_2x_3'x_4 + x_1x_2'x_3x_4'$.

***10.** Which of the following expressions denote threshold functions? In each case, either find a realization or prove that none exists.

(a) $f_1 = a'bc + bc'd + acd + ab'cd'$.
(b) $f_2 = a(b + c + d) + b(c + de)$.
(c) $f_3 = ab + cd$.
(d) $f_4 = a(b + c + deg) + bc(de + dg + eg) + (b + c)deg$.

***11.** Define *linear expression* over $x_1, \ldots, x_n$ inductively as follows:

Basis: $0, 1, x_1, \ldots, x_n$ are linear expressions.

Induction step: If E and F are linear expressions, then so is $E \oplus F$.

A Boolean function $f \in B_n$ is *linear* iff there exists a linear expression F such that $f = |F|$.

(a) How many linear functions are there in B_n?

(b) Let Λ_n be the set of all linear functions in B_n. Consider the poset $\langle \Lambda_n, \leq \rangle$. Can this poset be considered as a lattice? Can it be considered as a Boolean algebra?

(c) Is there a natural way in which one can associate a Boolean algebra with Λ_n? (*Hint:* Consider $x_1, \ldots, x_n$ and 1 to be atoms.)

12. Prove that every Boolean function $f \in B_n$ can be denoted by an expression of the form

$$F = P_1 \oplus P_2 \oplus \cdots \oplus P_{k_f},$$

where P_i is either a constant or a product of uncomplemented variables.

5 SIMPLIFICATION OF BOOLEAN EXPRESSIONS

ABOUT THIS CHAPTER

The main purpose of this chapter is to explain to the reader methods by which combinational gate networks may be economically designed out of gates. These methods are particularly applicable to the design of combinational networks by means of SSI packages and to the design of various MSI packages, but are of little help in the design of large digital networks by means of available IC packages. Also, they frequently offer only a first step toward the economical design of a network, whereas the remaining steps rely strongly on intuition. It follows that the importance of the formal minimization techniques developed here is not to be overemphasized, as is often done in conventional texts on switching theory.

The design examples of Section 5.8 illustrate how the formal methods developed here are applied to practical combinational problems. In Chapter 8 these methods will be applied to the design of sequential networks.

The simplification algorithms described in this chapter can be easily programmed for a computer. Some of the problems of this chapter are intended to guide the reader toward the design of such computer programs.

The reader of the unstarred material may omit the (iterated) consensus method, pp. 132–37, and the starred proofs.

5.1. INTRODUCTION

In Chapter 2 we considered a Boolean function $f \in B_6$ represented by the Boolean expression

$$F = x_1x_2'x_3 + x_1'x_3'x_4x_5x_6 + x_2x_3'x_4x_5'x_6'$$

and derived an implementation of f by means of 4 SSI packages of Table 2-1, (Chapter 2) as shown in Fig. 2-5. Assume now that f is given either by its table of combinations or by its canonical sum F_{CS}. One easily verifies that F_{CS} is the sum of 12 minterms. A direct implementation of F_{CS} by means of the SSI packages of Table 2-1 would require a 5-level network using 16 SSI packages as compared with the 3-level, 4-package implementation of Fig. 2-5. This example illustrates the importance of simplification techniques by which an expression such as F could be derived from F_{CS}. In this example we have used the number of SSI packages, as well as the number of levels of the gate network (which determines its overall delay time) as generalized cost criteria. In this particular case, F is better than F_{CS} from both points of view. In most real design problems the designer has to meet a number of partially contradicting requirements and must make decisions about various trade-offs.

In this chapter we discuss the simplification of two-level expressions. These techniques are useful in reducing the "cost" of a network for a wide range of cost criteria. In some cases, these methods can be used as the first step in obtaining absolutely minimal realizations.

5.2. SIMPLIFICATION OF TWO-LEVEL EXPRESSIONS

One can use various criteria to define minimal-cost expressions. For example, mathematically, one could minimize the total number of literals or the total number of binary operations or the total number of symbols in an expression. The minimization problem is difficult for all such cost criteria. For contact network applications, more relevant criteria would be the total number of contacts or the total number of relays with restrictions on the number of normally open and normally closed contacts per relay. Since there is no one-to-one correspondence between general (i.e., non series–parallel) contact networks and expressions, such minimal networks cannot be directly obtained from any simplified expression. Similar remarks apply to gate networks. Here one could minimize the total number of gates subject to such restrictions as fan-in, fan-out, number of levels, etc., or the total number of SSI packages, etc. In general, it is very difficult to find such minimal networks or to prove the minimality of a given network.

In spite of this, it is possible to solve a number of minimization problems using systematic techniques, provided that we are satisfied with less general solutions. In this chapter we describe the simplification of *sum-of-products* (SP) expressions and, dually, of *product-of-sums* (PS) expressions. (In view of Section 2.3 these methods are also applicable to two-level NAND and NOR networks.) This is a very severe restriction on the form of the expression since, in general, minimal expressions (according to some cost criterion) need not be of this form. However, the restriction makes the problem manageable, and the networks corresponding to simplified SP or PS expressions are usually far simpler than the canonical ones, according to most relevant cost criteria.

Unless otherwise stated, we now assume that each expression under consideration is of the form $F = P_1 + P_2 + \cdots + P_m$, where each P_i is a product of literals. We begin by making some rather weak assumptions about the concept of simplicity of such expressions:

A1. If $F_1 = P + G$, where P is a product of literals, and $F_2 = G$ are two Boolean expressions such that $|F_1| = |F_2|$, then F_2 is simpler than F_1.
A2. If $F_1 = \tilde{x}P + G$, where $\tilde{x}$ is a literal, P is a product of literals, and $F_2 = P + G$ are such that $|F_1| = |F_2|$, then F_2 is simpler than F_1.
A3. No expression is simpler than the expression 0 or 1.

These assumptions are certainly reasonable and most cost criteria satisfy the following relationship: Let $c(F)$ denote the cost of the expression F. If F_2 is simpler than F_1, then $c(F_2) \leq c(F_1)$, and frequently $c(F_2) < c(F_1)$. The reader can verify this for several criteria mentioned above.

Recall that by a *product* we mean an expression of the form $P = \tilde{x}_{i_1} \cdots \tilde{x}_{i_k}$, where the $\tilde{x}_{i_j}$ are literals and $k \geq 1$. By a *sum of products* we mean an expression of the form $E = P_1 + \cdots + P_m$, where the P_i are products and $m \geq 1$. Thus the expressions 0 and 1 are not sums of products but the expressions x_1, x_1x_1, x_1x_1', and $x_1 + x_2'$ are.

If $E = P_1 + \cdots + P_m$, $m \geq 2$, is a sum of products, then "removing a product P_i from E" means forming the sum $E_1 = P_1 + \cdots + P_{i-1} + P_{i+1} + \cdots + P_m$. If $E = P_1$, removing P_1 from E yields $E_1 = 0$. If $P = \tilde{x}_{i_1} \cdots \tilde{x}_{i_k}$ is a product and $k \geq 2$, then "removing a literal $\tilde{x}_{i_j}$ from P" means forming the product $\tilde{x}_{i_1} \cdots \tilde{x}_{i_{j-1}}\tilde{x}_{i_{j+1}} \cdots \tilde{x}_{i_k}$. If $P = \tilde{x}_{i_1}$, then removing $\tilde{x}_{i_1}$ from P yields $P_1 = 1$.

In view of assumption A1, if a product P appears in a sum of products expression E, and $|P| = 0$, then P can be removed, yielding a simpler expression. Therefore, *we assume in this chapter that products do not contain complementary literals*, i.e., that the corresponding functions are nonzero. In view of A2, we assume that *products do not contain repeated literals*. For example, at this point we will not have to consider such products as $x_1x_1'x_2$ or $x_1x_2x_2$.

If E is an expression, then $L(E)$ [respectively, $V(E)$] denotes the set of literals (respectively, variables) of E. For example, if $P = x_1'x_2x_3'x_5$, then $L(P) = \{x_1', x_2, x_3', x_5\}$ and $V(P) = \{x_1, x_2, x_3, x_5\}$. In view of our assumptions about products, each x_i in $V(P)$ appears exactly once in P as either x_i or x_i', and each $\tilde{x}_i$ in $L(P)$ appears exactly once in P.

We illustrate the concepts presented above by the following example. Let

$$E_1 = w'x'w' + wx + w'yz + xyz + x'w'zx.$$

Noticing that the product $x'w'zx$ contains both x and x', we consider E_2, where

$$E_2 = w'x'w' + wx + w'yz + xyz.$$

For E_2 we need 4 AND gates and one 4-input OR gate. Note also that E_1 would require a 4-input AND gate, but in E_2 the fan-in of the AND gates is at most 3. E_2 could be realized, for example, as

$$\hat{E}_2 = |(|(w', x', w'), |(w, x, 1, 1), |(w', y, z), |(x, y, z))$$

using one 7410 package with three 3-input NAND gates and one 7420 with two 4-input NAND gates.

Returning to E_2, we see that the first product contains a repeated literal (w'). Thus E_3 below is simpler, according to A2:

$$E_3 = w'x' + wx + w'yz + xyz.$$

If we are constrained to use only the packages of Table 2-1 (p. 31), E_3 is no better than E_2 in the sense that we still require one 7410 and one 7420. If the package count represents the cost c, then $c(E_2) = c(E_3)$, even though E_3 is simpler.

One can also verify algebraically or by making a table of combinations that E_3 is equivalent to

$$E_4 = w'x' + wx + w'yz.$$

This expression will also require at least two packages from Table 2-1. However, if we use two 7410 packages, we have two 3-input NAND gates left over. In this sense E_4 is preferable.

In the following sections we describe some algorithms related to the simplification of SP expressions. We assume that complemented inputs are available; hence each SP expression corresponds to a network of two levels—AND gates "feeding" an OR gate, or, in view of Section 2.3, two levels of NAND gates. Similarly, PS expressions correspond to OR gates "feeding" an AND gate or two levels of NOR gates. Our treatment deals mainly with SP forms, and we point out later how analogous techniques handle the PS problem.

5.3. PRIME IMPLICANTS [MCC, QUI2]

Every Boolean function other than $|0|$ can be represented by an SP expression, namely by the canonical sum. For example, for the OR function f we have $F_1 = xy' + x'y + xy$. However, $F_2 = x + y$ is another, simpler SP expression for f. In this section we examine "candidates" for products that may appear in simple SP expressions. Clearly, for the OR function f, the product $x'y'$ can never appear in any SP expression for f since when $x = y = 0$, we have $|x'y'| = 1$, but $f = 0$. Thus, if this product is included in an expression F, that expression cannot denote the OR function. In general, a product P can appear in an SP expression E for a function f only if $|P| \leq f$. This leads to the following definitions.

Definition 1

A Boolean function $p \neq |0|$ is a *product function* iff there exists a product P such that $|P| = p$. Note that p uniquely determines $L(P)$. Let f be a Boolean function. We say that a product function p is an *implicant* of f iff $p \leq f$. An implicant p is *prime* iff there does not exist an implicant q of f such that $p < q \leq f$.

(In terms of the poset of the algebra of Boolean functions, a prime implicant of f is a maximal product function $\leq f$.)

In the following we use the convention that if E, F, P, Q, etc., are Boolean expressions, then e, f, p, q, etc., are the corresponding Boolean functions.

Proposition 1

Let p and q be implicants of f. Then (a) $q < p$ iff $L(Q) \supset L(P)$ and (b) $q \leq p$ iff $L(Q) \supseteq L(P)$.

Proof

(a) If $L(Q) \supset \mathrm{L}(P)$ then $Q = PT$, where T is the product of all the literals that are in Q but not in P. Hence $|Q| = |P| \cdot |T|$ or (using our convention), $q = pt$. Clearly, $q = 1$ implies $p = 1$; hence $q \leq p$. Further, make all the literals of P equal to 1, and make one of the literals of T equal to 0. Then $p = 1$ and $t = 0$, implying $q = 0$. Hence $p \neq q$ and $q < p$ follows. Conversely, suppose that $q < p$ but $L(Q) \not\supset L(P)$. Then there must be at least one literal $\tilde{x}$ that is in P but not in Q. Assign the value 1 to all the literals of Q and let $\tilde{x} = 0$. Then $q = 1$, $p = 0$, contradicting $q < p$.

(b) One verifies that $q = p$ iff $L(Q) = L(P)$. This and (a) prove (b). ☐

Definition 2

An SP expression F is *literal-reduced* (*L-reduced*) iff whenever a literal is removed from a product of F yielding G, then $|F| \neq |G|$.

Theorem 1

An SP expression $F = P_1 + \cdots + P_m$ is L-reduced iff $p_i = |P_i|$ is a prime implicant of $f = |F|$, for $i = 1, \ldots, m$.

Proof

Suppose that each p_i is prime. If F is not L-reduced, there exists some product $P_j = \tilde{x}P$ such that $F_1 = P_1 + \cdots + P_{j-1} + P + P_{j+1} + \cdots + P_m$ also denotes f. By Proposition 1(a), $L(P_j) \supset L(P)$ implies $p_j < p$. But now $p_j < p \leq f$, contradicting that p_j is prime.

Conversely, suppose that F is L-reduced but p_j is not prime. Then there exists an implicant q of f such that $p_j < q \leq f$, and $L(P_j) \supset L(Q)$. Now let $F_1 = F + Q$, $F_2 = P_1 + \cdots + P_{j-1} + Q + P_{j+1} + \cdots + P_m$. Clearly, $f \geq q$ yields $f = f + q = f_1$. Also, $f_1 = f_2$ since $p_j + q = q$. However, F_2 is obtained from F by removing some literals from P_j. This contradicts the fact that F is L-reduced. ▯

We now illustrate the concepts introduced so far. Let $F = xy + z$ and $f = |F|$. Consider the product expression $Q = xyz'$. It defines a function q which is 1 iff $x = 1$, $y = 1$, $z = 0$. Since, for any input condition with $x = 1$ and $y = 1$ we have $f = 1$, it follows that $q \leq f$. Also $q \neq f$, because $f = 1$ and $q = 0$ when $z = 1$. Therefore, $q < f$. Because $q \leq f$, q is an implicant of f. It is not prime because $p = |P| = |xy|$ is also an implicant of f and $q < p$. The last statement follows from Proposition 1, since $L(Q) = \{x, y, z'\}$, $L(P) = \{x, y\}$, and clearly $L(Q) \supset L(P)$. Of course, the fact that $q < p$ can also be obtained from the tables of combinations for p and q.

Consider another expression for f, derived as follows:

$$f = |xy + z| = |xy(z + z') + z| = |xyz + xyz' + z|$$
$$= |xyz' + z|.$$

Let $\hat{F} = xyz' + z$. Now $\hat{F}$ is not L-reduced because it is possible to remove the literal z' without changing the function. However, the original expression F is L-reduced. This can be verified by removing each literal in turn and showing that the resulting function differs from f.

Given an SP expression $F = P_1 + \cdots + P_m$ that we wish to simplify, our first step is to remove unnecessary literals in the products P_i of F. It follows from Theorem 1 that the resulting expression is a sum of prime implicants. However, it may also happen that a prime implicant can be removed from an expression without changing the function. For example, one verifies that $xy + x'z + yz = xy + x'z$.

Definition 3

An SP expression F is *irredundant* iff F is a sum of prime implicants and whenever a product P is removed from F, the resulting expression E no longer denotes f.

Our next objective will be to find irredundant sums for f. In view of our assumptions A1 and A2, a simplified SP expression will always be an irredundant sum. In general, there may be more than one irredundant sum for a given function. For example, one can verify that the expressions below are both irredundant sums for the same function:

$$F_1 = w'x' + wx + wyz$$
$$F_2 = w'x' + wx + x'yz.$$

The simplification process is now divided into two steps: (1) generation of all the prime implicants and (2) selection of suitable prime implicants to find a minimal sum according to a specific cost criterion. Such a sum will be necessarily irredundant in view of our assumptions.

We now consider several techniques for finding the prime implicants of a Boolean function.

Tabular Method

The most direct method for finding the prime implicants of a Boolean function $f \in B_n$ is to generate all the product functions $\leq f$ and choose the maximal ones.

In what follows it will be convenient to represent a product function $p \in B_n$ by a ternary (i.e., three-valued) n-tuple $\mathbf{P} = (a_1, \ldots, a_n)$, where each $a_i \in \{0, 1, 2\}$. Namely, let $P = \tilde{x}_{i_1} \cdots \tilde{x}_{i_m}$ be a product expression such that $|P| = p$. We now set

$$a_j = 2 \quad \text{iff } x_j \notin V(P), \text{ i.e., } j \notin \{i_1, \ldots, i_m\}$$
$$a_j = 1 \quad \text{iff } j \in \{i_1, \ldots, i_m\} \text{ and } \tilde{x}_j = x_j$$
$$a_j = 0 \quad \text{iff } j \in \{i_1, \ldots, i_m\} \text{ and } x_j = x'_j.$$

For example, let $n = 5$ and $P = x'_1x_3x'_4$. Then $p = |P|$ is represented by $\mathbf{P} = (0, 2, 1, 0, 2)$. Usually, we omit the parentheses and commas and write

$$\mathbf{P} = 0 \ \ 2 \ \ 1 \ \ 0 \ \ 2.$$

For any ternary n-tuple $\mathbf{P}$, we denote by $|\mathbf{P}|$ the product function p represented by $\mathbf{P}$. We also set $|\mathbf{P}| = 1$ for $\mathbf{P} = (2, \ldots, 2)$.

Frequently, the symbols "—" or "×" are used instead of "2." However, in view of programming aspects (see Problem 22) we prefer the symbol set $T \triangleq \{0, 1, 2\}$ here. Later we also use $\{0, 1, -\}$.

Definition 4

Two product functions p and q are *adjacent* iff for some literal $\tilde{x}$ and some product function r or for $r = 1$, $p = |\tilde{x}| \cdot r$ and $q = |(\tilde{x})'| \cdot r$. Let $\mathbf{P}, \mathbf{Q} \in T^n$. Then $\mathbf{P}$ and $\mathbf{Q}$ are *adjacent*, iff $|\mathbf{P}|$ and $|\mathbf{Q}|$ are.

For example, $p = |x'_2x'_3x_5|$ and $q = |x'_2x_3x_5|$ are adjacent, by Definition 4. Consequently, the following ternary 5-tuples $\mathbf{P}$ and $\mathbf{Q}$ are adjacent, since

$p = |\mathbf{P}|$ and $q = |\mathbf{Q}|$:

$$\mathbf{P} = 2\ \ 0\ \ 0\ \ 2\ \ 1$$

$$\mathbf{Q} = 2\ \ 0\ \ 1\ \ 2\ \ 1.$$

Clearly, **P** and **Q** are adjacent iff they differ in exactly one coordinate and this discrepancy does not involve a 2-entry.

Definition 5

The *consensus* of two adjacent product functions $p = |\tilde{x}|\cdot r$ and $q = |(\tilde{x})'|\cdot r$ is the function r. The consensus of p and q will be denoted by $p \not\subset q$.

We also extend this notation to the corresponding ternary n-tuples. Thus $\mathbf{R} = \mathbf{P} \not\subset \mathbf{Q}$ iff $|\mathbf{R}| = |\mathbf{P}| \not\subset |\mathbf{Q}|$. **R** is obtained from **P** or **Q** by replacing by "2" the entry in the coordinate in which P and Q differ. For example,

$$\mathbf{P} = 1\ \ 2\ \ 0\ \ 0\ \ 2$$

$$\mathbf{Q} = 1\ \ 2\ \ 0\ \ 1\ \ 2$$

$$\mathbf{P} \not\subset \mathbf{Q} = 1\ \ 2\ \ 0\ \ 2\ \ 2.$$

Proposition 2

(a) If p and q are two adjacent product functions and f is a Boolean function such that $p \leq f$ and $q \leq f$, then $p \not\subset q \leq f$. (b) If $f \in B_n$ and R is a product expression of $m < n$ literals such that $|R| \leq f$, then there exist product expressions P and Q of $m + 1$ literals each, such that $|P| \leq f$, $|Q| \leq f$, and $|R| = |P| \not\subset |Q|$.

Proof

(a) Say $p = |x|\cdot r$ and $q = |x'|\cdot r$. Then

$$f \geq p + q = |x|\cdot r + |x'|\cdot r = r = p \not\subset q.$$

(b) Let x be a variable not in $V(R)$. Then $P = xR$ and $Q = x'R$ satisfy the conditions required. □

ALGORITHM TAB (*Tabular Method*)

Let f be a Boolean function in B_n, and let C_n be the set of all binary n-tuples $x \in \{0, 1\}^n$ such that $f(x) = 1$. Given C_n, the following algorithm yields the set C of all ternary n-tuples **P** such that $|\mathbf{P}| \leq f$.

Step 1. Construct $C_{n-r-1} = \{\mathbf{P} \not\subset \mathbf{Q} \mid (\mathbf{P} \in C_{n-r}) \wedge (\mathbf{Q} \in C_{n-r}) \wedge (\mathbf{P} \text{ and } \mathbf{Q} \text{ are adjacent})\}$ inductively, for $r = 0, 1, \ldots, n - 1$.

Step 2. Form $C = \bigcup_{r=0}^{n} C_{n-r}$.

Proposition 3

Let C be the set obtained by algorithm TAB. Then $\mathbf{P} \in C$ iff $\mathbf{P} \in T^n$ and $|\mathbf{P}| \leq f$.

Proof

By Proposition 2(a), if $|\mathbf{P}| \leq f$, $|\mathbf{Q}| \leq f$, then $|\mathbf{P} \not\subset \mathbf{Q}| \leq f$. Hence $\mathbf{P} \in C$ implies $|\mathbf{P}| \leq f$. Conversely, we use induction on the number r of 2-entries in $\mathbf{P}$. For $r = 0$ we have $\mathbf{P} \in \{0, 1\}^n$ and $|\mathbf{P}| \leq f$ implies that $\mathbf{P} \in C_n$ by construction of C_n. Assume now that for all n-tuples $\mathbf{P}$ with r 2-entries $|\mathbf{P}| \leq f$ implies $\mathbf{P} \in C_{n-r}$. By Proposition 2(b) and step 1 of the algorithm, all the n-tuples $\mathbf{P}$ with $r + 1$ 2-entries such that $|\mathbf{P}| \leq f$ will be in C_{n-r-1}. ☐

We now illustrate the algorithm by an example. It is convenient to rearrange the n-tuples in C_{n-r} into groups, where group G_i consists of all those n-tuples in C_{n-r} that have i 1's. One verifies that two n-tuples $\mathbf{P}$ and $\mathbf{Q}$ in C_{n-r} are adjacent only if one is in G_i and the other in G_{i+1} for some i. This means that a group G_i of n-tuples need be compared only with the next group G_{i+1} in order to find all adjacent n-tuples and thus generate all the consensus terms.

Table 5-1 shows the initial steps for constructing C_n grouped as above. For convenience we label each minterm by the corresponding decimal number and then each product by the set of minterms (decimal numbers) corresponding to it. Table 5-2 completes the algorithm.

Table 5-1 Construction of C_3

x	y	z	f
0	0	0	1
0	0	1	1
0	1	0	1
0	1	1	1
1	0	0	1
1	0	1	0
1	1	0	0
1	1	1	1

(a)

x	y	z
0	0	0
0	0	1
0	1	0
0	1	1
1	0	0
1	1	1

(b)

	x	y	z	
(0)	0	0	0	G_0
(1)	0	0	1	G_1
(2)	0	1	0	
(4)	1	0	0	
(3)	0	1	1	G_2
(7)	1	1	1	G_3

(c)

Let $\hat{C}_i$ denote the set of product expressions corresponding to the ternary n-tuples in C_i; i.e.,

$$\hat{C}_i = \{P \mid \mathbf{P} \in C_i\}.$$

From Tables 5-1 and 5-2 the set of all products P such that $|P| \leq f$ is $\hat{C}_3 \cup \hat{C}_2 \cup \hat{C}_1$, where

$$\hat{C}_3 = \{x'y'z', x'y'z, x'yz', xy'z', x'yz, xyz\}$$

$$\hat{C}_2 = \{x'y', x'z', y'z', x'z, x'y, yz\}$$

$$\hat{C}_1 = \{x'\}.$$

If one performs the algorithm TAB, modified as indicated in Tables 5-1 and 5-2, it is easy to find all the prime implicants. An n-tuple in C rep-

Table 5-2 Finding C_2 and C_1

(a)

	x	y	z	
(0, 1)	0	0	2	
(0, 2)	0	2	0	G_0
(0, 4)	2	0	0	
(1, 3)	0	2	1	
(2, 3)	0	1	2	G_1
(3, 7)	2	1	1	G_2

(b)

	x	y	z	
(0, 1, 2, 3)	0	2	2	G_0

resents a prime implicant of f iff the corresponding set of decimal labels is maximal, when these sets are partially ordered under inclusion.

In the example of Tables 5-1 and 5-2, the prime implicants of f are $|x'|$, $|yz|$, and $|y'z'|$, corresponding to the maximal sets $\{0, 1, 2, 3\}$, $\{3, 7\}$, and $\{0, 4\}$, respectively.

Every Boolean function f has a unique representation as the sum of all its prime implicants. This representation is called the *complete sum* of f and is denoted by F_C. In the example,

$$F_C = x' + yz + y'z'.$$

Alternatively, the prime implicants of $f \in B_n$ can be obtained by means of the following modification of the TAB algorithm.

ALGORITHM TABPI

Let f and C_n be as in algorithm TAB. This algorithm yields the set D of all ternary n-tuples $\mathbf{P}$ such that $|\mathbf{P}|$ is a prime implicant of f.

Step 1. Same as step 1 of algorithm TAB.
Step 2. For $r = 0, 1, \ldots, n$, form

$$D_{n-r} = \{\mathbf{P} \mid (\mathbf{P} \in C_{n-r}) \wedge (\mathbf{P} \text{ is not adjacent to any } \mathbf{Q} \in C_{n-r})\}.$$

Step 3. Form $D = \bigcup_{r=0}^{n} D_{n-r}$.

Actually, it is convenient to perform steps 1 and 2 of the TABPI algorithm concurrently.

Map Method

A convenient way of representing Boolean functions, their prime implicants, and other properties is obtained by using a geometrical interpretation. It is an intuitive visual representation in which one can recognize maximal products by inspection.

Consider, for example, the function f of Table 5-1. There are 8 possible input n-tuples. These n-tuples can be represented by the vertices of a unit cube in three-dimensional space, as shown in Fig. 5-1(a). A function f can

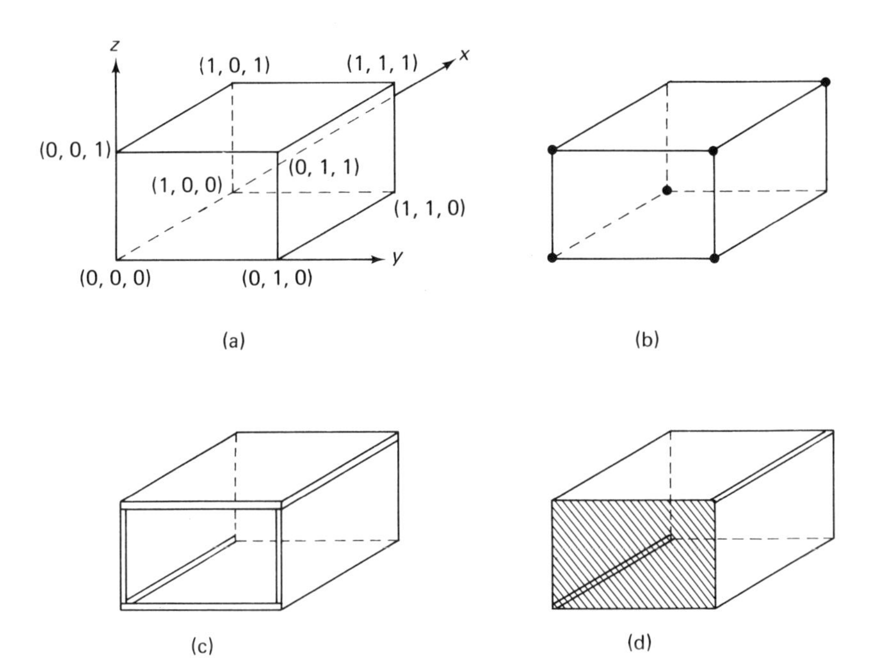

Fig. 5-1 (a) A cube; (b) f represented by points; (c) combining points into edges; (d) combining edges into faces.

then be represented by a subset of the 2^n points of the n-cube (the minterms), as shown in Fig. 5-1(b). The process of combining two points differing in a single coordinate corresponds to the formation of an *edge* to represent two points. All the edges of f are shown in Fig. 5-1(c). In the example, each product with 3 literals can be combined with at least one other product of 3 literals. Hence no product of 3 literals is a prime implicant, and Table 5-1(c) can be discarded. If there does exist a product p of n literals that is not adjacent to any other product of n literals, then p is a prime implicant of f. Now consider Table 5-2. The edge (0, 1) is represented by the product $x'y'$ and the edge (2, 3) by $x'y$. These two products can again be combined to yield x'. Thus the application of the consensus operation to two adjacent products represented by edges in the cube results in a *face*. It can be seen that finding prime implicants corresponds to finding maximal subcubes (i.e., points, edges, faces) in the cube.

Unfortunately, this model becomes awkward when we attempt to draw four-dimensional "cubes" in two dimensions. However, it can be successfully modified to extend the idea to 4 variables (or more, depending on the individual's visualizing ability) by using the map concept.

A map of 3 variables is shown in Fig. 5-2. Each square represents a minterm and the coordinates of the map are arranged in such a way as to

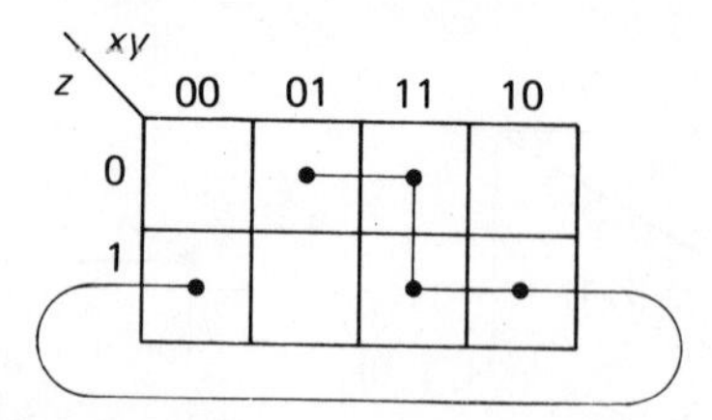

Fig. 5-2 A 3-variable map.

make as many pairs of adjacent minterms as possible adjacent as squares of the map. One verifies that all pairs of minterms that are adjacent are represented by adjacent squares except for the pairs in the first and last column. If we agree to identify the left and right borders of the map, this difficulty can be overcome.

Three-variable problems are usually too simple, but many useful concepts can be illustrated on a 4-variable map, as shown in Fig. 5-3. In order to

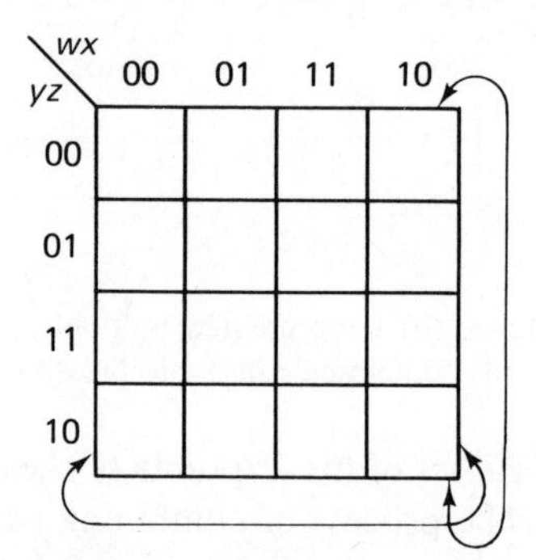

Fig. 5-3 A 4-variable map.

preserve minterm adjacency it is necessary to identify the left border with the right one and the top border with the bottom one. The problem now is to recognize subcubes, i.e., product functions, on the map. This comes easily with some practice for the case of 4 variables. In Fig. 5-4 we illustrate typical appearances of product functions, which are denoted by the corresponding expressions. The function of Table 5-1 and its prime implicants are shown in Fig. 5-5.

We shall use the map method later as a convenient means of presenting simple examples. For large problems this method becomes impractical.

*Consensus Method [QUI1]

Recall that to generate all the prime implicants using the tabular method we must start with the minterms of the function. It is possible to obtain the prime implicants from *any* sum of products expression by the method described below.

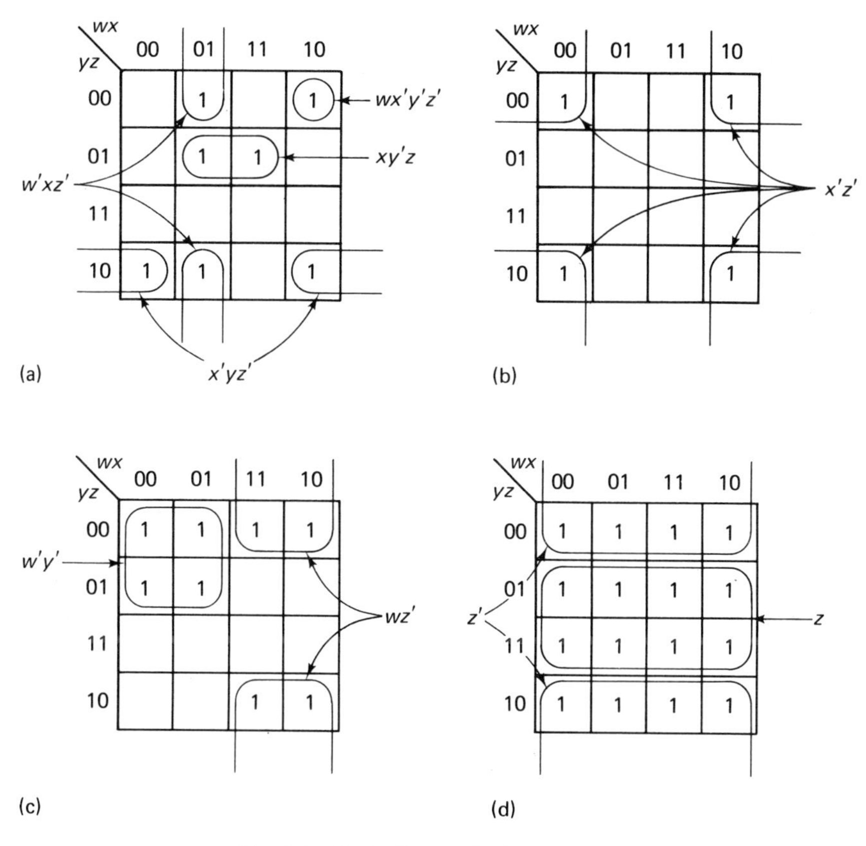

Fig. 5-4 Illustrating products on maps.

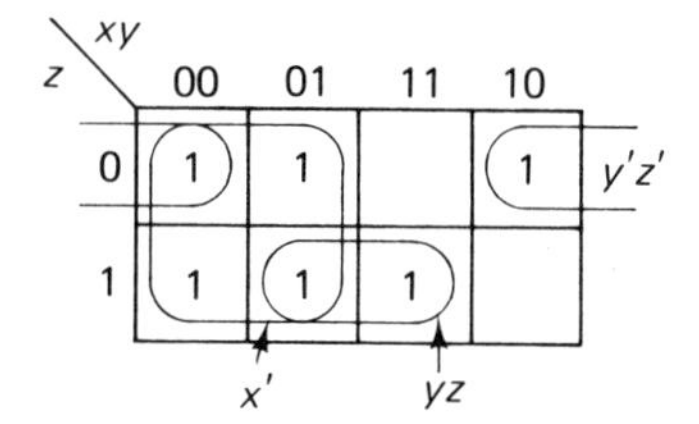

Figure 5-5

Definition 6

Two product functions p and q are *near* iff there exist a literal $\tilde{x}$ and product functions r and s such that $p = |\tilde{x}|\cdot r$, $q = |(\tilde{x})'|\cdot s$, and $r \cdot s \neq |0|$. The *consensus* of two near product functions $p = |\tilde{x}|\cdot r$ and $q = |(\tilde{x})'|\cdot s$ is defined to be $p \not{c}\, q = r \cdot s$.

We extend these definitions to the corresponding product expressions P and Q as well as to the ternary n-tuples $\mathbf{P}$ and $\mathbf{Q}$, representing p and q, in the evident way. Thus P and Q are *near* iff $|P|$ and $|Q|$ are, and similarly for **P** and **Q**. Also, $|P \not{c}\, Q| = |P| \not{c}\, |Q|$ and $|\mathbf{P} \not{c}\, \mathbf{Q}| = |\mathbf{P}| \not{c}\, |\mathbf{Q}|$. (Note, however, that $P \not{c}\, Q$ is defined only up to the order of its literals.)

The following are examples of this (generalized) consensus operation, on the variables v, w, x, y, and z.

			$v\ w\ x\ y\ z$
1.	$P = wx'yz'$	$\mathbf{P} =$	2 1 0 1 0
	$Q = vxy$	$\mathbf{Q} =$	1 2 1 1 2
	$P \not{c}\, Q = vwyz'$	$\mathbf{P} \not{c}\, \mathbf{Q} =$	1 1 2 1 0
2.	$P = x'$	$\mathbf{P} =$	2 2 0 2 2
	$Q = xyz$	$\mathbf{Q} =$	2 2 1 1 1
	$P \not{c}\, Q = yz$	$\mathbf{P} \not{c}\, \mathbf{Q} =$	2 2 2 1 1
3.	$P = x'$	$\mathbf{P} =$	2 2 0 2 2
	$Q = x$	$\mathbf{Q} =$	2 2 1 2 2
	$P \not{c}\, Q = 1$	$\mathbf{P} \not{c}\, \mathbf{Q} =$	2 2 2 2 2

Two ternary n-tuples **P** and **Q** are near iff there is exactly one coordinate i in which they are *opposed* (i.e., one coordinate is 0 and the other is 1). The consensus is then obtained by using the following coordinate operation:

¢	**0**	**1**	**2**
0	0	2	0
1	2	1	1
2	0	1	2

Proposition 4

If p and q are two near product functions such that $p \leq f$ and $q \leq f$, then $p \not{c}\, q \leq f$.

Proof

Say that $p = |xR|$ and $q = |x'S|$. Then

$$f \geq p + q = |xR + x'S| = |xR + xRS + x'S + x'RS|$$
$$= |xR + x'S + (x + x')RS| = |xR + x'S + RS|.$$

Thus $f \geq |RS| = p \not{c}\, q$. □

Definition 7

Let $F = P_1 + \cdots + P_k$ be an SP expression. If P_i and P_j are near, their consensus $P_i \not{c}\, P_j$ is *useless* for F if $|P_i \not{c}\, P_j| \leq |P_r|$ for some $r \in \{1, 2, \ldots, k\}$;

otherwise, $P_i \not{c} P_j$ is *useful* for F. F is *consensus-reduced* (*C-reduced*) iff for all i, j: (a) $|P_i| \leq |P_j|$ implies $i = j$, and (b) if P_i and P_j are near, then $P_i \not{c} P_j$ is useless for F.

We also introduce the notation $\mathbf{F} \triangleq \{\mathbf{P}_1, \ldots, \mathbf{P}_k\}$ and say that $\mathbf{P}_i \not{c} \mathbf{P}_j$ is useless (useful) for $\mathbf{F}$ iff $P_i \not{c} P_j$ is useless (useful) for F.

Theorem 2 (Quine)

An SP-expression $F = P_1 + \cdots + P_k$ is C-reduced iff it is the complete sum of $f = |F|$.

Proof

Suppose F is complete. Clearly condition (a) of Definition 7 holds, since no two products are comparable, both being maximal. Suppose now that P_i and P_j are near, and let $P = P_i \not{c} P_j$. By Proposition 4, $|P| \leq f$. Now there exists a prime implicant q such that $|P| \leq q \leq f$, and since F is complete there exists a product P_r in F such that $1 \leq r \leq k$ and $|P_r| = q$. It follows that $|P_i \not{c} P_j| \leq |P_r|$, i.e., $P_i \not{c} P_j$ is useless for F. Thus condition (b) of Definition 7 also holds.

Conversely, suppose that F is C-reduced and there exists a prime implicant p of f whose product P does not appear in F. Then P satisfies the following conditions: (a) $|P| \leq p$; (b) $|P| \leq |P_i|$ is false for all i; (c) $V(P) \subseteq V(F)$. The last condition follows because if $P = \tilde{x}Q$ and $x \notin V(F)$, then setting all the literals of Q alone to 1 will have the same effect on F as setting all the literals of P to 1; namely F will obtain the value 1. Hence $p < |Q| \leq f$, contradicting that p is prime. Now let R be a longest product (i.e., having the largest number of literals) that satisfies: (a) $r = |R| \leq p$, (b) $r \leq |P_i|$ is false for all i, and (c) $V(R) \subseteq V(F)$. If $V(F) - V(R) = \varnothing$, then by setting all the literals of R to 1 we assign a value to each variable of F. Hence each product of F becomes either 0 or 1. At least one product P_i must become 1 since $r \leq f$. This contradicts (b). Therefore, let $x \in V(F) - V(R)$. Now consider xR. Clearly, xR satisfies conditions (a) and (c) and so must violate (b) because R is the longest product satisfying (a), (b), and (c). Thus $|xR| \leq |P_i|$ for some i. If $x \notin L(P_i)$, then $r \leq |P_i|$. This is false and we must have $P_i = xS_i$, where $L(S_i) \subseteq L(R)$. By a similar argument, $|x'R| \leq |P_j|$ for some j, where $P_j = x'S_j$ and $L(S_j) \subseteq L(R)$. Now the products $P_i = xS_i$ and $P_j = x'S_j$ are near and have consensus $C = S_iS_j$. If $|C| \leq |P_t|$ for some P_t appearing in F, then $L(S_iS_j) \supseteq L(P_t)$. But now $L(R) \supseteq L(S_i) \cup L(S_j) = L(S_iS_j) \supseteq L(P_t)$. Hence $|R| \leq |P_t|$, contradicting condition (b). Therefore, if F is C-reduced, every prime implicant of f must appear in F. Finally, if F contains a product P_i that is not prime, then $|P_i| \leq |P_j|$ for some product P_j appearing in F, contradicting the fact that F is C-reduced. This concludes the proof that if F is C-reduced, then it is the complete sum. □

Theorem 2 yields the following algorithm.

ALGORITHM ITCO (*ITerated COnsensus*)

Let $F = P_1 + \cdots + P_k$ be an SP expression for $f \in B_n$. Given $\mathbf{F}$ (see Definition 7), the following algorithm yields $\mathbf{F}_C$, where F_C is the complete sum of f.

Step 1. Set $A = \mathbf{F}$.
Step 2. Form $B = A - \{\mathbf{P} \mid |\mathbf{P}| < |\mathbf{Q}| \text{ for some } \mathbf{Q} \text{ in } A\}$.
Step 3. If B contains two n-tuples $\mathbf{P}$ and $\mathbf{Q}$ such that $\mathbf{P} \not\subset \mathbf{Q}$ is useful for B (i.e., $|\mathbf{P} \not\subset \mathbf{Q}| \leq |\mathbf{R}|$ is false for all $\mathbf{R}$ in B), then set $A = \{\mathbf{P} \not\subset \mathbf{Q}\} \cup B$ and return to step 2. Otherwise, the algorithm terminates and $B = \mathbf{F}_C$.

It can be shown that the algorithm ITCO above always terminates [QUI1]. One also easily verifies that the SP expression F_C corresponding to the outcome $\mathbf{F}_C$ is C-reduced and that $|F_C| = f$. It follows, by Theorem 2, that F_C is indeed the complete sum of f.

The following example illustrates the ITCO algorithm. Let

$$F = w'xz + wy'z + x'yz + w'yz'.$$

Then

$$\mathbf{F} = \{0121, 1201, 2011, 0210\}.$$

Step 1. $A \leftarrow \mathbf{F}$. ($A \leftarrow B$ stands for "set $A = B$".)
Step 2. $B \leftarrow A$.
Step 3. $0121 \not\subset 1201 = 2101$ is useful for B.
$A \leftarrow \{2101, 0121, 1201, 2011, 0210\}$.
Step 2. $B \leftarrow A$.
Step 3. $0121 \not\subset 2011 = 0211$ is useful for B.
$A \leftarrow \{0211, 2101, 0121, 1201, 2011, 0210\}$.
Step 2. $B \leftarrow A$.
Step 3. $0211 \not\subset 0210 = 0212$ is useful for B.
$A \leftarrow \{0212, 0211, 2101, 0121, 1201, 2011, 0210\}$.
Step 2. $B \leftarrow \{0212, 2101, 0121, 1201, 2011\}$.
Step 3. $1201 \not\subset 2011 = 1021$ is useful for B.
$A \leftarrow \{1021, 0212, 2101, 0121, 1201, 2011\}$.
Step 2. $B \leftarrow A$.
Step 3. Since B contains no useful consensus, the algorithm terminates and $\mathbf{F}_C = B$.

The outcome is $\mathbf{F}_C = \{1021, 0212, 2101, 0121, 1201, 2011\}$. The complete sum of f is therefore

$$F_C = wx'z + w'y + xy'z + w'xz + wy'z + x'yz.$$

The maps of Fig. 5-6 show how the expressions F and F_C represent f.

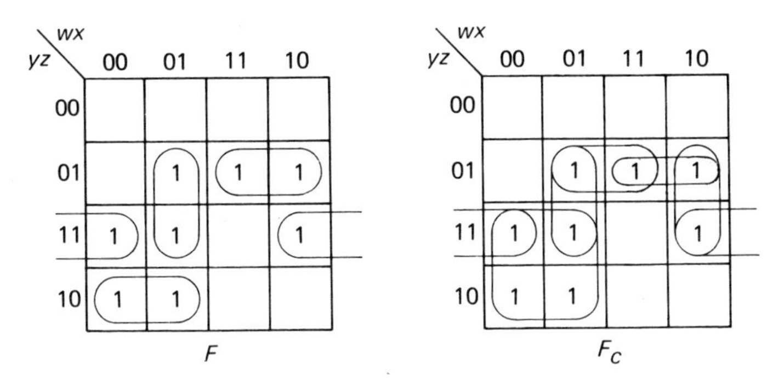

Fig. 5-6 Illustrating F and F_C.

5.4. IRREDUNDANT SUMS

If a is an atom function such that $a \leq f$, we call a an *atom of f*. If $f = p_1 + \cdots + p_k$ is any sum of prime implicants and a is an atom of f, then there exists at least one p_i such that $a \leq p_i$; otherwise, f will be 0 when $a = 1$. This means that if any subset of prime implicants is redundant, the remaining prime implicants must "cover" all the atoms of f, as above. This leads to the following approach for finding irredundant sums. Construct a table with prime implicants p_i corresponding to rows and atoms a_j to columns. Enter $\times$ in position i, j iff $a_j \leq p_i$. This table will be called the *PI table*. For brevity of notation we will use decimal numbers to represent atoms; the correspondence on a 4-variable map is shown in Fig. 5-7 for easy reference. The notation $f = \sum(2, 3, 5, 6, 7, 9, 11, 13)$ means that f consists of the atoms whose decimal equivalents are listed.

yz \ wx	00	01	11	10
00	0	4	12	8
01	1	5	13	9
11	3	7	15	11
10	2	6	14	10

Fig. 5-7 Numbering convention for map.

For the function of Fig. 5-6 we have the PI table of Fig. 5-8. In order to cover the atom 2, we require F. For 3 either C or F must be used, for 5 either A or D, etc. We introduce a *covering expression W* which is a Boolean expression over the prime implicant symbols. Here we have

$$W = F(C + F)(A + D)F(A + F)(B + E)(C + E)(B + D),$$

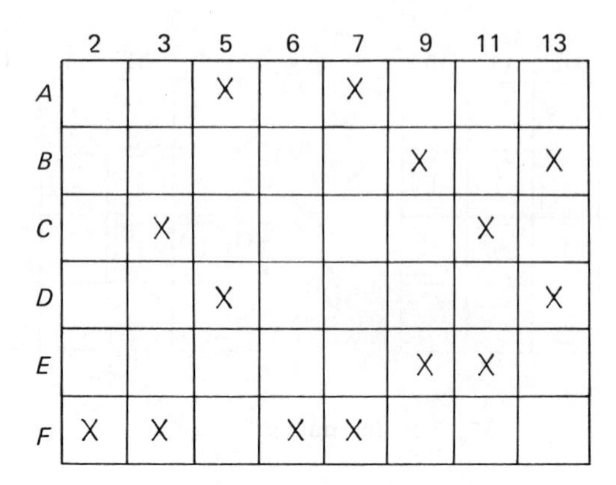

	2	3	5	6	7	9	11	13
A			X		X			
B						X		X
C		X					X	
D			X					X
E						X	X	
F	X	X		X	X			

Fig. 5-8 PI table for Fig. 5-6.

indicating that to cover all the atoms we require F *and* (C *or* F) *and* (A *or* D), etc.

Definition 8

Let f be a Boolean function with m atoms and r prime implicants. Let W be an expression over r variables (the r prime implicant symbols) formed as follows: W is a product of m sums (one for each atom), where the sum for atom a_i consists of all the prime implicants p_j such that $a_i \leq p_j$. Then W is the *covering expression* for f.

W is interpreted as follows: If p is a prime implicant symbol, let $p = 1$ if p is used in an SP expression for f and let $p = 0$ otherwise. A product π of prime implicant symbols corresponds to a sum s_π of prime implicants of f.

Proposition 5

$s_\pi = f$ iff $|\pi| \leq |W| = w$.

**Proof*

If $|\pi| \leq w$, then setting all the prime implicant symbols in π to 1 results in $w = 1$. This is only possible if each sum of W is 1 for this assignment. This, in turn, implies that at least one prime implicant symbol in each sum of W is 1. Therefore, the prime implicants in s_π have the property that for each atom a_i of f there is a prime implicant p_j in s_π such that $a_i \leq p_j$. Thus $s_\pi \geq f$. Since s_π is a sum of prime implicants of f, also $f \geq s_\pi$, and the claim follows. By reversing the argument, we have the converse. □

If we convert W to an SP form V and remove all repetitions as well as all the products π_i such that $|\pi_i| < |\pi_j|$ for some π_j in V, then the new expression U for w has the property that $|\pi_i| \leq |\pi_j|$ implies $i = j$. Since U does not contain any complemented literals, it cannot have any near products and hence no consensus terms exist. Therefore, U is C-reduced. By Theorem 2, U is the complete sum for w.

Theorem 3

Let f be a Boolean function and W its covering expression. The sum s_π is an irredundant sum for f iff $|\pi|$ is a prime implicant of w.

**Proof*

If s_π is irredundant but $|\pi|$ is not a prime implicant of w, then there exists another product π_1 such that $|\pi| < |\pi_1| \leq w$. By Proposition 5, $s_{\pi_1} = f$. However, the set of prime implicants of s_{π_1} is a proper subset of the set of prime implicants of s_π. Hence we have a contradiction of the fact that s_π is irredundant. Similar reasoning shows the converse. □

We now have the following algorithm for finding all irredundant sums for f. Construct the PI table and write the covering expression W. By "multiplying out" W, find an SP expression V for w. Finally, find the complete sum U for w by removing products π_i such that $|\pi_i| = |\pi_j|$ and $i > j$, or $|\pi_i| < |\pi_j|$, for some j in V. For each π in U, s_π is an irredundant sum for f.

Continuing with our example, we have

$$\begin{aligned} w &= F(AB + D)(BC + E) \\ &= ABCF + ABEF + BCDF + DEF. \end{aligned}$$

The irredundant sums for f are

$$\begin{aligned} F_1 &= A + B + C + F = w'xz + wy'z + x'yz + w'y \\ F_2 &= A + B + E + F = w'xz + wy'z + wx'z + w'y \\ F_3 &= B + C + D + F = wy'z + x'yz + xy'z + w'y \\ F_4 &= D + E + F = xy'z + wx'z + w'y. \end{aligned}$$

The maps of Fig. 5-9 illustrate these expressions. One verifies by inspection that each map has the property that removing any prime implicant leaves at least one atom uncovered. This, in fact, is the basic property needed for constructing irredundant sums by the map method.

Sometimes it is possible to simplify the PI table by removing certain prime implicants. We now examine this process.

Definition 9

An atom a of f is *distinguished* if there exists only one prime implicant p such that $a \leq p$. If $p \geq a$ and a is distinguished, then p is called *essential*. The sum of all essential prime implicants of f is called the *core* of f.

It is clear that all essential prime implicants must be included in every valid sum of prime implicants of f. In the PI table an atom is distinguished

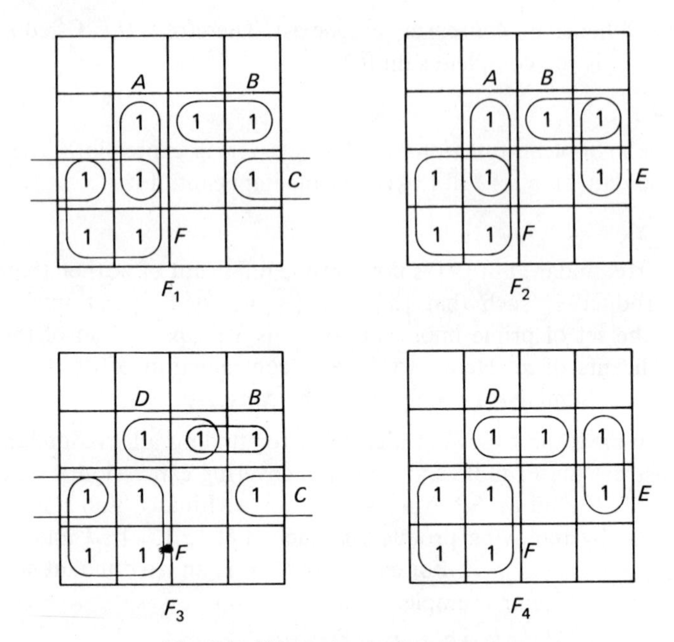

Fig. 5-9 Illustrating irredundant sums.

iff its column contains only one $\times$. The prime implicant in which that $\times$ appears is essential. Thus in Fig. 5-8, atoms 2 and 6 are distinguished and F is essential. If we remove all the essential prime implicants and all the atoms covered by them, we obtain a reduced PI table. In our example, we obtain the table of Fig. 5-10. The corresponding covering expression is a simpler one. Of course, when the covering problem for the reduced table is solved, we must add the essential prime implicants to the final solution. For additional methods of simplifying PI tables we refer the reader to the literature [MCC].

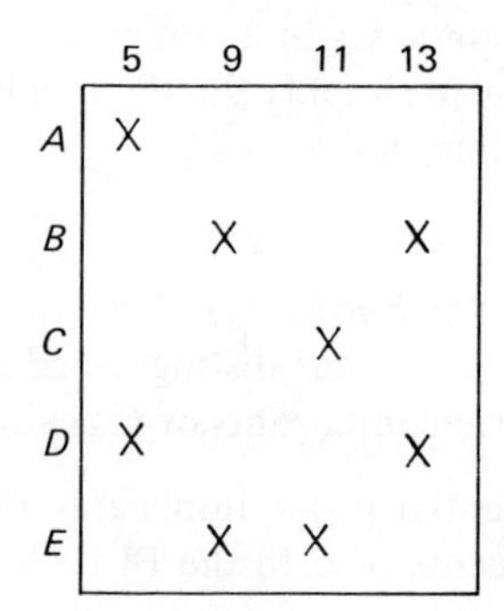

	5	9	11	13
A	X			
B		X		X
C			X	
D	X			X
E		X	X	

Fig. 5-10 Reduced PI table.

We illustrate some irredundant sums in Fig. 5-11. For the function f_1 all prime implicants are essential. In such a case there is always a unique irredundant sum, the core. In the function f_2 there is a "big" prime implicant that is not used. In general, whenever a prime implicant is not essential and is less than the core, that prime implicant will not appear in any irredundant sum because all its atoms are already covered by the core. In f_3 there are no essential prime implicants. There are altogether 10 irredundant sums. Among them are two sums each consisting of 4 prime implicants: the "horizontal" prime implicants and the "vertical" ones.

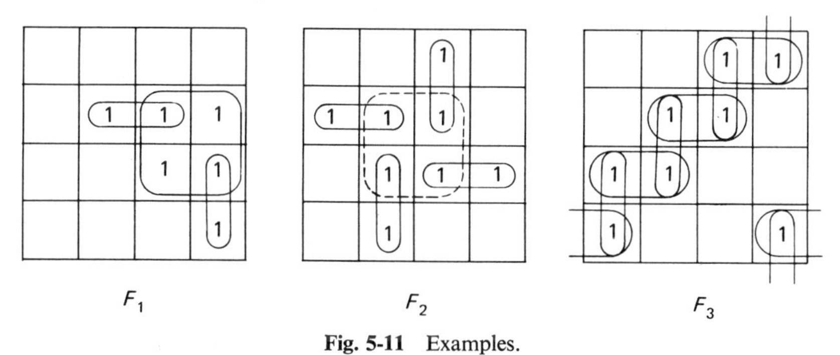

Fig. 5-11 Examples.

The method described above required that both the complete sum and the canonical sum of atoms be available. This is unfortunate because one can use the consensus method on any SP form to find the complete sum. It is then laborious to find all the atoms and construct the prime implicant table. It is possible to avoid this by using different methods (see [BRE]).

To obtain the irredundant products of sums for a given function f, one can proceed as follows. First, find all the irredundant sums for f' and then use the generalized De Morgan's laws to get $(f')' = f$, as we have done in Section 4.3.

5.5. PARTIAL BOOLEAN FUNCTIONS

Frequently, combinational networks to be designed are specified only for some of all the possible input combinations. This holds, for example, for the binary-to-BCD converter module BDM, specified in Section 2.7.

In this connection, we introduce the following.

Definition 10

A *partial Boolean function on n variables* is a function $f: X \longrightarrow \{0, 1\}$, where $X \subseteq \{0, 1\}^n$.

Since we also admit $X = \{0, 1\}^n$, the concept of a partial Boolean function generalizes the concept of a (complete) Boolean function, defined earlier.

A partial Boolean function may be specified by a partial table of combinations such as that of Table 5-3. The function f is undefined for $x = 010$ and $x = 111$.

Table 5-3 Table of Combinations of Partial Boolean Function

x_1	x_2	x_3	$f(x)$
0	0	0	1
0	0	1	1
0	1	1	0
1	0	0	0
1	0	1	1
1	1	0	1

It is both customary and convenient to associate with a partial Boolean function $f: X \longrightarrow \{0, 1\}$ on n variables another function $f_d: \{0, 1\}^n \longrightarrow \{0, 1, d\}$, where the symbol d may be taken to stand for "don't care", and f_d is determined as follows:

1. If $x \in X$, then $f_d(x) = f(x)$.
2. If $x \notin X$, then $f_d(x) = d$.

The function f_d associated with the partial function f of Table 5-3 may be specified by the table of combinations shown in Table 5-4.

Table 5-4 Table of Combinations for Function f_d Associated with f of Table 5-3

x_1	x_2	x_3	$f_d(x)$
0	0	0	1
0	0	1	1
0	1	0	d
0	1	1	0
1	0	0	0
1	0	1	1
1	1	0	1
1	1	1	d

Note: In examples, we often use the symbol – for don't cares.

Intuitively, the concept of realizing a partial Boolean function is rather obvious. However, to treat such realizations and their simplification mathematically, we have to introduce the necessary formal terminology.

Definition 11

Let $f: X \longrightarrow \{0, 1\}$ be a partial Boolean function on n variables. A Boolean expression F *realizes* f iff

$$|F|(x) = f(x) \qquad \text{for every } x \in X.$$

Definition 12

Let f be a partial Boolean function. An SP expression F is an *L-reduced realization* of f iff

(a) F realizes f.

(b) If a literal is removed from some product of F, yielding G, then G does not realize f.

It may not be immediately clear how to extend Theorem 1 to partial functions. To formulate this extension, we introduce the following notation. Let $f: X \longrightarrow \{0, 1\}$ be a partial Boolean function on n variables. Then $\hat{f} \in B_n$ is defined by

$$\hat{f}(x) = f(x) \qquad \text{for every } x \in X$$
$$\hat{f}(x) = 1 \qquad \text{for every } x \notin X.$$

Thus the table of combinations for $\hat{f}$ is obtained from the table for f_d by replacing all d-entries by 1-entries.

Theorem 4

Let f be a partial Boolean function. An SP expression $F = P_1 + \cdots + P_m$ is an L-reduced realization of f iff $p_i = |P_i|$ is a prime implicant of $\hat{f}$, for $i = 1, \ldots, m$.

Proof

This proof is rather easy and is left to the reader. □

Irredundant realizations of partial Boolean functions, to be defined next, play a similar role as the irredundant sums of (complete) Boolean functions. Namely, a simplified expression that realizes a partial Boolean function f will always be an irredundant realization of f.

Definition 13

An expression F is an *irredundant* realization of a partial Boolean function f iff: (a) F is an L-reduced realization of f, and (b) whenever a product P is removed from F, the resulting expression no longer realizes f.

The PI-table approach discussed in Section 5.4 is easily modified, in order to find irredundant realizations of a given partial function f. Namely, a modified PI table is prepared such that its rows correspond to the prime implicants of $\hat{f}$, and the columns to the atoms of f, in the obvious sense.

For example, for the function f of Table 5-3 we obtain the PI table of Fig. 5-12.

	0	1	5	6
$A = x_1'x_2'$	X	X		
$B = x_1'x_3'$	X			
$C = x_1x_2$				X
$D = x_1x_3$			X	
$E = x_2'x_3$		X	X	
$F = x_2x_3'$				X

Fig. 5-12 PI table for function of Table 5-3.

The covering expression becomes

$$\begin{aligned} W &= (A + B)(A + E)(D + E)(C + F) \\ &= (A + BE)(D + E)(C + F) \\ &= (AD + AE + BE)(C + F) \\ &= ACD + ACE + BCE + ADF + AEF + BEF. \end{aligned}$$

Thus the irredundant realizations of f are

$$\begin{aligned} F_1 &= A + C + D = x_1'x_2' + x_1x_2 + x_1x_3 \\ F_2 &= A + C + E = x_1'x_2' + x_1x_2 + x_2'x_3, \text{ etc.} \end{aligned}$$

5.6. MINIMAL SUMS

We have already pointed out that the general minimization problem is very difficult for all cost criteria of either theoretical or practical interest. In this section we discuss the minimization problem in a severely restricted form. First, we only consider SP (sum-of-products) expressions (or dually PS expressions). Second, we make the following assumptions about the cost function: (a) The cost of an SP expression is the arithmetic sum of the costs of the products. (b) The cost $\$(P)$ of a product P depends only on k, the number of literals of P; we set $\$(P) = \bar{\$}(k)$. (c) $k_1 < k_2$ implies $\bar{\$}(k_1) \leq \bar{\$}(k_2)$.

Thus our cost criterion is compatible with the simplicity assumptions A1–A3 of Section 5.2, in the sense that if the expression F is simpler than G, then the cost $\$(F)$ of F cannot be greater than $\$(G)$.

It follows that it suffices to consider only irredundant sums (for complete Boolean functions) or irredundant realizations (for partial Boolean functions) if we wish to find a minimal-cost SP expression.

One method of finding a minimal-cost SP expression for a given partial function f consists of determining all irredundant realizations of f, as discussed in Section 5.5. Among all irredundant realizations we select one with minimal cost. If f is complete, we first find all irredundant sums and then select a minimal-cost SP expression.

However, it turns out that even for Boolean functions with few variables, this method may be very cumbersome. For example, the 4-variable function shown in the map of Fig. 5-13 has 58 irredundant sums.

yz \ wx	00	01	11	10
00	0	1	1	1
01	1	1	1	1
11	1	1	0	1
10	1	1	1	1

Fig. 5-13 A Boolean function with 58 irredundant sums.

We now proceed to discuss alternative methods of determining a minimum-cost SP expression for a given partial function f. Our starting point is the modified PI table mentioned above. An example of such a table is given in Fig. 5-14, where a 1-entry now corresponds to a mark ($\times$) and a 0-entry to the absence of a mark. The cost of each product is also included.

		Atoms					
		a_1	a_2	a_3	a_4	a_5	$\$(P_i)$
Products	P_1	1	1	0	0	0	2
	P_2	1	0	1	0	0	3
	P_3	0	1	0	0	0	2
	P_4	0	0	1	1	0	1
	P_5	0	0	0	1	1	3
	P_6	0	0	0	0	1	1

Fig. 5-14 Example of PI table (with product costs shown).

In general, let the table have m rows, corresponding to m products, and n columns, corresponding to n atoms. Let K be the $m \times n$ matrix of 0's and 1's thus defined, and let c_i denote the cost of the ith product. Then our task can be stated mathematically as follows: Determine an m-bit binary word $y = y_1, \ldots, y_m$ such that

$$\sum_{i=1}^{m} y_i k_i^j \geq 1 \qquad \text{for every } j = 1, \ldots, n.$$

and

$$\sum_{i=1}^{m} y_i c_i \text{ is minimal.}$$

The first equation guarantees that atom j is covered by a prime implicant. If P_i is the ith product in question, the expression $y_1 P_1 + \cdots + y_m P_m$, with 0-terms omitted, is evidently a minimal sum.

In the formulation above, our task becomes a well-known problem of mathematical programming, for which a variety of algorithms have been proposed. For further details the reader is referred to the literature [MU-IB].

5.7. OTHER ASPECTS OF MINIMIZATION

So far we have considered the minimization problem for a single expression. In practice, one usually constructs multi-output networks, and this implies that several expressions must be minimized simultaneously. The example of Fig. 5-15 illustrates this problem for two functions f and g.

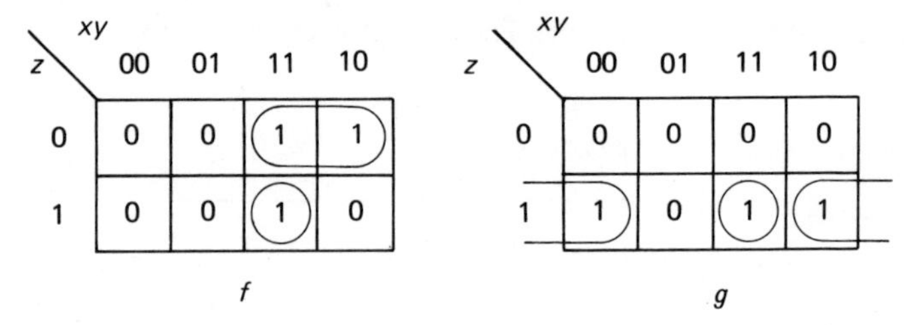

Fig. 5-15 Illustrating product sharing.

If we find minimal SP expressions for f and g we obtain

$$F = xz' + xy$$
$$G = xz + y'z.$$

This realization requires a total of 6 gates. On the other hand, we notice that f and g contain a common product xyz. If we use the expressions

$$F_1 = xyz + xz'$$
$$G_1 = xyz + y'z,$$

a realization with 5 gates can be obtained, since xyz need be computed only once. The network corresponding to these expressions is shown in Fig. 5-16.

It now appears that we have to abandon the notion of prime implicants when solving multi-output problems, since xyz is neither a prime implicant

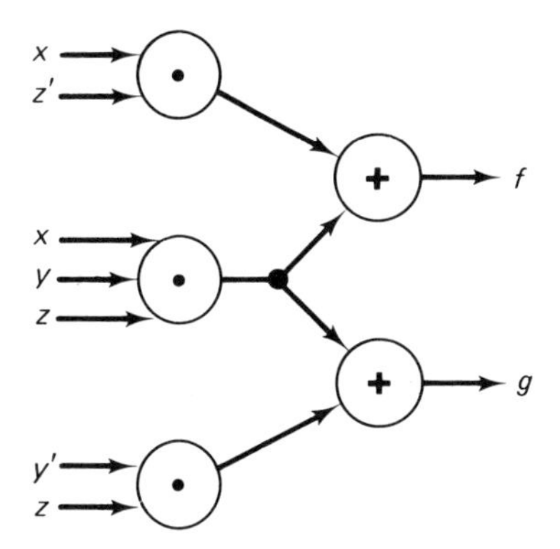

Fig. 5-16 Illustrating gate sharing.

of f nor of g. Fortunately, this notion can be recovered, but we must also consider prime implicants of the function $f \cdot g$ (xyz *is* a prime implicant of $f \cdot g$), as candidates for gates to be shared. For 3 functions f, g, and h we must look at the prime implicants of $f, g, h, f \cdot g, f \cdot h, g \cdot h$, and $f \cdot g \cdot h$. One can see that these problems can become very lengthy.

We will not pursue this topic here since the extensions of the notions of prime implicants, irredundant sums, minimal sums, prime implicant tables, etc., to the multiple-output minimization problem are rather straightforward, although laborious [MCC]. Furthermore, many books contain extensive treatments of this subject, as if it were the central problem in digital network design. In the present technology, the minimization problem has a relatively small role.

It should be pointed out that a minimal-product realization may be significantly better than the minimal-sum realization, and vice versa. For example, for the functions of Fig. 5-15, we can write

$$F_2 = x(y + z')$$
$$G_2 = z(x + y').$$

This realization requires only 4 gates. There does not exist any simple relation between a minimal sum and a minimal product. Thus one must find both and choose the better of the two.

In practical applications one should view the two-level minimization method as an aid to network design but not necessarily as the final solution. Often it may be convenient to use a minimal sum as a starting point. Then by heuristic methods such as factoring, one can look for good three- or four-level solutions which may include the sharing of some gates. We remind the reader again that a minimal SP(PS) expression corresponds to a minimal two-level NAND (NOR) realization, as explained in Section 2.3.

Other problems, such as minimization of multilevel NAND gate networks, decomposition of switching functions, etc., have been studied and are adequately covered in the literature [BRE].

5.8. DESIGN EXAMPLES

We close this chapter with some design examples in which the minimization techniques are used. Other such examples will be found in Chapter 8 in connection with sequential networks.

Because our examples are relatively simple, we apply the map method. The reader is encouraged to try other algorithms for solving the same problems. For larger problems the map method is no longer suitable. It is then convenient to use computer programs to implement the algorithms described here.

Binary-to-BCD Converter Module

The input–output characteristics of the binary-to-BCD converter module (BDM) were described in Section 2.7. For convenience we repeat the block diagram of the module in Fig. 5-17. Recall that the 5-bit binary number (y_1, y_2, y_3, y_4, c) is to be converted to BCD form, where $\bar{y}$ is to represent the binary code for the units digit and $\bar{c}$ corresponds to the tens digit. See Fig.

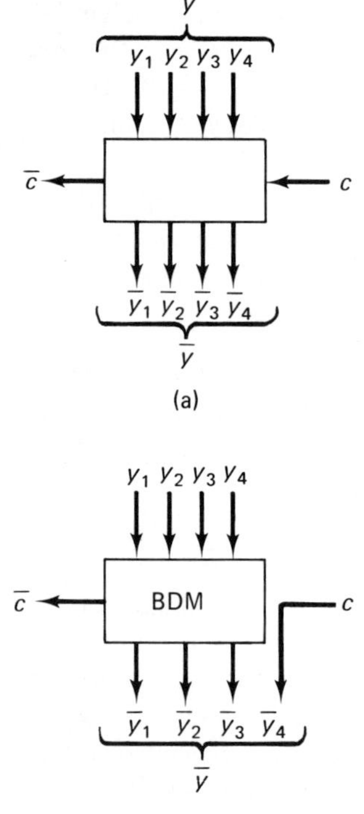

Fig. 5-17 Block diagram of BDM.

5-17(a). Now $\perp(y, c)$ is at most $2 \times 9 + 1 = 19$ since $\perp y \leq 9$. Therefore, the tens digit is either 0 or 1, and one bit ($\bar{c}$) is sufficient to represent it. In summary, we have the following specification:

$$\perp(y, c) = 2 \times \perp y + c = 10\bar{c} + \perp\bar{y}, \qquad \perp y \leq 9, \quad \perp\bar{y} \leq 9.$$

We can obtain a simplified version of the problem by noticing that $\perp(y, c)$ is odd iff $c = 1$, whereas $10\bar{c} + \perp\bar{y}$ is odd iff $\bar{y}_4 = 1$. We conclude therefore that $c = \bar{y}_4$. We are now left with a somewhat simpler problem, as shown in Fig. 5-17(b). Let $\hat{y} = (\bar{y}_1, \bar{y}_2, \bar{y}_3)$. We have

$$\begin{aligned} 2 \times \perp y + c &= 10\bar{c} + \perp\bar{y} \\ &= 10\bar{c} + 2 \times \perp\hat{y} + \bar{y}_4 \\ 2 \times \perp y &= 10\bar{c} + 2 \times \perp\hat{y} \\ \perp y &= 5\bar{c} + \perp\hat{y}, \end{aligned}$$

where $\perp y \leq 9$ and $\perp\hat{y} \leq 4$, since $\perp\bar{y} \leq 9$. Now as long as $\perp y$ is less than 5, we must have $\bar{c} = 0$ and $\perp\hat{y} = \perp y$. For $5 \leq \perp y \leq 9$, $\bar{c} = 1$ and $\perp\hat{y} = \perp y - 5$. This is summarized in Table 5-5, where –'s represent don't cares.

Table 5-5 Table of Combinations for BDM

$\perp y$	y_1 y_2 y_3 y_4	$\bar{c}$	$\bar{y}_1$ $\bar{y}_2$ $\bar{y}_3$
0	0 0 0 0	0	0 0 0
1	0 0 0 1	0	0 0 1
2	0 0 1 0	0	0 1 0
3	0 0 1 1	0	0 1 1
4	0 1 0 0	0	1 0 0
5	0 1 0 1	1	0 0 0
6	0 1 1 0	1	0 0 1
7	0 1 1 1	1	0 1 0
8	1 0 0 0	1	0 1 1
9	1 0 0 1	1	1 0 0
10–15		–	– – –

We have the problem of minimizing 4 partial functions of 4 variables. The maps are shown in Fig. 5-18, in which the prime implicants used for each function are circled.

The corresponding expressions are

$$\begin{aligned} \bar{c} &= y_1 + y_2 y_3 + y_2 y_4 \\ \bar{y}_1 &= y_1 y_4 + y_2 y_3' y_4' \\ \bar{y}_2 &= y_1 y_4' + y_2' y_3 + y_3 y_4 \\ \bar{y}_3 &= y_1 y_4' + y_1' y_2' y_4 + y_2 y_3 y_4'. \end{aligned}$$

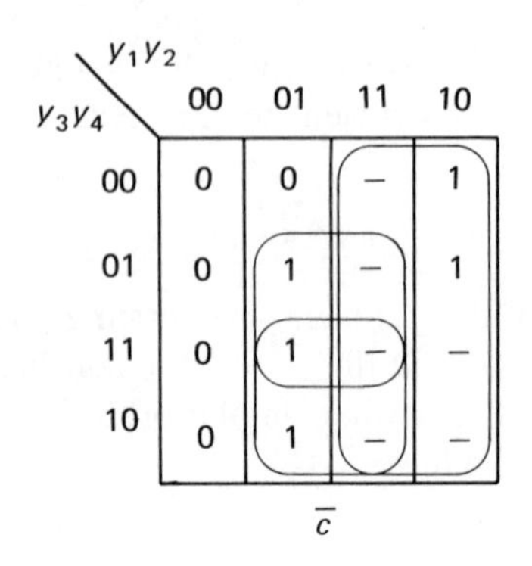

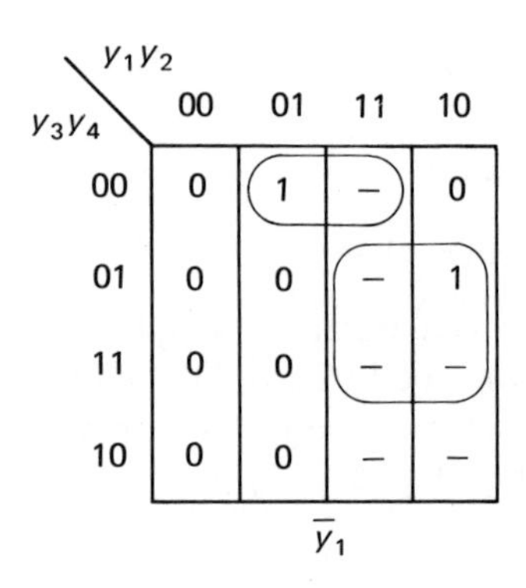

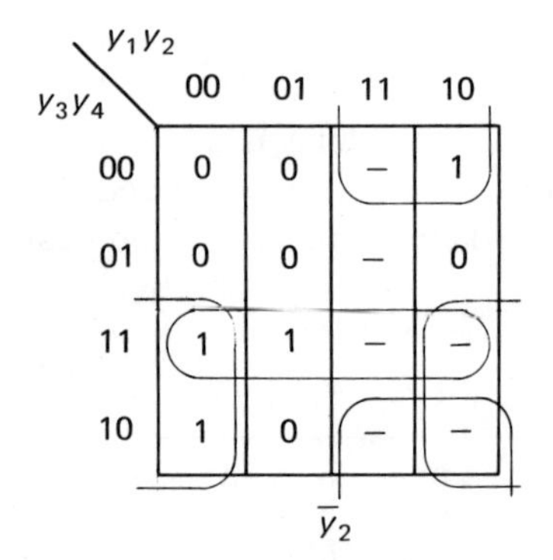

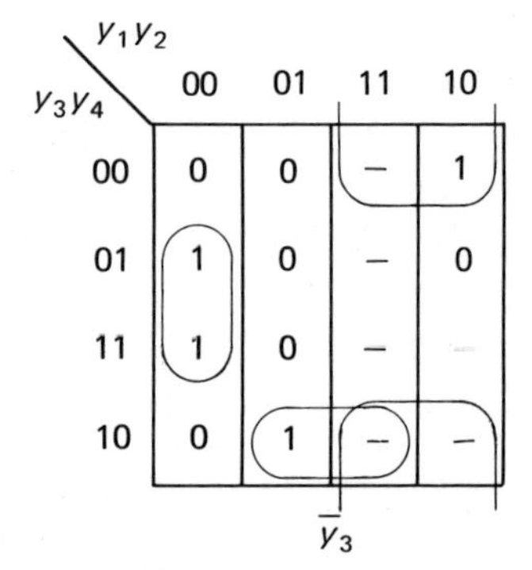

Fig. 5-18 Maps for BDM functions.

The realization involves 10 AND gates (if no sharing is used) and 4 OR gates. One AND gate can be saved by sharing y_1y_4' between $\bar{y}_2$ and $\bar{y}_3$. A more careful examination of the sharing leads to a solution corresponding to the expressions below where the first appearance of each product is underlined. We recommend that the reader draw the maps for this solution to see how this sharing was accomplished.

$$\begin{aligned}
\bar{c} &= \underline{y_1y_4'} + \underline{y_1y_4} + \underline{y_2y_3y_4'} + \underline{y_2y_4} \\
\bar{y}_1 &= y_1y_4 + \underline{y_2y_3'y_4'} \\
\bar{y}_2 &= y_1y_4' + \underline{y_3y_4} + \underline{y_2'y_3} \\
\bar{y}_3 &= y_1y_4' + y_2y_3y_4' + \underline{y_1'y_2'y_4}.
\end{aligned}$$

This solution uses 8 AND gates, 4 OR gates, and 4 inverters. The corresponding NAND realization requires 12 NAND gates and 4 inverters, as shown in Fig. 5-19.

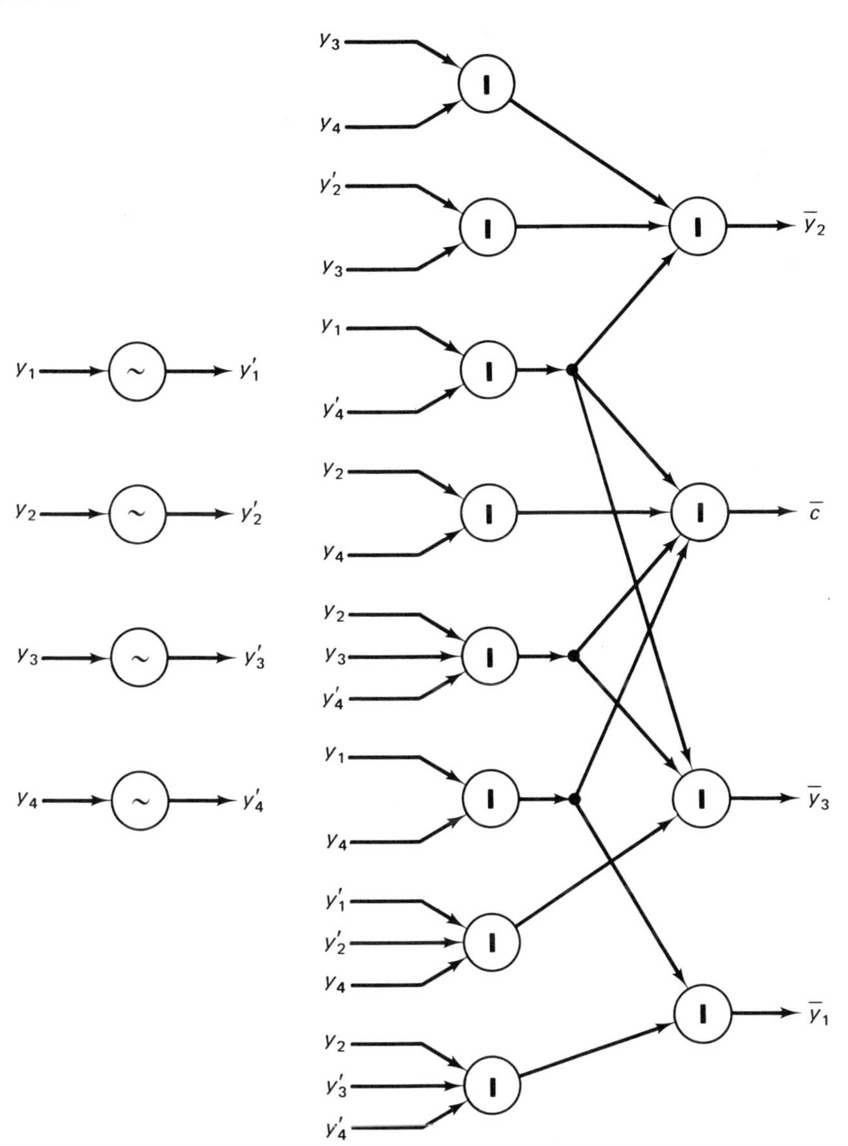

Fig. 5-19 NAND network for BDM.

Seven-Segment Display Network

The scheme shown in Fig. 5-20 can be used to display any digit (0–9). Altogether there are seven lamps, corresponding to the seven segments. The 10 digits are represented by the segments according to Table 5-6. The

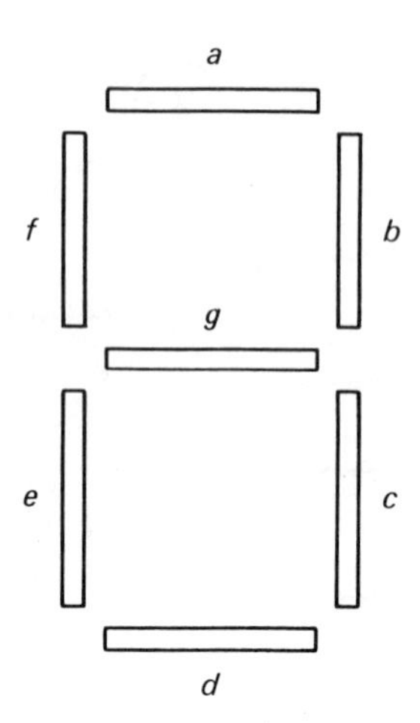

Fig. 5-20 Seven-segment display.

Table 5-6

Digit	Segments Used
0	a, b, c, d, e, f
1	b, c
2	a, b, d, e, g
3	a, b, c, d, g
4	b, c, f, g
5	a, c, d, f, g
6	c, d, e, f, g
7	a, b, c
8	a, b, c, d, e, f, g
9	a, b, c, f, g

problem is to design a code converter accepting $x \triangleq x_1, x_2, x_3, x_4$, the binary representation of the digit, as input and producing outputs $a, b, \ldots, g$, where $a = 1$ means that segment a is ON, etc. Such a code converter is called a "BCD-to-7-segment decoder" and is commercially available in TTL MSI packages such as 7446A, 7447A, 7448, and 7449.

We begin by constructing maps for the seven functions $a, b, \ldots, g$. As in our first example, the input combinations corresponding to $\perp x > 9$ represent don't cares. From Table 5-6 we see that segment a should be on for digits 0, 2, 3, 5, 7, 8, and 9, off for 1, 4, and 6 and a don't care for 10–15. This leads to the first map of Fig. 5-21; the other maps are constructed similarly. (Asterisks indicate distinguished minterms.)

We minimize each function separately first. There are many essential prime implicants, as shown in Fig. 5-21. To keep the diagram clear, we do not show the other prime implicants. Also it is evident that much sharing can be used. For example, the minterm $x'_1x'_2x_3x_4$ of function a can be covered

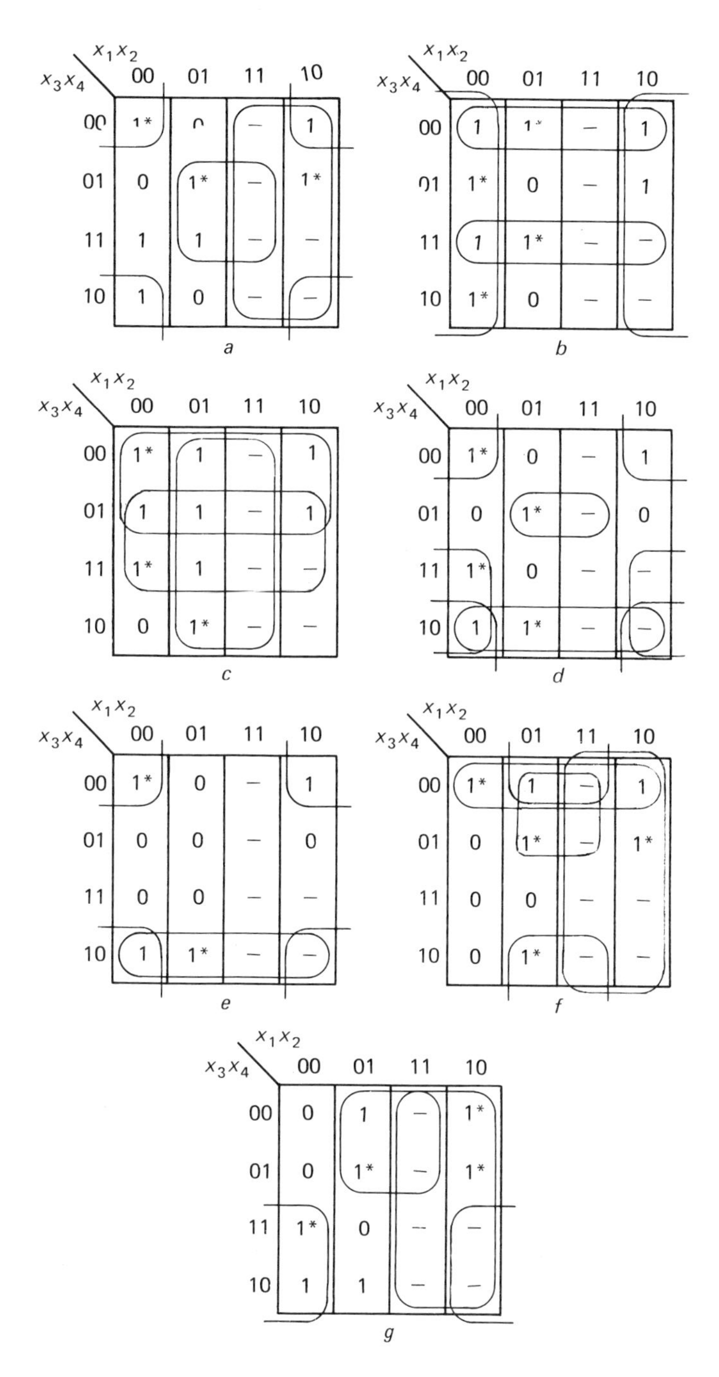

Fig. 5-21 Maps for segment functions.

by either x_3x_4 or x'_2x_3. Since x_3x_4 is essential for b, we can also use it for the above-mentioned minterm of a. In this way we arrive at the following expressions, where the first appearance of each product (of more than one variable) is underlined.

$$\begin{aligned}
a &= \underline{x'_2x'_4} + \underline{x_2x_4} + \underline{x_3x_4} + x_1 \\
b &= \underline{x'_3x'_4} + x_3x_4 + x'_2 \\
c &= x_2 + x'_3 + x_4 \\
d &= \underline{x_2x'_3x_4} + \underline{x_3x'_4} + \underline{x'_2x_3} + x'_2x'_4 \\
e &= x'_2x'_4 + x_3x'_4 \\
f &= \underline{x_2x'_3} + \underline{x_2x'_4} + x'_3x'_4 + x_1 \\
g &= x_2x'_3 + x'_2x_3 + x_3x'_4 + x_1.
\end{aligned}$$

We require 9 AND gates for the underlined products and 7 OR gates, or 16 NAND gates. We do not claim that this is minimal. Also, inverters are needed for x_1, x_2, x_3, and x_4, if a two-level NAND realization is used. Note that x'_1 *is* required because

$$\begin{aligned}
a &= x'_2x'_4 + x_2x_4 + x_3x_4 + x_1 \\
&= ((x'_2x'_4)' \cdot (x_2x_4)' \cdot (x_3x_4)' \cdot x'_1)'.
\end{aligned}$$

Therefore, x'_1 must be supplied to the final NAND gate, producing output a.

Sometimes a more economical solution can be obtained using products of sums instead of sums of products. Let us try this approach here, by finding expressions for $a', b', \ldots, g'$. The maps are shown in Fig. 5-22. The products used correspond to the solution given by the equations

$$\begin{aligned}
a' &= \underline{x_2x'_3x'_4} + \underline{x_2x_3x'_4} + \underline{x'_1x'_2x'_3x_4} \\
b' &= \underline{x_2x'_3x_4} + x_2x_3x'_4 \\
c' &= \underline{x'_2x_3x'_4} \\
d' &= x_2x'_3x'_4 + \underline{x_2x_3x_4} + \underline{x'_2x'_3x_4} \\
e' &= x_4 + x_2x'_3x'_4 \\
f' &= \underline{x_3x_4} + x'_1x'_2x'_3x_4 + x'_2x_3x'_4 \\
g' &= x_2x_3x_4 + \underline{x'_1x'_2x'_3}.
\end{aligned}$$

This solution requires 9 AND gates and 6 OR gates, or 15 NAND gates. We do not claim that this solution is minimal. The products enclosed in rectangles in Fig. 5-22 correspond to products that are *not* prime implicants of a given function. For example, a' has an essential prime implicant $x_2x'_4$. It costs one AND gate to produce this. However, $x_2x_3x'_4$ is needed for b'

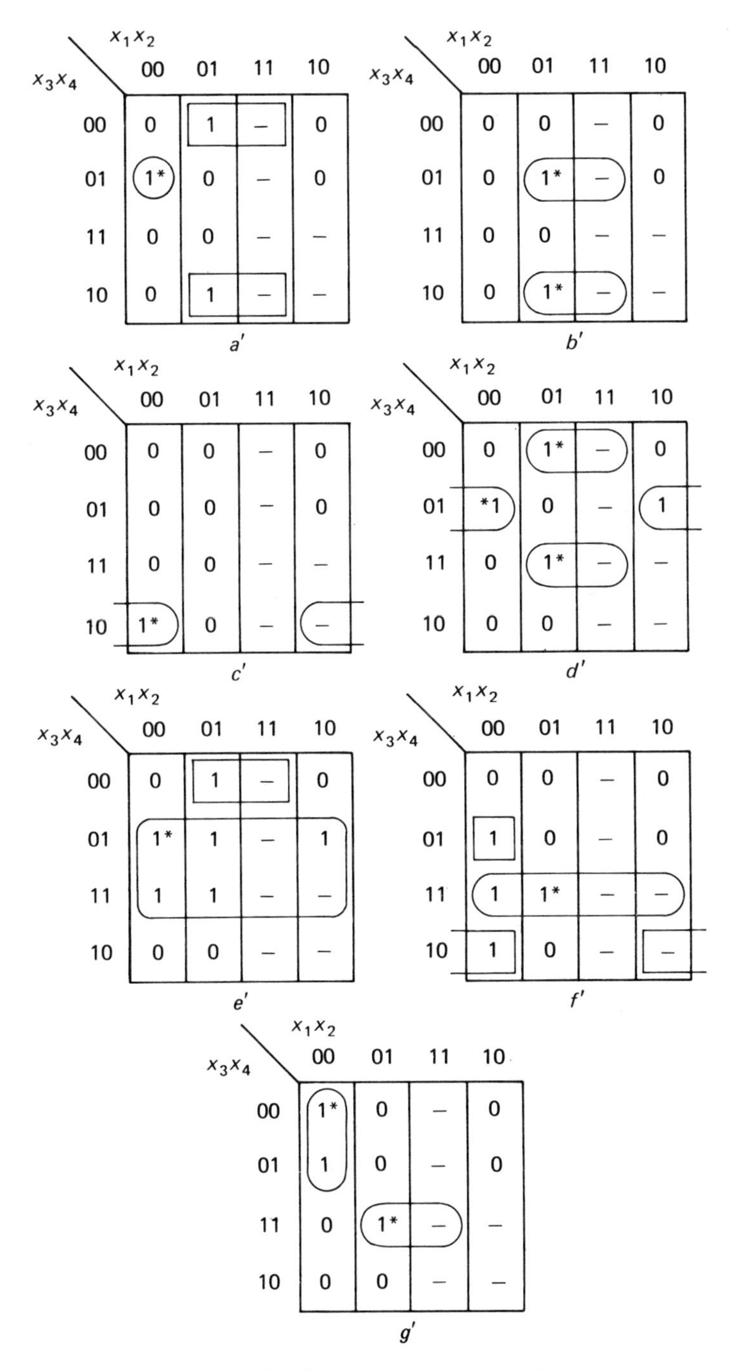

Fig. 5-22 Complements of segment functions.

and $x_2x_3'x_4'$ for d'. These have to be produced regardless of a'. By not using the prime implicant x_2x_4', we save one AND gate.

In practice, errors may occur and the input x may violate the condition $\perp x \leq 9$. In this case, a solution that uses don't cares as we did above can produce strange patterns for $\perp x > 9$ or can show a digit, misleading the viewer. For example, the solution of Fig. 5-21 yields $a = b = c = f = g = 1$ for $\perp x = 15$. This corresponds to the digit 9. For this reason it may be preferable to turn off all the lamps when $\perp x > 9$, and our don't cares become 0's. This feature of "false-data rejection" is sometimes incorporated in commercial units.

Priority Encoder

We will now design a 4-line priority encoder as described in Section 2.6. The table of combinations is shown in Table 5-7 in compact form and is derived as follows. When the enable input e is 0, all the outputs should be 0. When $e = 1$ and there are no requests, we should have $z_0 = 0$, and z_1 and z_2 are irrelevant. Finally, if $e = 1$ and the highest-priority request is on line p_i, we have $z_0 = 1$ and $\perp(z_1, z_2) = i$, as shown in Table 5-7.

Table 5-7 Specification of Encoder

e	p_0	p_1	p_2	p_3	z_0	z_1	z_2
0	–	–	–	–	0	0	0
1	0	0	0	0	0	–	–
1	1	–	–	–	1	0	0
1	0	1	–	–	1	0	1
1	0	0	1	–	1	1	0
1	0	0	0	1	1	1	1

We have 5 inputs $e, p_0, \ldots, p_3$ and 3 outputs. The problem can be simplified if we deal with the enable input separately, by introducing auxiliary variables w_i as follows. Let

$$z_i = e \cdot w_i \qquad \text{for } i = 0, 1, 2.$$

Now the w_i's depend only on $p_0, \ldots, p_3$, as shown in the map of Fig. 5-23. Routine minimization yields the following expressions:

$$w_0 = p_0 + p_1 + p_2 + p_3$$
$$w_1 = p_0'p_1'$$
$$w_2 = p_0'p_1 + p_0'p_2' = p_0'(p_1 + p_2').$$

Alternatively, the tabular method could be used to generate the prime implicants and then the prime implicant table could be solved to find a

p_2p_3 \ p_0p_1	00	01	11	10
00	0– –	101	100	100
01	111	101	100	100
11	110	101	100	100
10	110	101	100	100

w_0, w_1, w_2

Fig. 5-23 Map for auxiliary outputs.

minimal solution. Note that z_1 and z_2 are partial functions. (The reader of the starred material will note that the consensus method is more appropriate here, because of the form of Table 5-7.)

A realization of the 4-line priority encoder based on the preceding Boolean equations is shown in Fig. 5-24. If an SSI implementation is required,

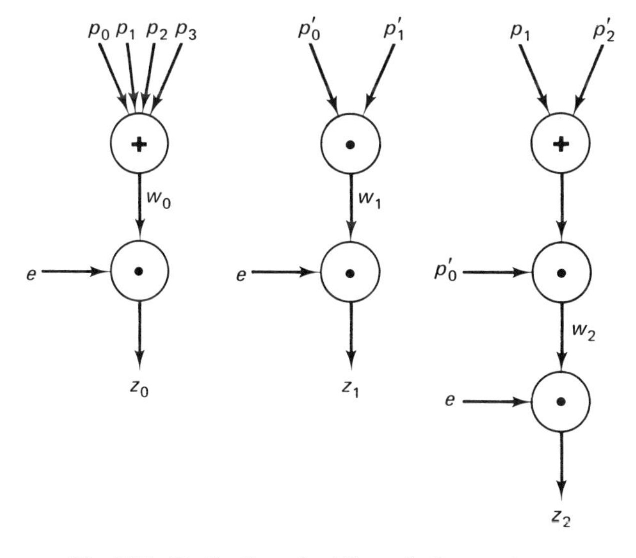

Fig. 5-24 Realization of a 4-line priority encoder.

this realization is easily modified by trial and error to suit the availability of SSI packages, as listed in Table 2-1, for example. One such modification is shown in Fig. 5-25.

We now proceed to show how two n-line priority encoders can be interconnected to provide an encoder for $2n$ priority lines. Figure 5-26 indicates how an 8-line priority encoder may be obtained from two 4-line priority encoders. The inputs to the auxiliary network are not all independent, because

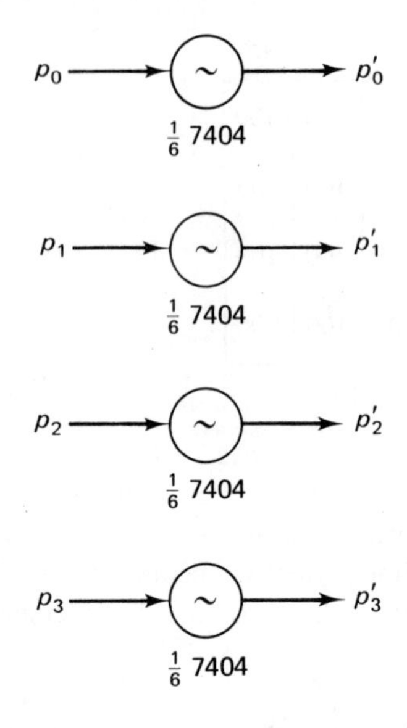

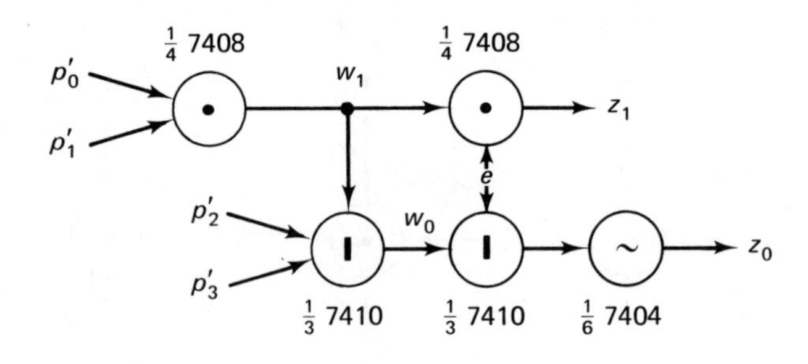

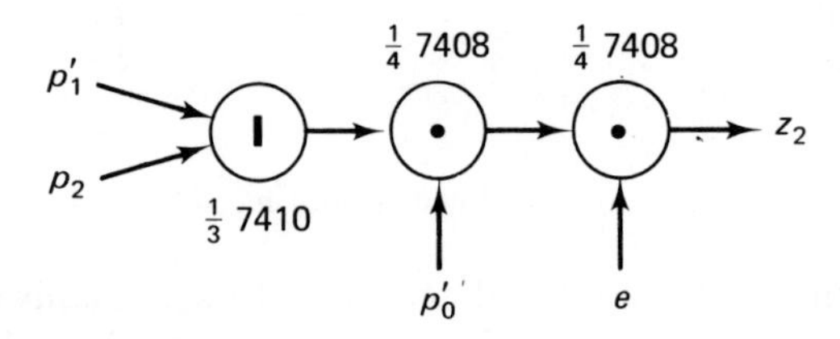

Fig. 5-25 SSI-package implementation of 4-line priority encoder.

the z outputs depend on e. Therefore, not all the input possibilities can occur. The allowed input combinations are shown in Table 5-8; the others become don't cares. Note that we must use the 4-line encoder as we have designed it and there are no longer any don't cares for its outputs.

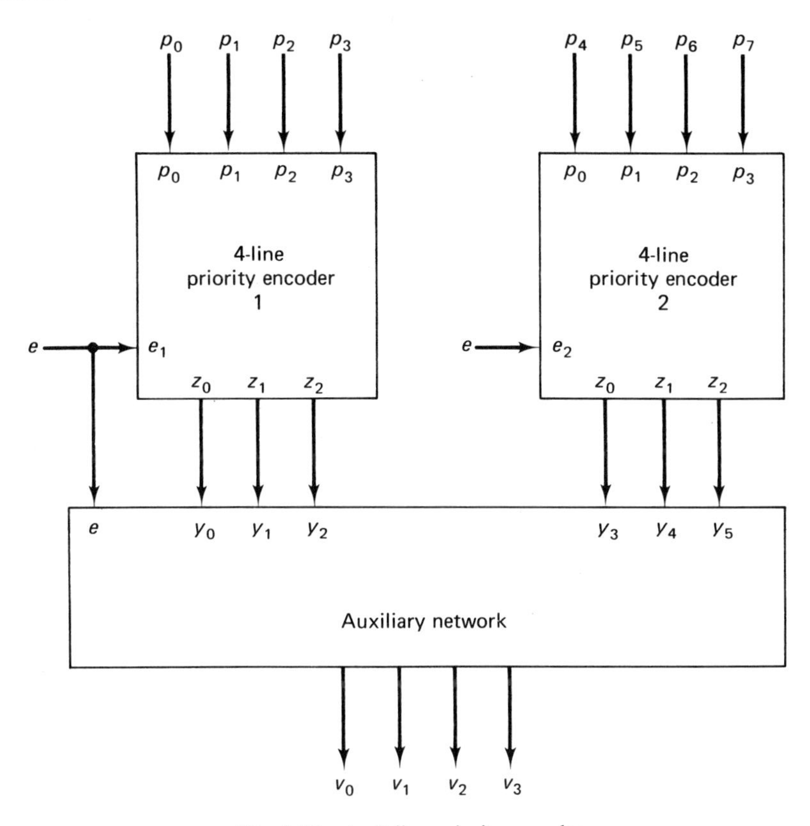

Fig. 5-26 An 8-line priority encoder.

There are 7 inputs, and the map method is not practical. The tabular method could be used if a computer program is available. One could also use the consensus method with some intuitive modifications. In any case, our solution is

$$v_0 = ey_0 + ey_3 = e(y_0 + y_3)$$
$$v_1 = y'_0 y_3$$
$$v_2 = y'_0 y_4 + y_0 y_1$$
$$v_3 = y'_0 y_5 + y_0 y_2$$

as can be verified intuitively. NAND-gate or IC-package implementation is now straightforward.

In view of the maximal propagation delays given in Table 2-1 the maximal delay of the 4-line priority encoder of Fig. 5-25 is ≤ 115 ns (5 levels are used for z_0). The auxiliary network can be implemented with a propagation delay of ≤ 66 ns, for a total propagation delay of about 180 ns for the 8-line encoder. We refer to such an encoder in Section 9.2, where it is used as a component of a larger digital network.

Table 5-8 Specification of Auxiliary Network

e	y_0	y_1	y_2	y_3	y_4	y_5	v_0	v_1	v_2	v_3
0	0	0	0	0	0	0	0	0	0	0
1	0	1	1	0	1	1	0	–	–	–
1	0	1	1	1	0	0	1	1	0	0
1	0	1	1	1	0	1	1	1	0	1
1	0	1	1	1	1	0	1	1	1	0
1	0	1	1	1	1	1	1	1	1	1
1	1	0	0	0	1	1	1	0	0	0
1	1	0	1	0	1	1	1	0	0	1
1	1	1	0	0	1	1	1	0	1	0
1	1	1	1	0	1	1	1	0	1	1
1	1	0	0	1	–	–	1	0	0	0
1	1	0	1	1	–	–	1	0	0	1
1	1	1	0	1	–	–	1	0	1	0
1	1	1	1	1	–	–	1	0	1	1
All other combinations							–	–	–	–

PROBLEMS

1. Use the tabular method to find all the prime implicants of

$$f(w, x, y, z) = \sum(2, 3, 7, 8, 9, 10, 11, 13).$$

2. Let $f = wy'z' + xy'z' + wyz + x'yz' + xy'z$.
(a) Show f on a map.
(b) Find all essential prime implicants.
(c) Find the complete sum.
(d) Find all irredundant sums.
(e) Find all sums using the minimal number of gates.

***3.** For $f = vwx' + wxy' + xyz'$, use the iterated consensus method to obtain all the prime implicants of f. Prove that vwx' is essential without constructing the prime implicant table.

4. The prime implicants of $f(w, x, y, z) = \sum(2, 4, 5, 8, 10, 11, 12, 13, 15)$ are $A = \sum(4, 5, 12, 13)$, $B = \sum(2, 10)$, $C = \sum(8, 12)$, $D = \sum(11, 15)$, $E = \sum(13, 15)$, $F = \sum(8, 10)$, $G = \sum(10, 11)$.
(a) Form the prime implicant table.
(b) Find all the essential prime implicants.
(c) Find all irredundant sums.

5. For the Boolean function $f = xz' + w'x'y' + wxz + wy'z + w'yz'$:
(a) Find a minimal sum using the total number of gates as the cost criterion.
(b) Find one irredundant sum that is not minimal as in part (a).

6. For the function $f(v, w, x, y, z) = \sum(1, 3, 7, 12, 13, 14, 15, 17, 21, 23, 29, 31)$:
(a) Find all prime implicants using the tabular method.
(b) Construct the prime implicant table.
(c) Find all essential prime implicants.
(d) What is the smallest number of gates required to realize f as a sum of products? Draw the corresponding gate network.

7. Assume that both complemented and uncomplemented inputs are available. Let $f = w'x'z' + xy'z + w'xz + wx'z'$.
(a) Find a minimal sum for f using the number of gates as the criterion of cost. Use the map method.
(b) Find an economical realization of f using only NAND gates.
(c) Repeat part (b) using NOR gates instead of NAND gates (4 NOR gates are sufficient).

8. Realize $f = (z + w'x')(w + x' + y)$ by a contact network using no more than 5 contacts.

9. Realize $f = wx' + wy' + x'z' + y'z'$ by a two-level gate circuit (AND gates and OR gates) using no more than a total of 7 gate inputs.

10. Realize $f(w, x, y, z) = \sum(3, 12, 13) + d(5, 6, 7, 15)$ by two-level SP and PS gate circuits, using the smallest number of gates in each case.

11. The Gray code for integers 0–15 is given in Table 2-3. Design a combinational gate network that converts the Gray code representation (g_1, g_2, g_3, g_4) of a digit from 0 to 9 to its BCD representation (b_1, b_2, b_3, b_4). Minimize SP expressions for each b_i, $i = 1, 2, 3, 4$, separately using don't cares.

12. The $(8, 4, -2, -1)$ code for the decimal digits 0–9 is defined as follows. If the code word is the binary word $c = c_1, c_2, c_3, c_4$, then the digit $d(c)$ denoted by c is

$$d(c) = 8c_1 + 4c_2 - 2c_3 - c_4.$$

(a) Write out the $(8, 4, -2, -1)$ code for 0–9.
(b) Design a gate network that translates the usual binary code word b_1, b_2, b_3, b_4 for d into the $(8, 4, -2, -1)$ code. Assume that complemented inputs are available. The network is to be constructed using only INVERTERS and multi-input AND and OR gates. Use as few gates as possible.
(c) Extend the network above to also provide the "9's complement" of d, i.e., the $(8, 4, -2, -1)$ representation of $9 - d$.

13. Consider the following bit sequence, which repeats itself every 10 bits:

00001011110000101111 . . .

Design a 4-input gate network that meets the following requirements:
(1) Whenever the input combination corresponds to four consecutive bits in the sequence, then the output is the immediately following bit of the sequence (e.g., the input combination 0111 yields output 1, since 01111 is a substring of the given sequence).
(2) The network should be a two-level gate network (of AND gates and OR gates) using a minimal number of gate inputs.

14. Let each decimal digit N, $0 \leq N \leq 9$, be represented by its BCD code (b_1, b_2, b_3, b_4). It is desired to add a "check" bit c to this code so that the resulting 5-tuple $(b_1, b_2, b_3, b_4; c)$ has an even number of 1's. For example, the new code word for 2 is $(0, 0, 1, 0; 1)$. Find a network using AND gates and OR gates that generates the check bit c from the binary code word (b_1, b_2, b_3, b_4). Assume that complemented inputs are available. Use the total number of gates as a criterion to obtain a minimal network in a sum-of-products form.

15. A Boolean function f is defined to be positive iff there exists a sum-of-products expression F for f which contains no complemented variables. For example, $f_0 = |vwx + wxy + xyz + vwyz + vwxy|$ is a positive function.
(a) Prove that the complete sum of f_0 above is an irredundant sum.
(b) Prove that for any positive function, the complete sum is an irredundant sum.

16. Below you are given a statement called *Theorem* and four statements, A–D, called *Proof*. The *Theorem* and *Proof* statements may or may not be correct.
(a) Which statements in the *Proof* are correct? If you believe a statement is incorrect, give your reasoning.
(b) If you believe the *Theorem* is correct but not the *Proof*, give a proper proof of the Theorem.
(c) If you believe the *Theorem* is incorrect, find a counterexample.

Theorem: If a Boolean function f has only one irredundant sum, then its complete sum (the sum of all prime implicants) is irredundant.

Proof: (1) Suppose that f has only one irredundant sum F.
A. If a prime implicant p of f is not essential, then there exists an irredundant sum G for f, where P does not appear in G.
B. By A and the hypothesis (1), every prime implicant p appearing in F is essential for f.
C. If every prime implicant of a function f is essential, then the complete sum is irredundant.
D. By B and C, it follows that the complete sum of f is irredundant.

17. Every Boolean function can be realized using only AND and XOR gates, since $x' = 1 \oplus x$. Thus AND and complementation can be performed and we know that these operations are functionally complete. Below you are given three statements, S1–S3, and the corresponding "proofs," P1–P3. For each pair Sk, Pk, decide whether
(1) Both Sk and Pk are true; or
(2) Sk is true but Pk is false. In this case produce a correct proof of Sk; or
(3) Sk is false. In this case give your reasoning briefly.

S1. $x + y + z = x \oplus y \oplus z \oplus xy \oplus xz \oplus yz \oplus xyz.$ (1)
P1. This can be verified by a table of combinations.
S2. In expression (1) no product with two or more literals can be removed without changing the function.
P2. Verify that $x + y + z \neq x \oplus y \oplus z \oplus xz \oplus yz \oplus xyz$. By symmetry, neither xz nor yz can be removed. Finally, verify that $x + y + z \neq x \oplus y \oplus z \oplus xy \oplus xz \oplus yz$. S2 now follows.

S3. Any network of XOR and AND gates that realizes $f = x + y + z$ must use at least 4 AND gates.

P3. By S1, f can be denoted by expression (1). By S2 no product can be removed from expression (1). Hence any expression for f using only XOR and AND gates must have at least the four products xy, xz, yz, and xyz.

18. Design the DBM specified in Section 2.7. The network is to be constructed using only INVERTERS and multi-input AND and OR gates. Use as few gates as possible.

19. Redesign the contact network of Section 1.6, Example 2. Prepare a table of combinations and then apply the simplification methods of this chapter.

20. Redesign the gate network of Problem 12 of Chapter 1, using the methods of this chapter. Implement the network by means of SSI packages, using as few packages as possible.

21. Design a BCD-to-7-segment decoder (see Section 5.8, pp. 151–56, for terminology) with false-data rejection, i.e., for any illegal input combination, all lamps should be blanked out. The networks are to be constructed as two-level NAND networks. Use as few NAND gates as possible.

22. This problem is concerned with the programming of the TABPI and ITCO algorithms. In this connection we represent a set C of m ternary n-tuples (i.e., $C \subseteq \{0, 1, 2\}^n$) by a ternary $m \times n$ matrix M. Let ROWS$(M) \triangleq \{M_1, \ldots, M_m\}$, where M_i denotes the ith row of M. Then M *represents* the set C iff ROWS$(M) = C$.

(a) In your favorite programming language write a program that implements the TABPI algorithm. The program is to accept as input a binary matrix M representing the set C_n of the TABPI algorithm. The output is to be a ternary matrix representing the set D.

*(b) Repeat part (a) for the ITCO algorithm.

6 INTRODUCTION TO SEQUENTIAL NETWORKS

ABOUT THIS CHAPTER

The combinational networks considered in Chapters 1, 2, 4, and 5 are very useful in digital network design. However, another dimension is added when memory is present. Without the ability to store information, a digital network cannot do more than realize Boolean functions.

Networks with memory are called sequential. In this chapter we introduce sequential networks and develop models for describing their behavior. Since the material is very fundamental, no topic should be omitted, except possibly the description of the AED model in Section 6.5, on first reading.

Our model differs from those described in most switching theory texts: We uniformly take into account the propagation delays of all gates, whereas many conventional models arbitrarily select some of those delays and treat the others differently. Such assumptions are somewhat misleading because important information about internal states and races may be lost. Our model is more realistic and is used to explain the detailed behavior of commercially used IC packages in Chapter 7.

We introduce two new models for the analysis of races and demonstrate their applicability. The GSW model is a modification and formalization of intuitively clear concepts. The AED model is completely original and leads to a realistic treatment of races.

6.1. GATE NETWORKS WITH PROPAGATION DELAYS

In Section 2.1 we mentioned the propagation delays of gates and pointed out the fact that, in actual networks, transitions between the signals 0 and 1 do not occur instantaneously.

We now investigate further the effects of propagation delays. To keep our gate model as simple as possible for our present purposes, we make the following two assumptions:

(1) Signal transitions are instantaneous.
(2) Consecutive transitions of a gate input signal occur at time intervals larger than the propagation delay of the gate.

In Section 6.5 we abandon assumption (2) and introduce a correspondingly more complex model which takes into account the appearance of short input pulses. Furthermore, in Chapter 10 we abandon assumption (1) and consider phenomena related to nonzero transition times.

The gate model we presently consider may be represented as shown in Fig. 6-1(a), namely, by (1) an ideal, delay-free gate with inputs $x_1, \ldots, x_n$ and output Y satisfying

$$Y(t) = f(x_1(t), \ldots, x_n(t))$$

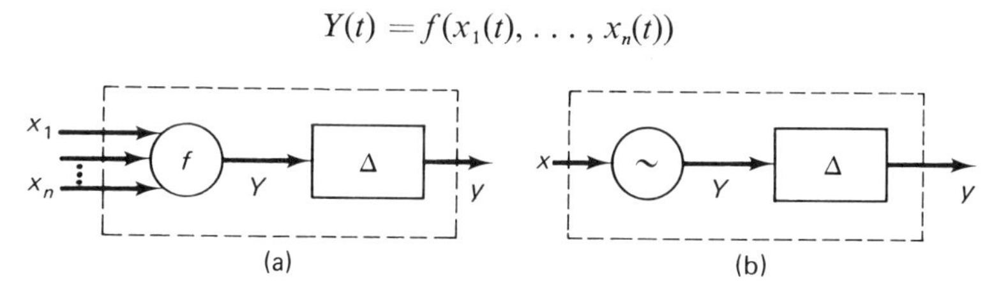

Fig. 6-1 (a) Model of a gate with delay; (b) inverter.

at any time t, where f is a Boolean function of n variables, and (2) an ideal delay of Δ units of time, whose input–output description is given by

$$y(t) = Y(t - \Delta) \qquad \text{for } t \geq \Delta$$
$$y(t) = y_0 \qquad \text{for } 0 \leq t < \Delta,$$

where y_0 is a given initial condition, $y_0 \in \{0, 1\}$. For example, the output $y(t)$ of the inverter of Fig. 6-1(b), corresponding to some input $x(t)$, will be as shown in Fig. 6-2. In the figure we also show the output $Y(t)$ of the delay-free inverter. Note that this output is not physically accessible. However, it is useful to have the concept of the "signal" $Y(t)$. In Fig. 6-1(a), the n-tuple $x(t) \triangleq (x_1(t), \ldots, x_n(t))$ is called the input state of the gate at time t, the

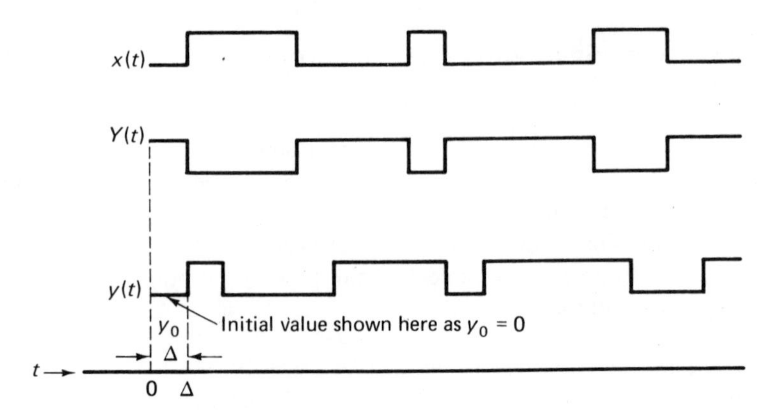

Fig. 6-2 Delayed output signal.

signal $y(t)$ is the output of the gate at time t, and $Y(t)$ is the fictitious signal, called the *excitation* of the gate, at time t. If we view t as the present time, then $x(t)$, $y(t)$, and $Y(t)$ are the present input, present output, and present excitation of the gate. Since the output at time $t + \Delta$ will be equal to the excitation at time t, we can think of $Y(t)$ as describing the next output of the gate.

We now make the following further assumptions about our gate model:

(3) We are neglecting the delays in the lines connecting gate outputs to gate inputs and in the input and output lines.
(4) The magnitude Δ_i of the delay associated with an individual gate varies from gate to gate and is not precisely known.

The general form of a gate network in our new model is shown in Fig. 6-3. Referring to Fig. 6-3, we define the following concepts:

input state, $x(t) \triangleq (x_1(t), \ldots, x_n(t))$

output state, $z(t) \triangleq (z_1(t), \ldots, z_m(t))$

internal state, $y(t) \triangleq (y_1(t), \ldots, y_s(t))$

excitation state, $Y(t) \triangleq (Y_1(t), \ldots, Y_s(t))$

total state $\triangleq$ (input state; internal state) $= (x(t); y(t))$.

The gate excitations satisfy the gate *excitation equations*:

$$Y_i(t) = f_i(x(t); y(t)) \qquad \text{for } i = 1, \ldots, s,$$

where the f_i are Boolean functions of $n + s$ variables.

We assume that each output z_j comes from the output of some gate. Thus $z_j = y_k$ for some k. Hence the external outputs correspond to a "portion" of the internal state; not necessarily all the gate outputs are of interest to the

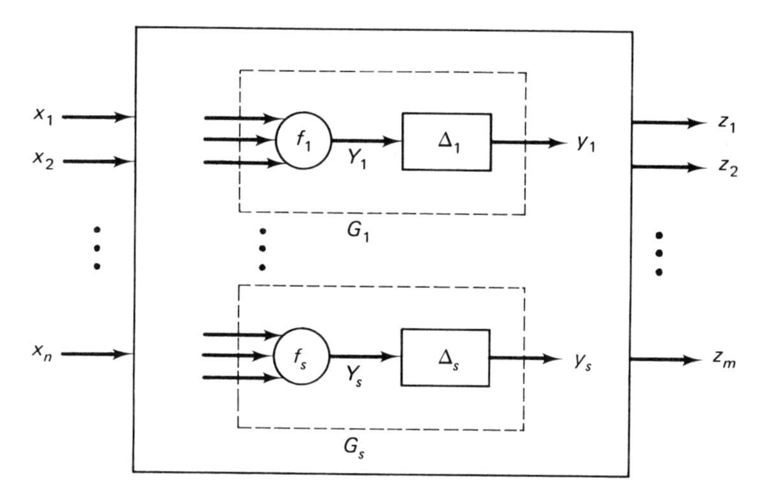

Fig. 6-3 General gate-delay model.

designer. Note that trivial outputs of the form $z_j = x_k$ for some k can be disregarded.

An important notion associated with network behavior is that of stability. A gate G_i is *stable* at time t iff $Y_i(t) = y_i(t)$, implying that G_i has no tendency to change, since its present output agrees with its present excitation. A given total state $(x; y)$ of a gate network is a *stable total state* iff

$$f_i(x; y) = y_i \qquad \text{for all } i = 1, \ldots, s.$$

In other words, the binary $(n + s)$-tuple $(x; y)$ is stable for a given network iff the excitation of each gate for the total state $(x; y)$ is equal to the gate output y_i, i.e., iff each gate is stable.

For most applications we will be mainly interested in the stable states of a gate network and well-behaved transitions from one stable state to another.

6.2. COMBINATIONAL NETWORKS AGAIN

Loop-Free Networks

If we apply our present gate model to a loop-free network of gates, how does the new model relate to our previous notions of combinational behavior?

Consider the network of Fig. 6-4. If $\Delta_1 = \Delta_2 = 0$, we have $z(t) = x_1'(t) + x_2(t)$, a combinational behavior. If the delays are nonzero, the network behavior is much more complex. The excitation equations are

$$Y_1 = x_1'$$

$$Y_2 = x_2 + y_1,$$

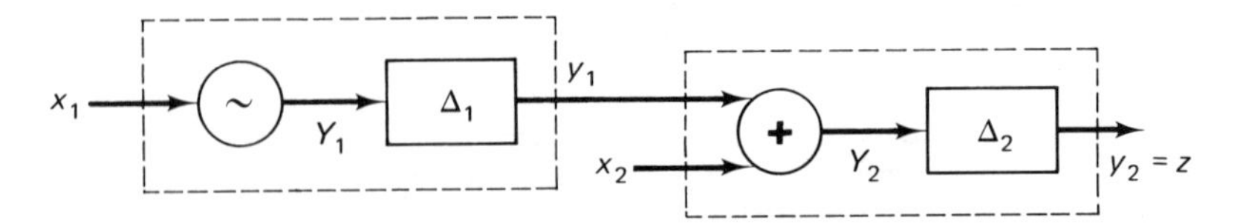

Fig. 6-4 Loop-free gate network.

where the argument t has been omitted everywhere for simplicity. Suppose that somehow the network finds itself in total state $(x_1, x_2; y_1, y_2) = (0, 0; 1, 1)$. We have

$$Y_1 = x_1' = 0' = 1 = y_1$$

$$Y_2 = x_2 + y_1 = 0 + 1 = 1 = y_2.$$

Thus the network is stable and no changes can take place unless an input changes. This total state is illustrated in Fig. 6-5, $t < 0$. If x_1 changes at

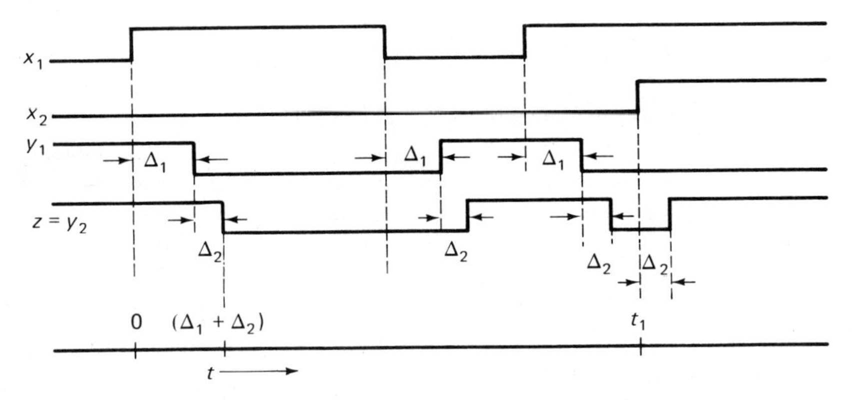

Fig. 6-5 Signals for network of Fig. 6-4.

$t = 0$, y_1 will change at $t = \Delta_1$. At this time the excitation of the OR gate changes and the OR gate responds at $t = \Delta_1 + \Delta_2$, etc. For $t < t_1$ we have $x_2(t) = 0$; hence the network should behave like an inverter $[z(t) = x_1'(t)]$ in the combinational, delay-free model. Clearly this is not so if gate delays are taken into account, and the behavior is not combinational. Consider, however, the case where the input is not allowed to change very frequently in comparison to $\Delta_1 + \Delta_2$; i.e., suppose that once the input state changes it must remain constant for at least τ units of time, where $\tau \gg \Delta_1 + \Delta_2$. Then the corresponding signals would have the appearance of those of Fig. 6-6, where $x_1(t)$ is as in Fig. 6-5, but $\Delta_1 + \Delta_2$ is assumed to be very small compared to the input "rate." In the case of Fig. 6-6, except for the short shaded intervals following input changes, the value of the output $z(t)$ is properly given by the function $x_1'(t) + x_2(t)$.

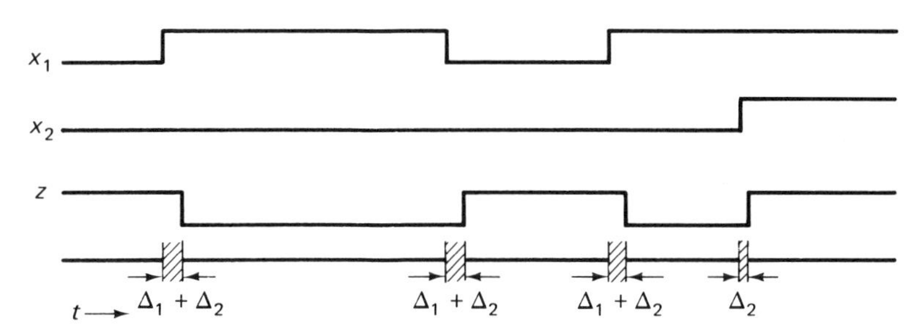

Fig. 6-6 Signals for "slower" inputs.

In summary, it is seen that, for a given gate network, if the inputs change infrequently, we are justified in using the combinational delay-free model. Otherwise, the gate delays must be taken into account.

These remarks apply to any loop-free gate network. Suppose that the delay of each gate is $\leq \Delta$ and that the network is a k-level network. Assume the input is kept constant, after changing at time t. Then, in the worst case, at time $t + k\Delta$, the network outputs agree with the values specified by the Boolean function of the combinational model. Note also that at time $t + k\Delta$ the total state is stable and the internal portion $y(t + k\Delta)$ of the state is uniquely determined by $x(t) = x(t + k\Delta)$.

For the analysis of the logical performance, the delay-free combinational model for loop-free networks will suffice. On the other hand, in many practical applications the overall propagation delay of the network is very important.

Combinational Networks with Loops

In general, gate networks with loops do not correspond to combinational behavior, as we shall see later. However, a network with loops can be combinational, as we now show.

Consider the network of Fig. 6-7. We will analyze it by a shortcut method which is fast but does not constitute a formal procedure. Basically, we would like to know the network behavior under all input conditions. However, one can often avoid a lengthy analysis involving the examination of all the input n-tuples, by considering sets of n-tuples at the same time. This turns out to be the case here.

Suppose that the inputs x_1, x_2, x_3 are held constant for $t \geq t_0$, for a long time, to permit the network to reach a stable state, if one exists for that input combination. (Later, we shall make these concepts more precise.) Suppose further that $x_2 = 1$. After delay Δ_2, the OR gate G_2 will become or remain stable, with $y_2 = 1$. Next consider the AND gate G_3. At time $t_0 + \Delta_2$, its inputs are $y_2 = 1$ and x_3. At time $t_0 + \Delta_2 + \Delta_3$, this gate is stable, with

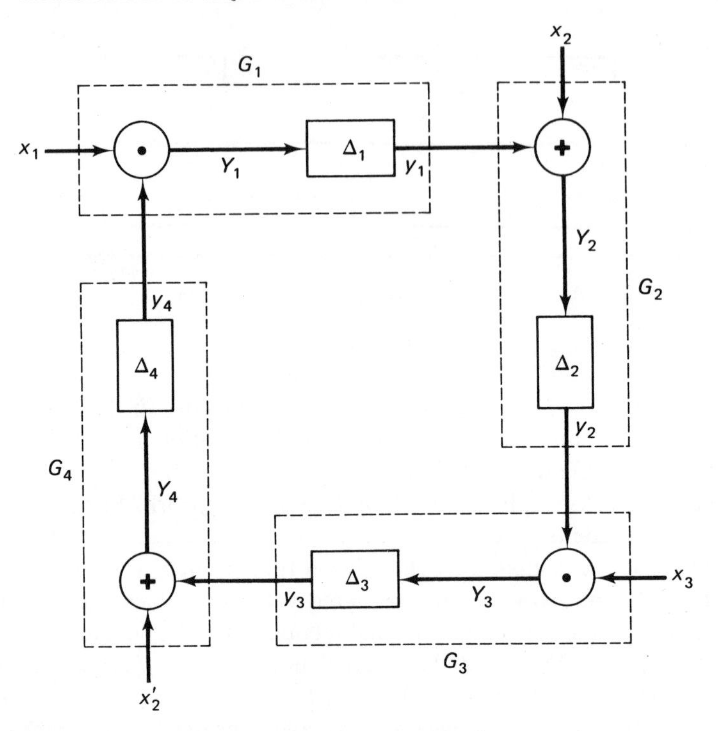

Fig. 6-7 Gate network with a loop.

$y_3 = x_3 \cdot 1 = x_3$. Similarly, since $x'_2 = 0$ (the input is held constant), the OR gate G_4 is stable at time $t_0 + \Delta_2 + \Delta_3 + \Delta_4$, with $y_4 = x_3$. Finally, at time $t_0 + \Delta$, where $\Delta = \Delta_1 + \Delta_2 + \Delta_3 + \Delta_4$, the AND gate G_1 is guaranteed to be stable, with $y_1 = x_1 x_3$. Note now that changes in y_1 cannot affect the OR gate G_2, because its input $x_2 = 1$ is a "forcing" input which, by itself, suffices to make $y_2 = 1$.

In summary, we have just shown that any input state of the form $x = (x_1, 1, x_3)$ [which of course represents the four 3-tuples (0, 1, 0), (0, 1, 1), (1, 1, 0), and (1, 1, 1)] forces the network to a corresponding unique internal state $y \triangleq (y_1, y_2, y_3, y_4) = (x_1 x_3, 1, x_3, x_3)$. Note also that this analysis is completely independent of the initial internal state—no matter where the network starts, the input $(x_1, 1, x_3)$ forces it to a unique stable total state $(x; y) = (x_1, 1, x_3; x_1 x_3, 1, x_3, x_3)$ after at most Δ units of time, where $\Delta = \Delta_1 + \Delta_2 + \Delta_3 + \Delta_4$.

A similar analysis takes care of the remaining input possibilities. Let $x = (x_1, 0, x_3)$. The input $x'_2 = 1$ is now forcing for the OR gate G_4, yielding $y_4 = 1$ (after $t_0 + \Delta_4$), $y_1 = x_1$ (after $t_0 + \Delta_4 + \Delta_1$), $y_2 = x_1$ (after $t_0 + \Delta_4 + \Delta_1 + \Delta_2$), and $y_3 = x_1 x_3$ (after $t_0 + \Delta$). Again, for each input of this type, the network reaches a unique stable total state $(x_1, 0, x_3; x_1, x_1, x_1 x_3, 1)$ after a time at most Δ, independently of the initial state.

It is now convenient to combine the two cases as follows:

$$y_1 = x_1x_3 \quad \text{when } x_2 = 1 \qquad \text{and} \qquad y_1 = x_1 \quad \text{when } x_2 = 0.$$

Thus

$$y_1 = x_1x_2x_3 + x_1x_2' = x_1(x_2x_3 + x_2') = x_1(x_2' + x_3). \tag{1}$$

Similarly,

$$y_2 = 1x_2 + x_1x_2' = x_1 + x_2 \tag{2}$$

$$y_3 = x_3x_2 + x_1x_3x_2' = x_3(x_2 + x_1x_2') = x_3(x_1 + x_2) \tag{3}$$

$$y_4 = x_3x_2 + 1x_2' = x_2' + x_3. \tag{4}$$

Therefore, all the gate outputs are uniquely determined by the external inputs only (after delay Δ), the network is combinational, and it can be easily replaced by an equivalent loop-free network.

Another approach to finding stable states consists of using the excitation equations. We have

$$Y_1 = x_1y_4$$
$$Y_2 = x_2 + y_1$$
$$Y_3 = x_3y_2$$
$$Y_4 = x_2' + y_3.$$

If there are any stable total states, we must have $Y_i = y_i$ for $i = 1, \ldots, 4$. Thus the stable states must satisfy the equations:

$$y_1 = x_1y_4$$
$$y_2 = x_2 + y_1$$
$$y_3 = x_3y_2$$
$$y_4 = x_2' + y_3.$$

By repeated substitution we can express all the y_i in terms of a single variable, say y_4, and the inputs:

$$y_1 = x_1y_4 \tag{5}$$

$$y_2 = x_2 + y_1 = x_2 + x_1y_4 \tag{6}$$

$$y_3 = x_3y_2 = x_3x_2 + x_3x_1y_4 \tag{7}$$

$$y_4 = x_2' + y_3 = x_2' + x_2x_3 + x_1x_3y_4. \tag{8}$$

Now it appears that y_4 depends on itself, and hence its value may not be uniquely determined by the inputs. However, this dependence is degenerate because

$$y_4 = (x_2' + x_2x_3) + x_1x_3y_4 = x_2' + (x_3 + x_1x_3y_4) = x_2' + x_3.$$

By substituting this in equations (5), (6), and (7) we find again the equations (1)–(4).

This network could also be analyzed by the "excitation table" method to be discussed later. However, in this case the shortcut method is simpler.

6.3. SEQUENTIAL NETWORKS: INTRODUCTORY EXAMPLES

Shortcut Analysis

The simple network of Fig. 6-8 will serve to illustrate a number of properties of gate networks which are not combinational. We first analyze it by using an intuitive approach.

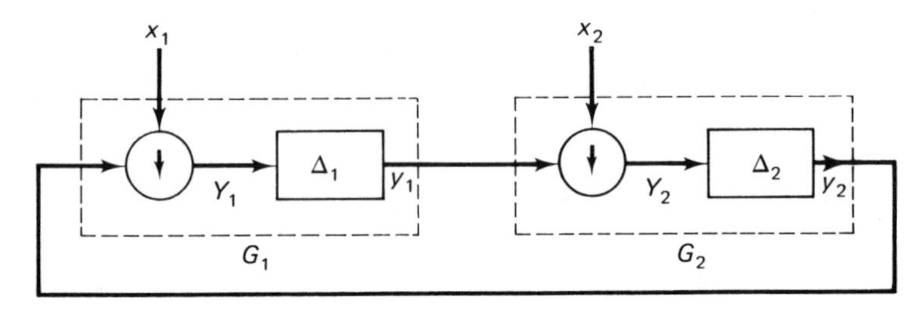

Fig. 6-8 Network which is not combinational.

We consider all four input possibilities in turn:

1. $x_1 = 1$, $x_2 = 0$. The input $x_1 = 1$ is forcing for G_1; after Δ_1 units of time, $y_1 = 0$. Gate G_2 now has as inputs $y_1 = x_2 = 0$. Hence (after Δ_2 units of time) $y_2 = 1$. The reader now verifies that no further changes will take place since both gates are stable. In this case the input is forcing to the unique stable total state (1, 0; 0, 1).
2. $x_1 = 0$, $x_2 = 1$. The network is symmetric with respect to x_1 and x_2, y_1 and y_2. Therefore, this input is also forcing, this time to the stable total state (0, 1; 1, 0).
3. $x_1 = 1$, $x_2 = 1$. This is again a forcing situation to (1, 1; 0, 0).

For these three input combinations, the network behaves like a combinational network. However, this is not so for the remaining case.

4. $x_1 = 0$, $x_2 = 0$. If the network reaches a stable state, the NOR gates behave like inverters, since $y_1 = y_2'$ and $y_2 = y_1'$. Nothing else can be said about the network using only the input information. The behavior is not combinational. There are two possible stable states $y_1 = 0$, $y_2 = 1$ and $y_1 = 1$, $y_2 = 0$, as can be easily verified.

It is now convenient to summarize our analysis in the form of a table that we will call a *flow table*. The rows of the flow table correspond to all those internal states that are stable for some input combination. In other words, enter y as a row iff there exists x such that the total state $(x; y)$ is stable. The columns of the table correspond to all the input n-tuples. The location (column x, row y) corresponds to the total state $(x; y)$.

The information discovered so far is summarized in Fig. 6-9. The circled entries correspond to stable total states. For example, if the network finds itself in total state (0, 0; 0, 1), we enter 01 to indicate that the internal state will not change, provided that the input remains constant. If the input then changes to 10, the network is again stable, etc. There are, altogether, 5 stable total states.

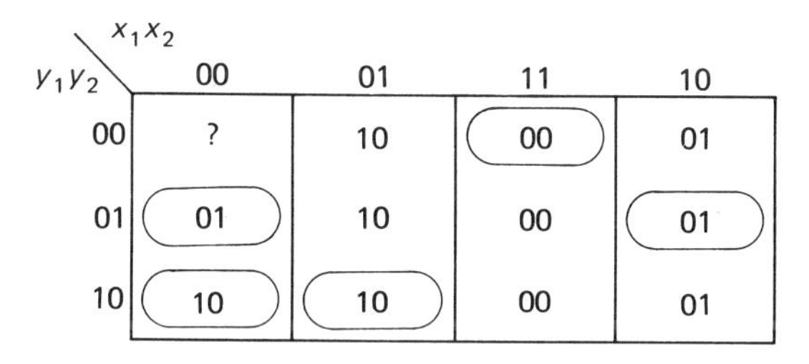

y_1y_2 \ x_1x_2	00	01	11	10
00	?	10	00	01
01	01	10	00	01
10	10	10	00	01

Fig. 6-9 Flow table for the network of Fig. 6-8.

The uncircled entries correspond to unstable total states; i.e., one verifies that in each case either $Y_1 \neq y_1$ or $Y_2 \neq y_2$ or both. From our shortcut analysis, we know that for $x_1 = 0$, $x_2 = 1$, the final destination of the network is the stable state (0, 1; 1, 0), no matter where it started. The internal state component (1, 0) is therefore entered for all the unstable states in column 01. (Remember that we agreed to keep the input constant until the network has a chance to settle.) By the same argument we complete the entries in columns 11 and 10, as shown in Fig. 6-9.

The entry marked "?" requires a careful analysis and will be considered later. For the time being, we notice that the only way the total state (0, 0; 0, 0) can be reached is by starting in the stable total state (1, 1; 0, 0) and changing both inputs simultaneously. It is convenient to use the approximation that two physical phenomena do not take place at exactly the same time. If x_1 becomes 0 first, the total state (0, 1; 0, 0) is reached. It is then possible for the network to reach the total state (0, 1; 1, 0), which is stable. When x_2 becomes 0 next, the network reaches the stable total state (0, 0; 1, 0). On the other hand, if x_2 changes first, the network can eventually reach the total state (0, 0; 0, 1). This behavior is therefore not well defined in our model, since the final state reached depends on the relative magnitudes of various "stray" delays in the network.

We therefore leave this entry with a "?", which will be used to indicate that the corresponding behavior is unreliable. More will be said about this later.

Excitation Table Analysis

We now describe another method of analyzing a gate network based on its excitation equations. If we again refer to the network of Fig. 6-8, we have

$$Y_1 = (x_1 + y_2)' = x_1' y_2'$$
$$Y_2 = (x_2 + y_1)' = x_2' y_1'.$$

The excitation state of a network is conveniently represented by a table called the *gate excitation table*, where the row coordinates are all the 2^s internal states $(y_1, \ldots, y_s)$ and the column coordinates are all the 2^n input states $(x_1, \ldots, x_n)$. Hence each square (column x, row y) in the table defines a total state. The entry in the square is the excitation state Y corresponding to the total state $(x; y)$. We arrange the excitation table for the network of Fig. 6-8 in the form of a 4-variable map as shown in Fig. 6-10. The table can be interpreted as follows. Suppose that the network somehow finds itself in total state $(x; y)$. The entry Y in column x, row y of the excitation table specifies what the internal state is trying to become in the future. As we shall see, there are several types of entries and several different types of behavior.

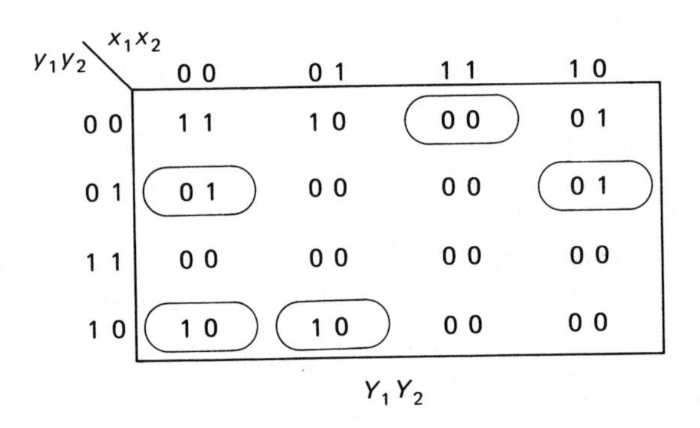

$y_1 y_2$ \ $x_1 x_2$	0 0	0 1	1 1	1 0
0 0	1 1	1 0	(0 0)	0 1
0 1	(0 1)	0 0	0 0	(0 1)
1 1	0 0	0 0	0 0	0 0
1 0	(1 0)	(1 0)	0 0	0 0

Fig. 6-10 Gate excitation table.

1. *Stable entries.* If $Y(x; y) = y$, then the total state $(x; y)$ is stable. No changes can take place unless the input changes. Stable states are of particular interest to us since they constitute the "steady-state" behavior of the network. The entries corresponding to stable states are shown circled in the excitation table. In our example of Fig. 6-8, there are 5 stable total states: (0, 0; 0, 1), (0, 0; 1, 0), (0, 1; 1, 0), (1, 1; 0, 0), and (1, 0; 0, 1).

By examining the various entries in the excitation table, we shall derive the flow table of the network obtained previously by the shortcut analysis. For now, copy all the stable entries from the excitation table into the flow table and omit rows without any stable entries, as shown in Fig. 6-11(a).

2. *Race-free transitions; single-step.* If we have a condition where for two or more gates the gate excitation differs from the gate output, such a condition is called a *race*. For example, in total state (1, 1; 1, 1) there is a race, since $Y = (0, 0)$ but $y = (1, 1)$. If Y and y differ in only one coordinate, only one gate is unstable and there is no race.

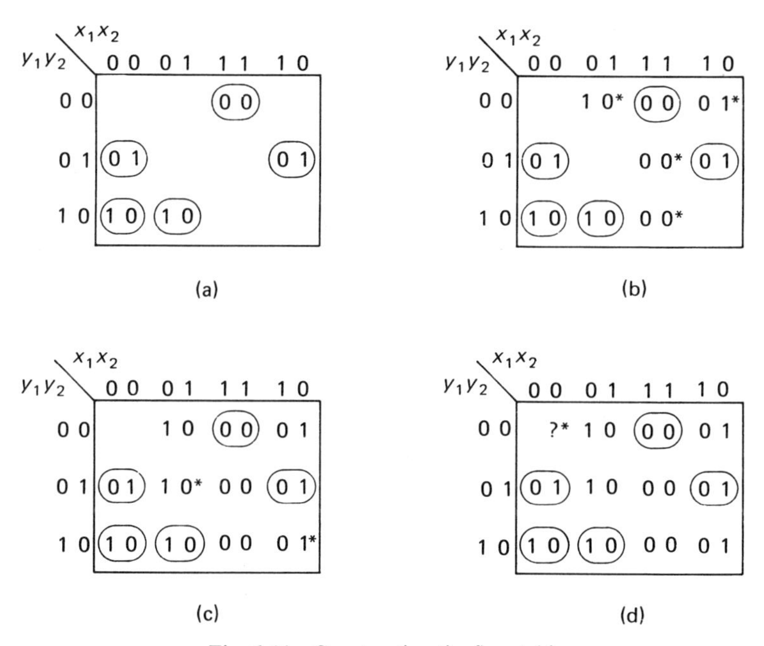

Fig. 6-11 Constructing the flow table.

Consider the total state (0, 1; 0, 0). Only gate G_1 is unstable since $y_1 = 0$ and $Y_1 = 1$. According to our gate model, y_1 becomes 1 Δ_1 units of time after the last change that put the network into the state (0, 1; 0, 0). Assuming that the input remains constant at (0, 1), the network reaches stable total state (0, 1; 1, 0). Thus no further changes will occur unless the input changes. The network has made a transition from one total state to another. This transition was race-free and only one gate changed state. Hence we classify it as a race-free, single-step transition.

Excitation entries resulting in race-free, single-step transitions are copied directly into the flow table. See Fig. 6-11(b), where the latest entries are marked with *.

3. *Race-free transitions; multistep.* A generalization of the previous case occurs if the network starts in state $(x, y) = (0, 1; 0, 1)$. The excitation is $Y = (0, 0)$. There are no races and after Δ_2 units of time the total state

(0, 1; 0, 0) is reached. However, this state is not stable and y_1 will also change without a race, resulting in the stable total state (0, 1; 1, 0). Any entry that eventually leads to a unique stable state after several changes and does not involve races will fall into this category.

For such transitions we enter in the flow table the internal component of the final stable state reached. See Fig. 6-11(c), where the most recent entries have a *.

4. *Critical races.* The possible outcomes if (0, 0; 0, 0) is the initial state are shown in Fig. 6-12. Here if $\Delta_1 < \Delta_2$, stable state (0, 0; 1, 0) is reached, and if $\Delta_1 > \Delta_2$, the final state is (0, 0; 0, 1). Since the outcome of this race depends on the relative magnitudes of Δ_1 and Δ_2, such a race is called *critical*.

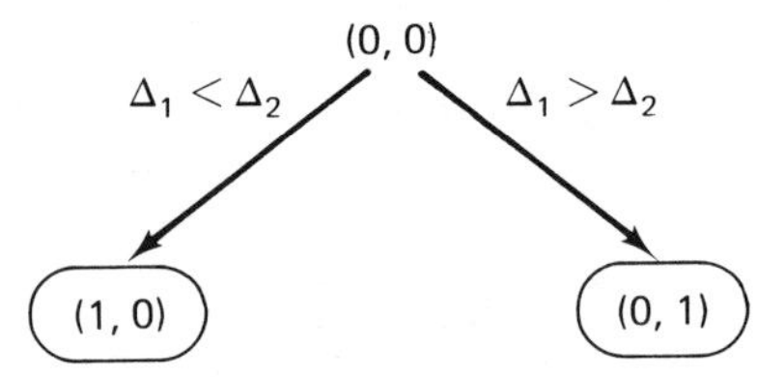

Fig. 6-12 A critical race.

In many applications such behavior is undesirable and should be avoided. We enter a ? in the flow table entries corresponding to critical races. See Fig. 6-11(d), which now coincides with Fig. 6-9.

A careful scrutiny of our analysis above for the initial state (0, 0; 0, 0) will show that this analysis is not consistent with our present gate model. The reader is encouraged to find out why! We shall return to this problem in Section 6.4.

Let us assume that the outputs for the network of Fig. 6-8 are $z_1 = y_1$ and $z_2 = y_2$. It is sometimes convenient to suppress the detailed values of the internal states. Suppose that we let $A \triangleq (0, 1)$, $B \triangleq (1, 0)$, and $C \triangleq (0, 0)$. Considering only stable states and reliable transitions among them, we obtain a new version of the flow table, called the *state table*, shown in Fig. 6-13. Here we also show the outputs associated with stable total states. The state table is a more convenient form for analysis of network behavior.

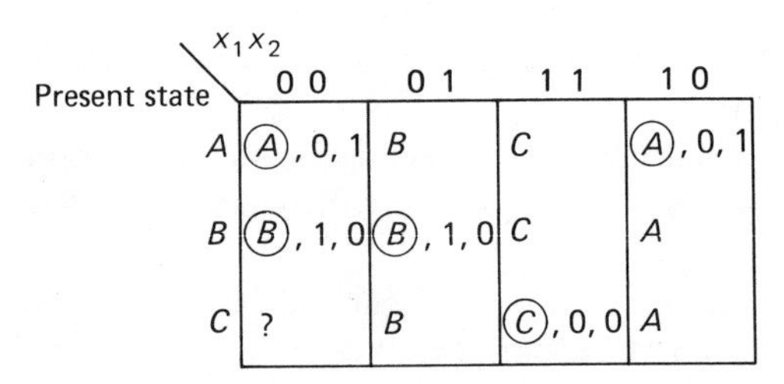

Present state \ x_1x_2	0 0	0 1	1 1	1 0
A	(A), 0, 1	B	C	(A), 0, 1
B	(B), 1, 0	(B), 1, 0	C	A
C	?	B	(C), 0, 0	A

Next state, z_1, z_2

Fig. 6-13 State table.

Note that the given network has an input combination [namely (0, 0)], with more than one stable output combination. We shall refer to such networks as *sequential*. However, we postpone the precise definition of sequential networks until later.

If we specify a given initial state and an input sequence, the state table can be used to predict the corresponding output sequence. All of this is under the assumption that the gate delays are small compared to the periods of constant input.

Typical signals for the network of Fig. 6-8 are shown in Fig. 6-14. In what follows, propagation delays between inputs and outputs will not be shown in waveform diagrams.

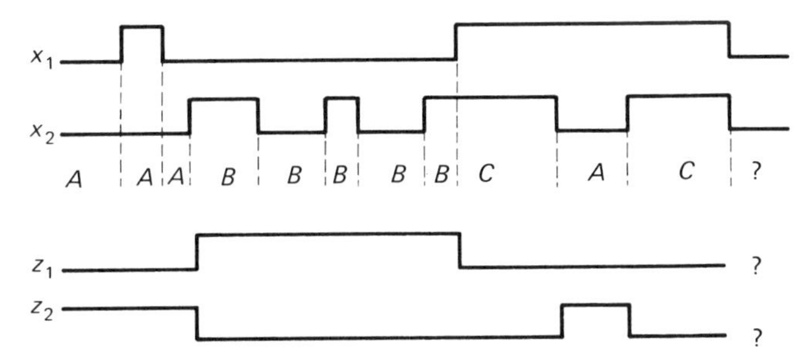

Fig. 6-14 Typical signals for network of Fig. 6-8.

At this point we wish to correct a widespread fallacy. This fallacy stems from the unjustified assumption that one delay element for the network of Fig. 6-8 (commonly known as a latch) is sufficient to describe its behavior. This assumption, in turn, was motivated by the erroneous model which associates delays with feedback loops rather than with gates. In the one-delay model there can be no races, in contradiction to reality.

Another misconception about the network of Fig. 6-8 is the frequently used phrase "the input condition $x_1 = x_2 = 1$ is not allowed." The fact is that this input not only can be applied without causing any damage but can be usefully exploited (see Section 9.2).

In all fairness, though, we point out that the network of Fig. 6-8 has many applications where the input $x_1 = x_2 = 1$ is never used. Under this condition the race will not occur.

5. *Noncritical races.* Consider now the gate network shown in Fig. 6-15. Assume this network to be in the total state $(x; y_1, y_2) = (1; 1, 1)$. This state is unstable since the excitation is $(Y_1, Y_2) = (0, 0)$. Both gates will change their outputs, and we have a race. We disregard the unlikely event that the gate delays Δ_1 and Δ_2 are precisely equal. If $\Delta_1 > \Delta_2$, the network first reaches the total state $(1; 1, 0)$. This is followed by a race-free transition to $(1; 0, 0)$. If $\Delta_1 < \Delta_2$, y_1 "wins the race," total state $(1; 0, 1)$ is first reached

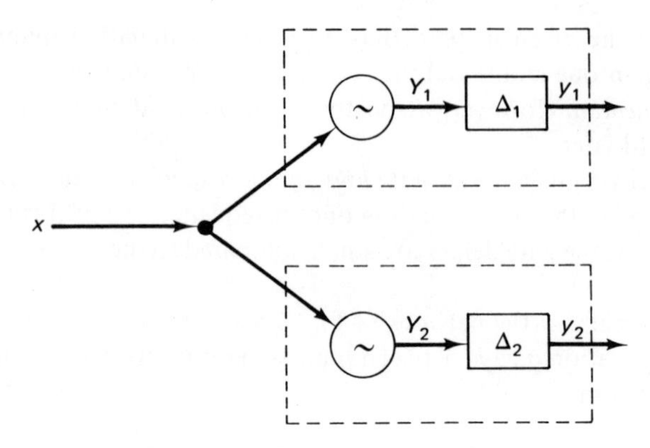

Fig. 6-15 Network with a noncritical race.

and is followed by a race-free transition to (1; 0, 0). Thus no matter what the relative magnitudes of Δ_1 and Δ_2 are, the final outcome is always the same, and a unique stable state is eventually reached. The possible successive values of y are shown in Fig. 6-16 for the initial states (1; 1, 1) and (0; 0, 0).

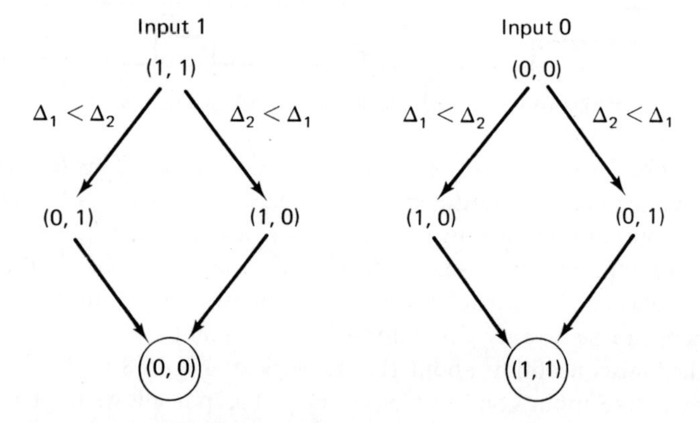

Fig. 6-16 Illustrating noncritical races.

The flow table for the network of Fig. 6-16 is shown in Fig. 6-17. In general, for any unstable state with a noncritical race, we simply enter the internal component of the final stable state reached.

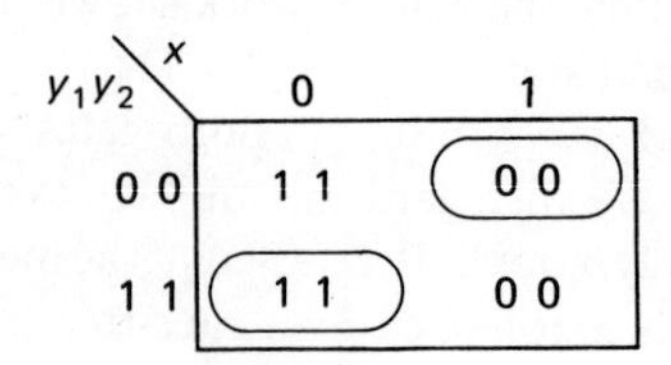

y_1y_2 \ x	0	1
0 0	1 1	(0 0)
1 1	(1 1)	0 0

Fig. 6-17 Flow table for network of Figure 6-16.

6. *Oscillations.* Next, we consider the network shown in Fig. 6-18. Its excitation table is shown in Fig. 6-19. Once this network enters the state $(x; y) = (1; 1)$, it will oscillate between this state and state $(1; 0)$ without reaching a stable state, as long as the input condition $x = 1$ persists. In such a case we shall enter "OSC" in the flow table. The principle of Fig. 6-18 can be usefully applied to the design of oscillators.

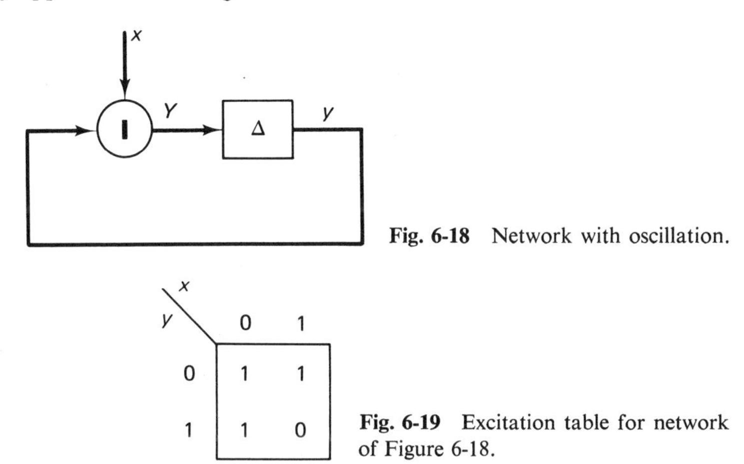

Fig. 6-18 Network with oscillation.

Fig. 6-19 Excitation table for network of Figure 6-18.

6.4. DIFFICULTIES WITH THE IDEAL-DELAY MODEL

In Section 6.1 we have made the assumption (2) that changes in the input signals to a gate occur at time intervals larger than the propagation delay Δ of the gate. In essence, we have ignored the question "How does a gate behave when its input is a very short pulse?"

We have not removed this assumption in Section 6.3 when we discussed the "?" transition in the network of Fig. 6-8. Nevertheless, the flow-table analysis is not consistent with the ideal-delay model of a gate for this particular condition. (We have stated this in Section 6.3). We now show how the problem arises. In Section 6.5 we present more realistic models for analyzing race conditions.

First we explain the difficulty with the ideal-delay model. The network model of Fig. 6-8 is reproduced in Fig. 6-20 with the additional simplifying assumption that $x_1 = x_2 = x$. This is sufficient for our purposes. The starting point is the stable total state $x = 1$, $Y_1 = y_1 = Y_2 = y_2 = 0$, as shown. Now suppose that the input changes from 1 to 0; this is denoted by $x: 1 \longrightarrow 0$. Both ideal (delay-free) gates H_1 and H_2 respond immediately, resulting in $Y_1: 0 \longrightarrow 1$, $Y_2: 0 \longrightarrow 1$, and both delay elements become unstable. There is

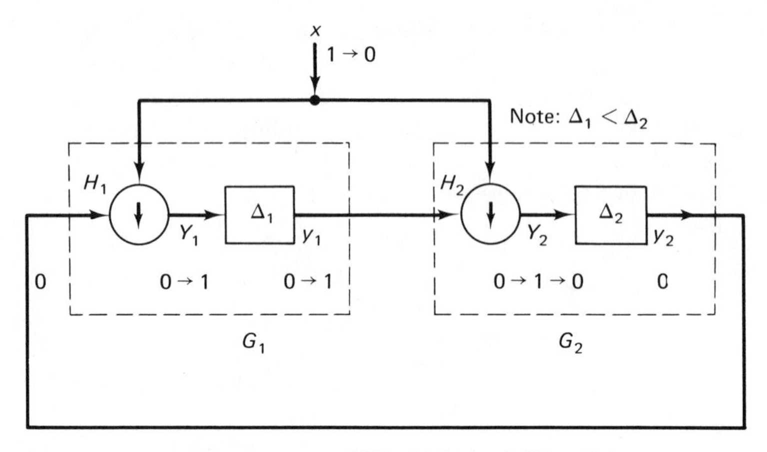

Fig. 6-20 Network of Fig. 6-8 during "?" transition.

now a race between the delays Δ_1 and Δ_2. To simplify, assume that Δ_1 is not exactly equal to Δ_2. Therefore, we consider first the case $\Delta_1 < \Delta_2$. The case $\Delta_2 < \Delta_1$ then follows by symmetry.

After delay Δ_1, y_1 responds to the excitation, i.e., $y_1: 0 \longrightarrow 1$. [See Fig. 6-21(a) for signal waveforms.] As soon as this happens, ideal gate H_2 responds immediately, $Y_2 = 1 \downarrow 0 = 0$. Hence we have the change $Y_2: 1 \longrightarrow 0$. In effect, a 1-pulse has appeared at the input to delay Δ_2. The duration of this pulse is $\Delta_1 < \Delta_2$. After time Δ_2 from the original input change, this pulse appears at the output y_2 and is immediately inverted into a 0-pulse by the ideal gate H_1, etc. Thus the ideal-delay model predicts a circulating pulse persisting forever. It will now become clear why the model fails to agree with the flow-table analysis. Although we have not violated assumption (2) of Section 6.1 with respect to the external input x, that assumption is violated with respect to input combination $(x; y_1)$ of G_2 which changes Δ_1 units of time after the change in x. Effectively an internally generated pulse of duration Δ_1 is applied to G_2, and the propagation delay Δ_2 is (by assumption) longer than Δ_1.

In the flow-table analysis we have the following process [see also Fig. 6-21(b)]:

(1) $x = 1,\ Y_1 = 0,\ y_1 = 0,\ \underline{Y_2 = 0},\ y_2 = 0$; stable.
(2) $x: 1 \longrightarrow 0,\ Y_1 = 1,\ y_1 = 0,\ \underline{Y_2 = 1},\ y_2 = 0$; unstable.

Assuming that $\Delta_1 < \Delta_2$, we reach:

(3) $x = 0,\ Y_1 = 1,\ y_1 = 1,\ \underline{Y_2 = 0},\ y_2 = 0$; stable!

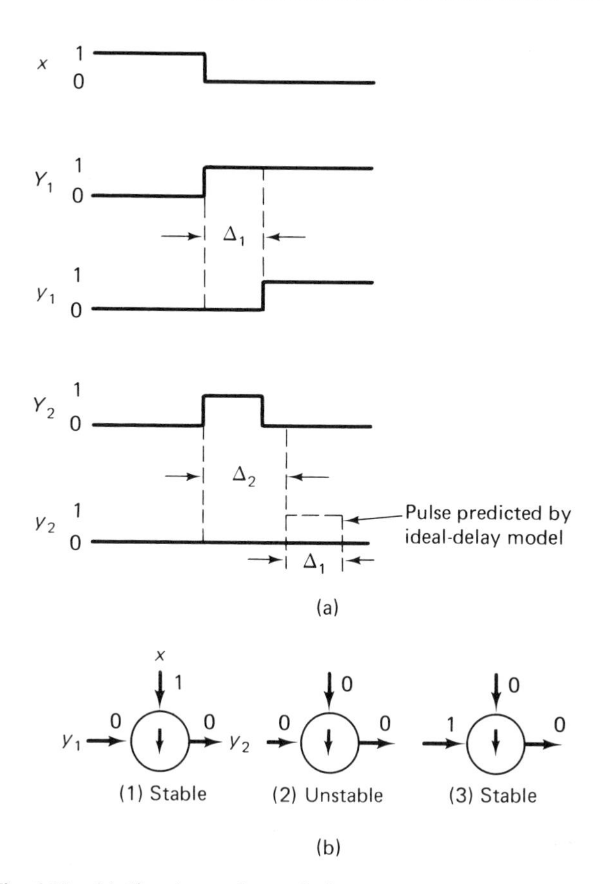

Fig. 6-21 (a) Signal waveforms during "?" transition; (b) gate G_2 during "?" transition.

The difference is that the change $Y_2: 0 \longrightarrow 1 \longrightarrow 0$ was not taken into account in the flow-table analysis because the 1-pulse in Y_2 of (2) above is hidden in the delay Δ_2 ("traveling through it but not yet seen") in (3). Since the delay Δ_2 is (by assumption) ideal, "what comes in must go out" and the conclusion in (3) is not consistent with the ideal-delay model.

What happens in a real network? One can easily verify experimentally that the forever circulating pulse predicted by the ideal-delay model does not exist! Thus we conclude that the ideal-delay model is not adequate for handling critical races. Nevertheless, it is a very simple model and is useful as long as we are aware of its limitations.

In the following section we introduce more realistic models for the analysis of sequential network behavior. In Chapter 10 we will present a further discussion of transient phenomena in gate networks.

6.5. MATHEMATICAL MODELS OF SEQUENTIAL NETWORKS

General Single-Winner Model [BR-YO1]

In the last section we pointed out a discrepancy between the ideal-delay model and the flow-table analysis. Also, we have made it clear that the flow-table analysis is more realistic than the ideal-delay model. In this section we want to formulate a precise mathematical procedure which is to correspond to the flow-table analysis. We shall refer to the corresponding mathematical model as the *general single-winner* (GSW) *model*. Its name is derived from the simplifying assumption that in any race there is always a single winner.

Consider the general gate network of Fig. 6-3. For every input state $x \in \{0, 1\}^n$, we define a binary relation R_x on $\{0, 1\}^s$ as follows. Let $(x; y)$ be some total state of the network and Y the corresponding excitation state. If $y = Y$, then yR_xy. If $y \neq Y$, consider each i such that $y_i \neq Y_i$. Then $yR_xy^{(i)}$, where

$$y^{(i)} \triangleq (y_1, \ldots, y_{i-1}, Y_i, y_{i+1}, \ldots, y_s).$$

For example, if $y = 1101$ and $Y = 1010$, we have 1101 R_x 1001, 1101 R_x 1111, and 1101 R_x 1100.

If one draws the relation diagram† for R_x, one can easily derive the flow table for column x from this relation diagram. If yR_xy, then $(x; y)$ is stable and we enter y for $(x; y)$. Otherwise, to obtain the entry for $(x; y)$, we follow all directed paths in the relation diagram for R_x, starting from node y. One easily verifies that any such path must reach a cycle after a finite number of steps. If, starting from y, we can reach more than one cycle, we enter "?" for $(x; y)$. If only one cycle can be reached and this cycle is of length greater than 1, we enter "OSC" for $(x; y)$. If only one cycle can be reached and this cycle is of length 1, we enter the corresponding node for $(x; y)$.

The application of the GSW model to the analysis of race conditions is illustrated for the network of Fig. 6-22. We wish to describe only one race

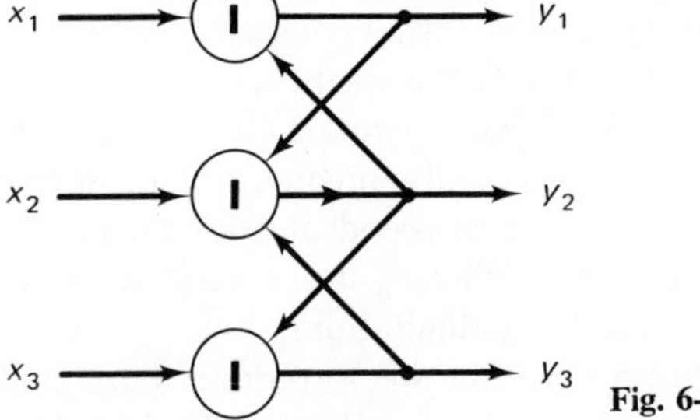

Fig. 6-22 Network N.

†In this diagram, the nodes correspond to internal states, and an arrow from node i to node j indicates that i is related to j (i.e., iR_xj).

condition; for this reason only the input combination $x \triangleq (0, 1, 1)$ will be of interest. The excitation equations for this case are

$$Y_1 = 1, \qquad Y_2 = (y_1 y_3)', \qquad Y_3 = y_2'.$$

We will not draw the complete relation diagram for R_x but only that portion relevant to the initial state $(x; 0, 1, 1)$. For the internal state $y = (0, 1, 1)$, the excitation state is $Y = (1, 1, 0)$. Therefore, y_1 and y_3 are racing. According to the GSW model, 011 R_x 111 and 011 R_x 010. Next, the excitation for 111 is 100, yielding 111 R_x 101 and 111 R_x 110. Similarly, we find 010 R_x 110, 101 R_x 101, and 110 R_x 110. The corresponding relation diagram is shown in Fig. 6-23, where unstable variables are underlined. Starting with $y = (0, 1, 1)$, we find that two cycles of length 1 can be reached. This corresponds to two possible stable states (101 and 110) and the result is interpreted in the GSW model as a critical race, i.e., a ? entry.

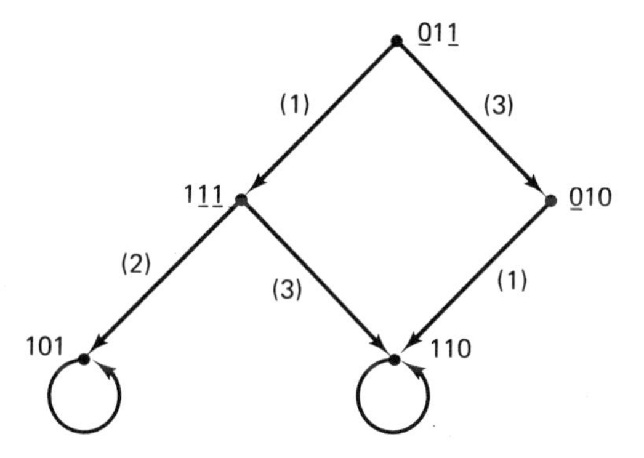

Fig. 6-23 Partial relation diagram for $x = (0,1,1)$.

In Fig. 6-23 we have shown in parentheses the variable winning each race, associated with each edge resulting from a race. Thus the edge from 111 to 101 is labeled (2), since y_2 is assumed to have won the race.

Almost-Equal-Delay Model [BR-YO1]

We have previously mentioned that the propagation delay Δ_i associated with gate G_i is not known precisely, may be different for two gates of the same type, and may vary with time. The GSW model is a very general model which describes the network behavior under all possible propagation delay distributions. For example, reconsider the diagram of Fig. 6-23. Starting in 011, if y_3 wins the race, the final result is stable state 110. This will happen under the condition $\Delta_1 > \Delta_3$. On the other hand, if $\Delta_1 < \Delta_3$, the network moves temporarily to 111, which is unstable. Note that, in 111, the state variable y_3 has not yet changed and is still unstable and racing with y_2. If y_3 wins the race, the result is stable state 110, as before. However, if y_2 wins the race, the result is stable state 101. In order for y_2 to win the race, we must have $\Delta_3 > \Delta_1 + \Delta_2$. In fact, y_3 has been unstable for $\Delta_1 + \Delta_2$ units of time, while y_2 has been unstable for only Δ_2 units.

In the *almost-equal-delay* (AED) *model*, we make the following assumptions:

1. The propagation delays of all the gates are approximately Δ units of time.
2. Gate G_i ignores all excitation conditions which persist for time less than Δ_i (approximately Δ).
3. If the excitation of gate G_i persists for time approximately 2Δ, then gate G_i must react. In other words, we assume that $\Delta_i < \Delta_j + \Delta_k$ for any propagation delays Δ_j and Δ_k in the network.

The analysis of the race condition starting with $y = (0, 1, 1)$ in the AED model is shown in Fig. 6-24. Under AED assumptions, the race is not critical!

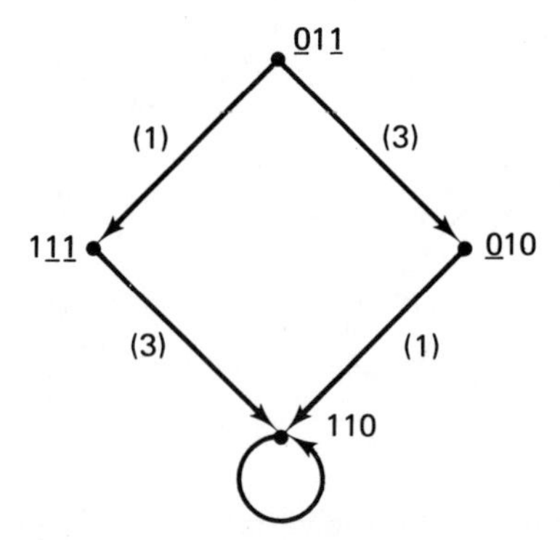

Fig. 6-24 Analysis in the AED model.

The ideas of the AED model are further illustrated in the example below. The network is actually that of Fig. 7-28 (Chapter 7) with ¢ $= J = K = C_0 = 1$. The gate excitation equations under these input conditions, which we denote by x, are

$$Y_1 = y_5', \quad Y_2 = (y_1 y_3 y_5)', \quad Y_3 = (y_1 y_2 y_4)',$$

$$Y_4 = (y_3 y_5)', \quad Y_5 = y_6', \quad Y_6 = (y_2 y_5)'.$$

Assume that the total state $(x; \hat{y} = 101011)$ can be reached from some stable state $(\bar{x}; \hat{y})$ by changing $\bar{x}$ into x. We wish to investigate the behavior of the network with $(x; \hat{y})$ as starting state, keeping the input x fixed. Consider Fig. 6-25, where we show a part of the corresponding relation diagram for the GSW model. Initially, in state $\hat{y} = \underline{1}010\underline{1}1$, variables 1 and 5 are unstable.

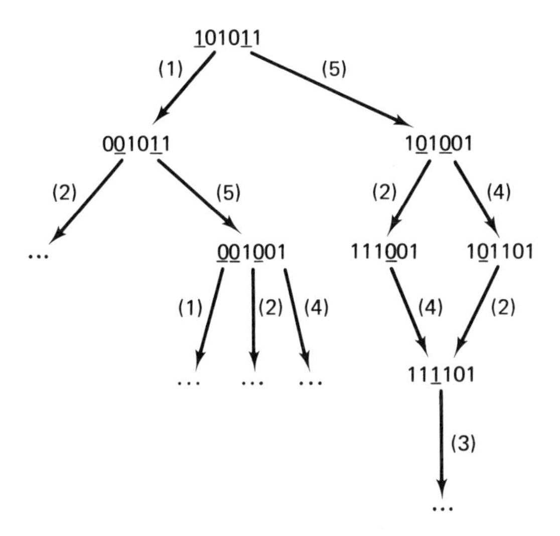

Fig. 6-25 Partial GSW analysis of a race.

If 1 wins the race, we reach state $0\underline{0}10\underline{1}1$. In this state, variables 2 and 5 are now racing. However, 5 has already been unstable for $\boldsymbol{\Delta}_1$ time units, whereas 2 has just become unstable. In the GSW model, variable 2 might still win the race. However, this is impossible in the AED model, since for 2 to win, we must have $\boldsymbol{\Delta}_5 > \boldsymbol{\Delta}_1 + \boldsymbol{\Delta}_2$. We shall indicate this situation in the AED model by replacing state $0\underline{0}10\underline{1}1$ by the *race state* $\langle 0\underline{0}10\underline{1}1, \{5\}\rangle$. Thus in the AED model a race state consists of an internal state y, together with the set of all variables that are unstable in y and are candidates for winning the race.

The complete AED analysis of the race under consideration is shown in Fig. 6-26. The diagram shows all race states reachable from the start state $\hat{y} = \underline{1}010\underline{1}1$ as well as all feasible transitions.

We now give a precise definition of the AED model. We introduce the following terminology. In a network of s gates, the set of all internal states is $Q \triangleq \{0, 1\}^s$. As before, given $y = (y_1, \ldots, y_{i-1}, y_i, y_{i+1}, \ldots, y_s) \in Q$, denote by $y^{(i)}$ the state y with the ith coordinate changed, i.e.,

$$y^{(i)} \triangleq (y_1, \ldots, y_{i-1}, y_i', y_{i+1}, \ldots, y_s).$$

Let $x \in \{0, 1\}^n$ be a fixed input, let $y \in \{0, 1\}^s$, and let $Y = Y(x; y)$ denote the excitation. The set of subscripts of variables unstable in the total state $(x; y)$ is denoted by $u_x(y)$ or simply $u(y)$, if x is understood; i.e.,

$$u(y) \triangleq \{i \mid y_i \neq Y_i\}.$$

Let $S \triangleq \{1, 2, \ldots, s\}$ and recall that $P(S)$ is the set of all subsets of S.

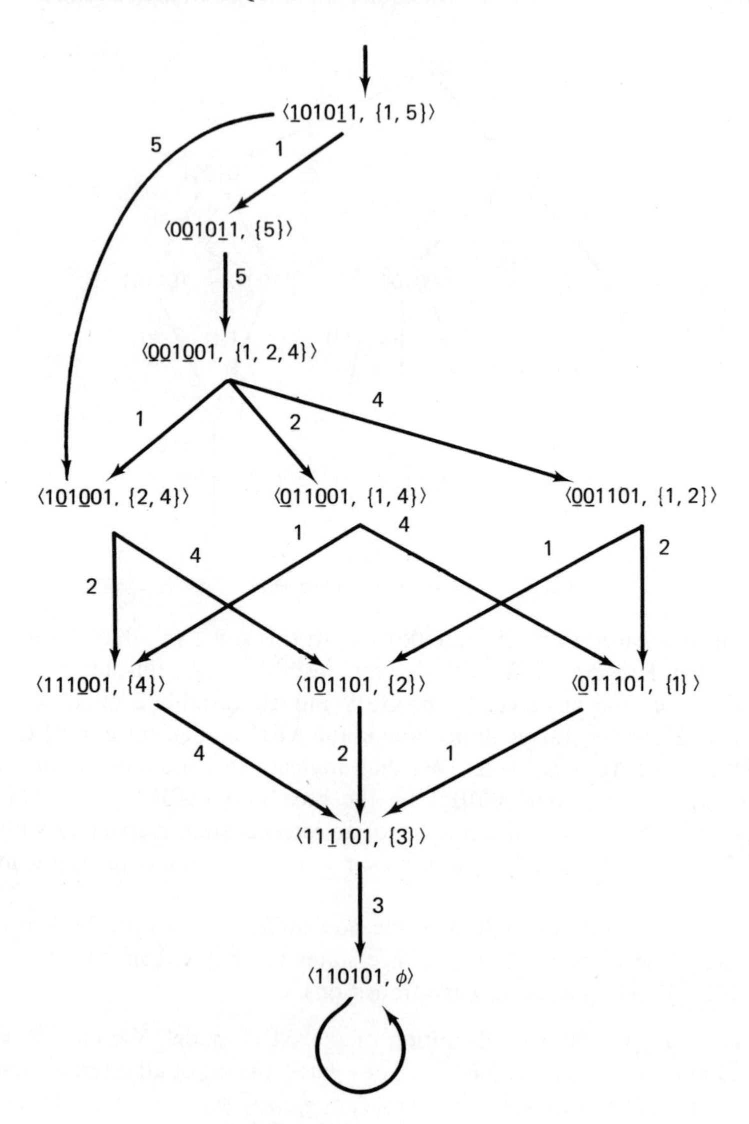

Fig. 6-26 AED analysis of a race.

Now define the set V of *race states* by

$$V \triangleq Q \times P(S) = \{0, 1\}^s \times P(\{1, \ldots, s\}).$$

We shall be interested in a subset T of V in the following discussion of races.

Definition 1

Given a network N with n inputs and s gates, we call the total state $(x; \hat{y})$ of N *primary* iff there exists an input $\bar{x}$ such that $(\bar{x}; \hat{y})$ is stable.

We are interested only in primary total states initially, because we want to avoid total states which are not reachable during the normal operation of the network. As will be seen, a total state does not have to be primary in order to be reachable, and other reachable total states will be introduced later. A primary total state simply represents an appropriate starting point. Suppose that the network has been stable in total state $(\bar{x}; \hat{y})$, and the input changes to x. The network enters the total state $(x; \hat{y})$, and we are interested in finding out what happens next. We will assume that the input remains fixed at x for a long time. We now define the set $T(x; \hat{y}) \subseteq V$ for a given primary total state $(x; \hat{y})$. Since x is assumed to be fixed and $\hat{y}$ is viewed as the initial internal state, we will write simply T for $T(x; \hat{y})$. T is to represent all race states reachable from $(x; \hat{y})$.

It turns out to be convenient to define T inductively and also to define a relation R on T at the same time. In our example, Fig. 6-26 is the relation diagram of R.

Definition 2

Let $(x; \hat{y})$ be primary. Define T and R as follows.
Basis: $\langle \hat{y}, u(\hat{y}) \rangle \in T$.
Induction step: Given $\langle y, v \rangle \in T$.

1. If $v = \varnothing$, then $\langle y, v \rangle \, R \, \langle y, v \rangle$.
2. If $v \neq \varnothing$, for each $i \in v$, compute

$$w_i \triangleq (v - \{i\}) \cap u(y^{(i)})$$

(a) If $w_i = \varnothing$, then

$$\langle y^{(i)}, u(y^{(i)}) \rangle \in T$$

and

$$\langle y, v \rangle \, R \, \langle y^{(i)}, u(y^{(i)}) \rangle.$$

(b) If $w_i \neq \varnothing$, then

$$\langle y^{(i)}, w_i \rangle \in T$$

and

$$\langle y, v \rangle \, R \, \langle y^{(i)}, w_i \rangle.$$

In the example shown in Fig. 6-26 we have $\hat{y} = \underline{1}010\underline{1}1$ and $u(\hat{y}) = \{1, 5\}$. Thus $\langle 101011, \{1, 5\} \rangle \in T$. Now $y^{(1)} = 0\underline{0}10\underline{1}1$, $u(y^{(1)}) = \{2, 5\}$ and $w_1 = (\{1, 5\} - \{1\}) \cap \{2, 5\} = \{5\}$. Therefore, $\langle 001011, \{5\} \rangle \in T$ and $\langle 101011, \{1, 5\} \rangle \, R \, \langle 001011, \{5\} \rangle$. Next $y^{(5)} = 1\underline{0}1\underline{0}01$, $u(y^{(5)}) = \{2, 4\}$, and $w_5 = \varnothing$. Thus $\langle 101001, \{2, 4\} \rangle \in T$ and $\langle 101011, \{1, 5\} \rangle \, R \, \langle 101001, \{2, 4\} \rangle$. If we continue in this manner we generate the relation R whose relation diagram is shown in Fig. 26.

As mentioned before, the model is based on the idea that all the gates have approximately equal delays ($\approx\Delta$). Figure 6-27 shows a timing diagram for part of the race of Fig. 6-26. The path we follow in the diagram of Fig. 6-26 is chosen as 1, 5, 1, 2, 4, 3. This transition takes approximately 3Δ units of time. It is useful to introduce the concept of *race units*. In the example of Fig. 6-27, we have three race units: $\{1, 5\}$, $\{1, 2, 4\}$, and $\{3\}$. A race unit starts either in the initial state $\hat{y}$ if that state is unstable, or when a previous race unit has finished.

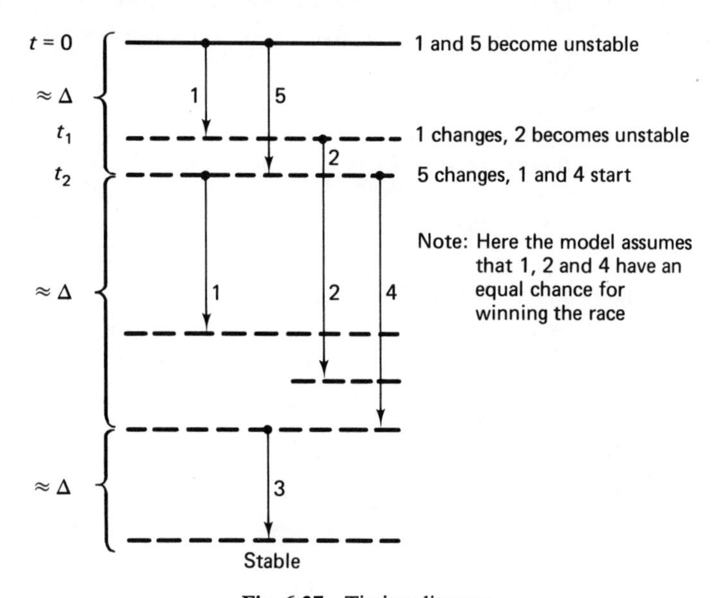

Fig. 6-27 Timing diagram.

If Δ_{max} (Δ_{min}) represents the maximum (minimum) value of gate delay, then it is clear that the duration of a race unit corresponds to at worst Δ_{max} units of time. Actually, a variable that takes part in a race unit can become unstable just before that race unit starts. This is the case of variable 2 in Fig. 6-27 for the race unit $\langle 001001, \{1, 2, 4\}\rangle$. However, the longest time for a variable to wait before the corresponding race unit starts is $\Delta_{max} - \Delta_{min}$. It seems unreasonable to assume that 2 will necessarily have priority over 1 and 4 just because it starts a "little" ($\leq(\Delta_{max} - \Delta_{min})$) earlier.

We point out that the concept of "internal state" is to be replaced by "race state" in the AED model, for it is not enough to know y ($\in \{0, 1\}^s$) in order to determine the subsequent behavior. We illustrate this by means of the following excitation equations:

$$Y_1 = xy_3, \quad Y_2 = xy_4, \quad Y_3 = (y_1 + y_4)', \quad Y_4 = (y_2 + y_3)'.$$

It is easy to verify that when $x = 0$, $y = 0001$ is a stable internal state. The relation diagram of R for the AED model for the total state (1; 0001) is shown in Fig. 6-28. Here the internal state 0000 appears twice. Of course, in both cases the unstable variables are the same. However, in the first instance, 3 must win the race, whereas in the second, variable 4 has priority.

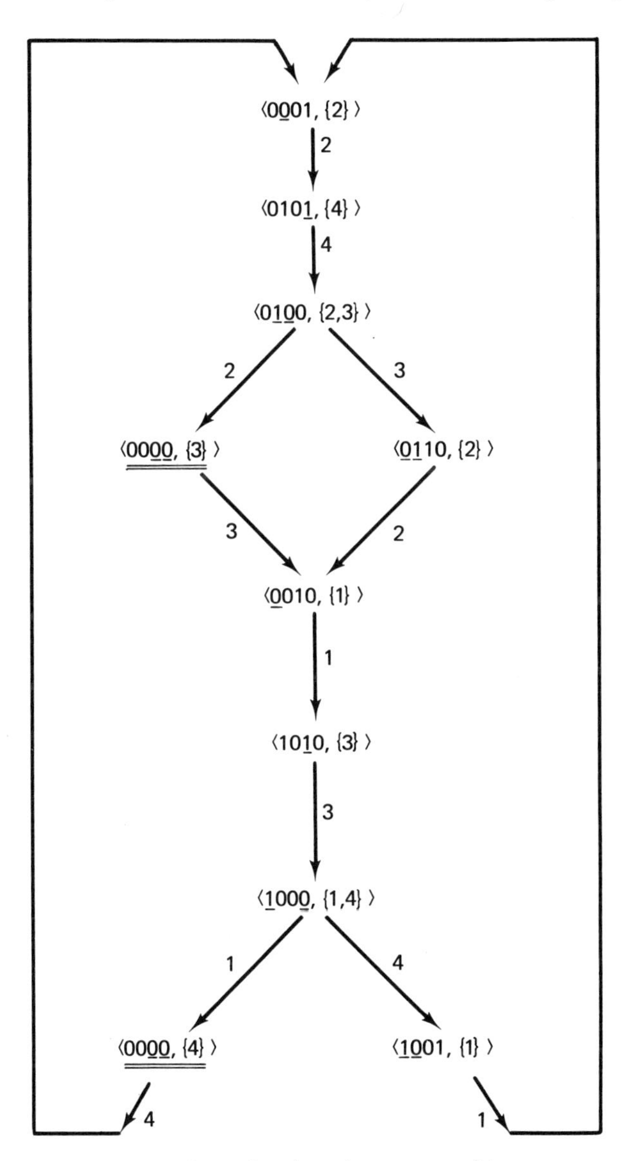

Fig. 6-28 Illustrating dependence on race history.

6.6. SUMMARIZING COMMENTS

We have seen that a number of interesting and complex phenomena may occur in sequential networks, even if we consider simplified and idealized models. One could easily postulate more complicated models which would provide a more accurate reflection of reality. For example, one could admit "multiple winners" or associate a delay with each line connecting a gate output to a gate input, or take into account the fact that the propagation delay of a gate while changing from 0 to 1 is different from the corresponding delay from 1 to 0. However, such models quickly become unmanageable. For this reason we compromise and accept a less accurate but relatively simple model. As we shall see in Chapter 7, this model is detailed enough to explain a large number of fundamental concepts.

In general, the networks that we will consider do not have oscillations. *We therefore assume that oscillations do not occur* in the networks to be analyzed. This leads to a simpler model, which we summarize below.

Starting with a logic diagram of a network, we want to obtain the flow table as the result of our analysis. The flow table will have rows corresponding only to those internal states that are stable for at least one input combination. To find these stable states, we may use the excitation table if the network is small and shortcut techniques if it is large.

Next we must examine the unstable entries. In view of our assumption that oscillations do not occur, each unstable entry can lead either to one stable state or to several stable states. In the latter case we have a critical race. We may first examine the critical races in the GSW model. In some applications, a critical race may be quite acceptable, as we shall see in Chapter 7. In any case, a critical race should be viewed as a warning that the network behavior may not be well defined. On the other hand, we can use the AED model to check whether the GSW model is not too pessimistic. It may be that certain critical races can occur only under very unusual delay distributions, and we may prefer to interpret the behavior as well defined, if certain possibilities are very unlikely. In practice a designer must be very well acquainted with the devices he is using, in order to decide which assumptions are valid.

Until now, we have not been stressing the concept of external outputs because we have tacitly assumed that all gate outputs are external outputs. If this is the case, the internal state contains both the state and output information. However, in many cases, only a few of the gate outputs are of interest. Then it is often preferable to suppress the details of the actual internal state coding and label each state by a suitable symbol, e.g., a, b, c, etc. However, now the outputs are of great importance. In a state table we normally list the values of all the outputs corresponding to stable states. Transient outputs are not considered.

We can now make precise the distinction between sequential and combinational behavior. We say that a column of a state table is a *combinational column* iff the same output word is associated with each stable state in that column. Otherwise, there must be at least two stable states with different output words in that column. In the latter case we call it a *sequential column*, since the output corresponding to the input word x of the column is not uniquely determined by x alone, but depends on the past sequence of inputs. A network is combinational iff all the columns in its state table are combinational. Otherwise, it is sequential. This definition can be modified to allow for oscillations, but we shall not need such general concepts here.

We close by illustrating these ideas with examples. The network of Fig. 6-29(a) is combinational. In fact, $z = x'$ under stable conditions. The network of Fig. 6-29(b) is sequential, although the $x = 1$ column of the state table is combinational.

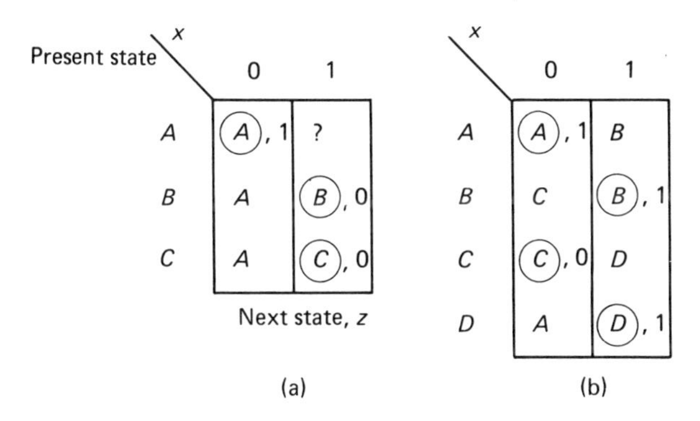

Fig. 6-29 (a) A combinational network; (b) a sequential network.

Further illustrations of the notions introduced above will be found in Chapter 7, where a number of practical networks are analyzed. See also [BR-YO2] for a further discussion of sequential network models.

PROBLEMS

1. Show that the network of Fig. P6-1 is combinational and find z as a function of x_1 and x_2.

2. [KAU] Analyze the network of Fig. P6-2. The inputs are x_1, x_2, x_3, the outputs z_1, z_2, z_3. Is the network combinational or sequential?

3. For the network of Fig. P6-3, find the flow table.
   (a) Use the GSW model.
   (b) Use the AED model.

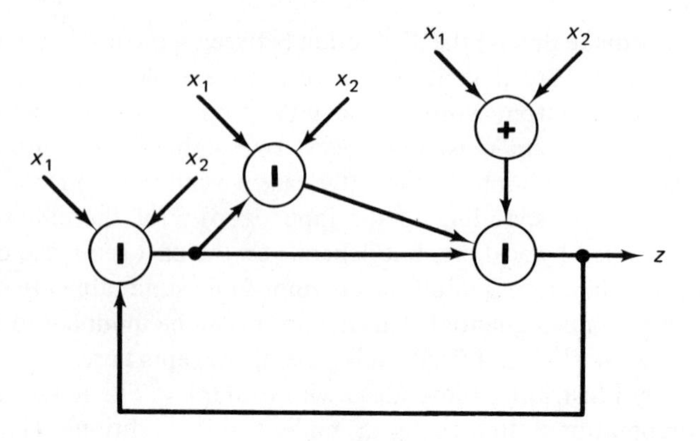

Figure P6-1

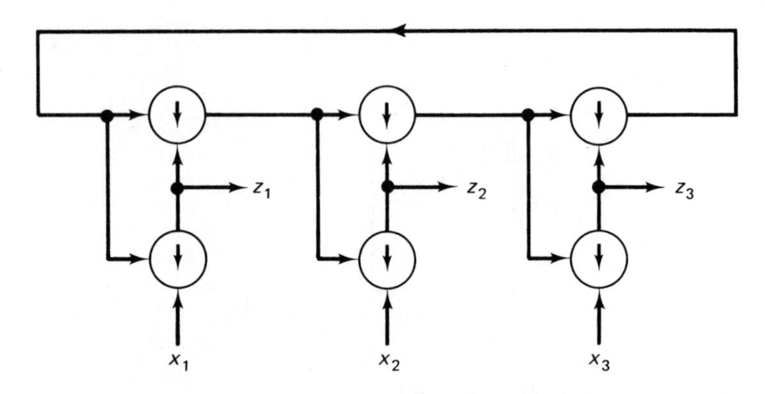

Figure P6-2

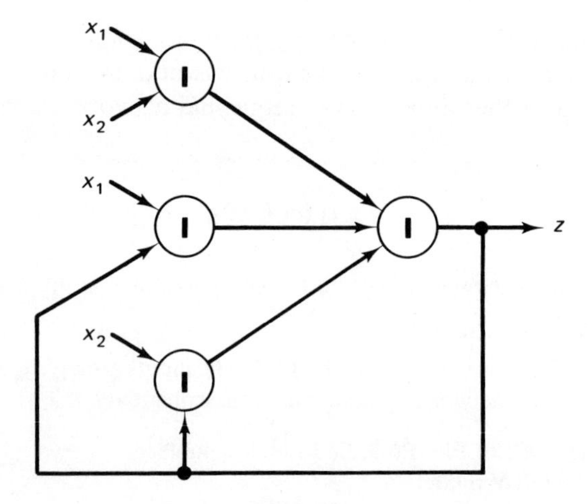

Figure P6-3

4. Repeat Problem 3 for the network of Fig. P6-4.

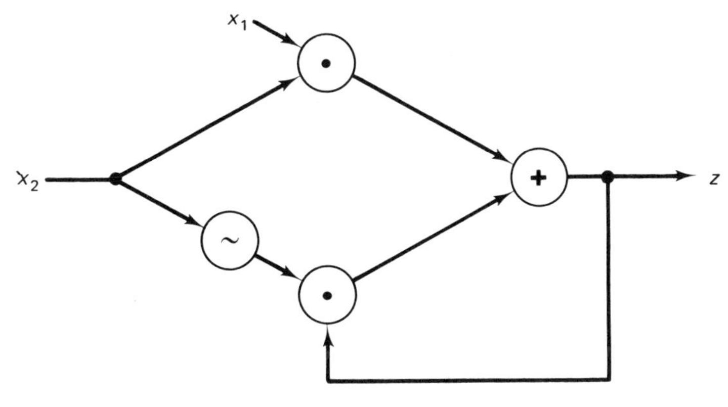

Figure P6-4

5. (a) Find the flow table for the network of Fig. P6-5.
(b) Repeat part (a) assuming that y_1 is an ideal delay-free gate. Compare with part (a).
(c) Repeat part (a) assuming that y_2 is delay-free. Compare with parts (a) and (b).

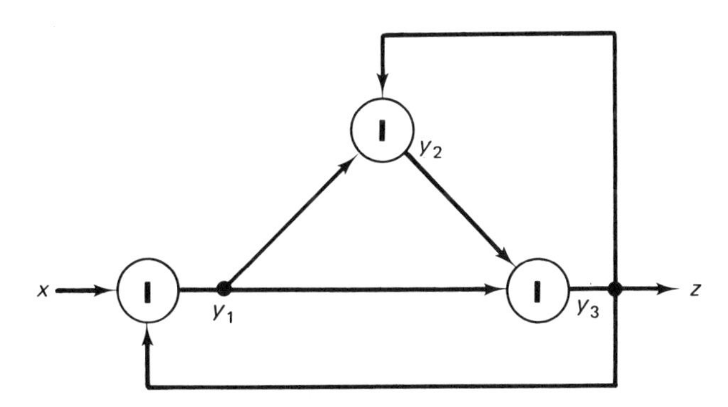

Figure P6-5

6. A network consisting of 8 NAND gates has the following excitation equations: $Y_1 = y_8'$, $Y_2 = 1$, $Y_3 = (y_1 y_4)'$, $Y_4 = (y_2 y_3)'$, $Y_5 = (y_1 y_3)'$, $Y_6 = (y_2 y_4)'$, $Y_7 = (y_5 y_8)'$, and $Y_8 = (y_6 y_7)'$. Find the entry in the flow table corresponding to the initial state $(y_1, y_2, \ldots, y_8) = (10011110)$.
(a) Use the GSW model.
(b) Use the AED model.

7 LATCHES AND FLIP-FLOPS

ABOUT THIS CHAPTER

In the previous chapter we developed sufficient techniques to tackle the analysis of practical, commercially used memory modules. We start with simple latches and one-latch flip-flops. We explain how they perform the function of memory. Next we point out their shortcomings and, to overcome them, introduce the master–slave principle. Edge-sensitive networks follow; rather sophisticated memory devices utilize this principle. Finally, we describe some commercial IC flip-flop packages.

The material of this chapter is fundamental, in that the understanding of flip-flops is assumed in Chapter 8, where they are treated as simple building blocks of more complex networks such as shift registers and counters. On the other hand, certain parts can be omitted. For example, the reader may omit the detailed analysis of complex flip-flops, such as those of Figs. 7-28 and 7-32, and accept the flow table in much the same way as one sometimes omits a proof and accepts a theorem. Also, although practically important, the material on edge-sensitive networks (Section 7.4) is not required in what follows; it is sufficient to understand the master–slave principle.

To the best of our knowledge, the systematic mathematical analysis of commercial IC flip-flops is original. These flip-flops are often sophisticated multistate gate networks, and we treat them as such. In contrast to this, conventional texts treat them very unrealistically as two-state black boxes.

It is worthwhile noting that we explain edge-sensitive behavior of flip-flops by flow tables derived from logic diagrams using our formal model. Whereas the flow-table method involves analysis at the gate level, the functional approach (pp. 225–29) considers higher-level analysis.

So far there are no general techniques for designing flip-flops with a minimal number of gates. The functional approach appears likely to lead to more systematic design methods.

7.1. NOR AND NAND LATCHES

The network of Fig. 6-8 is repeated in Fig. 7-1(a), where the delays are not shown and the input and output names are changed to correspond more closely to common usage ($C = x_2$, $P = x_1$, $Q = y_2$, $\bar{Q} = y_1$). Figure 7-1(b) shows a simplified symbol for the network. (It is customary to graphically

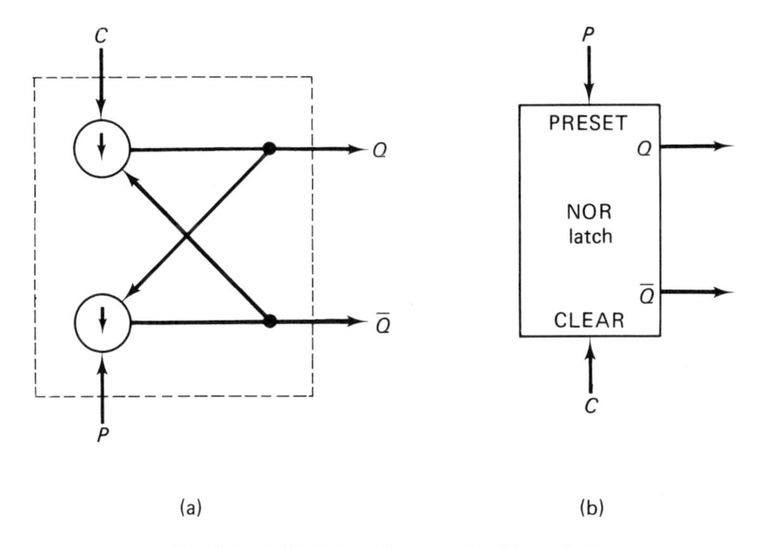

Fig. 7-1 (a) NOR latch network; (b) symbol.

associate the PRESET input with Q and the CLEAR input with $\bar{Q}$; hence the locations of P and C are interchanged.) For convenience, the flow table is repeated with the new symbols in Fig. 7-2(a). (Note that the inputs have been permuted.) Often the input combination $C = P = 1$ is not used. Under this assumption the state table of Fig. 6-13 reduces to that of Fig. 7-2(b). We can

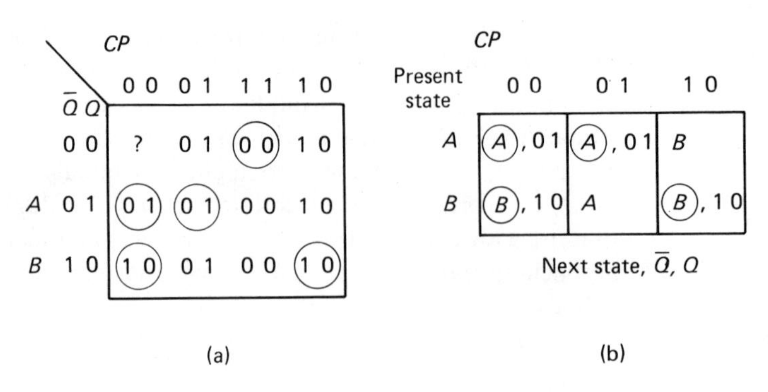

Fig. 7-2 (a) Flow table; (b) restricted state table.

now give a very simple word description of the state table of Fig. 7-2(b) and an explanation for the symbols used.

The NOR latch can be viewed as a binary storage cell, i.e., a memory device capable of storing one bit. We arbitrarily choose one of the outputs, Q (the circuit is symmetric), to represent the state of the memory cell. The input condition $C = P = 0$ is the passive condition where no change takes place and the latch simply remembers its state indefinitely. If a C input is applied, i.e., $C = 1$ (implying $P = 0$ since we agreed not to use the condition $C = P = 1$), Q becomes 0. We say that the cell is *cleared* and C is the CLEAR input. Similarly, if a P input is applied ($P = 1$, $C = 0$), Q becomes 1, the cell is *preset*, and P is the PRESET input. During the input condition $C = P = 0$, the latch remembers which input was applied last. If C was last, the state is B, and $Q = 0$; if P was last, the state is A, and $Q = 1$. Figure 7-3 illustrates the behavior of the latch under the assumption $CP = 0$.

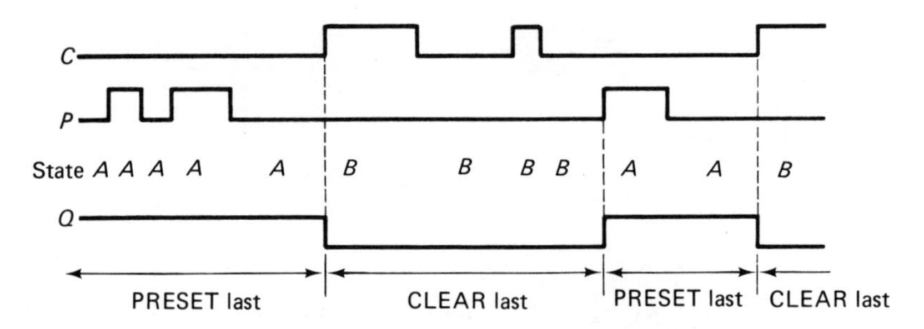

Fig. 7-3 Typical signals for the NOR latch.

Note that, under the restriction $CP = 0$, we have $\bar{Q} = Q'$. Thus Q represents the bit stored in the cell, and $\bar{Q}$ is its complemented version. For this reason, the $\bar{Q}$ output is sometimes not available externally, since it is not an independent output when $CP = 0$.

Sometimes the P input is called the *set* input, C is called the *reset* input, and the NOR latch is called an *RS flip-flop*, although we will use this name for a somewhat different network.

A similar memory device can be constructed with NAND gates as shown in Fig. 7-4. We leave to the reader the detailed analysis, which is similar to

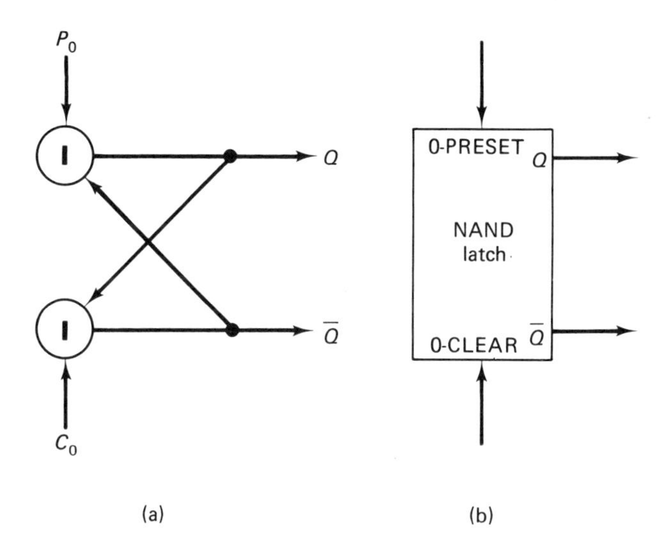

Fig. 7-4 (a) NAND latch; (b) symbol.

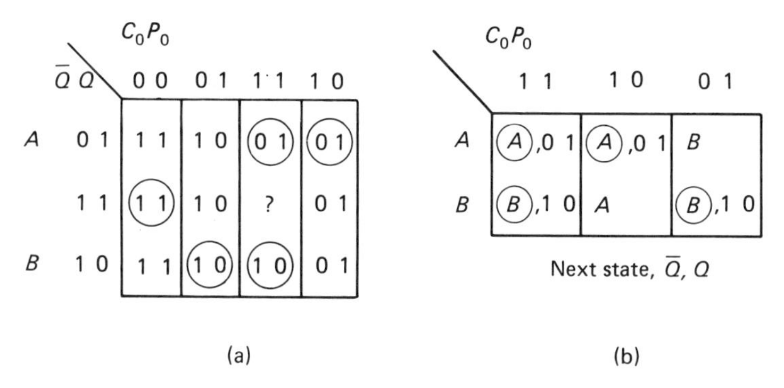

Fig. 7-5 NAND latch tables: (a) flow table; (b) restricted state table.

that for the NOR latch. In Fig. 7-5(a) we give the flow table and, in Fig. 7-5 (b), the state table restricted to the condition $C_0 + P_0 = 1$; i.e., $C_0 = P_0 = 0$ is not used. The following shortcut analysis provides another way of deriving the flow table. If $P_0 = 0$, then Q is forced to 1 regardless of the previous state. Such an input is called 0-PRESET. If also $C_0 = 1$, $\bar{Q}$ will be forced to 0. This accounts for the $C_0 = 1, P_0 = 0$ column of the transition table. By symmetry, $C_0 = 0, P_0 = 1$ yields $Q = 0, \bar{Q} = 1$ as the final stable state. The C_0 input is called 0-CLEAR. The condition $C_0 = P_0 = 0$ forces the outputs of both gates to become 1. Finally, suppose $C_0 = P_0 = 1$. Both NAND gates behave like inverters. Clearly, the only stable states possible are $Q = 0$, $\bar{Q} = 1$ and $Q = 1, \bar{Q} = 0$. It is easily verified that the network has a critical race in the unstable state corresponding to $C_0 = P_0 = 1$.

(We will often use such analysis, specialized for a given circuit, because the excitation table of the previous chapter can become very large.)

Note that the NAND latch operates exactly like the NOR latch, if the "passive" input state is taken as $C_0 = P_0 = 1$, and presetting and clearing occurs when the corresponding input goes to 0 (rather than 1 as in the NOR latch).

7.2. SIMPLE ONE-LATCH FLIP-FLOPS

We have seen that a latch has the ability to store a bit of information. It can be made more flexible, if the storing of information takes place under the control of another input, which we shall call ¢. If ¢ $= 0$, the latch remembers the last state; if ¢ $= 1$, new information is entered. These ideas are developed in this section.

RS Flip-Flop

Consider the network of Fig. 7-6(a). It is like a NOR latch except that there are two external inputs for each gate. Since

$$Y_1 = (R + C + \bar{Q})' = ((R + C) + \bar{Q})'$$
$$Y_2 = ((S + P) + Q)',$$

the network is logically equivalent to that of Fig. 7-6(b), where the OR gates are delay-free. In other words, this network acts like a NOR latch, with CLEAR input $(C + R)$ and PRESET input $(S + P)$, as shown in Fig. 7-6(c).

Suppose now that we *construct* the network of Fig. 7-6(c) by using one NOR latch and two *actual* OR gates. In our model each gate has some propagation delay. However, we now show that the delays associated with the OR gates can be essentially ignored and that we do not need to carry out a lengthy analysis with 4 internal state variables.

Suppose $R + C = 0$, $S + P = 1$, and that this input combination is held constant. After some delay we have CLEAR $= 0$, PRESET $= 1$. This condition forces the latch to the state $Q = 1, \bar{Q} = 0$. Similarly, $R + C = 1$, $S + P = 0$ permits only one stable total state of the whole network with $Q = 0, \bar{Q} = 1$, and $R + C = S + P = 1$ results in $Q = \bar{Q} = 0$.

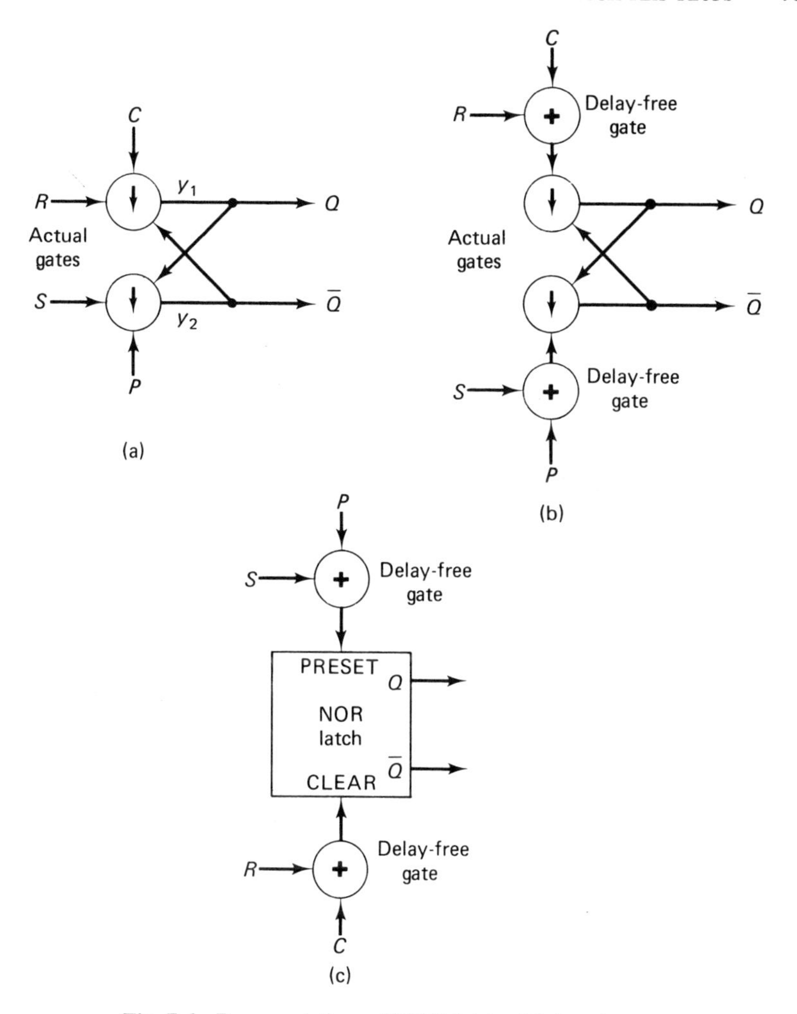

Fig. 7-6 Representations of NOR latch with four inputs.

Suppose now that we start in the stable total state $R + C = 0$, $S + P = 1$, $Q = 1$, $\bar{Q} = 0$. If S becomes 0, there will be no change if $P = 1$. However, if $P = 0$, the excitation of the bottom OR gate [Fig. 7-6(b)] will become 0. After the OR-gate delay the PRESET signal becomes 0. Clearly, the latch remains in the state $Q = 1$, $\bar{Q} = 0$.

Such arguments show that the delays present in the input OR gates affect the operation of the latch only to the extent that the latch will respond a little later. Otherwise, they can be ignored.

In Fig. 7-7(a) we show a network of an "*RS* flip-flop" similar to that of Fig. 7-6(c). The ¢ input can be thought of as a control input for the inputs R and S. When $C = P = 1$, we have CLEAR = PRESET = 1 and $Q = \bar{Q}$

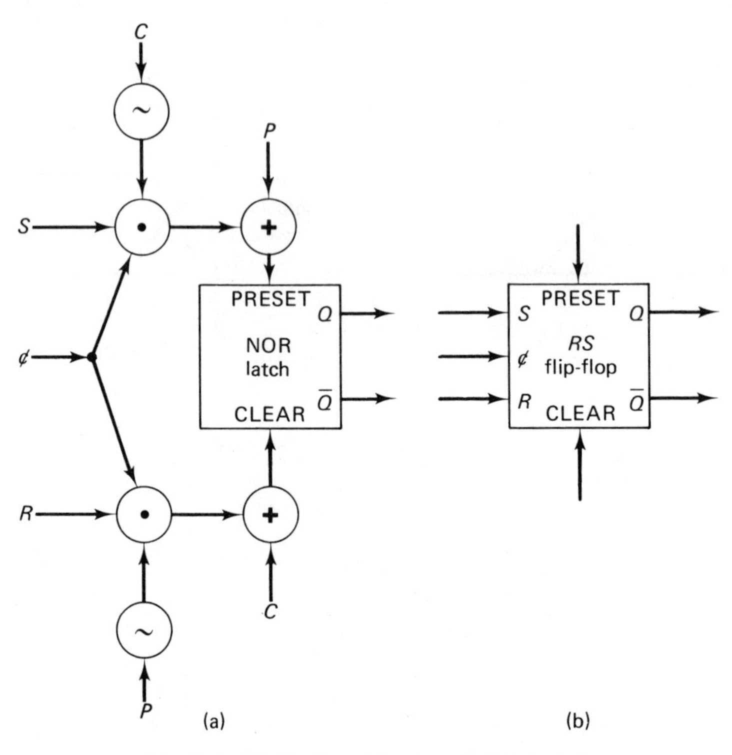

Fig. 7-7 RS flip-flop: (a) network; (b) symbol.

$= 0$. When $C = 1, P = 0$, we have CLEAR $= 1$ because of the bottom OR gate. The top AND gate has output 0 because of $C' = 0$; hence PRESET $=$ 0. Similarly, $C = 0$, $P = 1$ implies CLEAR $= 0$, PRESET $= 1$. Thus for any input condition satisfying $C + P = 1$, the network of Fig. 7-7 acts like a simple NOR latch with CLEAR $= C$, PRESET $= P$. $C + P = 1$ implies C and P are the inputs in control, overriding the remaining 3 inputs.

When the C and P inputs are inactive (i.e., $C = P = 0$), we have

$$\text{CLEAR} = R¢$$
$$\text{PRESET} = S¢.$$

Thus if $¢ = 0$, the inputs R and S are disabled. If $¢ = 1$, the *set* input S performs the preset function and the *reset* input R performs the clear function.

The reader may find this terminology somewhat confusing. We introduce it here because it is in common use. Usually, the overriding inputs are called *PRESET* and *CLEAR*, the input ¢ is called the *clock* input, and the inputs controlled by the clock are called the *Reset* (corresponding to CLEAR) and *Set* (corresponding to PRESET). We have purposely avoided the term "clock input" since the ¢ terminal is just an input terminal and one can apply any signal to it. For example, Fig. 7-8 illustrates a possible operation of the flip-flop. To simplify such diagrams, delays in the output waveforms caused by gate-propagation delays are not shown, as usual.

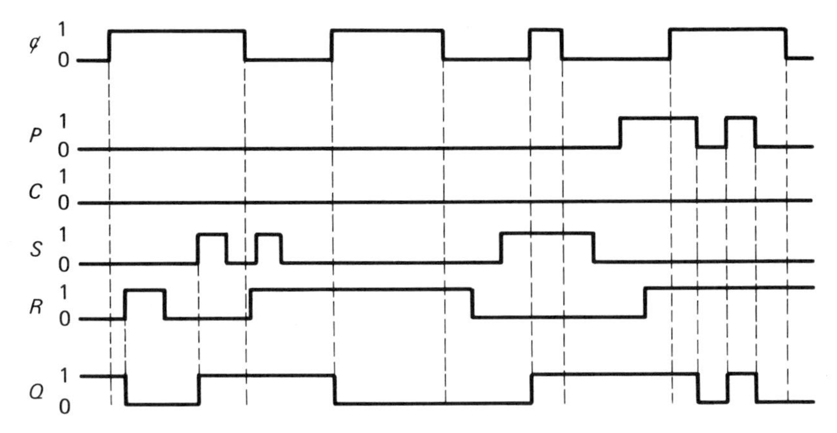

Fig. 7-8 Signals for RS flip-flop.

A regular waveform consisting of a sequence of equally spaced pulses of fixed duration is usually called a *clock*. Suppose we assume that the ¢ input is a clock, as shown in Fig. 7-9. The C and P inputs are still the overriding inputs unaffected by the clock. For this reason, they are called *asynchronous inputs*. If these inputs are "not used," i.e., $C = P = 0$, and, furthermore, if we agree not to change the R and S inputs when the clock is present ($¢ = 1$) and avoid $¢ = R = S = 1$, we get the behavior illustrated in Fig. 7-9. In effect, R and S look like "asynchronous" signals in the sense that they can be

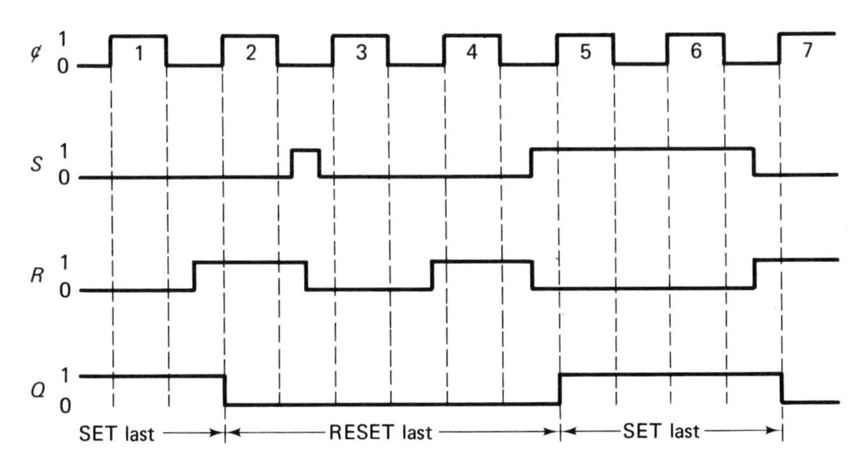

Fig. 7-9 Synchronous behavior of RS flip-flop.

constant for any period of time and can change at any time, *except during a clock pulse*. However, both R and S are connected to AND gates controlled by the clock. Hence the latch "sees" only the values of R and S during the clock pulse, and these are assumed to be constant. Thus the clock "samples" the R and S signals. For these reasons, R and S are called *synchronous* inputs. These distinctions will become more evident later when typical applications are discussed. Incidentally, it is the "synchronous" inputs that usually determine the name of the flip-flop type—in this case an *RS* (or *SR*) flip-flop.

Note what happens to the *RS* flip-flop if both R and S are 1 during a clock pulse. When the clock becomes 1, the latch goes to the stable state CLEAR = PRESET = 1, $\bar{Q} = Q = 0$ [see Fig. 7-2(a)]. When the clock signal becomes 0, the network finds itself in total state CLEAR = PRESET = 0, $\bar{Q} = Q = 0$. This, however, is a critical race, as we know. Therefore, the operation of an *RS* flip-flop is not reliable under these conditions, since the next state cannot be predicted. For this reason, we have introduced the restricted state table of Fig. 7-2(b).

A very simple model is available for the synchronous operations of an *RS* flip-flop. Assume that $C = P = 0$ (i.e., the asynchronous inputs are inactive) and that $R = S = 1$ is not allowed when ¢ = 1. When the clock is absent, no change can take place. When it arrives, it permits the latch to look at the R and S input values. This possibly produces a change of state, and then the clock is removed.

The table of Fig. 7-10 summarizes this. Let Q^n denote the value of Q just before the nth clock pulse and similarly define R^n and S^n. The *next state*, Q^{n+1}, just before the $(n + 1)$st clock pulse is shown as the entry in the table. More will be said about this type of state table later.

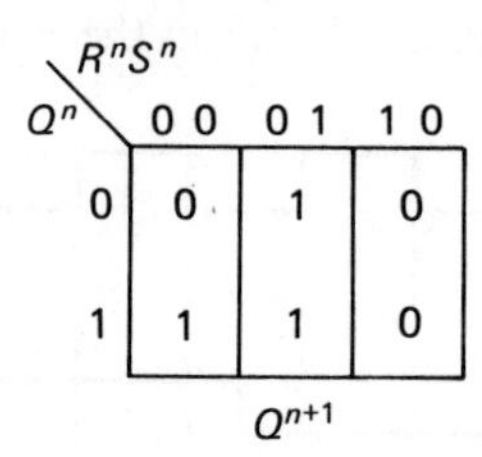

Fig. 7-10 Synchronous state table for RS flip-flop.

D Flip-Flop

Another simple latch with input logic, the "*D* flip-flop" (where D is sometimes called the *data input*), shown in Fig. 7-11. Notice that

$$\text{PRESET} = D¢$$

$$\text{CLEAR} = D'¢.$$

Thus PRESET and CLEAR are never both 1 in any stable state, and hence $\bar{Q} = Q'$. Note also that this network is equivalent in behavior to that of

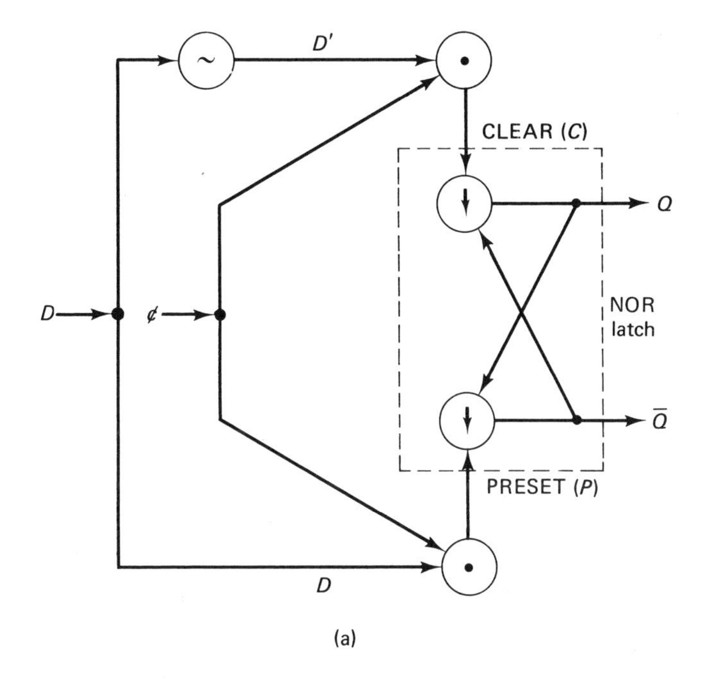

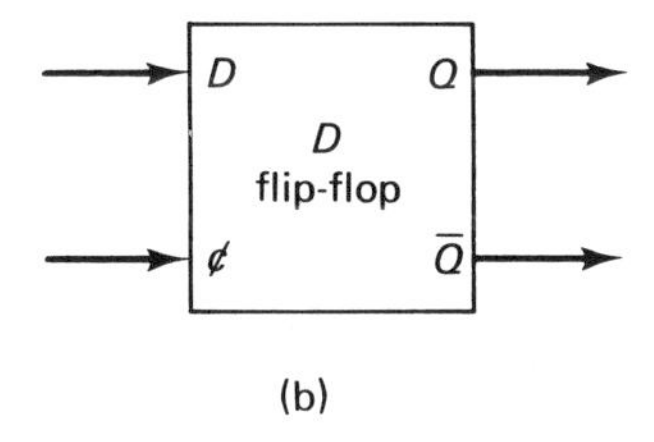

Fig. 7-11 (a) Basic D flip-flop; (b) symbol.

Fig. 7-7, if we set $P = C = 0$ and $S = D$, $R = D'$. Because of the delay of the inverter it is possible for both D and D' to be 1 during transient conditions. When $¢ = 1$ this might cause $Q = \bar{Q} = 0$ temporarily. However, we neglect such transient outputs here; more is said about this in Chapter 10.

A simplified flow table of the D flip-flop is shown in Fig. 7-12(a) and the synchronous state table in Fig. 7-12(b). Figure 7-13 illustrates the behavior of the D flip-flop. In the absence of the clock, D has no effect. During a clock pulse, Q follows D. After the clock falls, Q remembers the last value of D present when the clock was 1.

In the synchronous mode, assuming that D does not change during a clock pulse, the Q signal just before the $(n + 1)$st clock pulse is equal to the

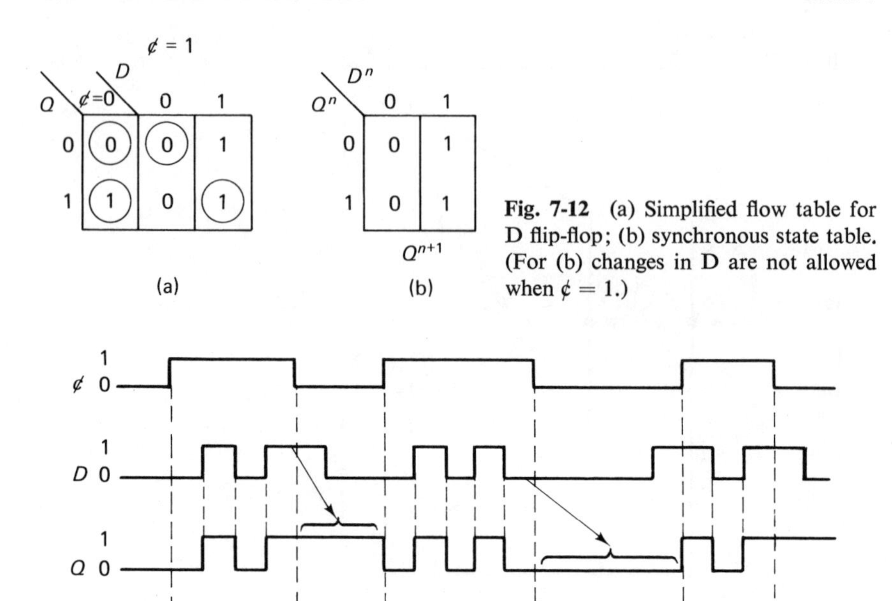

Fig. 7-12 (a) Simplified flow table for D flip-flop; (b) synchronous state table. (For (b) changes in D are not allowed when $¢ = 1$.)

Fig. 7-13 Waveforms for D flip-flop.

D signal just before the *n*th clock pulse. If the interval between clock pulses is considered to be a unit of time, the *D* flip-flop can be viewed as a *unit delay*.

The type of network described above is available in various packages [SIG, TI2]. For example, the TTL package 7475 is a 16-pin package consisting of 4 *D* flip-flops. Two clock terminals are supplied, and each is shared by two flip-flops. This package is called a "4-bit latch". An "8-bit latch" is available in package 74100. This 24-pin package contains 8 flip-flops, with *Q* as the single output from each flip-flop. Each of the two clock inputs is common to 4 flip-flops. A "4-bit latch" may be used to hold one BCD digit, for example.

We should point out that the terminology is not standardized and "latch" and "flip-flop" are often used interchangeably. In this book we adopt the convention that memory cells without a clock input are called latches and those with a clock input are flip-flops.

The *D* flip-flop is perhaps more natural than the *RS* flip-flop, when viewed as a memory cell. If we have one bit *D* of data to be stored, then a cell with a single input is natural. However, the latch is inherently a two-input device. Also, it is convenient to have three possibilities for data representation. For the NOR latch, $S = 0$, $R = 0$ represents "no data to be stored"; $S = 0$, $R = 1$ represents "write a 0"; and $S = 1$, $R = 0$ represents "write a 1." The fourth possibility is not used. From this point of view the two-input *RS* flip-flop is also quite natural. Notice also that, once a bit is stored in a latch, it is available both in its true form (Q output) and in its complemented form ($\bar{Q}$ output). When this information is used as an input to another memory cell, the two-line representation of a bit is quite convenient.

7.3. MASTER–SLAVE FLIP-FLOPS

Transfer Between Latches

It is usually not enough to be able to store information, but it is also important to be able to transfer it from one memory cell to another. We now consider this problem.

Figure 7-14(a) shows two NAND latches. Under the control of input x_1, we wish to transfer the contents of L_1 into L_2. We assume that L_1 can only be in the two states in which $\bar{Q}_1 = Q'_1$. A 1-pulse on the x_1 input is to cause Q_2 to become equal to Q_1. The requirements can be specified by the table:

Q_1	x_1	$\bar{P}_2$	$\bar{C}_2$
–	0	1	1
0	1	1	0
1	1	0	1

Hence we find

$$\bar{P}_2 = x'_1 + Q'_1 = (x_1 Q_1)'$$
$$\bar{C}_2 = x'_1 + Q_1 = x'_1 + \bar{Q}'_1 = (x_1 \bar{Q}_1)'.$$

This results in the design of Fig. 7-14(b). There are no difficulties here, but a slight variation of this problem does lead to difficulties, as described next.

For this problem, consider Fig. 7-14(c). Here when $x_1 = 1$ we would like to enter the bit from some external latch L_0 into L_1 and, at the same time, shift the old content of L_1 into L_2. Such networks are called *shift registers* and are described in detail in Chapter 8. One can easily verify that the proposed solution of Fig. 7-14(c) will not work, because if $x_1 = 1$ for a long enough period, both Q_1 and Q_2 will agree with Q_0; i.e., the old Q_1 information is lost instead of being stored in Q_2.

One could consider a solution in which the 1-pulse on x_1 is very narrow and is removed before the new value in L_1 has a chance to change Q_2. Such solutions require critical timing and are often unsatisfactory.

There is a solution that works, if we are willing to modify the problem statement. Suppose that we agree to perform the shift in two steps, by using two control inputs x_1 and x_2 as in Fig. 7-14(d). In the passive state, both x_1 and x_2 are 0. Both inputs to each latch are 1, ensuring that no change takes place in the state of the latches. Next apply a 1-pulse on x_2, keeping $x_1 = 0$.

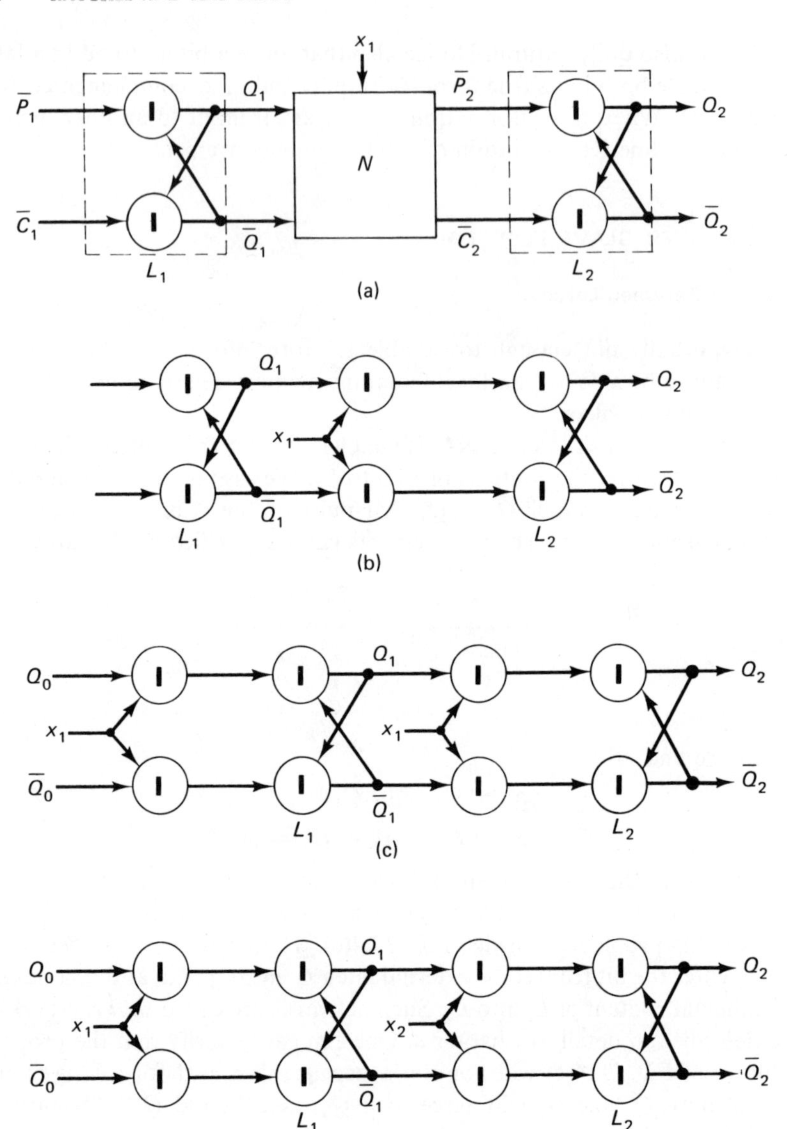

Figure 7-14

This causes Q_2 to agree with Q_1. Next, when x_2 becomes 0, L_2 is effectively isolated. We can now apply a 1-pulse on x_1, causing Q_1 to agree with Q_0. Note that the two pulses can be very wide, as long as they don't overlap. We will return to this solution later.

A third and commonly used solution is to use a more complex implementation of a memory cell as shown in Fig. 7-15, where (for purposes of illus-

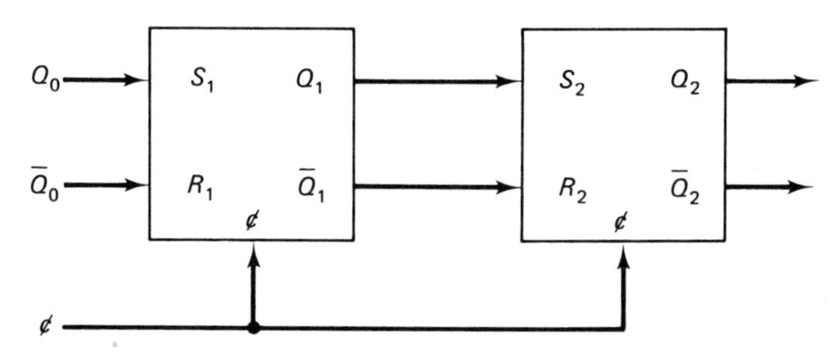

Fig. 7-15 Shift register arrangement.

tration) we have chosen an *RS* type of flip-flop. The reader can verify that the simple *RS* flip-flop of Fig. 7-7 cannot be used in Fig. 7-15 because we then have the same problem as in the network of Fig. 7-14(c). There are two quite different principles which can be used to make the arrangement of Fig. 7-15 possible. These are the master–slave principle and the edge-sensitive principle, which we describe later.

Master–Slave *RS* Flip-Flop

The network of Fig. 7-14(d) is redrawn in Fig. 7-16, where we have set $x_1 = ¢$ and $x_2 = ¢'$. This two-latch network is used to store one bit in the Q

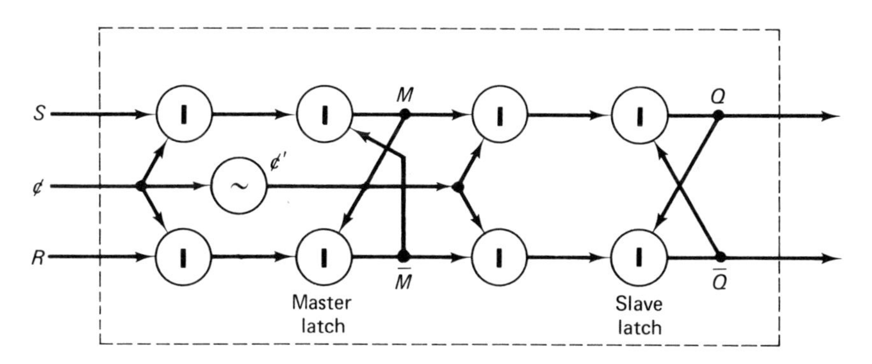

Fig. 7-16 M-S RS flip-flop.

output. When ¢ becomes 1, $¢' = 0$, causing the inputs to the slave latch to be both 1 and isolating the slave from the master. Thus the old value stored in the flip-flop remains in the slave as long as $¢ = 1$. At the same time, the *RS* information is entered into the master. When the clock returns to 0, the master becomes isolated from the *R* and *S* inputs and the slave gets connected to the master, causing Q to follow M. Now it is clear that the arrangement of Fig. 7-15 will work if master–slave *RS* flip-flops are used. The two-step transfer is shown in Fig. 7-17.

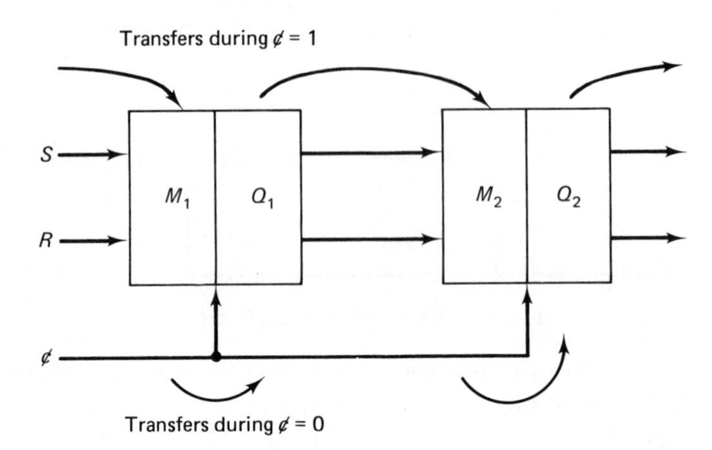

Fig. 7-17 M-S operation.

Because of the inverter in Fig. 7-16, it is possible for ¢ and ¢′ to be 1 at the same time for a short period following the $0 \longrightarrow 1$ change in ¢. This means that neither the master latch nor the slave latch is isolated during that period. This problem is sometimes overcome by electronic solutions which are beyond the scope of this book. Commercially available master-slave flip-flops will be described later.

Master–Slave *JK* Flip-Flop

We now introduce a version of the *RS* flip-flop in which the input *R* is usually called *K*, *S* is called *J*, and it is possible to have the input combination $J = K = 1$ when $¢ = 1$. In fact, this combination is used to perform a new function called *toggling*. Here, if $J = K = 1$, the state (Q) of the flip-flop will always change when the ¢-pulse arrives, or the flip-flop is said to toggle. As will be seen in Chapter 8, this toggling ability is convenient in the design of counters. We will also see later that the *JK* flip-flop is, in a sense, the most general two-state memory device. We now proceed to describe how the toggling function can be accomplished.

Consider Fig. 7-18, where we show an *RS* flip-flop with *J* and *K* inputs and a network *N*. We are not allowed to use $S = R = 1$. When $J = K = 1$ we will "transform" the *J* and *K* inputs by the network *N* so that the condition on *S* and *R* is not violated.

In Fig. 7-19(a) we show the information *M* that we want to store in the master latch, as a function of the slave *Q* and the *J* and *K* inputs. This information will be transferred to the slave *Q* when the clock is removed. Thus we will obtain in effect an *RS* flip-flop that toggles when $S = R = 1$.

When $¢ = 0$, $Q = M$; i.e., the slave follows the master (see Fig. 7-16). This is the initial condition. If *M* is presently 0 and we want it to remain 0

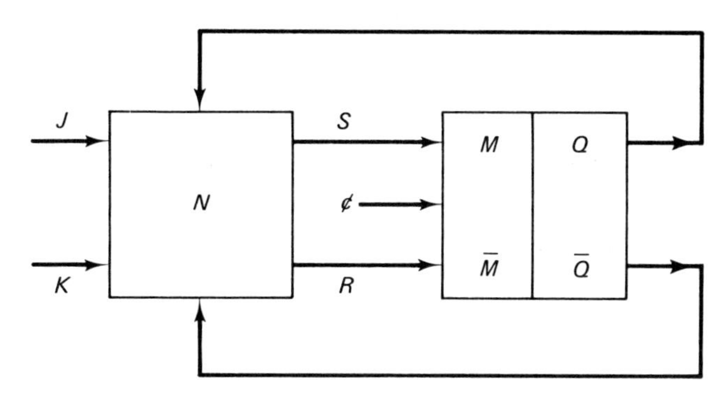

Fig. 7-18 Towards a JK flip-flop.

Q \ JK	00	01	11	10
0	0	0	1	1
1	1	0	0	1

M

(a)

Present $Q = M$	Next M	S	R
0	0	0	–
0	1	1	0
1	0	0	1
1	1	–	0

(b)

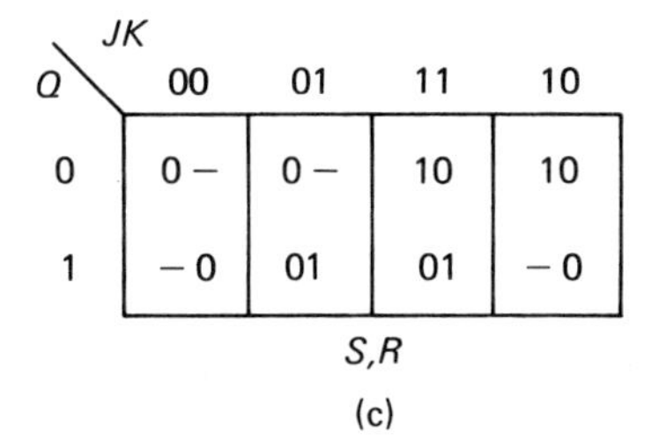

Q \ JK	00	01	11	10
0	0 –	0 –	10	10
1	– 0	01	01	– 0

S,R

(c)

Fig. 7-19 (a) Desired behavior; (b) S and R functions; (c) S and R table.

after the clock arrives, we can apply either $S = 0$, $R = 0$ or $S = 0$, $R = 1$; i.e., we can either do nothing or reset. This is summarized in the first row of Fig. 7-19(b), where "—" means that R is optional. Next, if $M = 0$ and we want M to become 1, the master latch must be set; i.e., $S = 1$, $R = 0$. The remaining two possibilities follow by a similar argument. It is preferable to use as many don't cares as possible because, in general, this will lead to simpler networks.

The values of S and R required to achieve the transitions of Fig. 7-19(a) are obtained using Fig. 7-19(b) and are shown in Fig. 7-19(c). Using the method of Chapter 5 we find

$$S = JQ' = J\bar{Q}$$
$$R = KQ.$$

This result can also be obtained by suitable intuitive arguments.

Figure 7-20 shows the completed design. If the RS flip-flop is that of Fig. 7-16, the AND gate can be eliminated by making J and $\bar{Q}$ (respectively, K

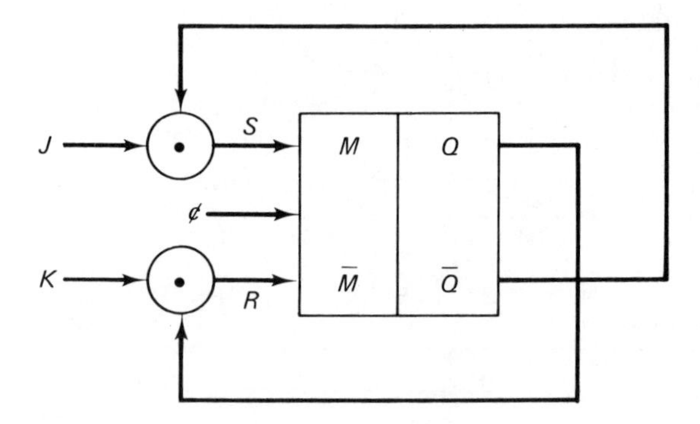

Fig. 7-20 JK flip-flop.

and Q) inputs to the NAND gate, which originally received the S (respectively, R) input.

In the next section we analyze a commercially available JK master–slave flip-flop by the methods of Chapter 6.

Flow-Table Analysis of a Master–Slave *JK* Flip-Flop

The network of Fig. 7-21 has 8 gates and 5 inputs. Its complete excitation table has 256×32 entries. Hence we prefer to analyze it by our shortcut approach. In Fig. 7-21 the reader will recognize two NAND latches. The one consisting of gates 3 and 4 will be called the *master* (latch), and the other (gates 7 and 8) will be called the *slave* (latch). First we consider the asynchronous 0-CLEAR (C_0) and 0-PRESET (P_0) inputs. Let the output of gate i be y_i.

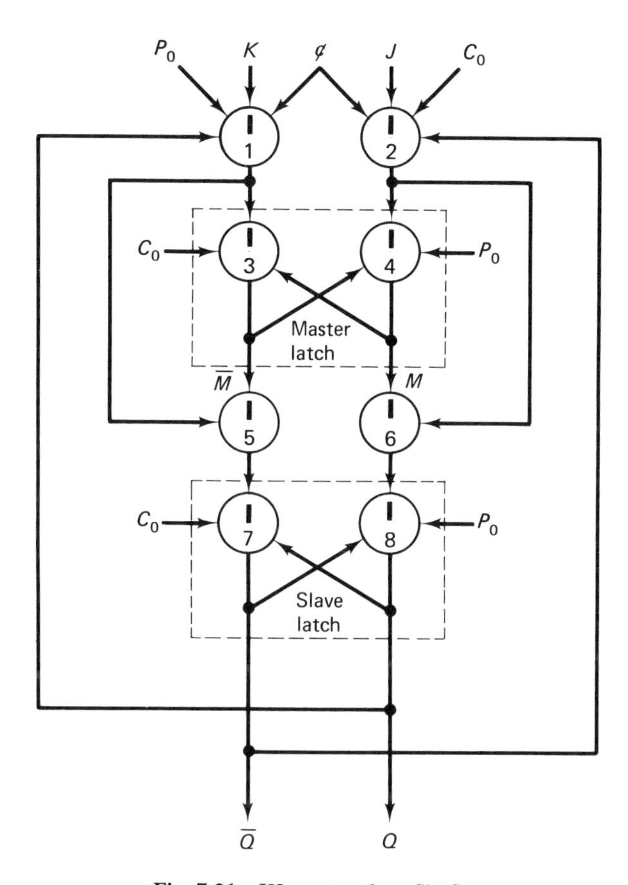

Fig. 7-21 JK master-slave flip-flop.

$\boldsymbol{C_0 = P_0 = 0}$. These inputs are forcing to 1 for gates 1, 2, 3, 4, 7, and 8. Next, $y_1 = 1$ and $y_3 = 1$ force $y_5 = 0$ and $y_2 = 1$, $y_4 = 1$ force $y_6 = 0$. Therefore, a unique stable internal state is reached:

$$y_{11} \triangleq (1\ 1\ \overset{M}{\overline{1\ 1}}\ \ 0\ 0\ \overset{Q}{\overline{1\ 1}})$$

where the "bars" over y_3, y_4 and y_7, y_8 help to locate the states of the master and slave, respectively. The symbol y_{11} is chosen to indicate that this is a state where both Q and $\overline{Q}$ are 1 for both the master and the slave.

$\boldsymbol{C_0 = 1, P_0 = 0}$. $P_0 = 0$ forces $y_1 = 1$, $y_4 = 1$, $y_8 = 1$. Next, $y_1 = 1$, $y_4 = 1$, and $C_0 = 1$ force $y_3 = 0$, which is followed by $y_5 = 1$. The inputs to gate 7 are now all 1, yielding $y_7 = 0$. This, in turn, forces $y_2 = 1$. Finally, $y_2 = y_4 = 1$ results in $y_6 = 0$. Therefore, for any input condition with $C_0 = 1$, $P_0 = 0$, there is a unique stable internal state:

$$y_{MQ} \triangleq (1\ 1\ \overline{0\ 1}\ \ 1\ 0\ \overline{0\ 1}),$$

where the M and Q in y_{MQ} indicate that $M = 1$ and $Q = 1$, i.e., that both the master and the slave are set.

$C_0 = 0, P_0 = 1$. By an argument symmetric to the last one, we find one stable state,

$$y_{M'Q'} \triangleq (1\ 1\ \overline{1\ 0}\quad 0\ 1\ \overline{1\ 0}),$$

where both the master and the slave are reset.

For the rest of the analysis we consider $C_0 = P_0 = 1$; i.e., these inputs are inactive.

$\not{c} = 0$. Gates 1 and 2 both have output 1. Thus the input to the master latch is (1, 1) and this latch could be either set or reset. Suppose that it is set; i.e., $\bar{M} = 0$ and $M = 1$. This results in $y_5 = 1, y_6 = 0$. This last condition is a set condition for the slave and will result in $\bar{Q} = 0, Q = 1$. Similarly, $\bar{M} = 1, M = 0$ implies $y_5 = 0, y_6 = 1$ and $\bar{Q} = 1, Q = 0$. Hence the two stable states possible when $\not{c} = 0$ are

$$(1\ 1\ \overline{0\ 1}\quad 1\ 0\ \overline{0\ 1}) = y_{MQ}$$

$$(1\ 1\ \overline{1\ 0}\quad 0\ 1\ \overline{1\ 0}) = y_{M'Q'}$$

Note that when $\not{c} = 0$, the $y_1 = 1, y_2 = 1$ inputs to the master ensure that the master is unaffected by the slave. In this sense, the master is *isolated* from the slave. On the other hand, the slave is forced to follow the master.

$\not{c} = 1$. We must now examine the effect of the J and K inputs. Clearly, $J = K = 0$ is logically equivalent to $\not{c} = 0$, the case we have just discussed.

Suppose that the flip-flop starts in state y_{MQ} with both latches set and $J = 0, K = 1$ when the clock arrives, $\not{c}: 0 \longrightarrow 1$. We get $y_1 = 0, y_2 = 1$, and so the master latch will reset to $\bar{M} = 1, M = 0$. This further yields $y_5 = 1$, $y_6 = 1$, which is the inactive input for the slave. Hence the slave remains set; i.e., it is isolated from the master. We have reached the stable state

$$y_{M'Q} \triangleq (0\ 1\ \overline{1\ 0}\quad 1\ 1\ \overline{0\ 1}).$$

Note that this state remains stable when J changes to 1, since $\bar{Q} = 0$ keeps $y_2 = 1$ regardless of J. See Fig. 7-22 for the corresponding flow-table entries.

A more formal verification that the transition above takes place reliably can be obtained with the aid of the excitation equations. For $C_0 = P_0 = 1$, $J = 0, K = 1, \not{c} = 1$, we have

$$Y_1 = y_8' \qquad Y_2 = 1$$

$$Y_3 = (y_1 y_4)' \qquad Y_4 = (y_2 y_3)'$$

$$Y_5 = (y_1 y_3)' \qquad Y_6 = (y_2 y_4)'$$

$$Y_7 = (y_5 y_8)' \qquad Y_8 = (y_6 y_7)'.$$

Using these excitation equations we can obtain the step-by-step description of the transition:

	$\not{c} = 0$	JK 0 0 ($\not{c} = 1$)	0 1	1 1	1 0
$y_{M'Q'}$	$y_{M'Q'}$	$y_{M'Q'}$	$y_{M'Q'}$		
y_{MQ}	y_{MQ}	y_{MQ}	$y_{M'Q}$	$y_{M'Q}$	
$y_{M'Q}$			$y_{M'Q}$	$y_{M'Q}$	
$y_{MQ'}$					

Fig. 7-22 Partial flow table.

$$
\begin{aligned}
y_{MQ} = {} & \underline{1}\ 1\ 0\ 1 \quad 1\ 0\ 0\ 1 \\
& 0\ 1\ \underline{0}\ 1 \quad 1\ 0\ 0\ 1 \\
& 0\ 1\ 1\ \underline{1} \quad 1\ 0\ 0\ 1 \\
& 0\ 1\ 1\ 0 \quad 1\ \underline{0}\ 0\ 1 \\
& 0\ 1\ 1\ 0 \quad 1\ 1\ 0\ 1 = y_{M'Q}
\end{aligned}
$$

where the changing variable is underlined.

At this stage it is also easy to compute the outcome if the network is started in y_{MQ} and $\not{c} = 1, J = 1, K = 1$. J affects only the excitation of gate 2: $Y_2 = (\not{c}JC_0y_7)' = y_7'$ here. All the other excitations are the same. One verifies that y_7 remains 0 during this transition and hence y_2 remains 1. Therefore, for $J = K = 1$, the same transition $y_{MQ} \longrightarrow y_{M'Q}$ takes place.

Let us complete the analysis of the case $J = 0, K = 1$. If $y_{M'Q'}$ is the initial state, where

$$y_{M'Q'} = (1\ \ 1\ \ \overline{1\ \ 0}\quad 0\ \ 1\ \ \overline{1\ \ 0}),$$

one verifies that it is stable.

A partial flow table is shown in Fig. 7-22. It summarizes all the entries that we have verified so far. These entries are also shown (unstarred) in Fig. 7-23. Symmetrically, we obtain the five starred entries, where

$$y_{MQ'} = (1\ \ 0\ \ \overline{0\ \ 1}\quad 1\ \ 1\ \ \overline{1\ \ 0}).$$

Consider now what happens when the present state is $y_{M'Q}$ and the clock falls, $\not{c}: 1 \longrightarrow 0$. We have $y_1 = y_2 = 1$ and the master is unaffected. The inputs to the slave, however, become $y_5 = 0, y_6 = 1$ and the slave resets, following the master. The state reached is therefore $y_{M'Q'}$. Similarly, when the clock falls in state $y_{MQ'}$, the state reached will be y_{MQ}, where the slave again follows the master. Since the condition $\not{c} = 0$ is logically equivalent to $J = K = 0$, we obtain the double-starred entries shown in Fig. 7-23.

	$\not{c} = 0$	JK 0 0 ($\not{c} = 1$)	0 1	1 1	1 0
$(1\,1\,\overline{1\,0}\;0\,1\,\overline{1\,0}) = y_{M'Q'}$	$(y_{M'Q'})$	$(y_{M'Q'})$	$(y_{M'Q'})$	* $y_{MQ'}$	* $y_{MQ'}$
$(1\,1\,\overline{0\,1}\;1\,0\,\overline{0\,1}) = y_{MQ}$	(y_{MQ})	(y_{MQ})	$y_{M'Q}$	$y_{M'Q}$	* (y_{MQ})
$(0\,1\,\overline{1\,0}\;1\,1\,\overline{0\,1}) = y_{M'Q}$	** $y_{M'Q'}$	** $y_{M'Q'}$	$(y_{M'Q})$	$(y_{M'Q})$	□ $y_{MQ'}$
$(1\,0\,\overline{0\,1}\;1\,1\,\overline{1\,0}) = y_{MQ'}$	** y_{MQ}	** y_{MQ}	□ $y_{M'Q}$	* $(y_{MQ'})$	* $(y_{MQ'})$

Fig. 7-23 Flow table of JK M-S flip-flop.

Finally, the two entries marked □ are found using the AED† model. (See Problem 6-6.) The GSW model differs here because it permits a critical race. However, this race will be critical only if one of the delays exceeds the sum of five other delays. Since this is unrealistic, we use the AED† model. The details are left to the reader as an exercise.

For most practical applications, a restricted mode of operation is suggested for the flip-flop. A *JK* master–slave flip-flop of the type we have just described is available, for example, in a 16-pin TTL package 74H76 [TI2] containing two such flip-flops ("dual *JK* master–slave flip-flops"). The manufacturer states that the *J* and *K* inputs must not change when the clock pulse is high. The application that the manufacturer has in mind is the following. The inputs *J* and *K* are to be "prepared" before the clock comes. The values J^n and K^n of *J* and *K* just before the *n*th clock pulse are to determine the function to be performed by the flip-flop. During the clock pulse *J* and *K* are not to be changed. When the clock falls again, the value Q^{n+1} of the slave will remember the action caused by J^n and K^n until the next clock pulse arrives. This is best illustrated by typical waveforms of Fig. 7-24. If $J^n = K^n = 0$, then Q^{n+1} is to be equal to Q^n; this is the "*remember*" input. If $J^n = 1$, $K^n = 0$, then Q^{n+1} is to become 1 regardless of Q^n; this is the *set* input. Similarly, $J^n = 0$, $K^n = 1$ results in $Q^{n+1} = 0$; this is the *reset* input. When $J^n = K^n = 1$, the flip-flop is to change state or *toggle*; i.e., $Q^{n+1} = (Q^n)'$.

Now we can easily derive the desired behavior that the manufacturer had in mind from our analysis, namely from Fig. 7-23. Suppose the state before the clock pulse is $y_{M'Q'}$; i.e., $Q^n = 0$ and $J^n = K^n = 0$. When $\not{c}$ arrives, this state remains stable. When $\not{c}$ disappears, no change takes place; this is the remember input. If $J^n = 0$, $K^n = 1$ and the starting state is $y_{M'Q'}$, again no change takes place and $Q^{n+1} = 0$. If $J^n = 1$, $K^n = 0$, then $y_{M'Q'}$ is unstable when $\not{c}$

†More precisely, this is a modified AED model where no delay exceeds the sum of 5 other delays.

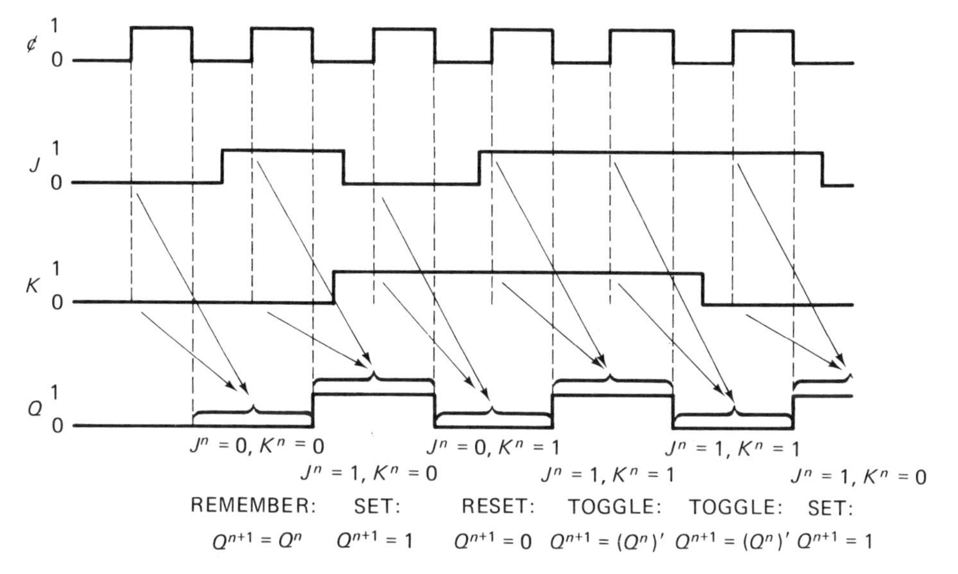

Fig. 7-24 Waveforms for JK M-S flip-flop.

arrives and changes to $y_{MQ'}$. Thus a 1 was stored in the master. When ¢ disappears, $y_{MQ'}$ is unstable and changes to y_{MQ}; i.e., the 1 from the master was transferred to the slave. Thus the overall effect of the ¢ pulse when $J^n = 1$, $K^n = 0$, and $Q^n = 0$ is to make $Q^{n+1} = 1$. The same happens when $J^n = K^n = 1$ and $Q^n = 0$.

By a similar analysis one verifies that the flip-flop indeed performs correctly when started in state y_{MQ}. The synchronous behavior of the master–slave *JK* flip-flop is conveniently summarized in Fig. 7-25.

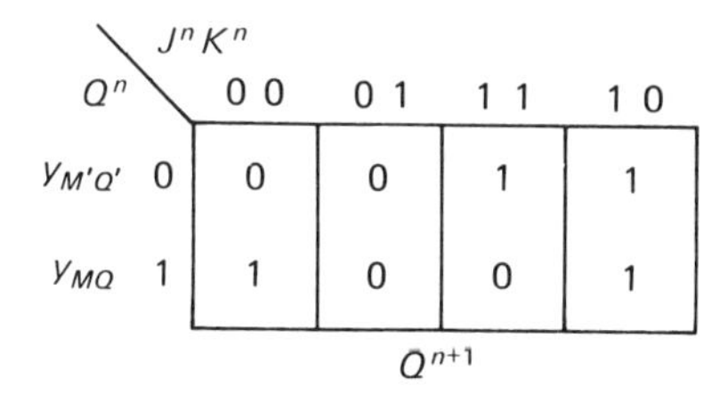

Q^n \ J^nK^n	00	01	11	10
$y_{M'Q'}$ 0	0	0	1	1
y_{MQ} 1	1	0	0	1
	Q^{n+1}			

Fig. 7-25 Synchronous state table for JK M-S flip-flop.

We have seen that when the clock is 0, both inputs to the master are 1, thus effectively isolating it from the slave. When the clock arrives and a change in Q is to take place, one verifies from Fig. 7-23 that the master changes first and the network stays in either $y_{MQ'}$ or $y_{M'Q}$, depending on where it started. In both of these states the inputs y_5 and y_6 to the slave latch are both 1, in effect isolating the slave from the master. Again, when the clock falls, the master is isolated from the slave.

The manufacturers explain this behavior by timing diagrams such as that shown in Fig. 7-26. Of course, our model cannot give any accurate timing

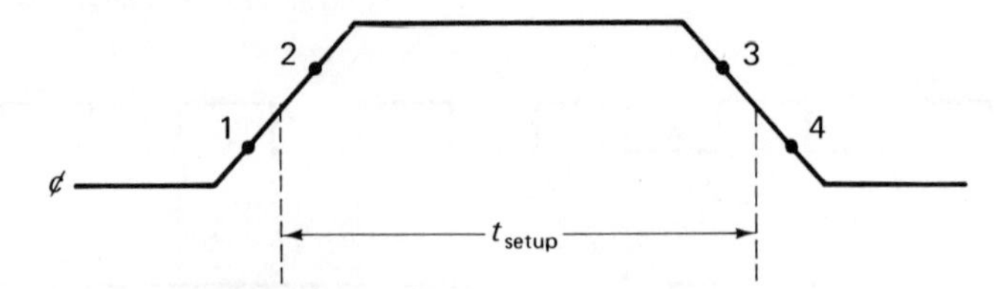

Figure 7-26 Flip-flop timing: (1) isolate slave; (2) enter J and K information to master; (3) isolate master; (4) transfer information from master to slave. *Note:* The inputs J and K should not change in the interval t_{setup}.

information, since we have assumed that the transition time of all signals is zero. Also the logic diagram is only an approximate representation of the actual circuit. Nevertheless, we were able to explain the isolation phenomena logically.

Our symbol for the master-slave *JK* flip-flop is given in Fig. 7-27.

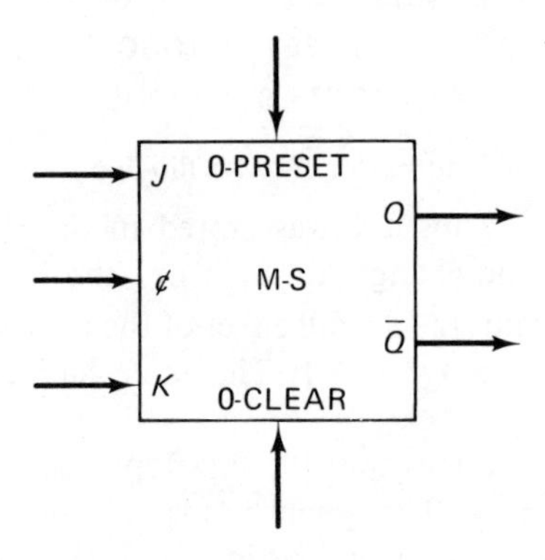

Fig. 7-27 JK flip-flop symbol.

Different Realization of a *JK* Flip-Flop

The 6-gate *JK* flip-flop of Fig. 7-28 is not of the master–slave type. Furthermore, it will not function properly according to the GSW model, as will be seen later. We leave the detailed analysis of this network as an exercise to the reader. We only indicate the main points. C_0 is a 0-clear input. Neither a 0-preset input nor the $\bar{Q}$ output is available. We first find the stable states using shortcuts as much as possible.

$\boldsymbol{C_0 = 0, \not{c} = 0.}$ This condition is forcing to $A \triangleq 110110$.

$\boldsymbol{C_0 = 0, \not{c} = 1, J = 0.}$ This is logically equivalent to the condition $C_0 = 0, \not{c} = 0$.

$\boldsymbol{C_0 = 0, \not{c} = 1, J = 1.}$ This condition is forcing to $B \triangleq 011110$.

This completes all the cases where $C_0 = 0$.

$\boldsymbol{C_0 = 1, \not{c} = 0.}$ Two stable states are possible: A and $C \triangleq 101011$.

$\boldsymbol{C_0 = 1, \not{c} = 1, J = 0, K = 0.}$ This is logically equivalent to $C_0 = 1$, $\not{c} = 0$.

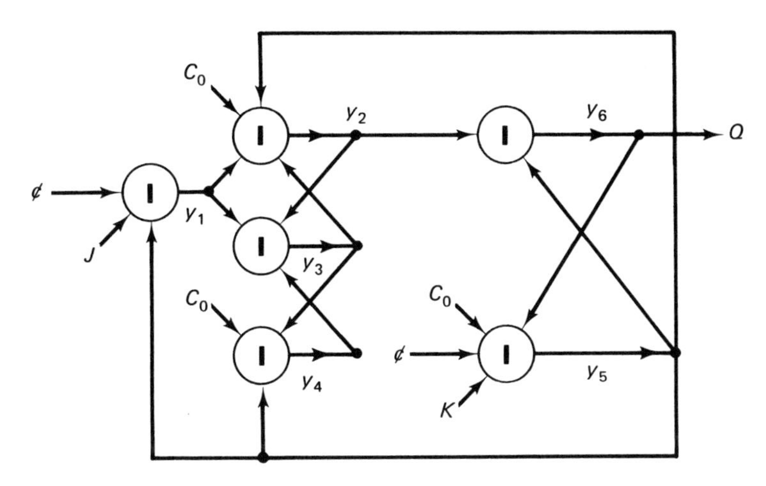

Fig. 7-28 A 6-gate JK flip-flop.

$\boldsymbol{C_0 = 1,\ ¢ = 1,\ J = 0,\ K = 1.}$ There are two stable states: A and $D \triangleq 110101$.

$\boldsymbol{C_0 = 1,\ ¢ = 1,\ J = 1,\ K = 1.}$ Two stable states: D and $E \triangleq 011010$.

$\boldsymbol{C_0 = 1,\ ¢ = 1,\ J = 1,\ K = 0.}$ This condition is forcing to E.

The results obtained above are summarized in the partially completed state table of Fig. 7-29.

Present state	$C_0 = 0$, ¢ = 0	$C_0 = 0$, ¢ = 1, $J = 0$	$C_0 = 0$, ¢ = 1, $J = 1$	$C_0 = 1$, ¢ = 0	$C_0 = 1$, ¢ = 1, JK = 00	$C_0 = 1$, ¢ = 1, JK = 01	$C_0 = 1$, ¢ = 1, JK = 11	$C_0 = 1$, ¢ = 1, JK = 10	
A	(A), 0	(A), 0	B	(A), 0	(A), 0	(A), 0		E	$A = 110110$
B	A	A	(B), 0					E	$B = 011110$
C	A	A	B	(C), 1	(C), 1			E	$C = 101011$
D	A	A	B			(D), 1	(D), 1	E	$D = 110101$
E	A	A	B				(E), 0	(E), 0	$E = 011010$

Next state, Q

Fig. 7-29 Partial flow table.

Next, we compute the unstable entries using both the AED and the GSW models. The resulting state tables are shown in Fig. 7-30, with the C_0 input removed for simplicity. One verifies that the synchronous state table corresponding to Fig. 7-30(a) is as shown in Fig. 7-31, where s denotes the state. The table of Fig. 7-30(b) shows that incorrect toggling may occur due to the

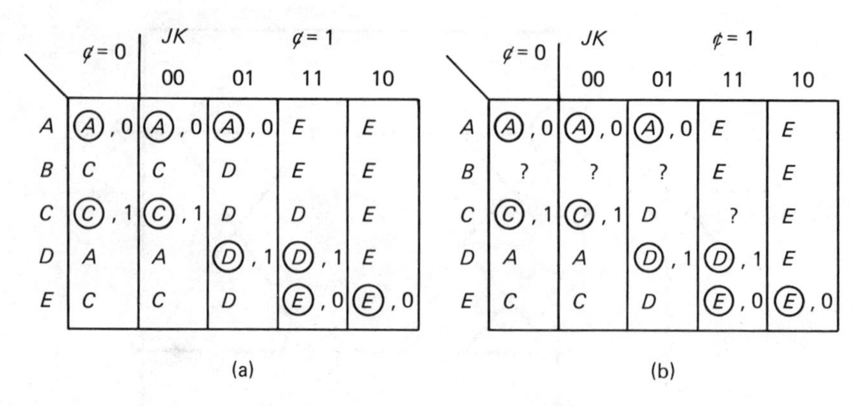

(a)

	¢ = 0	JK 00 (¢ = 1)	01	11	10
A	(A), 0	(A), 0	(A), 0	E	E
B	C	C	D	E	E
C	(C), 1	(C), 1	D	D	E
D	A	A	(D), 1	(D), 1	E
E	C	C	D	(E), 0	(E), 0

(b)

	¢ = 0	JK 00 (¢ = 1)	01	11	10
A	(A), 0	(A), 0	(A), 0	E	E
B	?	?	?	E	E
C	(C), 1	(C), 1	D	?	E
D	A	A	(D), 1	(D), 1	E
E	C	C	D	(E), 0	(E), 0

Fig. 7-30 State tables from (a) AED model and (b) GSW model.

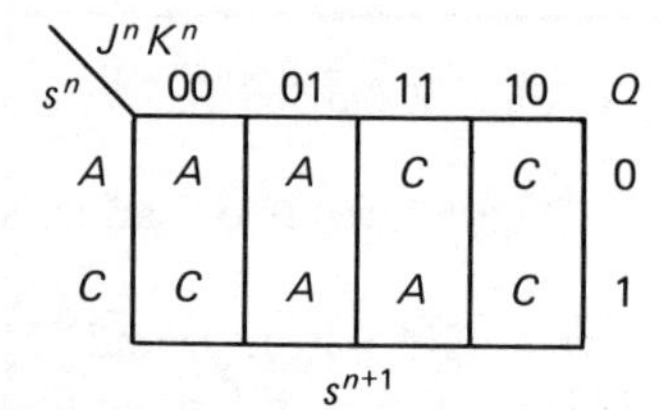

s^n \ J^nK^n	00	01	11	10	Q
A	A	A	C	C	0
C	C	A	A	C	1

s^{n+1}

Fig. 7-31 Synchronous state table.

critical race predicted by the GSW model in state $J = K = ¢ = 1$; $y = C$. Since the flip-flop is commercially used [MOT2], we conclude that the AED model is indeed appropriate.

Note that, if the flip-flop starts in state C and $J = 1$, $K = 0$, then the output Q goes to 0 while the clock pulse is on. This limits the use of such a flip-flop to situations where Q is observed only when $¢ = 0$. This problem is overcome in the edge-sensitive flip-flops of the next section. Note also that in the master-slave flip-flop of Fig. 7-21 the old Q value is kept until the clock falls (see Fig. 7-23).

7.4. EDGE-SENSITIVE FLIP-FLOPS

In this section we introduce another important concept, that of edge-sensitive devices. The master–slave flip-flops of the previous section overcome some of the difficulties present in the simple one-latch flip-flops. However, the master–slave flip-flops will not work properly if the synchronous inputs change while the clock is present. In this section we describe devices that respond to the values of the synchronous inputs present during a short time interval around the rising edge of the clock. This permits the synchronous (data) inputs to change soon after the clock arrives; such changes are ignored by an edge-sensitive device. Often both the edge-sensitive and the master–slave principles are incorporated into a single flip-flop.

Positive-Edge-Triggered D Flip-Flop

The network of Fig. 7-32 consists of 3 NAND latches. Its inputs are C_0 and P_0, the asynchronous 0-CLEAR and 0-PRESET inputs, the clock input ¢, and the synchronous data input D. Its outputs are Q and $\bar{Q}$.

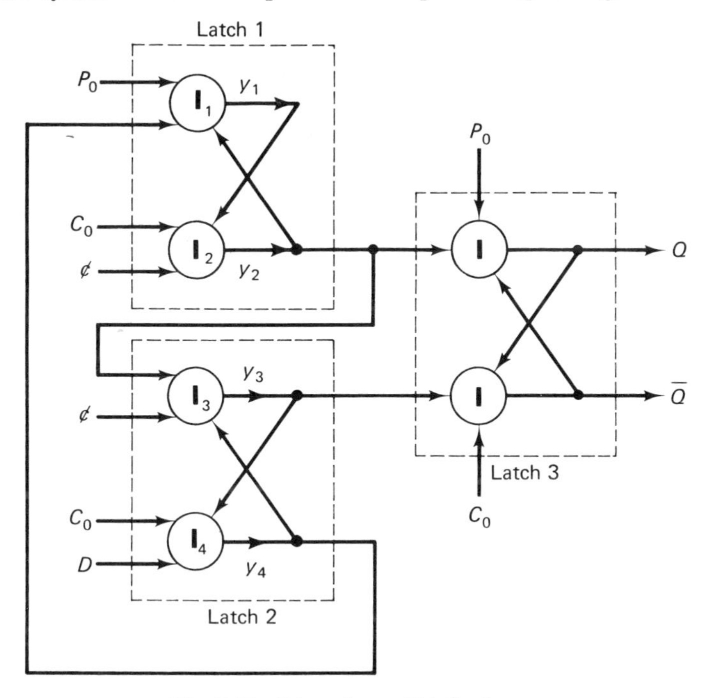

Fig. 7-32 Edge-triggered D flip-flop.

The first two latches form an interconnected subnetwork N_1. There is no easy way of separating them and we will analyze them together. On the other hand, the third latch, L_3, depends only on the outputs y_2 and y_3, and its outputs Q and $\bar{Q}$ do not affect the behavior of N_1. In effect, we have a serial connection of N_1 and L_3. The complexity of the analysis is reduced when N_1 is treated separately.

For the moment, however, we consider the entire network with respect to inputs P_0 and C_0, since these inputs turn out to be easily disposed of.

$\boldsymbol{C_0 = P_0 = 0}$. This is a forcing input resulting in $Q = \bar{Q} = 1$. Gates 1, 2, and 4 are also forced immediately to the 1 output. The output y_3 of gate 3 depends only on the clock, since $y_2 = y_4 = 1$ and $y_3 = (y_2 y_4 ¢)' = ¢'$. The variations in ¢ have no effect on Q or $\bar{Q}$ since P_0 and C_0 are forcing. The internal state of N_1 is therefore $(y_1, y_2, y_3, y_4) = (1, 1, ¢', 1)$, and for the entire flip-flop it is $(y_1, y_2, y_3, y_4, \bar{Q}, Q) = (1, 1, ¢', 1, 1, 1)$. The forcing inputs for $\bar{Q}$ and Q are illustrated in Fig. 7-33(a).

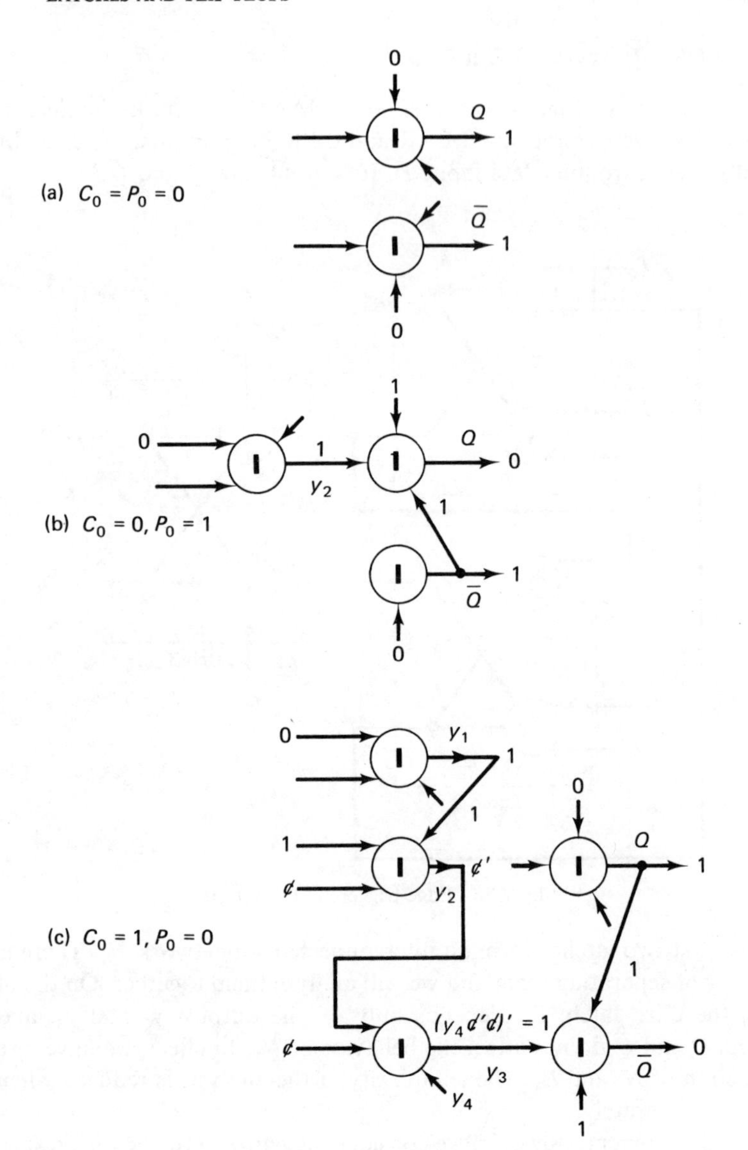

Fig. 7-33 Forcing conditions.

$\boldsymbol{C_0 = 0, P_0 = 1.}$ Here $\bar{Q} = 1$, but to determine Q we must know the y_2 input. Since $C_0 = 0$ is forcing for y_2, we have $y_2 = \bar{Q} = P_0 = 1$, resulting in $Q = 0$. [See Fig. 7-33(b).]

$\boldsymbol{C_0 = 1, P_0 = 0.}$ $P_0 = 0$ forces $Q = 1$ and $y_1 = 1$. Then $y_1 = 1$ and $C_0 = 1$ yield $y_2 = ¢'$. Next $y_3 = (¢'¢)' = 1$ and $\bar{Q} = 0$. [See Fig. 7-33(c).]

We have just discovered that the input y_3 to the third latch is $(¢¢')' = 1$.

When $\not{c}$ is constant, there is no difficulty. However, when $\not{c}$ changes from 0 to 1, it is possible that $\not{c}'$ changes from 1 to 0 after a short delay. Thus there may be a short period when $\not{c} = \not{c}' = 1$ and $y_4 = 1$, resulting in a 0-pulse at the input y_3 to the third latch. If the NAND gate with output $\bar{Q}$ ignores this pulse, $\bar{Q}$ remains 0. However, the pulse may be of sufficient duration to cause a 1-pulse on $\bar{Q}$. This, of course, is undesirable, for it may create additional errors.

More general techniques for discovering such situations will be treated in Chapter 10.

Let us return now to our analysis and let us assume that the asynchronous inputs C_0 and P_0 are inactive from now on; i.e., $C_0 = P_0 = 1$. We now confine our attention to the clocked operation and also analyze N_1 separately, as it appears in Fig. 7-34.

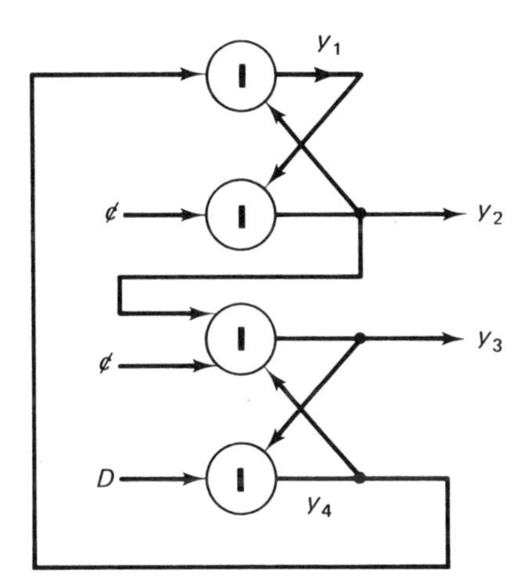

Fig. 7-34 Analysis of clocked operation of N_1.

$\not{c} = D = 0$. We immediately find $y_2 = y_3 = y_4 = 1$, from which $y_1 = 0$ follows. The input is therefore forcing to state $A = (y_1, y_2, y_3, y_4) = (0, 1, 1, 1)$.

$\not{c} = 0, D = 1$. Again, $y_2 = y_3 = 1$. In the second step we find $y_4 = 0$, and finally $y_1 = 1$. The input forces to $B = (1, 1, 1, 0)$.

$\not{c} = 1, D = 0$. Here $y_4 = 1$. To find the stable states possible for this input, notice that $y_1 = (y_4 y_2)' = y_2'$ and $y_3 = (\not{c} y_2 y_4)' = y_2'$. Thus we can have $C = (1, 0, 1, 1)$ when $y_2 = 0$, and $E = (0, 1, 0, 1)$ when $y_2 = 1$. One verifies that C and E are indeed both stable.

$\not{c} = 1, D = 1$. Suppose that $y_3 = 0$ and $y_2 = 0$. Then gate 3 is unstable since the y_2 input forces it to 1. Hence, if there exists a stable state with $y_3 = 0$, we must have $y_2 = 1$. We find $y_4 = 1$ and $y_1 = 0$. Thus we have state E again. If $y_3 = 1$, then $y_4 = 0$, $y_1 = 1$, and $y_2 = 0$. This is a new state $F = (1, 0, 1, 0)$. These states are shown in the state table of Fig. 7-35.

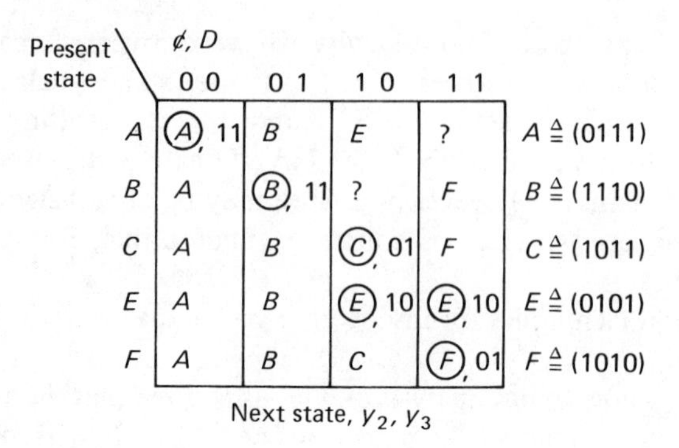

Present state	¢, D = 0 0	0 1	1 0	1 1	
A	(A), 11	B	E	?	$A \triangleq (0111)$
B	A	(B), 11	?	F	$B \triangleq (1110)$
C	A	B	(C), 01	F	$C \triangleq (1011)$
E	A	B	(E), 10	(E), 10	$E \triangleq (0101)$
F	A	B	C	(F), 01	$F \triangleq (1010)$

Next state, y_2, y_3

Fig. 7-35 State table for N_1.

The remaining entries can be computed with the aid of the excitation equations, and are shown in Fig. 7-35. The state A in column $¢ = 1$, $D = 1$ leads to a critical race in both the GSW and the AED models. State B in column $¢ = 1$, $D = 0$ leads to a critical race in the GSW model and to state C in the AED model.

The waveforms of Fig. 7-36 illustrate the operation of the flip-flop. Note that we have avoided transitions leading to critical races. From the figure it is clear that the value of D present just after the rising edge of the clock pulse is stored in the variables y_2 and y_3: $y_2 = 1$, $y_3 = 0$ if D was 0, and $y_2 = 0$,

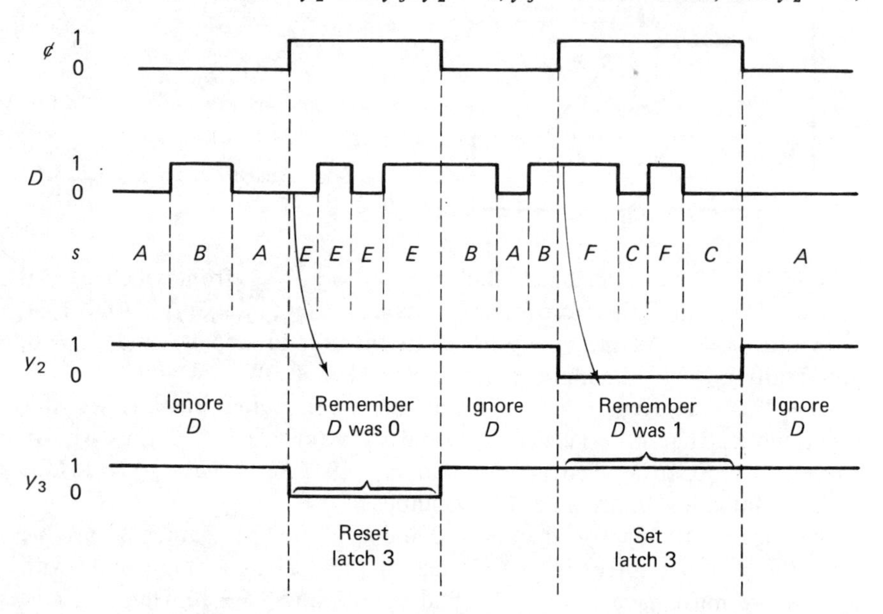

Fig. 7-36 Waveforms for positive-edge-triggered D flip-flop.

$y_3 = 1$ if D was 1. Values of D at other times are simply ignored. For this reason the flip-flop is called "positive-edge-triggered."

Return now to the complete flip-flop of Fig. 7-32. If D was 0, when ¢ arrived, $y_2 = 1, y_3 = 0$ forces the third latch during the clock pulse to the reset state $Q = 0, \bar{Q} = 1$. Similarly, if D was 1, when ¢ arrived, the latch is forced to the set state $Q = 1, \bar{Q} = 0$. This completes the analysis except for the critical race conditions. The critical races occur only when ¢ and D change simultaneously. More realistically, they may occur if ¢ and D change within a small time interval. Now the N_1 network is trying to memorize the value of D just after the clock arrives. If at this time the value of D is changing, our analysis shows that N_1 may interpret D as either a 0 or a 1. Therefore, the critical race does not have disastrous consequences. Usually, the user will avoid this condition. A more realistic treatment of this problem is given next.

Although it is beyond the scope of this book to explain in detail the electronic aspects of the networks described, we digress briefly to discuss a little more accurately some of the parameters affecting the flip-flop operation. The network just described is available in a 14-pin package 7474 which contains two edge-triggered D flip-flops each with inputs D, ¢, P_0, and C_0 and outputs Q and $\bar{Q}$. This package is labeled "dual, D-type, edge-triggered flip-flops" [SIG, TI2].

IC manufacturers specify precisely the conditions required for the proper functioning of the IC packages offered and provide typical and worst-case performance details. As an example, we describe some switching characteristics of the D-type flip-flop 7474.

The *clear propagation delay time* t_{pd0} is measured by applying a 0-pulse to the C_0 input having the characteristics shown in Fig. 7-37. The delay t_{pd0}

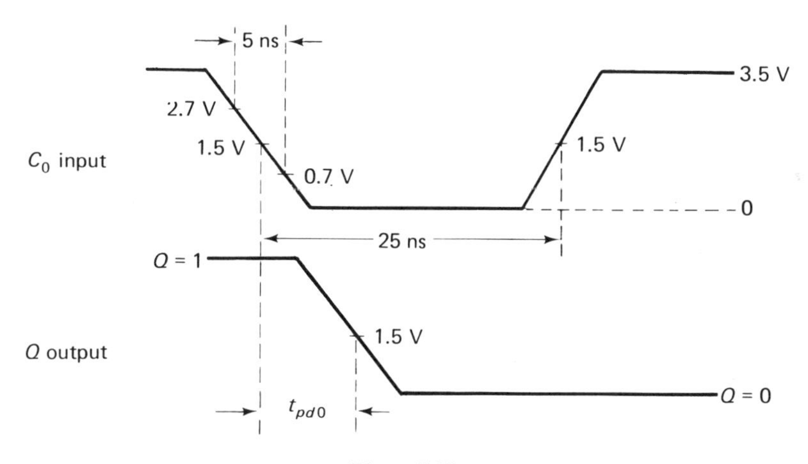

Figure 7-37

is then measured as indicated in Fig. 7-37. The *preset propagation delay time* is measured similarly.

As to the measurement of additional switching characteristics of the D-type flip-flop 7474, refer to Fig. 7-38, which indicates a change of the Q state from 0 to 1, caused by the application of a 1-signal to the D input. t_{setup} is a measure of the (setup) time the proper D input is applied, prior to the rising edge of the clock pulse. Table 7-1 quotes some switching characteristics. For further details and electrical characteristics, the reader is referred to [SIG, TI2].

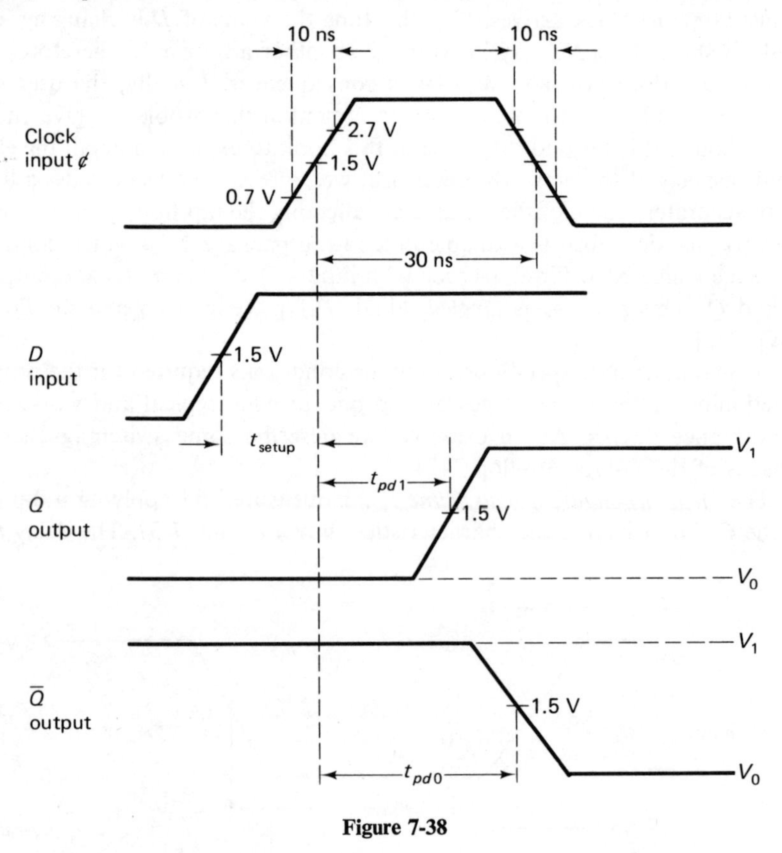

Figure 7-38

Table 7-1 Switching Characteristics of D-Type Flip-Flop 7474 [SIG, TI2]

Parameter	Minimum (ns)	Maximum (ns)
t_{pd0} (Fig. 7-37)		40
t_{pd1} (Fig. 7-38)	10	25
t_{pd0} (Fig. 7-38)	10	40
Minimum setup time		20

Functional Approach to Edge-Sensitive Networks

We now examine in more detail the principles involved in edge-sensitive (ES) networks. The simple NAND (or NOR) latch plays the key role of a basic memory module in sequential networks. In spite of the fact that we had already carried out a careful analysis of latches, it is worthwhile to reconsider latches as clocked devices.

In Fig. 7-39(a) we show the NAND latch with a data input *D* and a clock input ¢, and for convenience we repeat its transition table in Fig. 7-39(b). Typical waveforms are shown in Fig. 7-40.

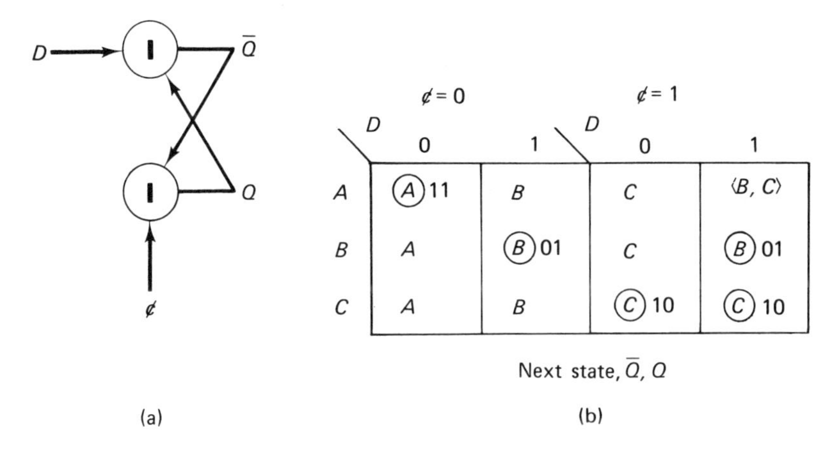

	¢ = 0, D = 0	¢ = 0, D = 1	¢ = 1, D = 0	¢ = 1, D = 1
A	(A) 11	B	C	⟨B, C⟩
B	A	(B) 01	C	(B) 01
C	A	B	(C) 10	(C) 10

Next state, $\bar{Q}$, Q

Fig. 7-39 (a) Clocked latch; (b) transition table.

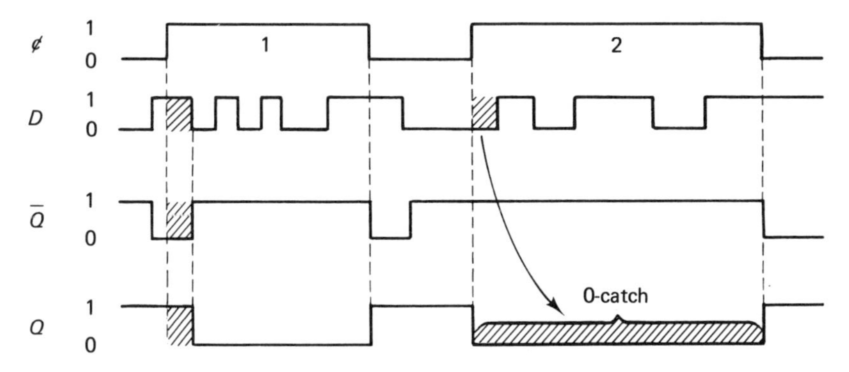

Fig. 7-40 Clocked latch waveforms.

For ES applications we are interested in the value of the *D* input during the rising edge of the clock. More precisely, we agree to set the value of *D* just before the clock pulse arrives and to keep it constant while the clock is changing. In an ES network we want to detect and store this value of *D*. After some minimum time following the clock change, the D input is allowed to change again any number of times. We shall designate as $x_{\uparrow_i}$ the value of a signal x from the time the rising edge of the ith clock pulse arrives until x changes. If a signal x is constant while the ith clock pulse is present, we shall designate its value as $x_{\wedge_i}$. If x is constant during the $¢ = 0$ period following the ith clock pulse, we shall write $x_{\vee_i}$ for its value. Refer to Fig. 7-40. With respect to clock pulse 1, we have

$$D_{\uparrow_1} = 1, \qquad \bar{Q}_{\uparrow_1} = 0, \qquad Q_{\uparrow_1} = 1,$$

and for clock pulse 2,

$$D_{\uparrow_2} = 0, \qquad \bar{Q}_{\uparrow_2} = 1, \qquad Q_{\uparrow_2} = 0$$

and

$$\bar{Q}_{\wedge_2} = 1, \qquad Q_{\wedge_2} = 0.$$

Note also that, regardless of D, we have

$$Q_{\vee_i} = 1 \qquad \text{for all } i.$$

$Q_{\wedge_i}$ is undefined if the value of Q changes during the ith clock pulse. Similar remarks apply to $\bar{Q}_{\vee_i}$. When there is no possibility of ambiguity, we will write $x_\uparrow$, $x_\wedge$, and $x_\vee$, without the subscript.

We also introduce the following terminology. If a network N with data input D, clock $¢$, and output Q has the property: For $a \in \{0, 1\}$ and for all i,

$$D_{\uparrow_i} = a \Longrightarrow Q_{\wedge_i} = a \qquad \text{where} \Longrightarrow \text{stands for "implies"},$$

then N is said to have (the) *a-catch* (property). In simple words, N has a-catch when the value a of the input D *during the rising edge of the clock* forces the output Q to have the value *a during the entire clock pulse*. The waveforms of Fig. 7-40 show clearly that the clocked latch has 0-catch but does not have 1-catch.

A network that has both 0-catch and 1-catch will be called a *catch*. We now discuss the design of such a network. We know that the NAND latch has 0-catch but not 1-catch. Notice, however, that when $D_\uparrow = 1$, we always have $\bar{Q}_\uparrow = 0$. If we used $\bar{Q}$ as the input to another NAND latch, the 0-pulse on $\bar{Q}$ could be caught by the second latch. [See Fig. 7-41(a).] We now have the following situation:

$$D_\uparrow = 0 \Longrightarrow Q_\wedge = 0$$

$$D_\uparrow = 1 \Longrightarrow P_\wedge = 0.$$

Note also that the first NAND latch "starts out" the correct way when $D_\uparrow = 1$, in the sense that $Q_\uparrow = 1$. The condition $D = 1$ when the clock pulse arrives is shown in Fig. 7-41(a). We would like to maintain Q constant during the entire clock pulse; i.e., we must now make Q independent of D. One way to solve this problem is as follows. The 0-pulse in $\bar{Q}$ can be viewed as being

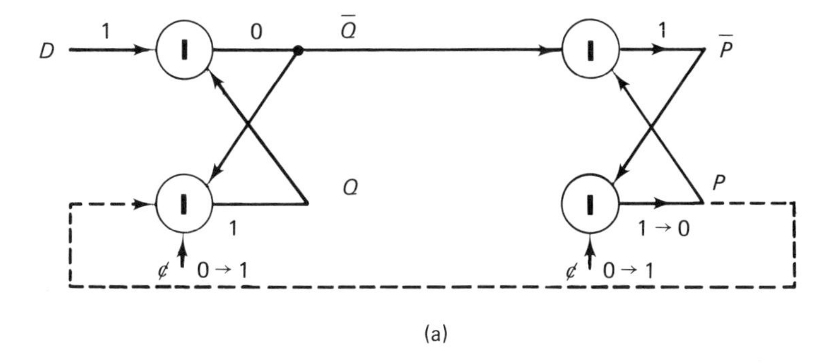

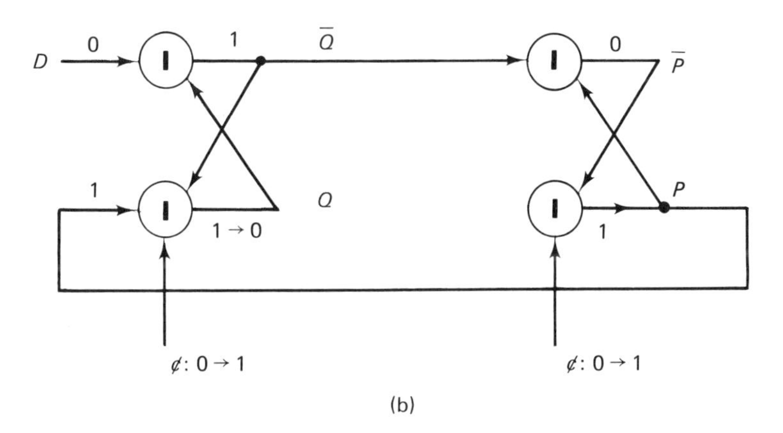

Fig. 7-41 (a) N_2; (b) catch network.

stored and held in P while the clock is 1. By feeding the P signal back to the Q-gate, as shown by the dashed connection, we force Q to be 1 as long as $¢ = 1$. Thus we have achieved our goal, since the output Q now has 1-catch for D.

We must, however, verify that the new feedback we have introduced does not destroy the other properties we wanted. This is easily done. Clearly, if $¢ = 0$, the feedback is ineffective because the Q-gate is forced to 1 by $¢ = 0$. Thus $¢ = 0 \Rightarrow Q = 1$, as before. It remains to be verified that the 0-catch property of Q has not been destroyed. Start with $D = ¢ = 0$. We find $\bar{Q} = 1$, $Q = 1$, $P = 1$, $\bar{P} = 0$. Now the change $¢: 0 \longrightarrow 1$ [see Fig. 7-41(b)] will not affect the P-gate, which is forced to 1 by $\bar{P} = 0$. The Q-gate will become 0 as required, and $Q = 0$ forces $\bar{Q} = 1$ independent of D. Thus D is now "locked-out" and the network of Fig. 7-41(b) is indeed a catch.

For completeness, we show in Fig. 7-42 the flow table of the catch of Fig. 7-41(b). The entries corresponding to double input changes are irrelevant

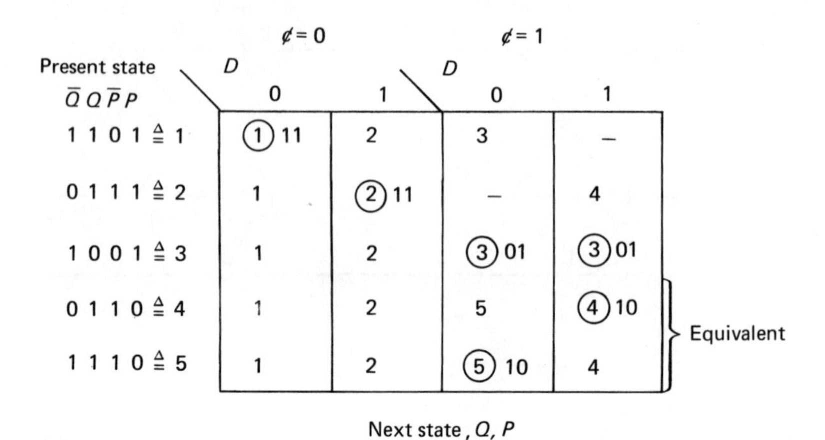

Present state $\bar{Q}Q\bar{P}P$	$¢ = 0$, $D = 0$	$¢ = 0$, $D = 1$	$¢ = 1$, $D = 0$	$¢ = 1$, $D = 1$
1 1 0 1 ≙ 1	(1) 11	2	3	–
0 1 1 1 ≙ 2	1	(2) 11	–	4
1 0 0 1 ≙ 3	1	2	(3) 01	(3) 01
0 1 1 0 ≙ 4	1	2	5	(4) 10
1 1 1 0 ≙ 5	1	2	(5) 10	4

Fig. 7-42 Flow table for catch.

for our purposes. Note that states 4 and 5 are equivalent as far as the outputs Q and P are concerned. The output P is included here because of its usefulness in the next example. Incidentally, the network we just designed is precisely the network N_1 of Fig. 7-34.

As we have said before, the ES D flip-flop is to store $D_{\uparrow}$ in Q *until the next rising edge of the clock*. The catch network stores $D_{\uparrow_i}$ only in $Q_{\wedge_i}$; we now also want the same value for $Q_{\vee_i}$. It is an easy matter to store the P and Q outputs of the catch in another latch as shown in Fig. 7-32 (ignoring the C_0 and P_0 inputs), as we have explained before.

The network of Fig. 7-43 represents another realization of an ES D flip-flop. Consider the front network without the dashed connections. Clearly

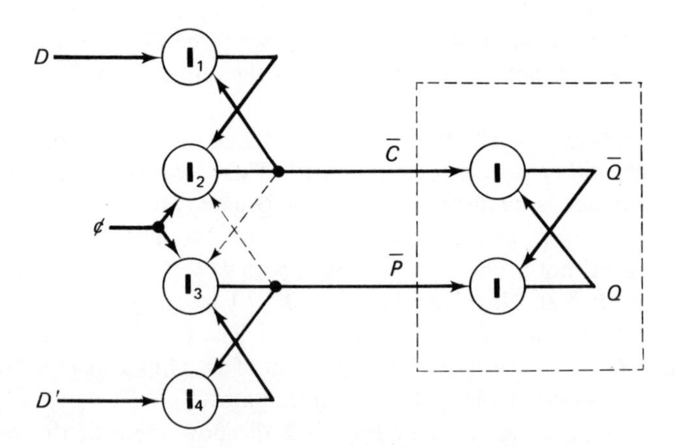

Fig. 7-43 ES D flip-flop #2.

$\bar{C}$ has 0-catch for input D, and $\bar{P}$ has 0-catch for input D', i.e., 1-catch for input D. Thus $D_\uparrow = 0$ implies that $\bar{C}_\wedge = 0$. However, $D'_\uparrow = 1$ gives only $\bar{P}_\uparrow = 1$, not $\bar{P}_\wedge = 1$. To lock the 1-value in $\bar{P}$, connect $\bar{C}$ to gate 3 as shown by the dashed connection. Then $\bar{C}_\wedge = 0$ will guarantee that $\bar{P}_\wedge = 1$. The second dashed connection yields

$$D_\uparrow = 1 \Longrightarrow \bar{P}_\wedge = 0 \qquad \text{and} \qquad \bar{C}_\wedge = 1.$$

Note also that $\not{c} = 0$ implies $\bar{P} = \bar{C} = 1$, which is the "remember" condition for a NAND latch. Thus we have the desired property of storing $D_\uparrow$ in $Q_\wedge$ and $Q_\vee$.

An example of an edge-sensitive *JK* flip-flop is given in Problem 14.

7.5. OTHER FLIP-FLOPS

Figure 7-44(a) summarizes the logical synchronous properties of four common types of flip-flops. If we view Q^{n+1} as a Boolean function of Q^n with J^n and K^n as control variables, it is evident that the *JK* flip-flop has the ability to realize all four functions of a single variable.

The characteristic equations for all four types of flip-flops are easily found from Fig. 7-44(a).

JK type: $Q^{n+1} = J^n(Q^n)' + (K^n)'Q^n$

RS type: $Q^{n+1} = S^n + (R^n)'Q^n$ (using $R = S = 1$ as a don't care)

D type: $Q^{n+1} = D^n$

T type: $Q^{n+1} = T^n \oplus Q^n$.

These equations will be used in Chapter 8 for the analysis of networks constructed with flip-flops as memory elements.

Briefly, the *RS* flip-flop can (1) remember, (2) set, and (3) reset. The *D* flip-flop can only (1) set or (2) reset. The *T* flip-flop can (1) remember or (2) toggle, i.e., change the state. All this is evident from the state tables.

We can also view the *RS*, *D*, and *T* flip-flops as restricted modes of operation of the *JK* flip-flop. In fact, this is commonly done. Figure 7-45 shows how to obtain flip-flops of the *RS*, *D*, and *T* types from a *JK* flip-flop. The verification that they work as claimed is trivial. As we have seen, each type of flip-flop has a variety of implementations.

We close this section by mentioning a few more examples of flip-flops. The basic concepts of NOR and NAND latches, master–slave operation, and edge-triggered operation should give the reader sufficient background to understand the logical behavior of these networks, without a detailed description.

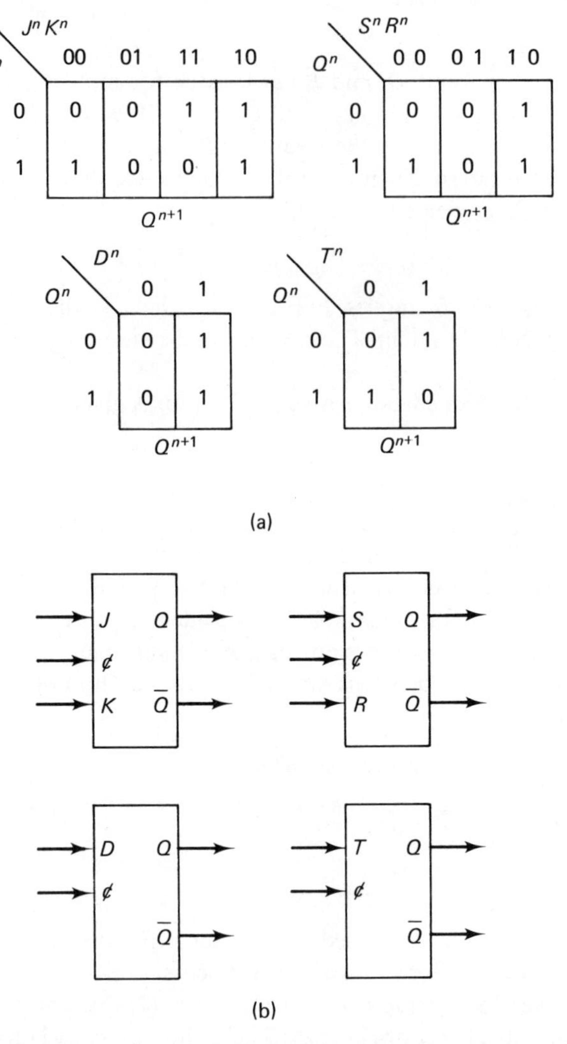

Fig. 7-44 (a) Synchronous state tables; (b) symbols for JK, RS, D and T flip-flops.

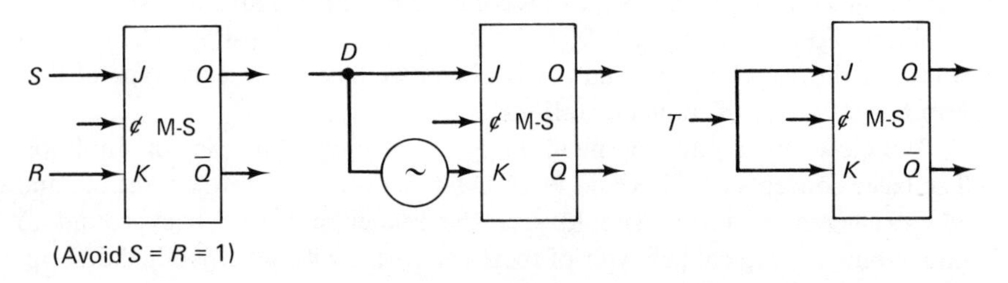

Fig. 7-45 Realization of RS, D and T flip-flops with JK M-S flip-flops.

Package 74111 (Dual *JK* Master–Slave Flip-Flops with Data Lockout) [TI1, TI2]

This is a 16-pin TTL package containing two flip-flops which are similar to the 74H76 type. They differ from the 74H76 type only in the clocking arrangements; namely, the *J* and *K* inputs are connected to the master latch only during a short period starting with the rising edge of the clock. The data stored in the master are transferred to the slave latch during the falling edge of the clock. The 74111 clocking arrangement has the advantage that the *J* and *K* inputs need not be held steady during the whole clock pulse, as is the case for the 74H76 type. The idea is similar to that in the edge-triggered *D* flip-flop.

Package 74110 (Gated *JK* Master–Slave Flip-Flop with Data Lockout) [TI1, TI2]

This is a 14-pin TTL package containing one 74111-type flip-flop with additional AND gates as shown in Fig. 7-46.

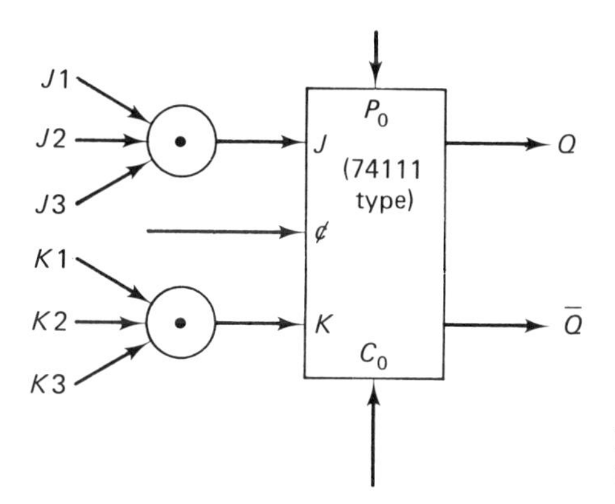

Fig. 7-46 Functional diagram of 74110-type flip-flop.

Package 7470 (Edge-Triggered *JK* Flip-Flop) [SIG, TI1, TI2]

This 14-pin TTL package contains one *JK* flip-flop with preset and clear inputs P_0 and C_0; *K*-type inputs K_1, K_2, and K^*; and *J*-type inputs J_1, J_2, and J^*. Input logic is provided so that the effective *J* and *K* signals are

$$J = J_1J_2(J^*)'$$
$$K = K_1K_2(K^*)'.$$

The input information present during the first part of the rising edge of the clock is stored and later data inputs are locked out.

Package 74L71 (Master–Slave *RS* Flip-Flop)
[TI1, TI2]

This 14-pin TTL package contains a single *RS* flip-flop of the master–slave type with P_0 and C_0 inputs. The effective R and S inputs are

$$R = R_1R_2R_3$$
$$S = S_1S_2S_3,$$

and both Q and $\bar{Q}$ outputs are available.

7.6. MONOSTABLE MULTIVIBRATORS

A *monostable multivibrator* (*mono* or *one-shot*) is a circuit capable of detecting a change in an input signal and producing an output pulse of fixed duration corresponding to this change. Although this function is logically quite different from that of a flip-flop, monostable multivibrators are usually grouped together with flip-flops (otherwise known as *bistable multivibrators*), because they are electronically similar.

Figure 7-47(a) shows our symbol for a basic monostable multivibrator. Normally, if $\tau = 0$, we have $Q = 0$, $\bar{Q} = 1$. When τ changes from 0 to 1, the mono is "triggered" and produces a 1-pulse on Q and a 0 pulse on $\bar{Q}$, as shown in Fig. 7-48. Once the mono is triggered, its outputs are independent of further input changes for the duration of the output pulse.

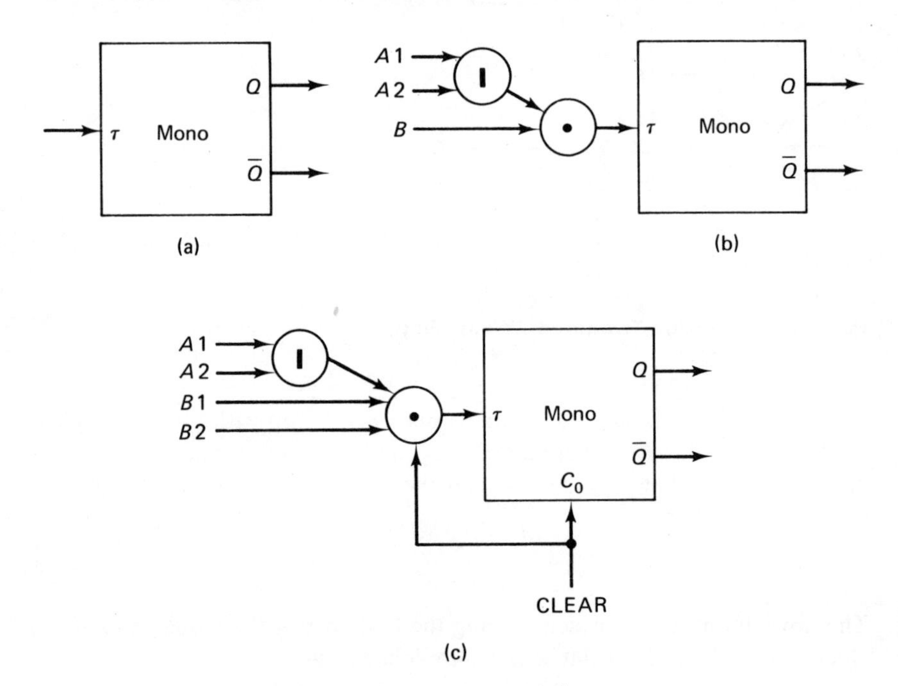

Fig. 7-47 Monostable multivibrators.

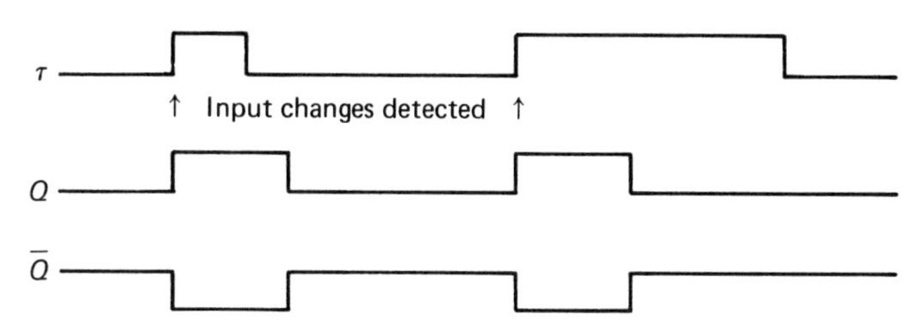

Fig. 7-48 Mono waveforms.

The mono of Fig. 7-47(b) is provided with additional input logic. If either of the A inputs is 0, positive changes in B are detected. If $B = 1$, $A1 = A2 = 1$, and one of the A inputs changes, the negative change in that input is detected. Such a unit is available as an IC TTL package 74121 [SIG, TI2]. Extra pins are provided for connecting external timing components (resistors and capacitors) to modify the output pulse width. For this unit the output pulse width can be varied from 40 ns to 40 seconds!

A still more flexible mono is shown in Fig. 7-47(c) and corresponds to package 74122 (*retriggerable monostable multivibrator with clear*) [SIG, TI2]. The waveforms of Fig. 7-49 illustrate the new feature.

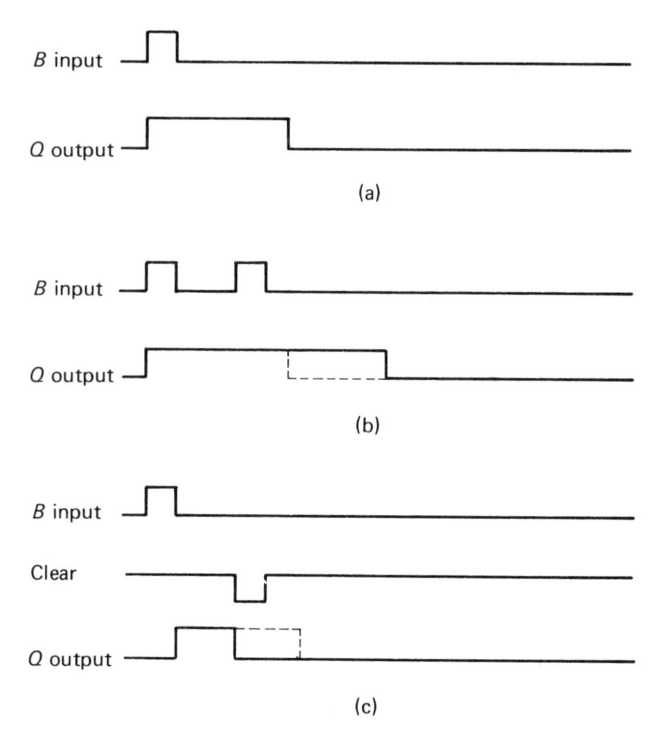

Fig. 7-49 Retriggerable mono: (a) Normal use; (b) extending output pulse by retriggering; (c) shortening output pulse by clearing.

An *astable multivibrator* is essentially an oscillator. Basically an oscillator can be constructed as shown in Fig. 6-18 (p. 179). If $x = 0$, the network is stable with $y = 1$. When $x = 1$, the network oscillates as discussed before. Thus the input x acts as the START/STOP control input for the oscillator. The frequency of oscillation can be changed by additional timing elements (resistors and capacitors).

If a more precise frequency and pulse waveform are required, commercially produced clocks (crystal oscillators) are available.

PROBLEMS

1. Find the flow table for the D flip-flop 7475 of Fig. P7-1 [TI2]. Use the shortcut approach. Find the stable states first and then find the missing unstable entries. Use both the GSW and the AED model to analyze all the races.

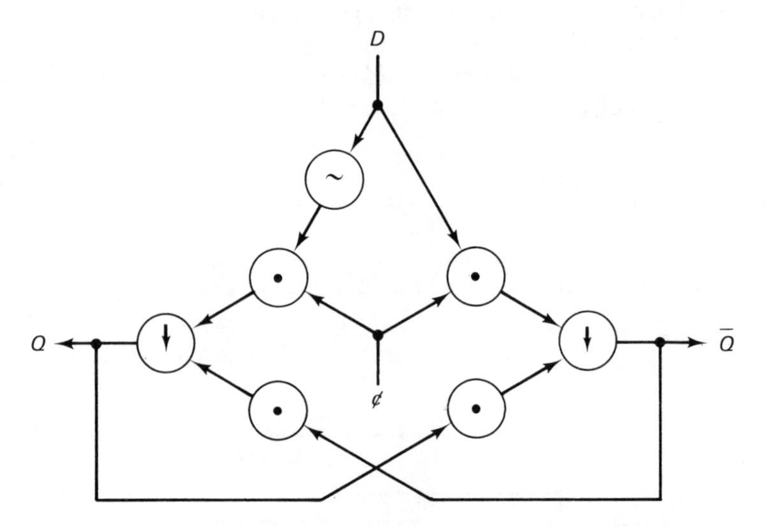

Figure P7-1

2. Repeat Problem 1 for the RS flip-flop RSN 54L71 of Fig. P7-2 [TI2]. For this problem set $a = b = 1$.

3. Repeat Problem 2, this time with $a = Q$, $b = \bar{Q}$. What type of flip-flop does this construction yield?

4. Repeat Problem 1 for the JK flip-flop of Fig. P7-3.

5. Analyze the flip-flop of Fig. P7-4. Discover which inputs correspond to the clock, clear, preset and data inputs. What type of device is this?

6. In practical applications it is often necessary to have several "J-type" and "K-type" inputs. One way to achieve this is by combinational networks external to the flip-flop package. However, there are also commercially available

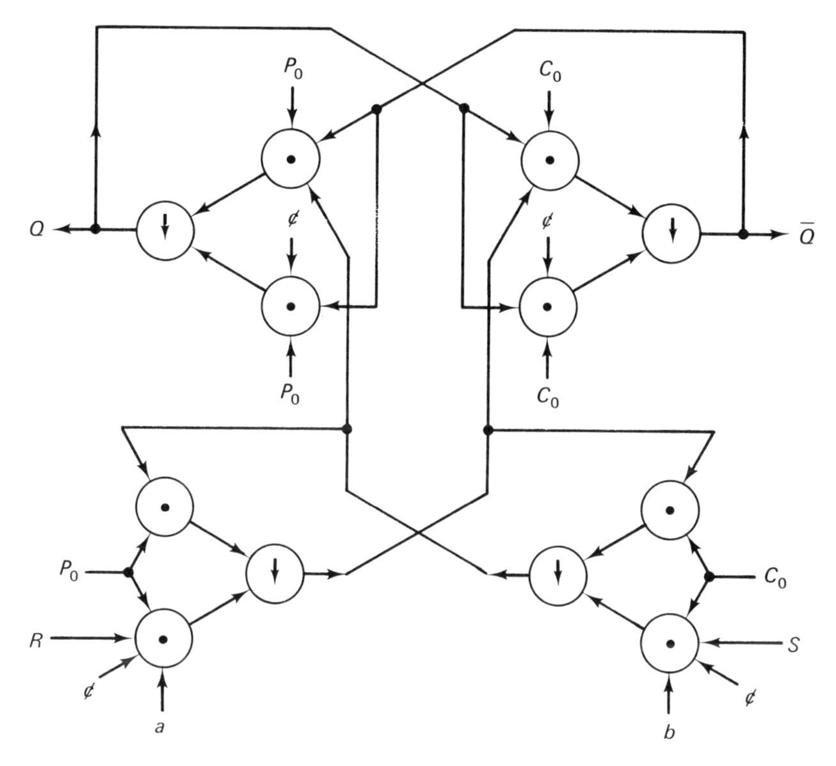

Figure P7-2

modules like the "gated *JK* with preset" flip-flop 74H71 [TI2]. Refer to Fig. 7-21. The top two NAND gates have been reproduced in Fig. P7-5(a) without the asynchronous inputs. In the 74H71, these two gates are replaced by the network of Fig. P7-5(b). What are the "effective" *J* and *K* functions for Fig. P7-5(b)?

7. A toggle flip-flop can be realized using an edge-triggered *D* flip-flop (Fig. 7-32) as shown in Fig. P7-6. Clearly no change takes place when ¢ = 0. When ¢ = $T = 1$, $\bar{Q}$ (representing the complement of Q) is effectively used as the data input *D*. Since the *D* flip-flop is edge-sensitive, the changes in the $\bar{Q}$ signal taking place after the rising edge of the clock will not have any further effect on the flip-flop. Verify that this network works as claimed, by constructing its flow table. Is the resulting *T* flip-flop edge-sensitive?

8. Try the approach of Problem 7, using the *D* flip-flop of Fig. 7-11 instead of the edge-triggered *D* flip-flop. The resulting network can be simplified to that of Fig. P7-7.

(a) Suppose that the network is stable, with ¢ = 0. If *T* is set to 1 and the clock pulse arrives, verify that the network will indeed change state. Complete the analysis and show that the network does not toggle properly.

(b) Could this network be made to operate properly if the clock pulse width could be accurately controlled?

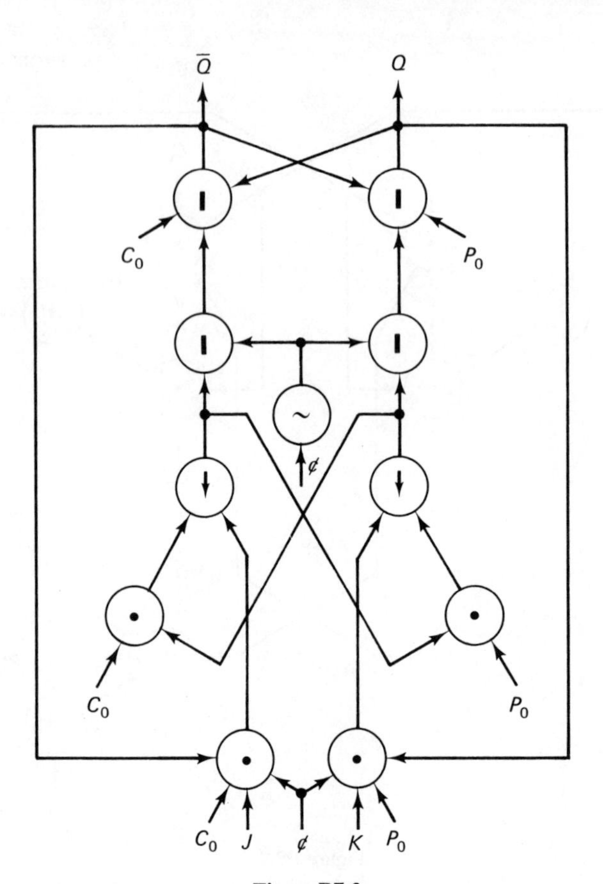

Figure P7-3

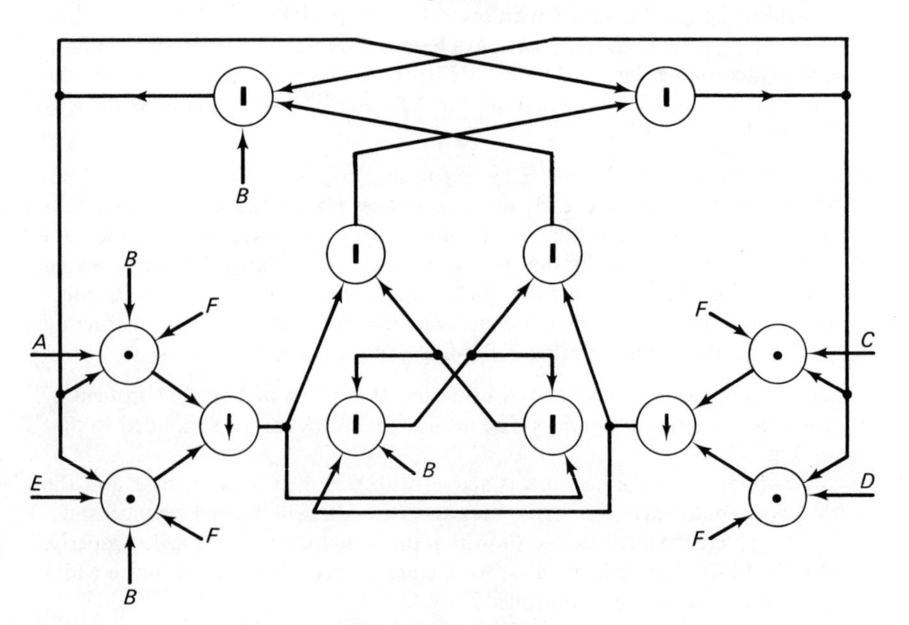

Figure P7-4

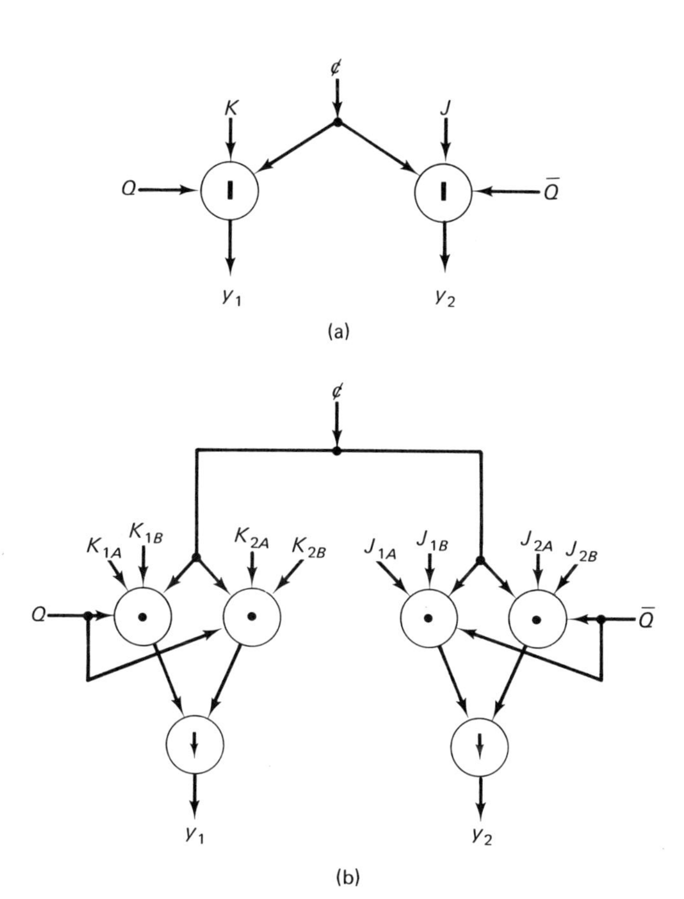

Figure P7-5

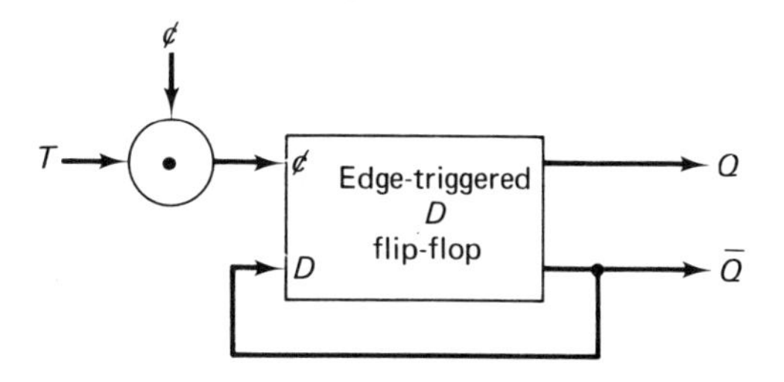

Figure P7-6

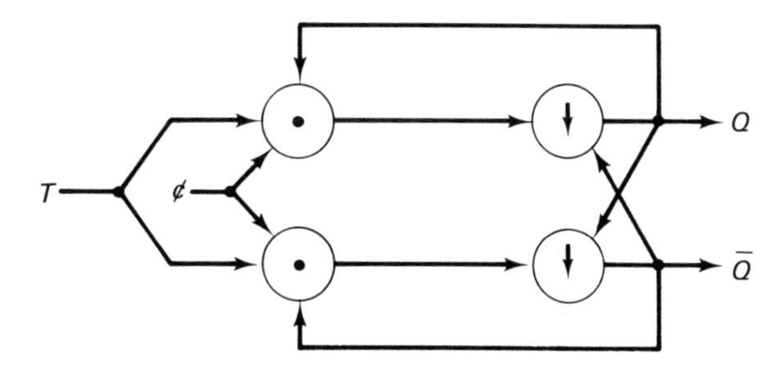

Figure P7-7

9. Repeat Problem 7 for the *JK* flip-flop realization of Fig. P7-8.

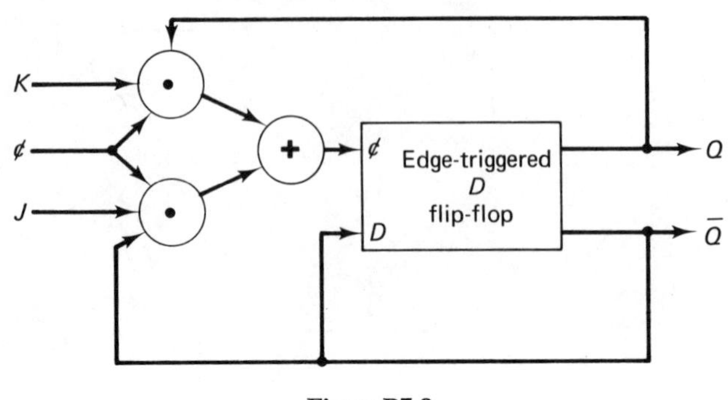

Figure P7-8

10. Analyze the network of Fig. P7-9. Give a word description of the device.

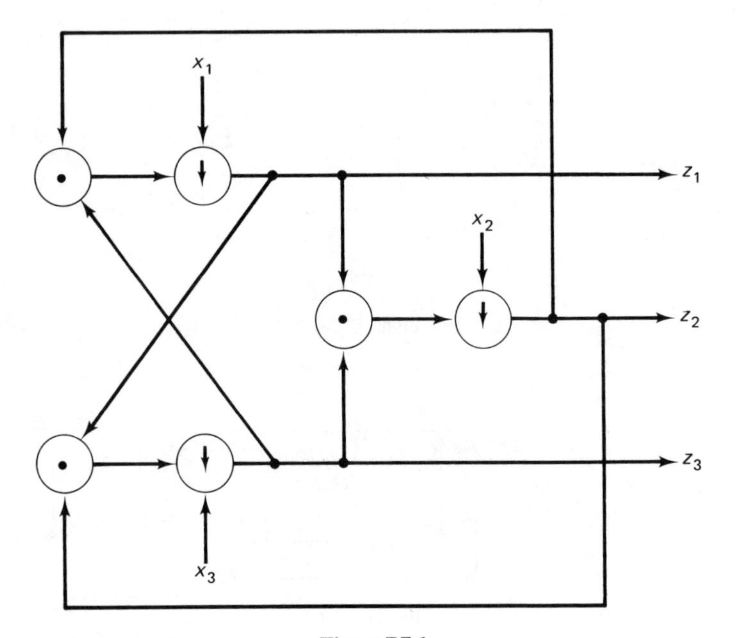

Figure P7-9

11. Find the flow table for the flip-flop of Fig. 7-16.

12. Find the flow table for the network resulting from Fig. 7-16 when $\bar{Q}$ is fed back to the *S*-gate and *Q* is fed back to the *R*-gate (See the discussion of Fig. 7-20, pp. 208-10.)

13. Find the flow table for the network consisting of the first 4 gates of Fig. 7-43, with only one of the dashed connections made, namely that pointing downward. Repeat with both dashed connections made.

14. Analyze the network described by the following excitation equations:

$$Y_1 = (Jy_5y_7)' \quad Y_2 = (Ky_6y_8)' \quad Y_3 = (y_1y_7)' \quad Y_4 = (y_2y_8)'$$

$$Y_5 = (¢y_1y_4y_6)' \quad Y_6 = (¢y_2y_3y_5)' \quad Y_7 = (y_5y_8)' \quad Y_8 = (y_6y_7)'$$

15. Modify the network of Problem 14 so as to provide asynchronous preset and clear facilities.

8 MORE COMPLEX SEQUENTIAL NETWORKS

About This Chapter In the previous chapter we considered relatively small sequential networks. The most complex network analyzed, the master–slave *JK* flip-flop, was just a memory cell capable of storing one bit of information. The gate-delay model and flow tables were useful for introducing sequential behavior and for describing some of the features and difficulties related to sequential operation. However, such a detailed point of view is not practical for larger sequential networks. We therefore adopt a new approach in this chapter. We now assume that latches and flip-flops are well-understood building blocks. Larger sequential networks are constructed using flip-flops to perform the basic function of memory, along with combinational logic to produce the desired changes in the memory.

There are no general algorithms for designing large asynchronous sequential gate networks. However, there does exist an elegant and useful theory of so-called "synchronous" sequential networks. We prefer to use the term "synchronous-mode" networks, because these are asynchronous (i.e., general) sequential networks operating under certain restrictions. A common example of synchronous-mode operation is provided by networks that use master–slave flip-flops (synchronized by a common clock input) as memory devices, and gates for performing logical operations. We discuss the analysis and synthesis of such networks in some detail.

We describe some commercially available sequential IC packages and illustrate design based on such modules. The approach is intuitive in view of the fact that no systematic methods exist for such problems. We also include a section on IC memories.

Comprehension of this chapter requires knowledge of the background material of Chapters 2, 5, and 7.

8.1. RIPPLE COUNTERS

Binary Ripple Counters

In this section we discuss several closely related networks. The network of Fig. 8-1(a) is available as package 7493 [SIG, TI2], a 14-pin MSI TTL-package consisting of 4 master–slave flip-flops interconnected as shown. Each flip-flop acts as a T-type flip-flop with T permanently set to 1. Such a T flip-flop can be obtained, for example, from the master–slave JK flip-flop 74H76 discussed in Section 7.3, as shown in Fig. 8-1(b). Thus each $1 \longrightarrow 0$ transition of the ¢-input will cause a change of the output. The counter is cleared (set to its 0-state) by the simultaneous application of a 1-pulse to both R_0 inputs. If a sequence of pulses is then applied to the B input, the outputs Q_B, Q_C, and Q_D will perform as shown in Fig. 8-2, provided that propagation delays are neglected. Figure 8-2 represents a modulo-8 counter. Its state is given by $\perp(Q_D, Q_C, Q_B)$. If the output Q_A is connected to input B, and A is used as the external input, a modulo-16 counter is obtained.

This type of counter is also called a *divide-by-eight* counter (or *divide-by-sixteen*, if A is used), because the frequency of the output Q_D is one eighth of the frequency of the input B. (See Fig. 8-2.)

If propagation delays are taken into consideration, the ripple-through counter of Fig. 8-1 will not perform as shown in Fig. 8-2. In Fig. 8-3 we illustrate the transition from state 3 to state 4, assuming a propagation delay time δ for each flip-flop. As is clear from Fig. 8-3, the transition from state 3 to state 4 passes temporarily through the undesired state combinations 2 and 0. To overcome this difficulty, state combinations have to be observed only at appropriate time slots. The additional input providing these time slots is usually referred to as *strobe pulse input*.

In distinction from ripple-through counters, the counters operating in the *synchronous mode* (to be discussed later) avoid the difficulty of undesired state transitions. Another advantage of the synchronous mode is higher speed. Consider the transition when the count changes from 11 . . . 1 to

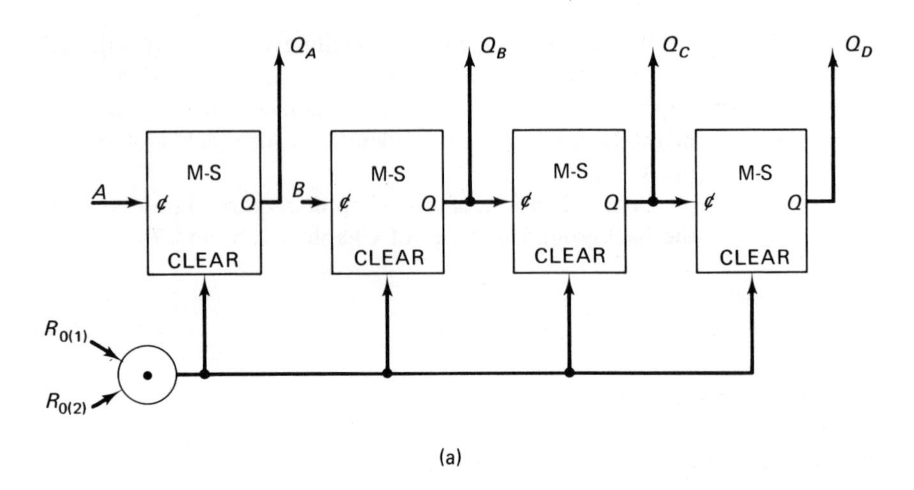

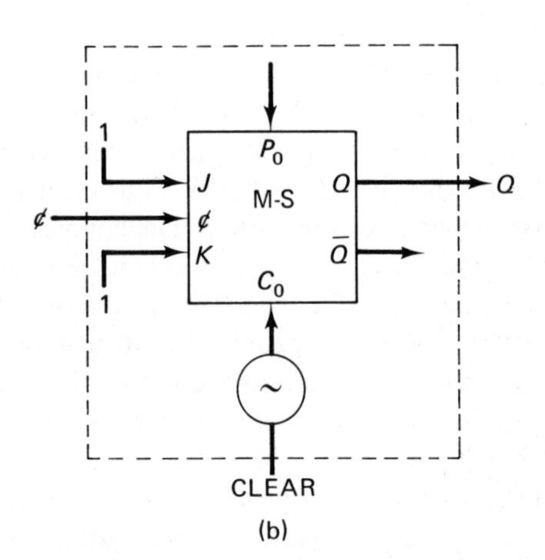

Fig. 8-1 (a) 4-bit binary ripple-through counter 7493; (b) flip-flop used in part (a).

00 . . . 0. In the ripple counter, a "ripple carry" has to propagate through all the stages. In the synchronous counter all flip-flops change in parallel.

We now give examples of counters that can be obtained by modifying the binary ripple counter.

Other Ripple Counters

It is possible to modify the basic binary ripple counter to obtain counters that count modulo 10 or modulo 12, for example. Consider the network of Fig. 8-4, consisting of four master–slave flip-flops as shown. The first and third flip-flops act like toggles, with inputs A and Q_B, respectively. The

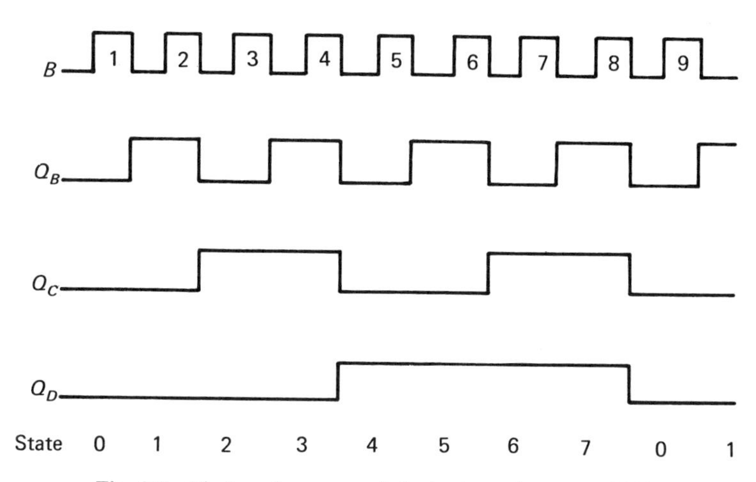

Fig. 8-2 Ideal performance of ripple-through counter 7493.

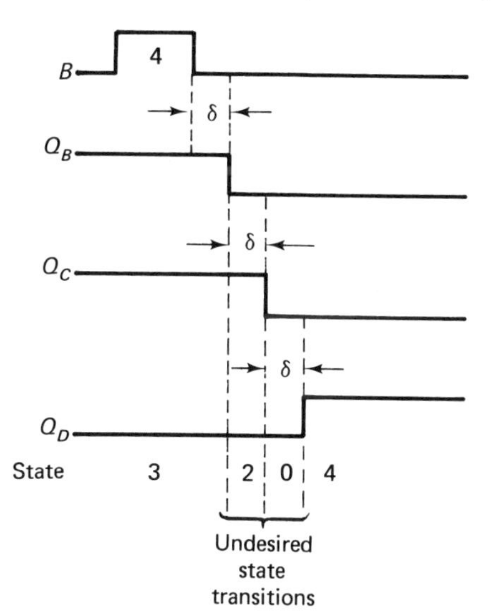

Fig. 8-3 Effect of propagation delays on ripple-through counters.

second acts like a toggle with input Q_A if $Q_D = 0$; if $Q_D = 1$, it resets after the next Q_A pulse. The last flip-flop is of the RS type with Q_A as the ȼ-input and $R = Q_D$, $S = Q_B Q_C$.

Typical waveforms for this decade counter are shown in Fig. 8-5. Assume that the counter starts at 0, i.e., $Q_D, Q_C, Q_B, Q_A = 0, 0, 0, 0$. As long as $Q_D = 0$ (i.e., until the count of 8), the first three flip-flops act like a 3-bit binary counter. The fourth flip-flop sets for the first time only after $Q_B = Q_C = 1$ and Q_A falls to 0, i.e., just after the count of 7. This explains the behavior up to the end of the eighth pulse. When the count is 8, $\bar{Q}_D = 0$ and the second flip-flop stays reset, since $J = 0$, $K = 1$ until Q_D becomes 0 again. Also, the reset input to the fourth flip-flop becomes 1, and $S = Q_B Q_C = 0$.

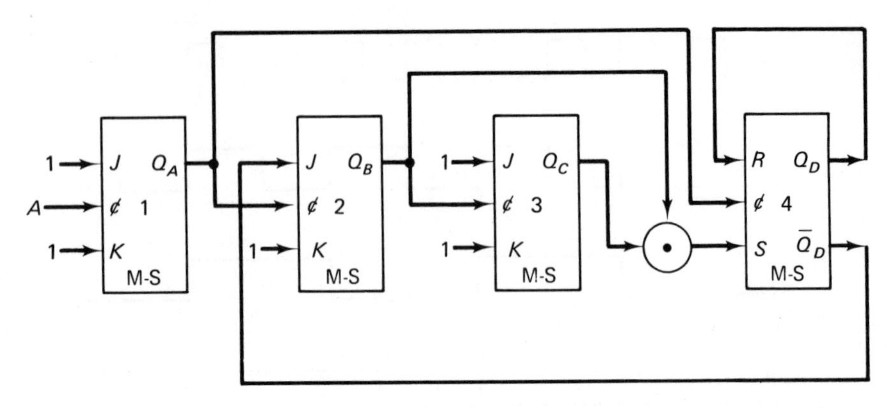

Fig. 8-4 Decade counter.

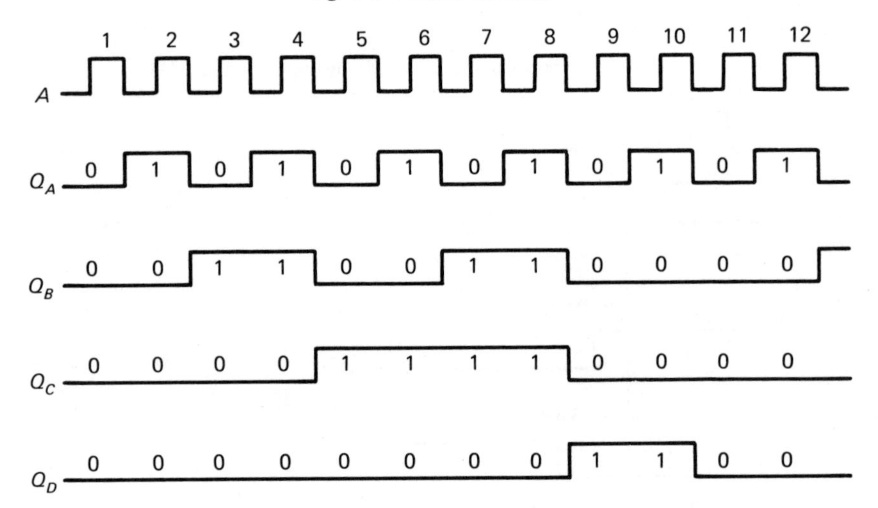

Fig. 8-5 Typical waveforms for a decade counter.

Thus when the Q_A-pulse (following the ninth A-pulse) arrives, it acts as a clock for the fourth flip-flop, which then resets to 0. The cycle then repeats. Clearly $\perp(Q_D, Q_C, Q_B, Q_A)$ counts the A-input modulo 10.

The network of Fig. 8-6 is a modified version of that of Fig. 8-4. An analysis (similar to that given for the decade counter) shows that Q_D, Q_C, Q_B count INPUT modulo 5. The Q_A-output is 0 for the first 5 input pulses and it is 1 for the remaining 5. Thus Q_A has one pulse for every 10 input pulses. This corresponds to dividing the input frequency by 10. (See Fig. 8-7 for waveform details.)

Both the counters of Figs. 8-4 and 8-6 are available in a single IC package, 7490 [SIG, TI2]. The modifications in the network are performed by external wiring. In fact, the network can act as a decade counter, a divide-by-10 counter, and also as a divide-by-2 counter (flip-flop A) and an independent divide-by-5 counter (flip-flops B, C, and D).

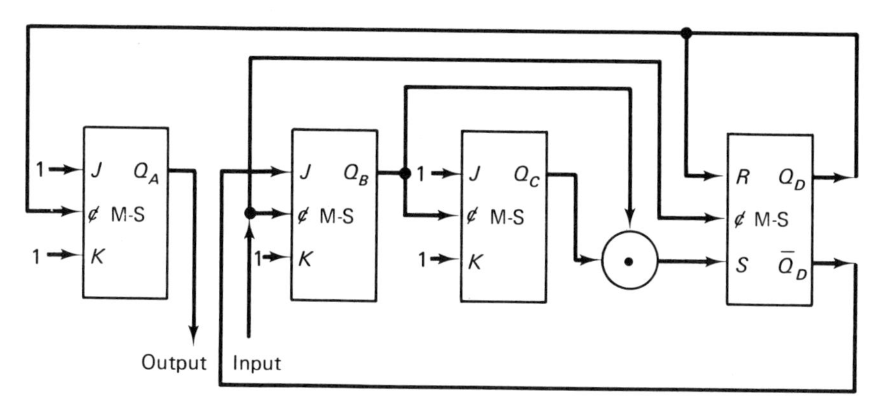

Fig. 8-6 Divide-by-ten counter.

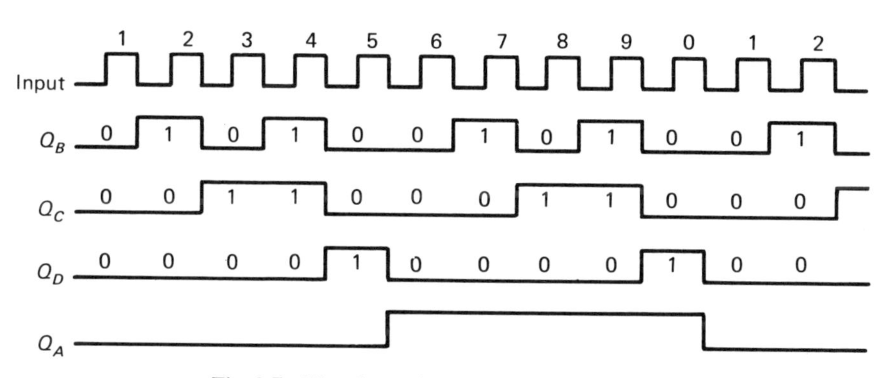

Fig. 8-7 Waveforms for divide-by-ten counter.

Additional facilities are provided for resetting the count to 0 and 9. The latter facility is useful for 9's-complement decimal applications.

A divide-by-12 counter is another example of a ripple counter modified by feedback to clear all the flip-flops to 0 when the count reaches 12. See, for example, package 7492 [SIG, TI2].

8.2. SHIFT REGISTERS

A *shift register* is a cascade of flip-flops together with facilities for shifting the bit string stored in one direction, or sometimes both to the right and to the left. Shift registers usually provide a *serial-in* feature, i.e., the ability to load input data into the register serially, through the input to the first flip-flop, as well as a *serial-out* feature, i.e., serial reading out, by means of the output of the last flip-flop. Shift registers frequently incorporate additional facilities, especially *parallel-out* (i.e., the outputs of all flip-flops are connected to outside terminals) and *parallel-in* (i.e., all flip-flops can be loaded simultaneously). The parallel-out feature can be used for serial-to-parallel conversion, and the parallel-in feature for parallel-to-serial conversion.

Master–slave flip-flops are frequently used in the design of shift registers. To provide for parallel-in facilities, the flip-flops are equipped with CLEAR and PRESET inputs. We now give two examples of shift registers available in IC packages.

Shift Register 7496 (5-Bit Shift Register) [SIG, TI2]

This 16-pin MSI package is a 5-bit (right)-shift register with parallel-in and parallel-out facilities. It consists of 5 *RS* master–slave flip-flops connected as shown in Fig. 8-8. Each clock pulse will cause the bit string stored in the

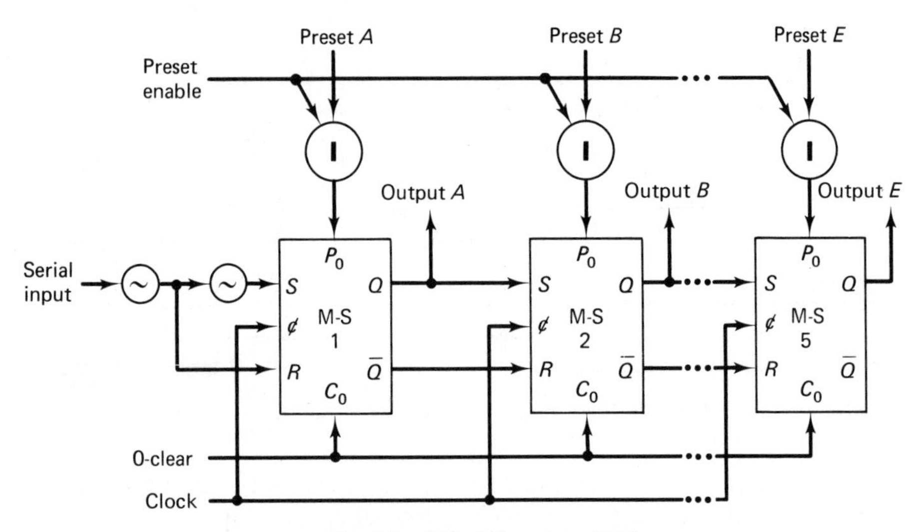

Fig. 8-8 5-bit shift register 7496.

register to be shifted once to the right. The bit appearing at the SERIAL INPUT is first stored in the master latch of the first flip-flop just after the clock rises. At the same time, the bit stored in the first slave latch S_1 is transferred into the second master M_2, the S_2-bit is stored in M_3, etc. Just after the clock falls, the bit stored in each master M_i is transferred to the corresponding slave S_i. The overall effect of one clock pulse is to perform the following shift. If x_0 is the SERIAL INPUT and $x_1x_2x_3x_4x_5$ is the binary word stored in the register before the clock pulse, then the register holds $x_0x_1x_2x_3x_4$ after the clock pulse.

Note that the CLEAR input is common to all 5 flip-flops. To achieve parallel loading a 0-pulse is first applied to the CLEAR input, clearing the register. The word to be loaded is then connected to the inputs PRESET *A–E* and a 1-pulse is applied to the PRESET ENABLE input. For serial operations the CLEAR input must be 1, and the PRESET inputs must be 0.

Shift Register 74194 (4-Bit Bidirectional Universal Shift Register) [SIG, TI2]

This 16-pin MSI package is a 4-bit bidirectional shift register with parallel-in and parallel-out facilities. It consists of 4 *RS* positive-edge triggered flip-flops with 0-CLEAR, and an additional gate network, as shown in Fig. 8-9. The register has four distinct modes of operation, controlled by the mode

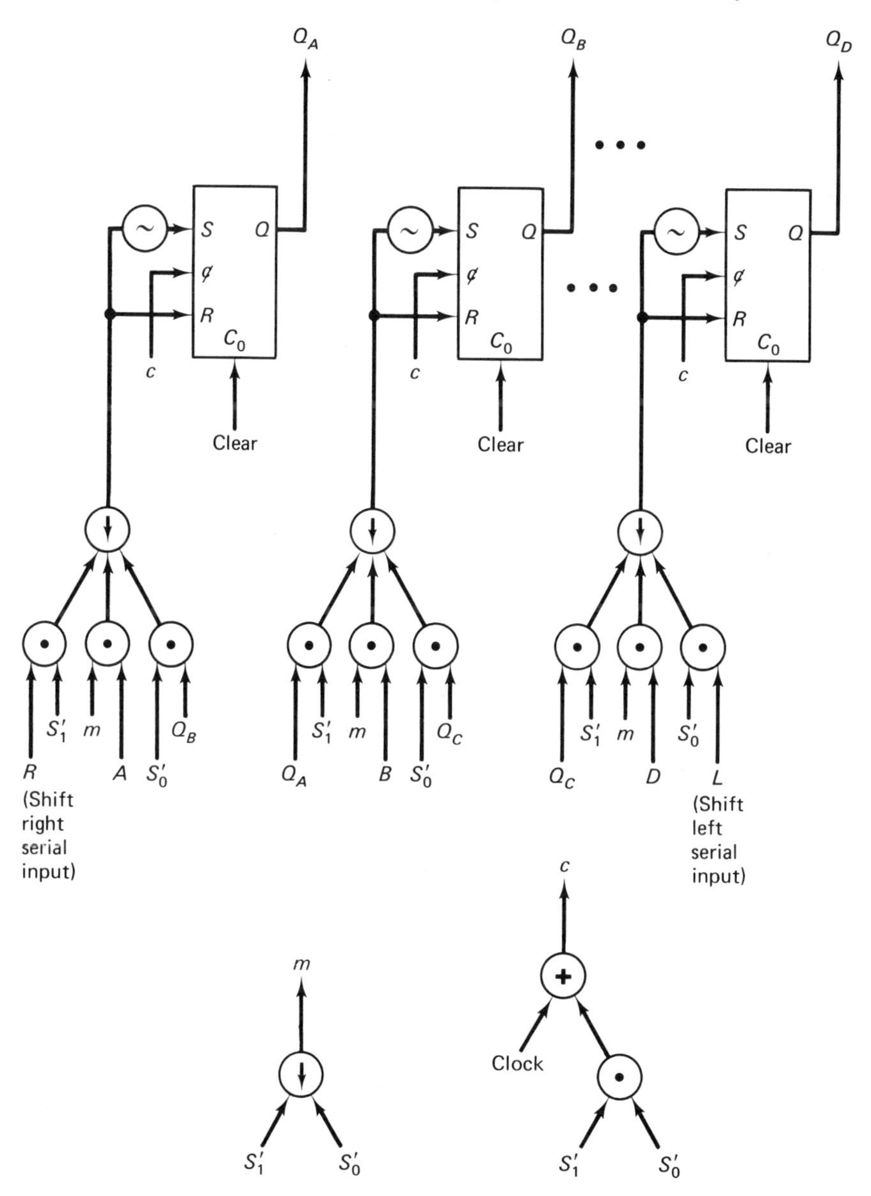

Fig. 8-9 4-bit bidirectional universal shift register 74194.

control inputs S_0 and S_1. If $S_0 = 1$ and $S_1 = 0$, the register performs as a right-shift register; if $S_0 = 0$ and $S_1 = 1$, it performs as a left-shift register. For parallel loading, the input word is connected to inputs A–D, and both S_0 and S_1 are set to 1. The proper output will appear after the clock transition from 0 to 1. If both S_0 and S_1 are set to 0, the "clock" input c to the flip-flops is inhibited (c is locked to 1). The mode control inputs should be changed only while the CLOCK input is 1. The application of a 0-pulse to the CLEAR input will clear the register, overriding all other inputs as usual.

8.3. SYSTEMATIC ANALYSIS OF SYNCHRONOUS-MODE NETWORKS

In many applications, sequential networks use master–slave flip-flops, controlled by a common clock, as memory devices. For example, this was the case for the shift registers of Section 8.2, which we have just analyzed by intuitive methods. We will now present systematic methods for the analysis and synthesis of such networks. Sequential networks in which all flip-flops are controlled (synchronized) by a common clock are said to operate in the *synchronous mode.*

Typically, no changes in the flip-flop states can take place when $¢ = 0$. When $¢$ becomes 1, those flip-flops which are supposed to change state will record this information in their master sections. It is assumed that all the masters will stabilize before the clock returns to 0. Since the state of the master of each flip-flop is independent of the states of the other masters (although it usually depends on the states of the slaves), any races that may be present among the various masters are noncritical. Similarly, when the clock is removed, each slave depends only on its own master, and all races among the various slaves are noncritical. For these reasons the analysis and design techniques for synchronous-mode networks are considerably easier than those of arbitrary sequential networks.

We develop the ideas by a series of examples.

Analysis Example 1 (*RS* Flip-Flops)

The network N_1 of Fig. 8-10 consists of two master–slave *RS* flip-flops and two gates, connected as shown. There is only one external input $¢$, which is assumed to be a periodic clock input. The outputs are y_1 and y_2.

Consider the clock waveform of Fig. 8-11. We assume that the clock downtime is long enough that all the transients in the flip-flops and gates settle some time before the clock rises again. We will refer to "time n" as the time immediately preceding the nth clock pulse. At this time all the signals in the network are stable. For any signal x, denote by x^n the value of x at time n. The signals must be related by the following equations:

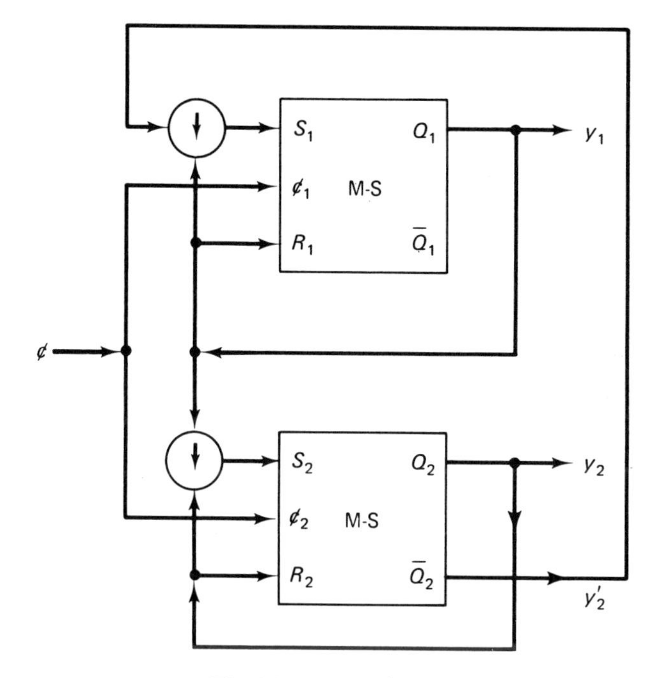

Fig. 8-10 Network N_1.

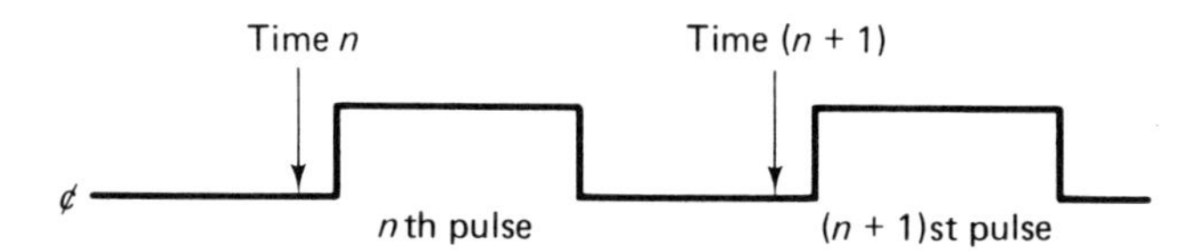

Fig. 8-11 Clock waveform.

$$S_1^n = (y_2')^n \downarrow y_1^n \tag{1}$$

$$R_1^n = y_1^n \tag{2}$$

$$S_2^n = y_1^n \downarrow y_2^n \tag{3}$$

$$R_2^n = y_2^n. \tag{4}$$

The values of y_1^n and y_2^n are not known because they depend on the previous history. However, whatever these values are, they uniquely determine the values of S_1, R_1, S_2, R_2 at time n. Now when the clock rises, the master section of each flip-flop will "see" the S and R values and react accordingly. We call the S and R signals the (flip-flop) *excitation signals*. Equations (1)–(4) are called the (flip-flop) *excitation equations*.

Some time after the clock has risen, the masters have both stabilized. When the clock falls, the slaves will follow the masters, and at time $n + 1$ the network is guaranteed to have completed its transition to the new state (which could be the same as the old state). Thus we are really interested in the signals y_1^{n+1} and y_2^{n+1} and we must compute them using (1)–(4).

To simplify the notation, we drop the superscript n and use Y_i for y_i^{n+1}. Thus the excitation equations become

$$S_1 = y_1 \downarrow y_2' = (y_1 + y_2')' = y_1'y_2$$
$$R_1 = y_1$$
$$S_2 = y_1'y_2'$$
$$R_2 = y_2.$$

In computing the values Y_1 and Y_2, it is convenient to represent the excitation equations in terms of a table called the (flip-flop) *excitation* table. This table for N_1 is shown in Fig. 8-12(a). From the excitation table we can

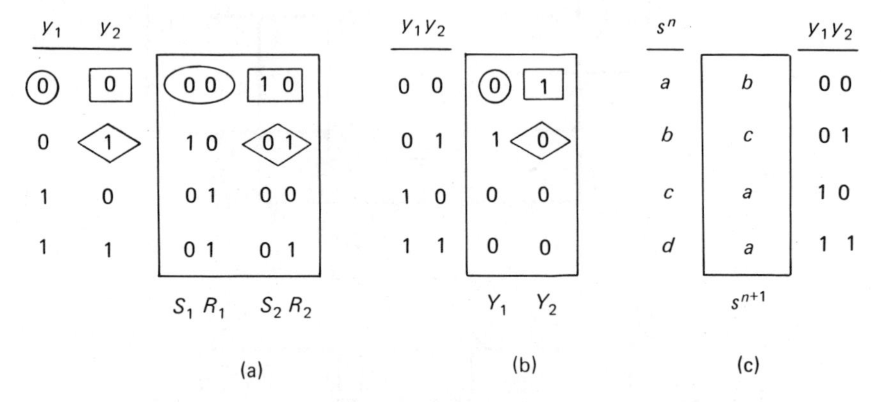

Fig. 8-12 (a) Flip-flop excitation table for N_1; (b) flip-flop transition table for N_1; (c) state table for N_1.

easily obtain the (flip-flop) *transition table*, which is a table listing the values of y_1 and y_2 at time $n + 1$ (denoted by Y_1 and Y_2) as functions of the values of y_1 and y_2 at time n. This is obtained by using the definition of an *RS* flip-flop. Recall that

when $S_i = 0$, $R_i = 0$, then $Y_i = y_i$ (REMEMBER condition)
when $S_i = 0$, $R_i = 1$, then $Y_i = 0$ (RESET condition)
when $S_i = 1$, $R_i = 0$, then $Y_i = 1$ (SET condition)
when $S_i = 1$, $R_i = 1$, then Y_i is not well defined.

Now examine the first row of the excitation table of Fig. 8-12(a). We have $y_1 = y_2 = 0$. Since $S_1 = R_1 = 0$, we find $Y_1 = y_1 = 0$ (circled entries). Since $S_2 = 1$, $R_2 = 0$, we have $Y_2 = 1$ (square entries). In the next row, $S_2 = 0$, $R_2 = 1$; hence $Y_2 = 0$ (diamond-shaped entries). Repeating this for all the possibilities we find the transition table of Fig. 8-12(b). Sometimes it is convenient to suppress the details of the actual coding of flip-flop states and refer to them simply by letters a, b, c, etc. This leads to the table of Fig. 8-12 (c), called the *state table*. Along with each state symbol s, we usually list the corresponding output values.

It is sometimes easier to understand the behavior of a network if its transition or state table is represented in graphical form by a *state graph* or *state diagram*. Introduce a node for each flip-flop state s and a directed edge from s to its successor. This is done in Fig. 8-13. Notice from this figure that if the network is started in state 00, it performs like a modulo-3 counter.

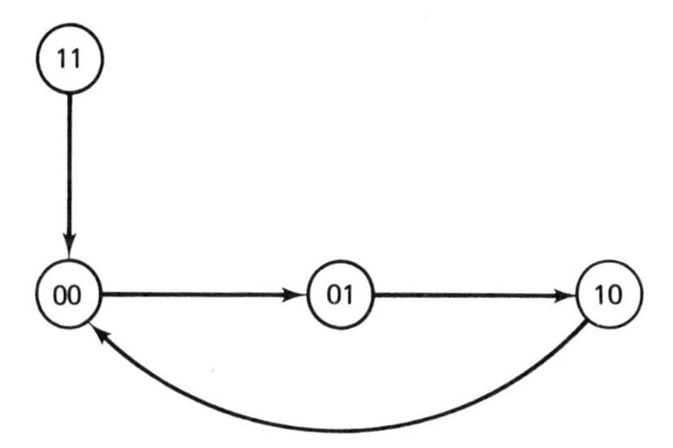

Fig. 8-13 State graph for N_1.

Finally, we may complete the analysis procedure by showing typical waveforms. This is done in Fig. 8-14, where we have assumed that initially $y_1 = y_2 = 0$. As usual, propagation delays are not shown in these waveforms.

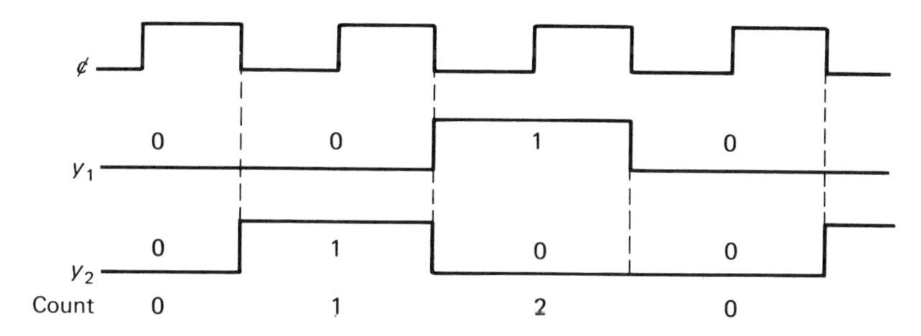

Fig. 8-14 Waveforms for N_1.

Analysis Example 2 (*T* Flip-Flops)

Figure 8-15 shows a network N_2 consisting of three toggle flip-flops. We follow an analysis sequence very similar to that of Example 1. First the flip-flop excitation equations are

$$T_1 = 1$$
$$T_2 = y_1$$
$$T_3 = y_1 y_2.$$

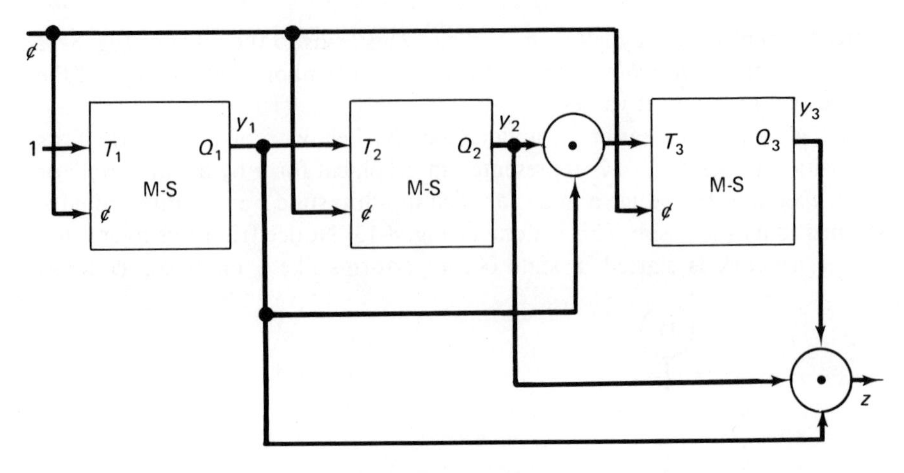

Fig. 8-15 Network N_2.

These are represented in the excitation table of Fig. 8-16(a). This time we must use the basic property of the toggle flip-flop in order to compute the flip-flop transitions; namely, the flip-flop stays in the same state if $T = 0$ and changes state if $T = 1$. In this way we obtain the transition table of Fig. 8-16(b). Examine the specially marked entries to follow the detailed computation. In fact, we have for T flip-flops,

$$Y = y \oplus T$$

as the fundamental equation for obtaining the transition table.

The state table for N_2 is shown in Fig. 8-16(c), where the output z was obtained from the *output equation*:

$$z = y_1 y_2 y_3.$$

(a)

y_1	y_2	y_3	T_1	T_2	T_3
0	0	0	1	0	0
0	0	1	1	0	0
0	1	0	1	0	0
0	1	1	1	0	0
1	0	0	1	1	0
1	0	1	1	1	0
1	1	0	1	1	1
1	1	1	1	1	1

(b)

y_1	y_2	y_3	Y_1	Y_2	Y_3
0	0	0	1	0	0
0	0	1	1	0	1
0	1	0	1	1	0
0	1	1	1	1	1
1	0	0	0	1	0
1	0	1	0	1	1
1	1	0	0	0	1
1	1	1	0	0	0

(c)

s^n	s^{n+1}	z
a	e	0
b	f	0
c	g	0
d	h	0
e	c	0
f	d	0
g	b	0
h	a	1

Fig. 8-16 Tables for N_2: (a) excitation table; (b) transition table; (c) state table.

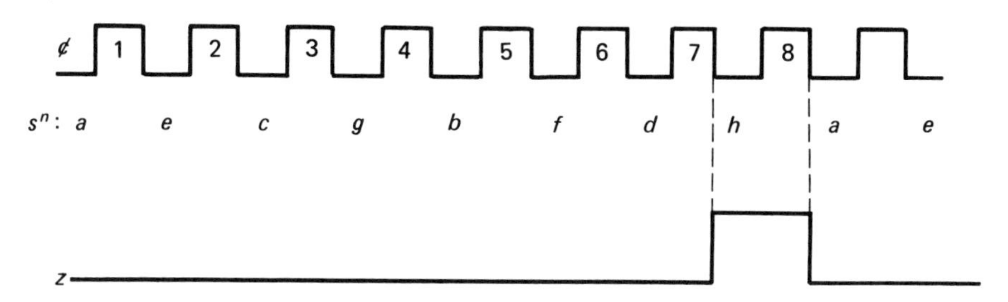

Fig. 8-17 Waveforms for N_2.

The waveforms of Fig. 8-17 explain the behavior of N_2. The network cycles through the 8 states in the order $a, e, c, g, b, f, d, h, a, e, c$, etc. For every 8 clock pulses, there is one output pulse, as shown.

Analysis Example 3 (Clock Gating)

In the network N_3 of Fig. 8-18, all the toggle functions are 1. Note, however, that the clock inputs of flip-flops 2 and 3 are *not* connected directly to the network clock. Hence we cannot proceed as before. Notice, however, that flip-flop i toggles when $T_i = 1$ and $¢ = 1$. Hence we should look at $T_i¢_i$:

$$T_1¢_1 = ¢$$
$$T_2¢_2 = ¢y_1$$
$$T_3¢_3 = ¢y_1y_2.$$

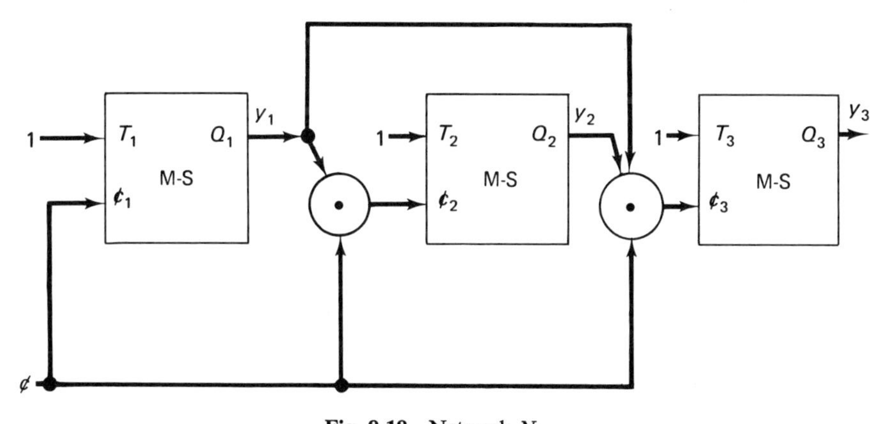

Fig. 8-18 Network N_3.

Compare this with the excitation equations for N_2. Clearly both networks will have identical transition tables. For N_3 the *effective excitation functions* $\hat{T}_i$ are obtained by setting $¢ = 1$ in $T_i¢_i$. Thus

$$\hat{T}_1 = 1$$
$$\hat{T}_2 = y_1$$
$$\hat{T}_3 = y_1y_2.$$

The technique of gating the clock to achieve effective excitation functions is often used since it provides more flexibility in the design. This will be further illustrated in later examples.

Analysis Example 4

We have introduced some analysis techniques for synchronous-mode networks by means of examples. However, these ideas are quite general. The flip-flops used can be of the *JK*, *RS*, *D*, or *T* type. In any case, we can always express the excitation inputs in terms of the external inputs and the *Q*-variables, which act here as internal state variables. Knowing these excitation inputs we can find the next state of each flip-flop by using the following equations:

$$D \textit{ flip-flop:} \quad Q^{n+1} = D^n \tag{5}$$

$$T \textit{ flip-flop:} \quad Q^{n+1} = T^n \oplus Q^n \tag{6}$$

$$SR \textit{ flip-flop:} \quad Q^{n+1} = S^n + (R^n)'Q^n \tag{7}$$

$$JK \textit{ flip-flop:} \quad Q^{n+1} = J^n(Q^n)' + (K^n)'Q^n. \tag{8}$$

These equations were obtained in Chapter 7.

The modules available in IC packages usually use the same type of flip-flop throughout the module. However, to illustrate the generality of our method, we present an example below that uses three different types of flip-flops. (See Fig. 8-19.)

The excitation equations are found to be

$$D = M'AB + MAB' + AC + BC$$

$$\begin{cases} J = MC + A'C \\ K = MAC' + A'C \end{cases}$$

$$\begin{cases} S = A'C' + M'BC' + MB'C' \\ R = A'C + M'BC + MB'C. \end{cases}$$

Let A^{n+1}, B^{n+1}, and C^{n+1} denote the flip-flop states at time $n + 1$ and let A, B, and C denote the state at time n for simplicity of notation. Then

$$A^{n+1} = D^n = D = M'AB + MAB' + AC + BC \tag{9}$$

$$\begin{aligned} B^{n+1} &= JB' + K'B \qquad \text{[from (8)]} \\ &= (MC + A'C)B' + (MAC' + A'C)'B \\ &= MB'C + A'B'C + (M' + A' + C)(A + C')B \\ &= MB'C + M'AB + M'BC' + A'B'C + A'BC' + ABC \end{aligned} \tag{10}$$

$$\begin{aligned} C^{n+1} &= S + R'C = A'C' + M'BC' + MB'C' \\ &\quad + (A + C')(M + B' + C')(M' + B + C')C \\ &= A'C' + M'BC' + MB'C' + MABC + M'AB'C. \end{aligned} \tag{11}$$

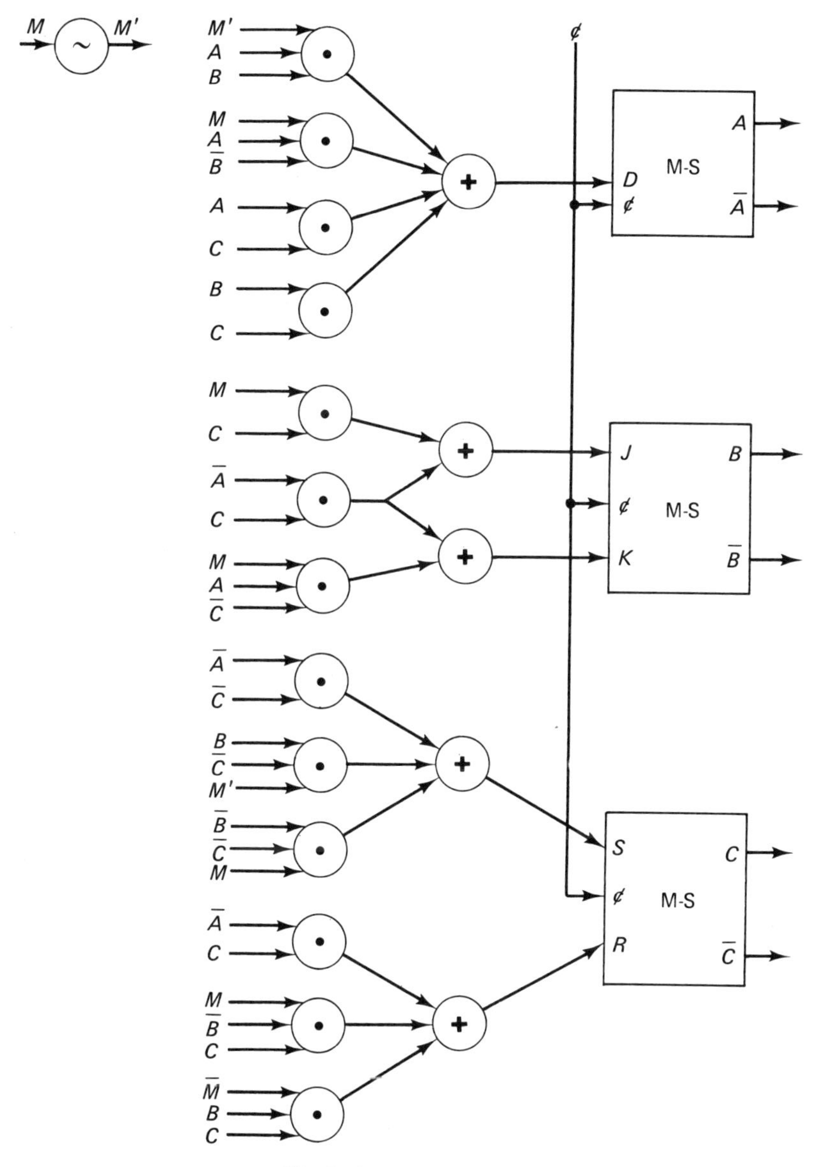

Fig. 8-19 Network N_4.

The transition table is shown in Fig. 8-20. Let us represent each state by the corresponding decimal number and let us construct the state graph as shown in Fig. 8-21. As before, the states are represented by nodes and transitions by directed branches. This time there is an external input M and the transitions are labeled by the corresponding value of M. Thus, if $s^n = 3$,

s	A B C \ M	0	1
(0)	0 0 0	0 0 1	0 0 1
(1)	0 0 1	0 1 0	0 1 0
(2)	0 1 0	0 1 1	0 1 1
(3)	0 1 1	1 0 0	1 0 0
(4)	1 0 0	0 0 0	1 0 1
(5)	1 0 1	1 0 1	1 1 0
(6)	1 1 0	1 1 1	0 0 0
(7)	1 1 1	1 1 0	1 1 1
		A^{n+1} B^{n+1} C^{n+1}	

Fig. 8-20 Transition table for N_4.

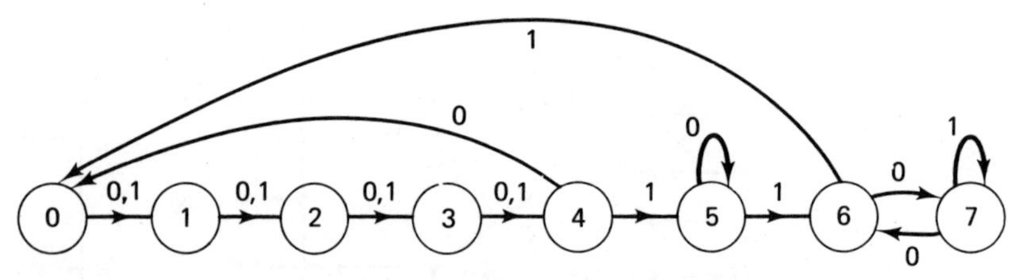

Fig. 8-21 State graph for N_4.

then $s^{n+1} = 4$, independently of M. However, when $s^n = 4$, $s^{n+1} = 5$ if $M = 1$ and $s^{n+1} = 0$ if $M = 0$.

It is seen from Fig. 8-21 that if $M = 0$ and the network is started in state 0, it acts as a modulo-5 counter. If started in state 6, it will act as a modulo-2 counter. On the other hand, if $M = 1$, the network started in state 0 is a modulo-7 counter. Thus the input M can be used as a *mode* control to get two different behavior modes from the network. If M is not constant, then the best description of the network behavior is the state graph of Fig. 8-21. Given any input (M) sequence and an initial state, the graph permits us to predict the state of the flip-flop corresponding to that sequence.

Analysis Example 5 (Decade Counter)

The network N_5 of Fig. 8-22 has four toggle flip-flops obtained from master–slave JK flip-flops as in Fig. 8-1(b). It has only one external input ¢, the clock input.

We obtain the following effective excitation equations:

$$T_A = 1$$
$$T_B = Q_A\bar{Q}_D$$
$$T_C = Q_AQ_B$$
$$T_D = Q_AQ_D + Q_AQ_BQ_C.$$

Columns I and II of Fig. 8-23 form the excitation table. Columns I and III constitute the transition table.

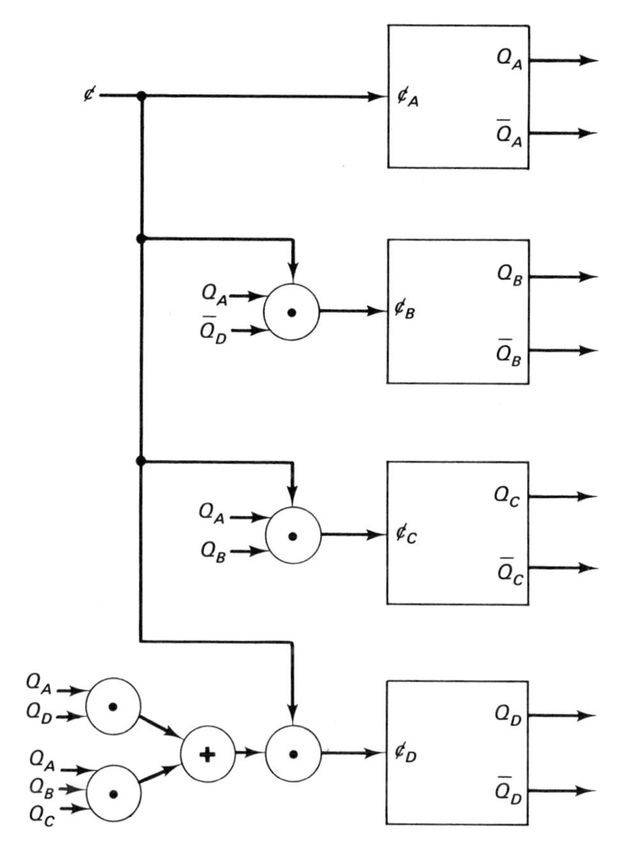

Fig. 8-22 Network N_5.

For convenience we have assigned letters to the states (s) in Fig. 8-23. This is shorthand for the state of the flip-flops. It is more instructive to represent the information in columns I′ and III′ (equivalent to I and III) by the state graph of Fig. 8-24. It is now clear from Fig. 8-24 that the network is basically a modulo-10 counter of the clock pulses, if states $a, b, c, e, g, i, j, k, m$, and o are used. Typical waveforms are shown in Fig. 8-25, assuming that the network starts in state d. Ignore the RC waveform for the time being. Except possibly for initial transient situations, states h, p, d, l, f, and n will never be reached. The present count modulo-10 is represented in binary by the flip-flop outputs Q_D, Q_C, Q_B, and Q_A.

Asynchronous Facilities

A network of the type analyzed above is available in package 74160 [SIG, TI2] (*synchronous decade counter*). There are a number of asynchronous facilities provided to make the counter more versatile. The complete logic

I′	I				II				III				III′
s^n	Q_A^n	Q_B^n	Q_C^n	Q_D^n	T_A^n	T_B^n	T_C^n	T_D^n	Q_A^{n+1}	Q_B^{n+1}	Q_C^{n+1}	Q_D^{n+1}	s^{n+1}
a	0	0	0	0	1	0	0	0	1	0	0	0	*i*
b	0	0	0	1	1	0	0	0	1	0	0	1	*j*
c	0	0	1	0	1	0	0	0	1	0	1	0	*k*
d	0	0	1	1	1	0	0	0	1	0	1	1	*l*
e	0	1	0	0	1	0	0	0	1	1	0	0	*m*
f	0	1	0	1	1	0	0	0	1	1	0	1	*n*
g	0	1	1	0	1	0	0	0	1	1	1	0	*o*
h	0	1	1	1	1	0	0	0	1	1	1	1	*p*
i	1	0	0	0	1	1	0	0	0	1	0	0	*e*
j	1	0	0	1	1	0	0	1	0	0	0	0	*a*
k	1	0	1	0	1	1	0	0	0	1	1	0	*g*
l	1	0	1	1	1	0	0	1	0	0	1	0	*c*
m	1	1	0	0	1	1	1	0	0	0	1	0	*c*
n	1	1	0	1	1	0	1	1	0	1	1	0	*g*
o	1	1	1	0	1	1	1	1	0	0	0	1	*b*
p	1	1	1	1	1	0	1	1	0	1	0	0	*e*

Fig. 8-23 Excitation and transition tables.

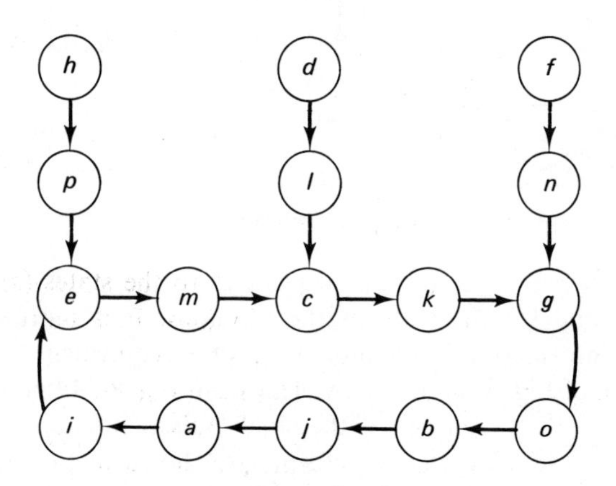

Fig. 8-24 State graph.

network is as shown in Fig. 8-26. By inspection of the diagram and some algebraic manipulation, we find

$$J_A = L + D_A \qquad K_A = L + D'_A \qquad ¢_A = ¢(E + L')$$
$$J_B = L + D_B \qquad K_B = L + D'_B \qquad ¢_B = ¢(E\, Q_A\bar{Q}_D + L')$$
$$J_C = L + D_C \qquad K_C = L + D'_C \qquad ¢_C = ¢(E\, Q_A Q_B + L')$$
$$J_D = L + D_D \qquad K_D = L + D'_D \qquad ¢_D = ¢(X + Y + L').$$

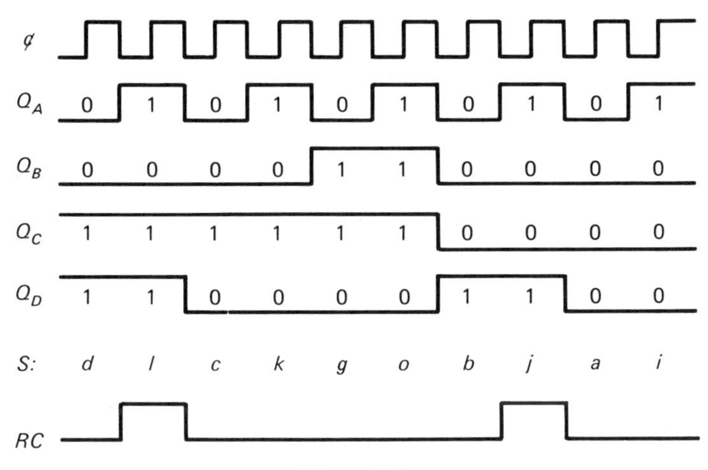

Figure 8-25

where $X = E\,Q_AQ_BQ_C$ and $Y = E\,Q_AQ_D$. Thus

$$¢_D = ¢E(Q_AQ_BQ_C + Q_AQ_D) + ¢L'.$$

From these equations we can describe the additional facilities. Normally $L = 1$. The input E is called the *count enable* input. If $E = 0$, the clock input ¢ is effectively disabled; thus we must have $E = 1$ for the normal counting operation. This counter has a presetting facility provided by the inputs $D_A, \ldots, D_D$ (the *data inputs*) under the control of the *load input L*. When $L = 1$, we have $J_N = K_N = 1$ for each N. This is the passive condition, blocking the data inputs and forcing each *JK* flip-flop to behave like a toggle flip-flop. When $L = 0$, we have $J_N = D_N$, $K_N = D'_N$. After the next clock pulse, the outputs will agree with the data word; for this reason, the presetting is called *synchronous*. An asynchronous *0-clear* input C_0 is provided for setting the count to 0 (regardless of the state of the clock).

Figure 8-26(b) shows that the enable input E is actually derived from two inputs P and T. The input T enables the ripple-carry output RC. A pulse will be produced at RC when the count is 9 since $Q_A = Q_D = 1$ (see Fig. 8-25). The RC can be used to trigger another counter of the same type without additional gating.

The analysis given above illustrates that for practical applications it is desirable to have a number of asynchronous facilities like clearing and presetting along with the basic synchronous counting ability. However, in order to focus our attention on the synchronous aspects of the operation, in this chapter we will often omit the discussion of asynchronous facilities.

This example also points out the versatility of the *JK* flip-flops. The J and K inputs are used here for a presetting function when $L = 0$. Otherwise,

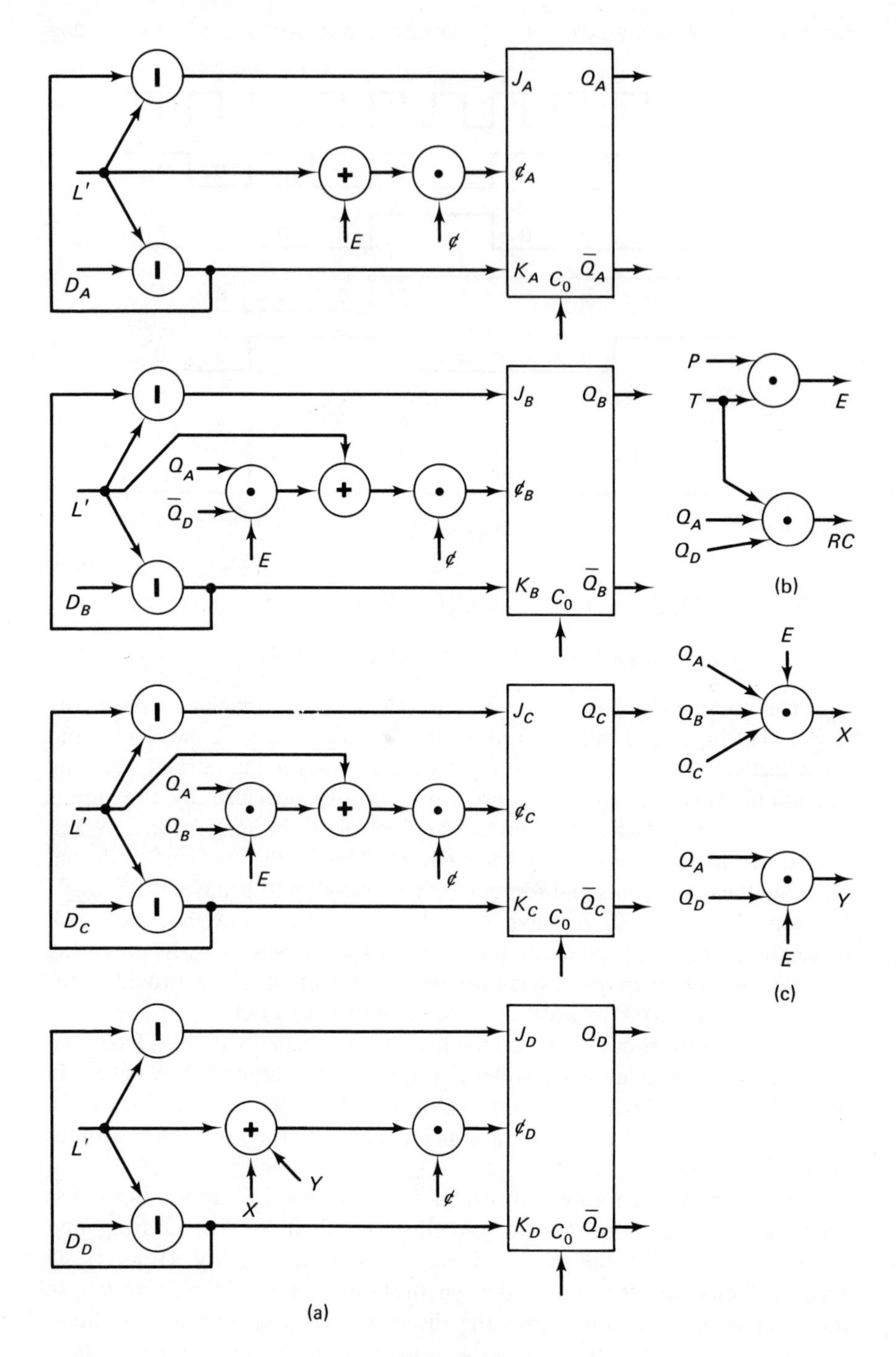

J_A
Q_A
L'
E
$¢$
$¢_A$
D_A
K_A
C_0
$\bar{Q}_A$
J_B
Q_B
Q_A
$\bar{Q}_D$
L'
E
$¢$
$¢_B$
D_B
K_B
C_0
$\bar{Q}_B$
P
T
E
Q_A
Q_D
RC
(b)
J_C
Q_C
Q_A
Q_B
L'
E
$¢$
$¢_C$
D_C
K_C
C_0
$\bar{Q}_C$
E
Q_A
Q_B
Q_C
X
Q_A
Q_D
Y
E
(c)
J_D
Q_D
L'
X
Y
$¢$
$¢_D$
D_D
K_D
C_0
$\bar{Q}_D$
(a)

Figure 8-26

$J = K = 1$ and the flip-flops act like toggles. Note that the toggle function here is implemented at the ¢-input.

8.4. SYNTHESIS OF SYNCHRONOUS-MODE SEQUENTIAL NETWORKS

We now consider the design of sequential networks operating in the synchronous mode. Although the methods are presented by means of examples, they are quite general. We assume that master–slave flip-flops and gates are available as the basic building blocks.

Synthesis Example 1 (Shift Registers)

We illustrate some steps in the design process by using the simple and already familiar example of a shift register.

WORD DESCRIPTION

The problem can be described as follows: Design a two-bit shift register with data input d and control input x. The contents of the register should remain unchanged if $x = 0$. If $x = 1$, the contents are to shift to the right by one bit, and the leftmost flip-flop of the register is to receive the data bit d. The network is to use clocked master–slave flip-flops. We will illustrate the design using D, RS, and JK flip-flops.

TRANSITION TABLE

Let y_1 and y_2 represent the shift register state at time n. The state at time $n + 1$ is represented by Y_1 and Y_2. The transition table corresponding to the word description just given is easily obtained as shown in Fig. 8-27. For convenience, we use the 4-variable map form for the transition table.

$y_1 y_2$ \ xd	00	01	11	10
00	00	00	10	00
01	01	01	10	00
11	11	11	11	01
10	10	10	11	01

$Y_1 Y_2$

Fig. 8-27 Transition table for shift register.

The transition table defines precisely how the two flip-flops (that are to be used as memory devices) are to behave. Now we must find suitable excitation functions to cause the flip-flops to follow the transition table. We now describe the procedure for D, RS, and JK master–slave flip-flops. Since the excitation table is different for each type of flip-flop, we consider the three cases separately. We leave to the reader the design of this shift register with T flip-flops, as an exercise.

EXCITATION TABLE

D Flip-Flops

The problem is reduced to the following one. If the present state of a flip-flop is y^n and the next state is to be $y^{n+1} \triangleq Y^n$, what should be the excitation D^n? From (5) we know that $y^{n+1} = Y^n = D^n$. Hence *for D flip-flops, the excitation table is identical to the transition table.* Therefore, the excitation equations can be found directly from the transition table. The reader will verify that

$$D_1 = Y_1 = x'y_1 + xd$$

$$D_2 = Y_2 = x'y_2 + xy_1.$$

The network diagram can now be drawn as shown in Fig. 8-28. Notice that the two stages have identical logic, as one would expect.

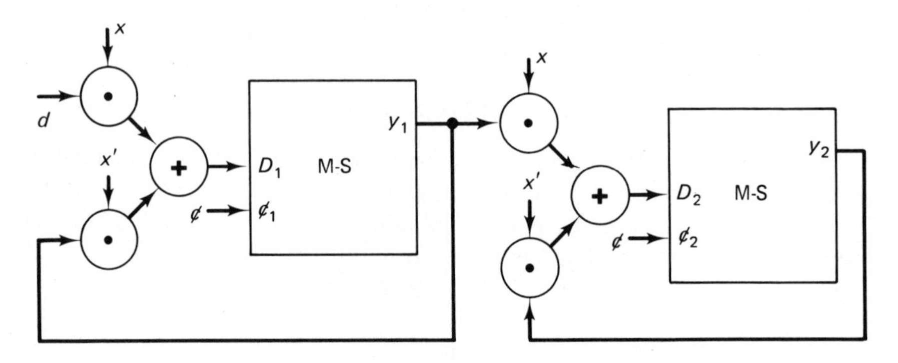

Fig. 8-28 2-bit shift register with D flip-flops.

RS Flip-Flops

We must return to the transition table and find the proper R and S inputs. The RS "design table" is shown in Fig. 8-29. If y is 0 and is to remain 0, we cannot set, but we may reset. Thus $S = 0$ and R is irrelevant, i.e., a don't care. If $y = 0$ and $Y = 1$, we must set but not reset, etc. We now repeatedly apply the RS design table to the transition table of Fig. 8-27, obtaining the excitation table shown in Fig. 8-30. The don't cares are often helpful in simplifying the excitation logic. In Fig. 8-31 the excitation table is separated

y	Y	S	R
0	0	0	–
0	1	1	0
1	0	0	1
1	1	–	0

Fig. 8-29 RS design table.

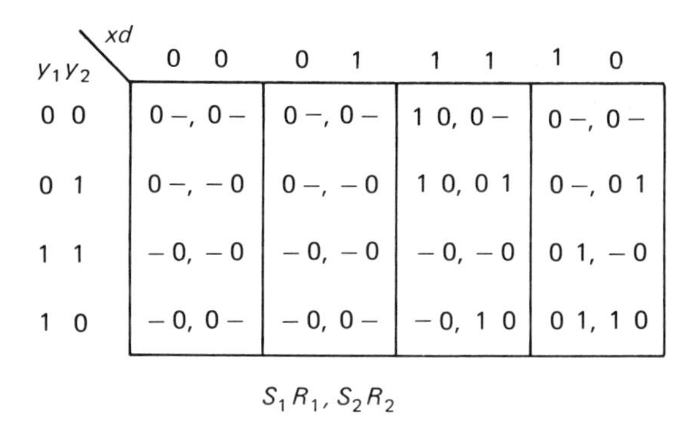

y_1y_2 \ xd	0 0	0 1	1 1	1 0
0 0	0 –, 0 –	0 –, 0 –	1 0, 0 –	0 –, 0 –
0 1	0 –, – 0	0 –, – 0	1 0, 0 1	0 –, 0 1
1 1	– 0, – 0	– 0, – 0	– 0, – 0	0 1, – 0
1 0	– 0, 0 –	– 0, 0 –	– 0, 1 0	0 1, 1 0

S_1R_1, S_2R_2

Fig. 8-30 Excitation table for RS flip-flops.

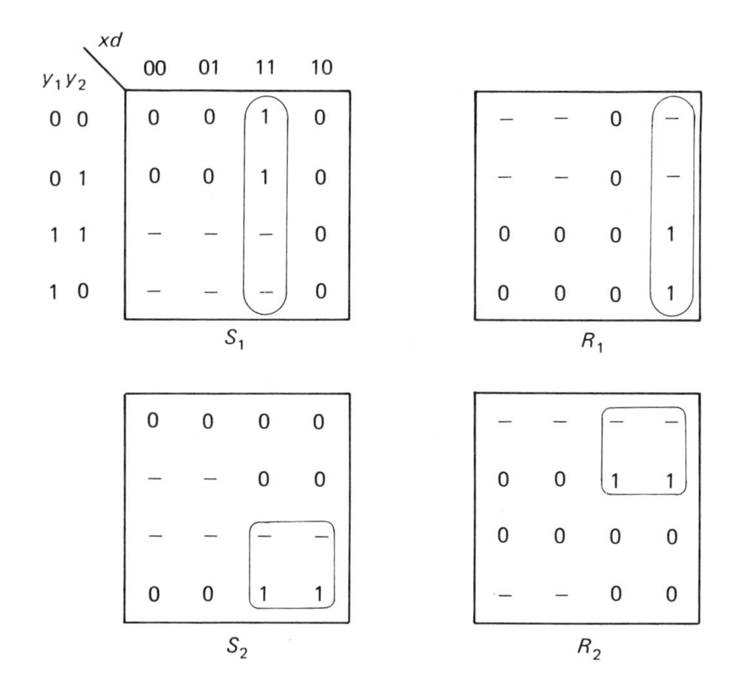

S_1:

y_1y_2 \ xd	00	01	11	10
0 0	0	0	1	0
0 1	0	0	1	0
1 1	–	–	–	0
1 0	–	–	–	0

R_1:

–	–	0	–
–	–	0	–
0	0	0	1
0	0	0	1

S_2:

0	0	0	0
–	–	0	0
–	–	–	–
0	0	1	1

R_2:

–	–	–	–
0	0	1	1
0	0	0	0
–	–	0	0

Fig. 8-31 Maps for excitation variables.

into four maps, from which we find

$$S_1 = xd \qquad R_1 = xd'$$

$$S_2 = xy_1 \qquad R_2 = xy_1'.$$

The corresponding network is shown in Fig. 8-32.

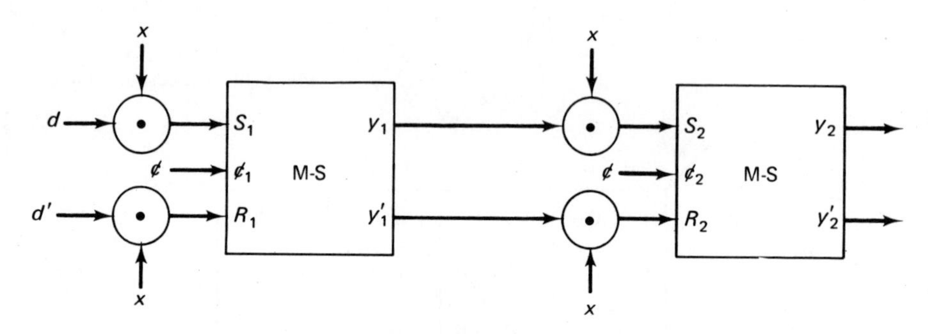

Fig. 8-32 2-bit shift register with RS flip-flops.

The method given above assumes that no logic will be associated with the clock input. If this assumption is removed, a simpler realization can be obtained as shown (for one stage) in Fig. 8-33.

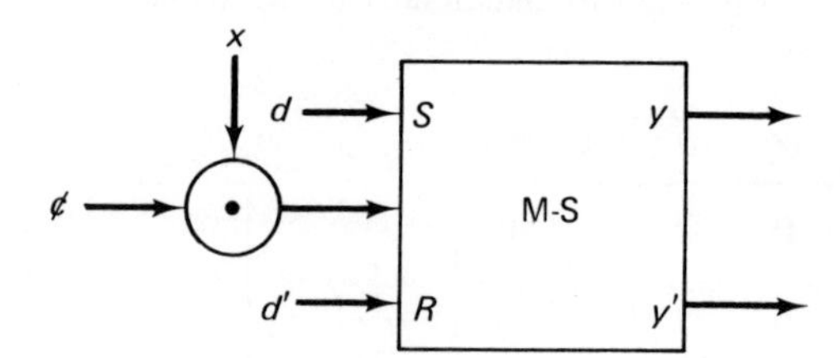

Fig. 8-33 A shift register stage with clock gating.

JK Flip-Flops

The design table for *JK* flip-flops is shown in Fig. 8-34. Notice that there are two more don't cares than in the corresponding *RS* table. This is because the transition $y = 1$ to $Y = 0$ can be accomplished by either resetting or toggling. Similarly, for $y = 0$, $Y = 1$ we can either set or toggle.

y	Y	J	K
0	0	0	–
0	1	1	–
1	0	–	1
1	1	–	0

Fig. 8-34 JK design table.

y_1y_2 \ xd	00	01	11	10
00	0–, 0–	0–, 0–	1–, 0–	0–, 0–
01	0–, –0	0–, –0	1–, –1	0–, –1
11	–0, –0	–0, –0	–0, –0	–1, –0
10	–0, 0–	–0, 0–	–0, 1–	–1, 1–

J_1K_1, J_2K_2

Fig. 8-35 Excitation table for JK flip-flops.

The excitation table is shown in Fig. 8-35. The reader can verify that the excitation equations are

$$J_1 = xd \qquad K_1 = xd'$$

$$J_2 = xy_1 \qquad K_2 = xy'_1.$$

These are identical to the equations for the *RS* flip-flops, in this case. Hence the network is as shown in Fig. 8-32, with S_1, R_1, S_2, and R_2 replaced by J_1, K_1, J_2, and K_2, respectively. Clock gating as in Fig. 8-33 can also be used. This completes the synthesis procedure. The interested reader can carry out the design using *T* flip-flops, for completeness.

In this simple example, the designs are also easily arrived at by an informal, intuitive approach. However, for more complex design problems, the formal methods above might prove easier than the intuitive approach.

We also point out that actual design problems are hardly ever restricted to the synchronous mode but usually involve some asynchronous features. We illustrate this in the next example.

Synthesis Example 2 (Up/Down Decade Counter)

Our second example is the design of an "up/down decade" counter. The counter is to have an up/down control input x, in addition to the clock input $\cent$. When $x = 0$, the counter should count as usual or "up"; i.e., it should produce the BCD representation of the sequence 0, 1, . . . , 9, 0, 1, etc., as consecutive clock pulses arrive. When $x = 1$, the counter should count "down"; i.e., it should produce the BCD representation of the sequence 0, 9, 8, . . . , 1, 0, 9, etc.

The above word description of the desired behavior is rather simple, and one easily converts it into the state table as shown in Fig. 8-36.

We will use master–slave flip-flops to represent the states of the network. We must therefore encode the 10 states (0–9) by a set of binary variables. In general, n binary variables can represent at most 2^n distinct states, so at least

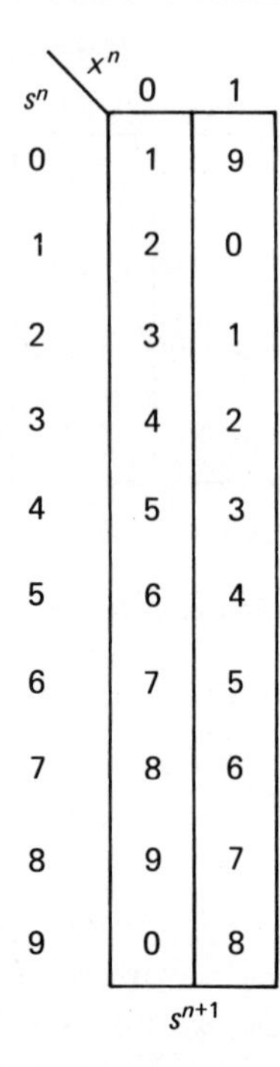

s^n \ x^n	0	1
0	1	9
1	2	0
2	3	1
3	4	2
4	5	3
5	6	4
6	7	5
7	8	6
8	9	7
9	0	8

s^{n+1}

Fig. 8-36 State table to be realized.

4 (internal) state variables are required here. In this case, we have no reason for using any more than 4 variables, so we let $y \triangleq y_1, y_2, y_3, y_4$ be the state representation. It is possible to assign any 10 distinct 4-tuples to the 10 states to realize the required behavior. However, in the present case there is a "natural" assignment. Recall that the counter is to have an output word $z \triangleq z_1, z_2, z_3, z_4$ such that $\perp z$ represents the count. If the state variables $y_1, \ldots, y_4$ are assigned so that the count is represented by $\perp y$ then $z = y$ and there is no need for any output logic.

The encoded state table using the "natural" assignment becomes the transition table of Fig. 8-37. Any transition table can be implemented using any one of the 4 types of flip-flops. For the present problem we will use toggle flip-flops. Each y_i will correspond to the Q-output of a toggle flip-flop, and we must now find suitable excitation functions T_i to ensure that the next state of the flip-flop will be Y_i. This is particularly simple for toggle flip-flops because

$$Y_i = y_i \oplus T_i.$$

Hence

$$T_i = Y_i \oplus y_i.$$

We now apply this repeatedly as follows:

$$\text{if } Y_i = y_i, \text{ set } T_i = 0$$
$$\text{if } Y_i = y_i', \text{ set } T_i = 1.$$

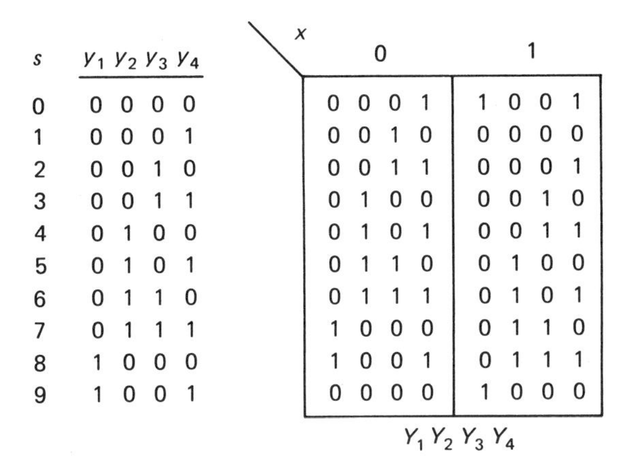

s	$y_1 y_2 y_3 y_4$	$x = 0$	$x = 1$
0	0 0 0 0	0 0 0 1	1 0 0 1
1	0 0 0 1	0 0 1 0	0 0 0 0
2	0 0 1 0	0 0 1 1	0 0 0 1
3	0 0 1 1	0 1 0 0	0 0 1 0
4	0 1 0 0	0 1 0 1	0 0 1 1
5	0 1 0 1	0 1 1 0	0 1 0 0
6	0 1 1 0	0 1 1 1	0 1 0 1
7	0 1 1 1	1 0 0 0	0 1 1 0
8	1 0 0 0	1 0 0 1	0 1 1 1
9	1 0 0 1	0 0 0 0	1 0 0 0
		$Y_1 Y_2 Y_3 Y_4$	

Fig. 8-37 Transition table.

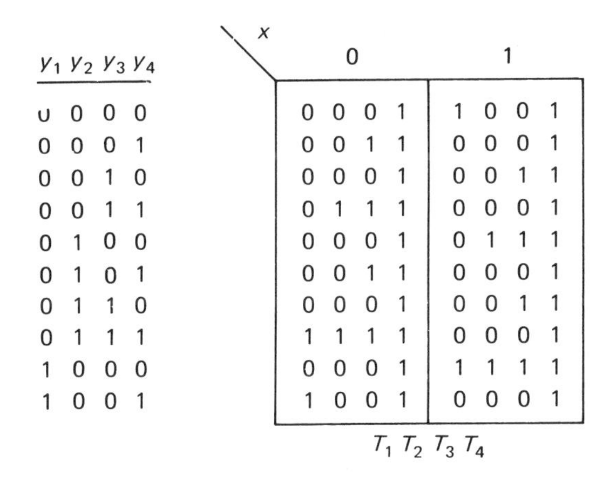

$y_1 y_2 y_3 y_4$	$x = 0$	$x = 1$
0 0 0 0	0 0 0 1	1 0 0 1
0 0 0 1	0 0 1 1	0 0 0 1
0 0 1 0	0 0 0 1	0 0 1 1
0 0 1 1	0 1 1 1	0 0 0 1
0 1 0 0	0 0 0 1	0 1 1 1
0 1 0 1	0 0 1 1	0 0 0 1
0 1 1 0	0 0 0 1	0 0 1 1
0 1 1 1	1 1 1 1	0 0 0 1
1 0 0 0	0 0 0 1	1 1 1 1
1 0 0 1	1 0 0 1	0 0 0 1
	$T_1 T_2 T_3 T_4$	

Fig. 8-38 Excitation table.

The resulting excitation table is shown in Fig. 8-38. From this table we can now find the excitation equations.

Notice that the network we will construct will actually have 16 states, not 10, because binary devices are used. The assumption that the states y with $\perp y > 9$ are of no interest to us implies that the excitation functions for these states are optional, i.e., represent don't cares. These don't care entries can be used advantageously to simplify the combinational logic generating the excitation functions. For convenience, we represent the excitation functions on 5-variable maps in Fig. 8-39. (The 5-variable map consists of two 4-variable maps side by side. On one half of the map one can find 4 points adjacent to a given point. To find the fifth adjacent point, simply consider the right 4-variable map to be on top of the left 4-variable map.) Using the minimiza-

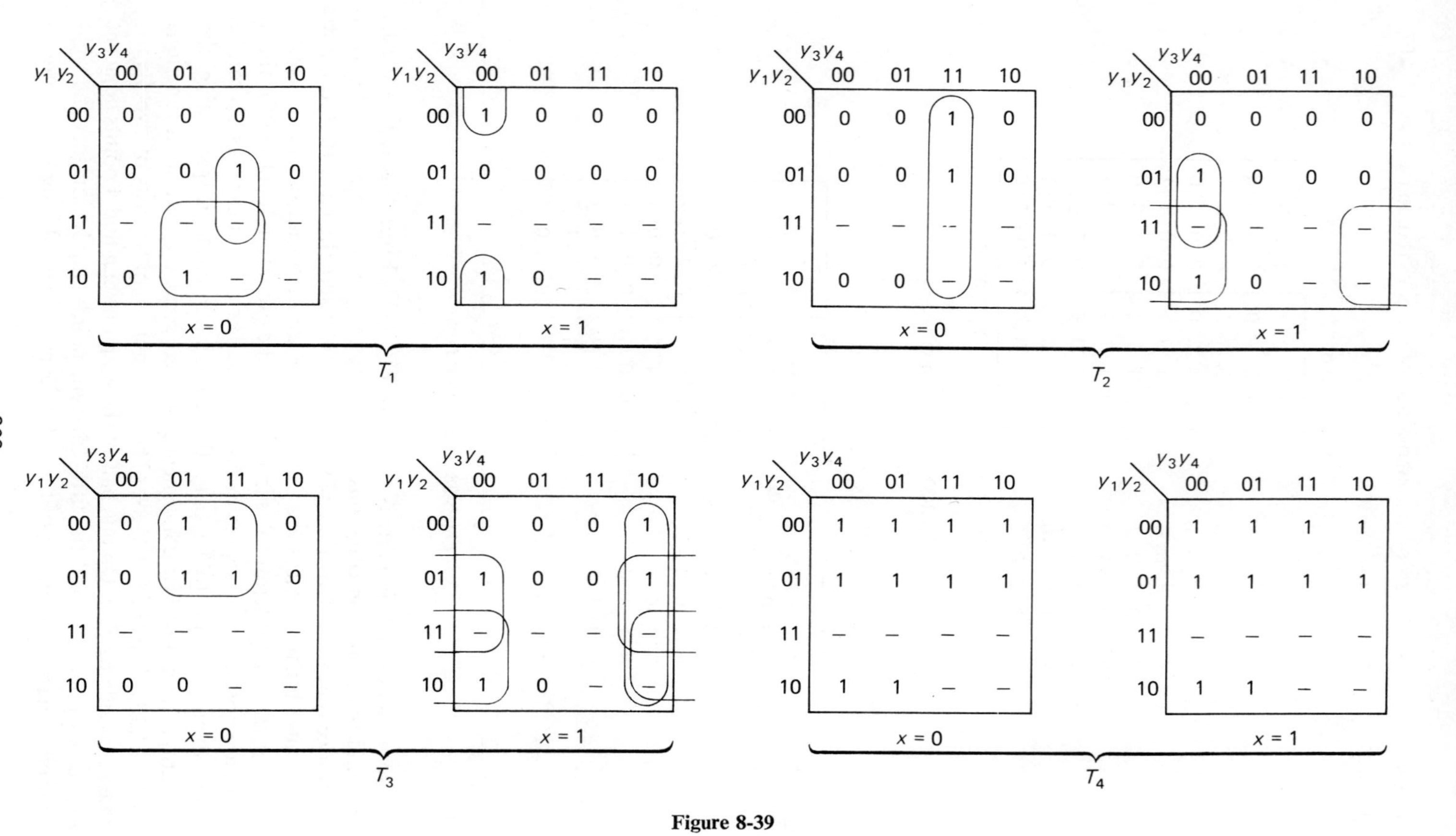

Figure 8-39

tion techniques of Chapter 5, we find the following minimal sums:

$$T_1 = x'y_1y_4 + x'y_2y_3y_4 + xy_2'y_3'y_4'$$
$$T_2 = x'y_3y_4 + xy_1y_4' + xy_2y_3'y_4'$$
$$T_3 = x'y_1'y_4 + xy_3y_4' + xy_1y_4' + xy_2y_4'$$
$$T_4 = 1.$$

Often minimal sums are used only as an intermediate step, to be followed by some heuristic techniques such as factoring and sharing. For instance, one can rewrite T_3,

$$\begin{aligned} T_3 &= x'y_1'y_4 + xy_4'(y_1 + y_2 + y_3) \\ &= x' \cdot y_1' \cdot y_4 + x \cdot y_4' \cdot (|(y_1', y_2', y_3')). \end{aligned}$$

Now T_2 can be modified as follows: $T_2 = x'y_3y_4 + E$, where

$$\begin{aligned} E &= xy_4'(y_1 + y_2y_3') \\ &= xy_4'(y_1y_3 + y_1y_3' + y_2y_3'). \end{aligned}$$

Noting that $xy_1y_3y_4'$ is optional for T_2 and can be dropped, we can write F instead of E, where

$$\begin{aligned} F &= xy_4'(y_1y_3' + y_2y_3') \\ &= xy_3'y_4'(y_1 + y_2) \\ &= xy_3'y_4'(y_1 + y_2 + y_3). \end{aligned}$$

The sum $y_1 + y_2 + y_3$ can be generated by one NAND gate and shared by T_2 and T_3 as shown in Fig. 8-40.

The particular logic arrangement of Fig. 8-40 is actually used in the IC TTL package 74190 [TI2]. The clock is actually inverted so that state changes take place on the rising edge. This package has a number of additional facilities. The general block diagram of the counter is shown in Fig. 8-41.

The counter can be loaded with an arbitrary data word (initial count) D_1, D_2, D_3, D_4 by setting the external LOAD input L to 0. This presets or clears the appropriate flip-flop as shown in Fig. 8-42. If the external COUNT ENABLE input G is set to 0, the normal counting operation can take place. Otherwise, counting is inhibited.

Two additional outputs, RC and MM, are available. These are used when two or more counters of this type are cascaded. We have

$$\mathrm{MM} = x'y_1y_4 + xy_1'y_2'y_3'y_4'$$
$$\mathrm{RC} = (\mathrm{MM})' + \not{c} + G.$$

Clearly, MM is 1 only if the counter is counting up ($x = 0$) and the count reaches 9, indicating an *overflow*, or it is counting down ($x = 1$) and the count reaches 0, indicating an *underflow*. The duration of the MM pulse is one complete cycle of the clock.

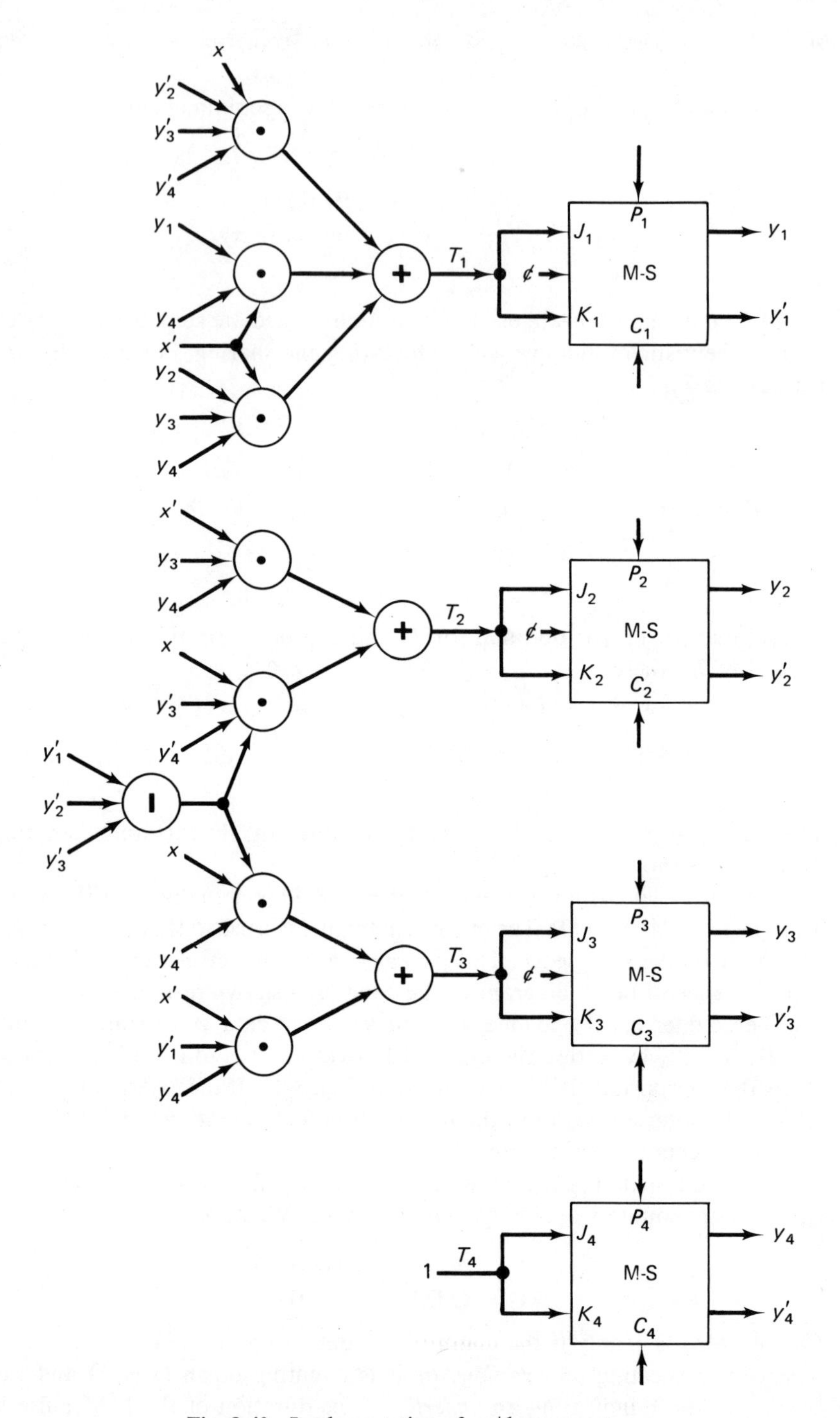

Fig. 8-40 Implementation of up/down counter.

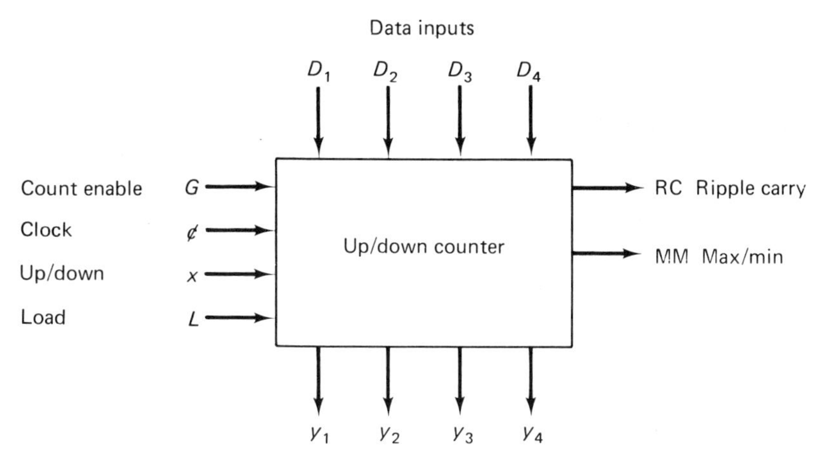

Fig. 8-41 Block diagram of up/down counter.

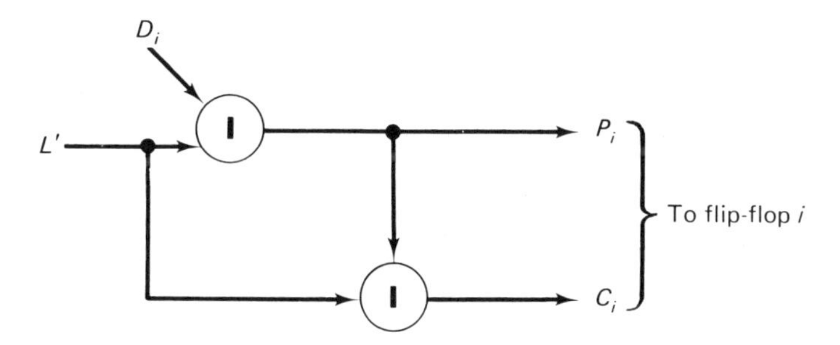

Fig. 8-42 Data loading facility.

The RC or RIPPLE CLOCK output will have a 0-pulse equal in duration to the 0-pulse of the clock ¢, provided the counter is enabled ($G = 0$) and overflow or underflow occurs (MM $= 1$).

More about the cascading of counters will be said in the next section.

Synthesis Example 3 (Serial Comparator) [MO-MI]

We wish to compare two binary words $x = x_1, x_2, \ldots, x_m$ and $z = z_1, z_2, \ldots, z_m$ and produce outputs $[X = Z]$, $[X > Z]$ and $[X < Z]$, where $X = \perp x$, $Z = \perp z$, as in Chapter 2. (Recall that $[X = Z] = 1$ iff $X = Z$, etc., as in Section 2.4.) However, this time the comparison is to be done "serially" in the following sense. The comparator will examine the numbers one bit at a time beginning with the most significant bit. Thus x_1 is compared with z_1 first. If $x_1 > z_1$, we have $X > Z$ and the computation stops. Similarly, $x_1 < z_1$ implies $X < Z$. If $x_1 = z_1$, examine x_2 and z_2, etc. The network to be designed is to operate in the synchronous mode, under the control of clock ϕ.

From this description we can derive the state table of Fig. 8-43, as follows. Having examined the first $i - 1$ bits, we need to remember only the relative magnitudes of $\hat{X} \triangleq x_1, \ldots, x_{i-1}$ and $\hat{Z} \triangleq z_1, \ldots, z_{i-1}$. Therefore, our state table will have 3 states, which we will call $\langle \hat{X} = \hat{Z} \rangle$, $\langle \hat{X} < \hat{Z} \rangle$, and $\langle \hat{X} > \hat{Z} \rangle$. When the clock pulse arrives, we examine x_i and z_i and modify the state according to Fig. 8-43.

Present state	$x_i z_i$ = 0 0	0 1	1 1	1 0	Outputs $[X = Z]$	$[X < Z]$	$[X > Z]$
$\langle \hat{X} = \hat{Z} \rangle$	$\langle \hat{X} = \hat{Z} \rangle$	$\langle \hat{X} < \hat{Z} \rangle$	$\langle \hat{X} = \hat{Z} \rangle$	$\langle \hat{X} > \hat{Z} \rangle$	1	0	0
$\langle \hat{X} < \hat{Z} \rangle$	$\langle \hat{X} < \hat{Z} \rangle$	$\langle \hat{X} < \hat{Z} \rangle$	$\langle \hat{X} < \hat{Z} \rangle$	$\langle \hat{X} < \hat{Z} \rangle$	0	1	0
$\langle \hat{X} > \hat{Z} \rangle$	$\langle \hat{X} > \hat{Z} \rangle$	$\langle \hat{X} > \hat{Z} \rangle$	$\langle \hat{X} > \hat{Z} \rangle$	$\langle \hat{X} > \hat{Z} \rangle$	0	0	1

Next state

Fig. 8-43 State table for serial comparator.

Next we must assign (at least) two binary variables y_1 and y_2 to represent the 3 states by flip-flop outputs. There are many choices for this. Note, however, that we can reduce the output logic by choosing the variables in such a way that two of the outputs coincide with y_1 and y_2. For example, we may choose $y_1 = [X < Z]$ and $y_2 = [X > Z]$. This leads to the transition table of Fig. 8-44.

If we choose *JK* flip-flops for the implementation, we obtain the excitation functions shown below. (The detailed construction of the excitation table is left as an exercise.)

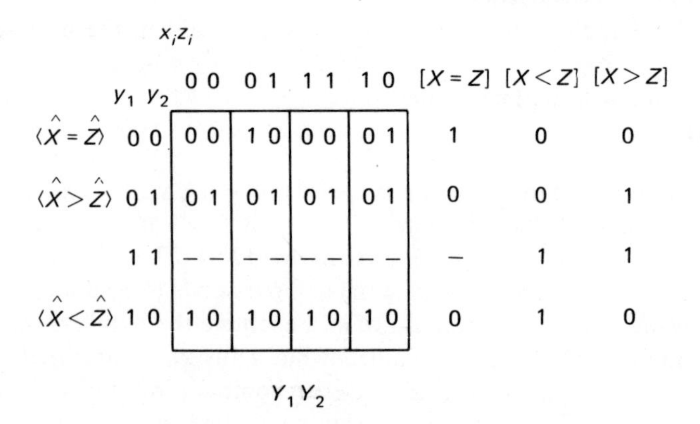

	$y_1 y_2$	$x_i z_i$ = 0 0	0 1	1 1	1 0	$[X = Z]$	$[X < Z]$	$[X > Z]$
$\langle \hat{X} = \hat{Z} \rangle$	0 0	0 0	1 0	0 0	0 1	1	0	0
$\langle \hat{X} > \hat{Z} \rangle$	0 1	0 1	0 1	0 1	0 1	0	0	1
	1 1	– –	– –	– –	– –	–	1	1
$\langle \hat{X} < \hat{Z} \rangle$	1 0	1 0	1 0	1 0	1 0	0	1	0

$Y_1 Y_2$

Fig. 8-44 Transition table for comparator.

$$J_1 = x_i' z_i y_2' \qquad K_1 = 0$$
$$J_2 = x_i z_i' y_1' \qquad K_2 = 0.$$

For the outputs we have

$$[X = Z] = y_1' y_2'$$
$$[X < Z] = y_1$$
$$[X > Z] = y_2.$$

These equations lead to the network of Fig. 8-45. (The COMPARE input is explained below.)

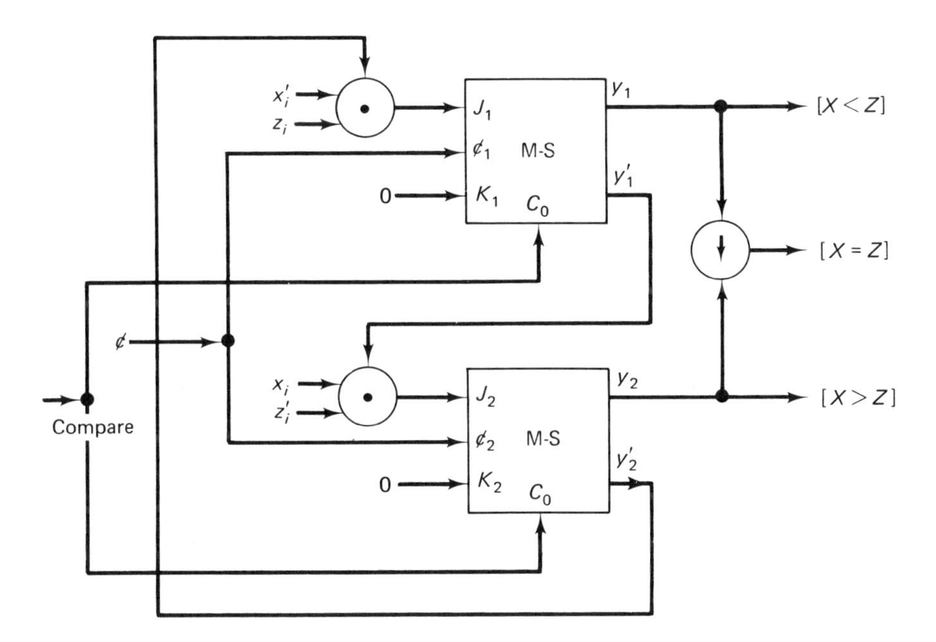

Fig. 8-45 Network for serial comparator.

In order to implement the design, it is convenient to choose commercially available modules. As we have mentioned before in Chapter 7, many flip-flops are provided with some logic preceding the J and K inputs. For example, the 74105 gated master–slave JK flip-flop has the following logic:

$$J = J_a J_b' J_c \qquad K = K_a K_b' K_c.$$

If we choose such a flip-flop for Fig. 8-45, the AND gates become part of the package. Thus we only need one NOR gate for the $[X = Z]$ output.

Finally, notice that we must be able to clear the flip-flops before the compare operation can start. This is conveniently done by applying a COMPARE 0-pulse to the C_0 terminals of each flip-flop before the operation is to start.

8.5. ITERATIVE NETWORKS

In Section 4.7 we have shown an iterative realization of the symmetric function s_2 (see Fig. 4-15). For a variety of combinational design problems, iterative networks offer a convenient solution. In this section we briefly discuss the analysis and synthesis aspects of iterative networks. The reason for doing this presently is the fact that iterative networks are closely related to sequential synchronous-mode networks. For example, the 4-bit comparator of Fig. 8-46(b) is the iterative analog of the serial comparator of Fig. 8-45.

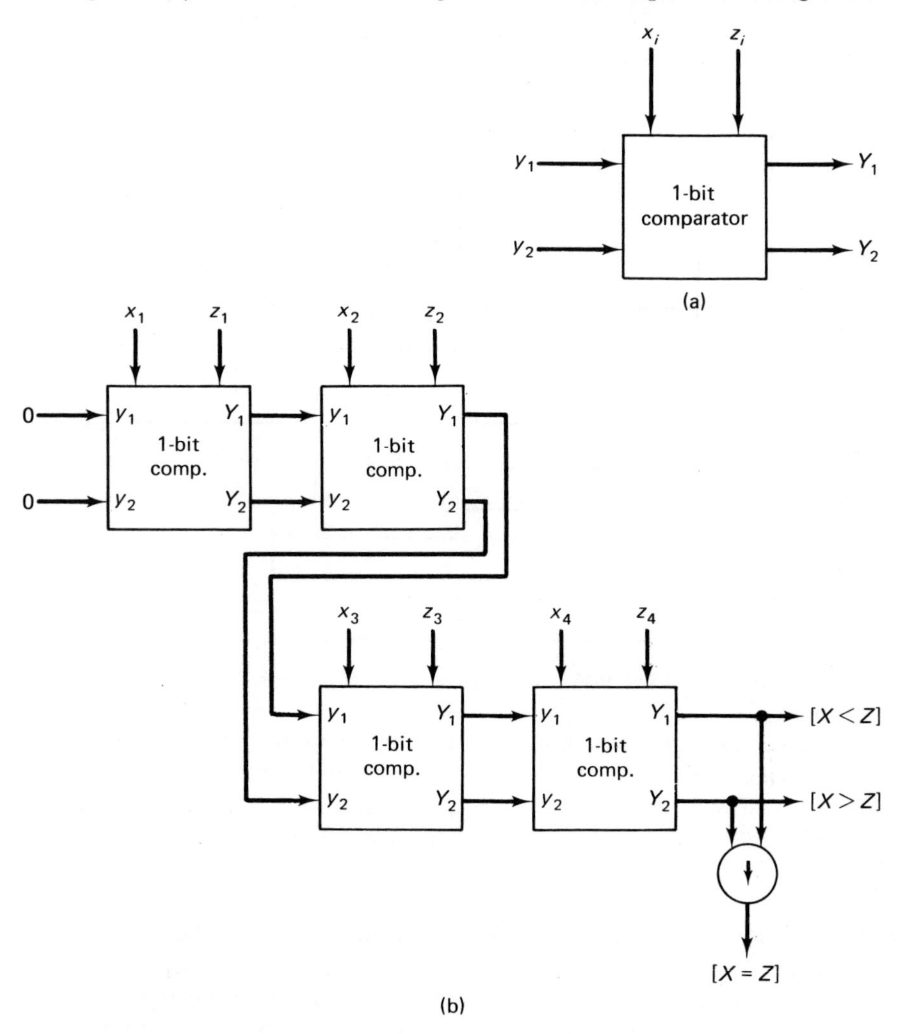

Fig. 8-46 (a) 1-bit comparator cell; (b) 4-bit iterative comparator.

Figure 8-46(a) shows the typical (1-bit comparator) cell of the iterative network. This cell is designed as a combinational network in accordance with the table of Fig. 8-44, where now x_i, z_i, y_1, and y_2 are the inputs and Y_1 and Y_2 are the outputs. An application of the map method yields the following:

$$Y_1 = y_1 + x_i'z_iy_2'$$
$$Y_2 = y_2 + x_iz_i'y_1'.$$

From the considerations which led to Fig. 8-45, one easily derives the fact that Fig. 8-46(b) indeed realizes a 4-bit comparator. Note that the horizontal inputs ("carry-in") to each cell correspond to the present state; and the outputs from each cell ("carry-out") correspond to the next state of the analogous sequential network.

In Section 2.4 we have designed a 4-bit comparator by realizing the required input–output functions directly. The iterative approach is more systematic and reduces the design problem to that of designing a typical cell. However, the overall iterative network will usually have a greater propagation delay than an equivalent network designed by realizing the required input–output behavior directly.

The reader is now advised to review the design of the symmetric function s_2 discussed in Section 4.7. In the analogous sequential problem, Table 4-10 represents the required transition table. The corresponding state table is given in Fig. 8-47, where state 3 corresponds to the carry-in "3 or more."

For further examples of iterative networks and their relation to synchronous-mode sequential networks see Problem 10.

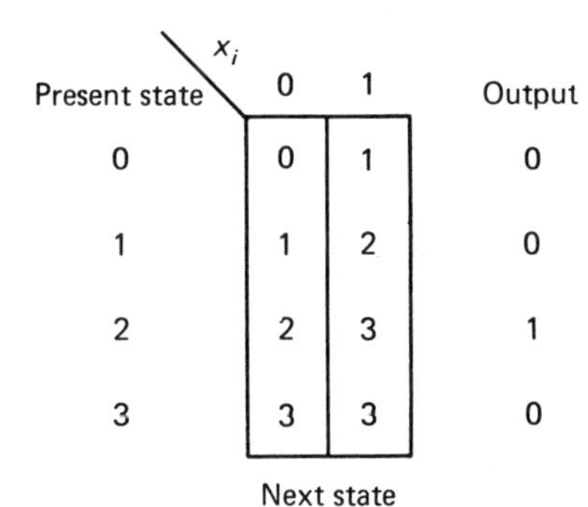

Present state \ x_i	0	1	Output
0	0	1	0
1	1	2	0
2	2	3	1
3	3	3	0
	Next state		

Fig. 8-47 State table for symmetric function s_2.

8.6. MODULAR DESIGN OF SYNCHRONOUS NETWORKS

In Section 8.4 we discussed some basic techniques for synthesizing synchronous networks. In our examples we used gates and flip-flops as basic building blocks. In practice, one usually designs synchronous networks with commercially available IC packages as building blocks. This modular approach to synchronous network design will be discussed in this section and also in Chapter 9.

Mod-*N* Counters [MO-MI]

Our first example is the modular design of a mod-N counter, with N presettable in the range, say, $2 \leq N \leq 9$. When started in its initial state, the mod-N counter will produce an output pulse upon receiving the Nth clock pulse and will then return to its initial state. One easily designs such a mod-N counter, using the 74190 counter of Fig. 8-41 as the main building block. Indeed, the 74190 counter may be converted into a mod-N counter, $2 \leq N \leq 9$, by simply providing the connections shown in Fig. 8-48.

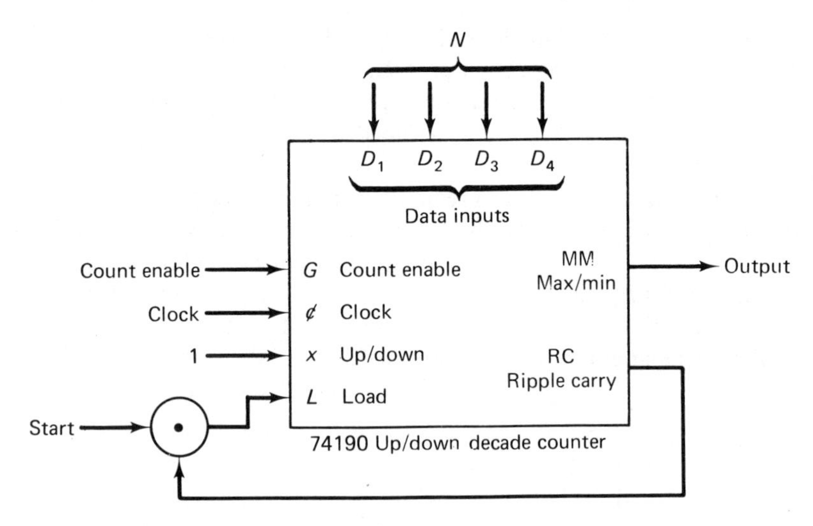

Fig. 8-48 Divide-by-N counter ($2 \leq N \leq 9$) using 74190 up/down decade counter.

Note that the 74190 counter is used as a down counter. If, for example, N is set to 7 (i.e., $D_1 = 0$, $D_2 = D_3 = D_4 = 1$) and a 0-pulse is applied to START, the word 0111 will be loaded as DATA INPUTS into the counter. The counter will then go through the (binary equivalent of) the following counting sequence: 6, 5, 4, 3, 2, 1, 0, 6, 5, 4, ... Whenever the counter reaches the 0-state, a 0-pulse on the *RC*-output occurs and the word 0111 is again loaded into the counter, restarting the counting sequence 6, 5, 4, ... An output pulse is provided whenever the counter enters state 0.

Cascading of Counters [MO-MI, PEA]

In Section 8.4 we have mentioned the cascading of 74190 counters. This concept is further illustrated in Fig. 8-49, where three 74190 counters are

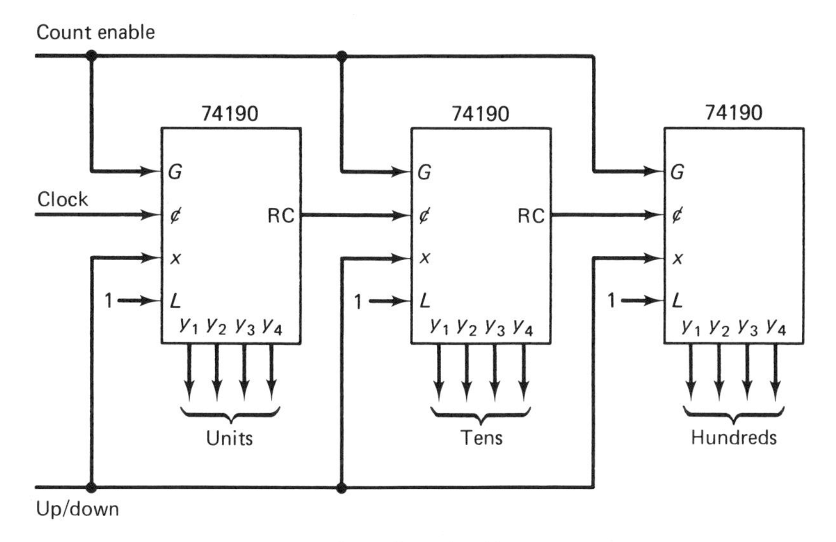

Fig. 8-49 Cascading of 74190 counters.

shown cascaded to form an up/down mod-1000 counter. Note that the *RC*-output is used as a ripple-carry input to the next stage.

To avoid the disadvantages of a ripple carry, a synchronous-mode mod-1000 counter with parallel carry can be obtained by cascading the three 74190 counters as shown in Fig. 8-50. The parallel-carry scheme allows more stages to be cascaded without having to reduce the counting frequency.

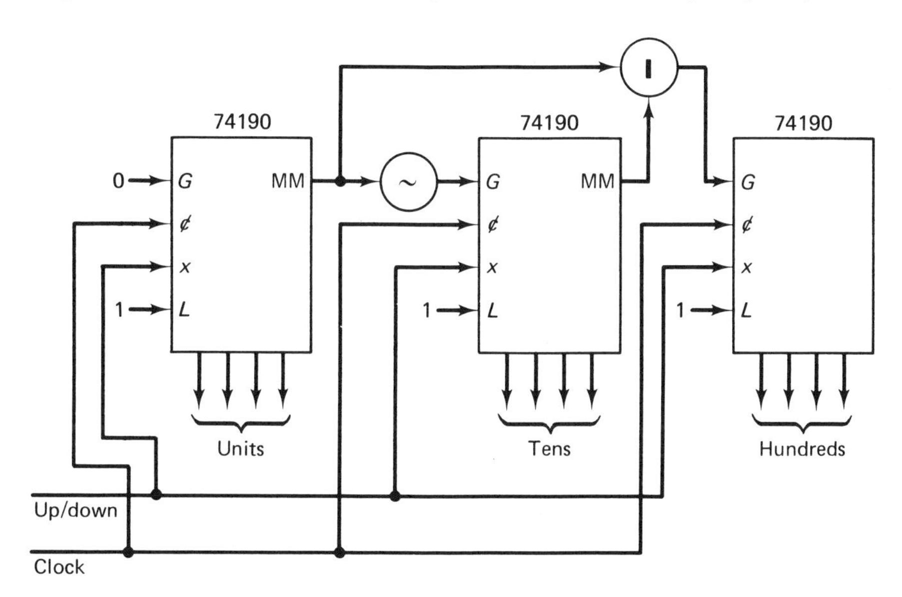

Fig. 8-50 Cascading of 74190 counters with parallel carry.

Shift Register Counters [MO-MI]

It is frequently convenient to use shift registers for counting purposes. Such shift register counters are especially applicable if one output is required for each state of the count sequence. As an example, Fig. 8-51 shows a 5-bit shift register modified to form a decade counter. This counter is obtained by feeding back the complement of the last output (E) to the serial input of the shift register. The count sequence obtained is shown in Fig. 8-52.

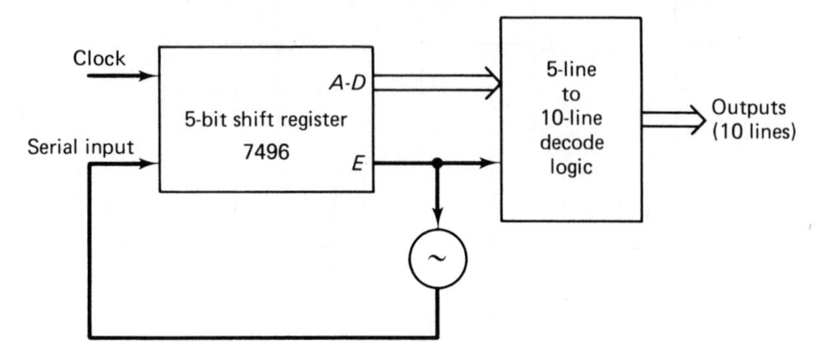

Fig. 8-51 Shift register decade counter.

	A	B	C	D	E
0	0	0	0	0	0
1	1	0	0	0	0
2	1	1	0	0	0
3	1	1	1	0	0
4	1	1	1	1	0
5	1	1	1	1	1
6	0	1	1	1	1
7	0	0	1	1	1
8	0	0	0	1	1
9	0	0	0	0	1

Fig. 8-52 Count sequence for counter of Fig. 8-51.

Generally, shift register counters consist of a feedback shift register and a decode logic as shown in Fig. 8-53. Usually this decode logic is quite simple. Extensive research has been devoted to the design of shift register counters. For further details, the reader is referred to the literature [GOL, YOE].

Universal ROM Logic

In Section 2.8 we discussed the ROM and some of its applications. We now show how ROMs can be applied to design synchronous-mode sequential networks. The ROM shown in Fig. 2-17 can be used to implement k arbitrary Boolean functions $f_1, \ldots, f_k$ as explained in Section 2.8 (p. 47).

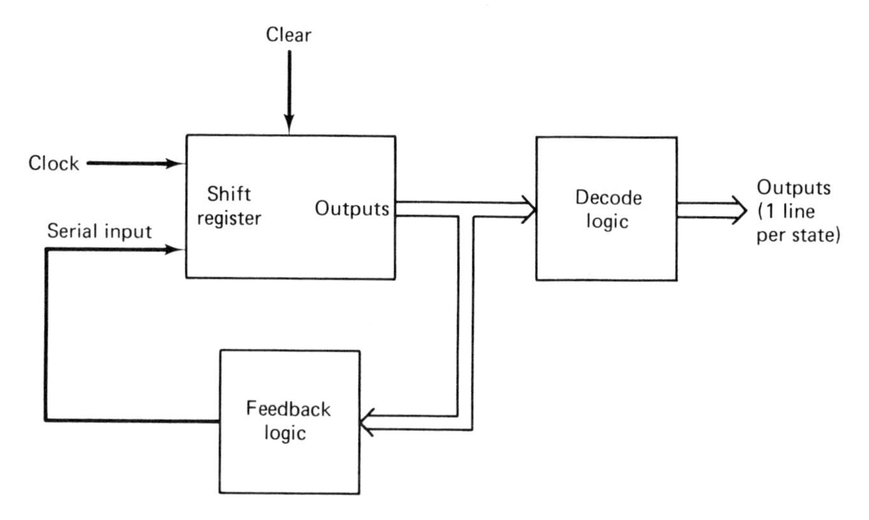

Fig. 8-53 Shift register counter.

The synthesis procedures for synchronous-mode sequential networks always lead to a design consisting of a number of flip-flops together with combinational logic which provides the various flip-flop excitation and output functions. This combinational logic can be implemented by means of ROMs, as explained earlier.

8.7. IC MEMORIES

Basic Principles

A *memory* is a special type of digital network used for storing digital information. In this book we shall be concerned only with integrated-circuit memories. Other types of memories have been widely described in the computer literature [RE-CO] and will not be discussed here. IC memories are becoming cheaper and their use is increasing. Furthermore, they are often compatible directly, without any interface equipment, with suitable combinational and sequential IC packages described previously. Their discussion here is therefore quite relevant.

Frequently, the information to be stored is represented by a set of n-bit words. Most often it is convenient to fix the length of all words to be the same for a particular set of problems. For example, for applications that handle binary-coded integers in the range 0–999, one requires 10-bit words, since $2^{10} = 1{,}024 > 1000 > 512 = 2^9$. A memory, then, can be thought of as a network capable of storing m n-bit words as shown in Fig. 8-54.

Normally there are no facilities for transferring the contents of one cell into another or for transferring words from one register to another. (The

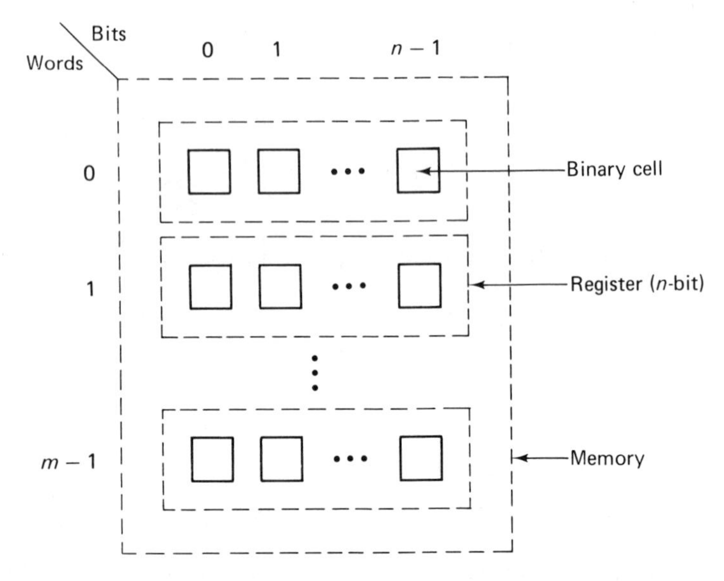

Fig. 8-54 Memory organization.

typical register is not, for example, a shift register.) Operations such as these must be performed outside the memory. Thus the basic function of a memory is to store information, not to transform it.

RAMs

A *random access memory* (RAM) has the facility of (1) *writing* a word from an external register called MDR (memory data register) into a location prescribed by another external register called MAR (memory address register; see Fig. 8-55), and (2) *reading* the word in the location specified by the MAR. Normally, the number n of words in a memory is a power of 2, $n = 2^k$, and in order to specify one out of 2^k words, k bits are sufficient for the MAR. The ability to select one out of n words by specifying a k-bit address ($k = \log_2 n$) is certainly convenient to the user. However, to perform the actual selection, n control lines are used. This calls for k-line to 2^k-line decoder (1-out-of-2^k decoder) as shown in Fig. 8-55(b). The "random access" label means that the time required to read out a word is the same regardless of the location of the word. This is to be contrasted with "sequential access" memories such as tapes or drums. Furthermore, the time required to write a word into memory is usually of the same order of magnitude as the time required for reading.

We now proceed to explain how a RAM can be constructed using components that we are already familiar with. A 4-word by 4-bit memory will be used to illustrate the basic principles.

The typical cell is shown in Fig. 8-56. The inputs are:

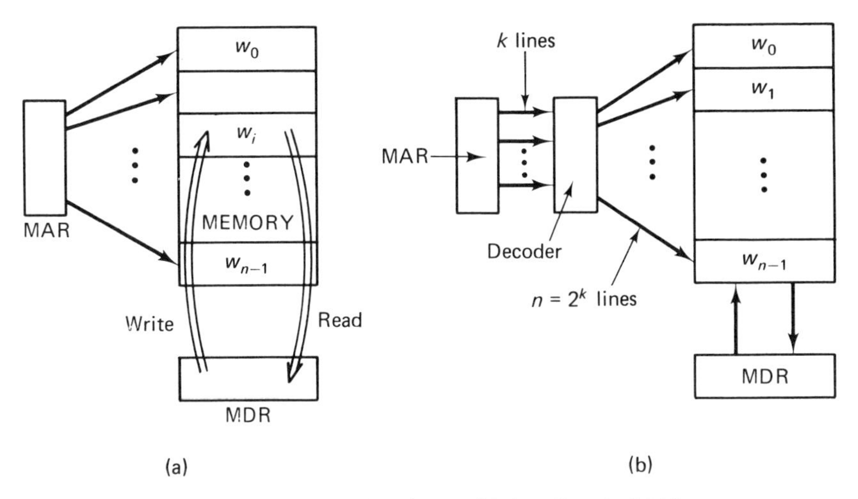

Fig. 8-55 (a) External registers; (b) decoding the MAR.

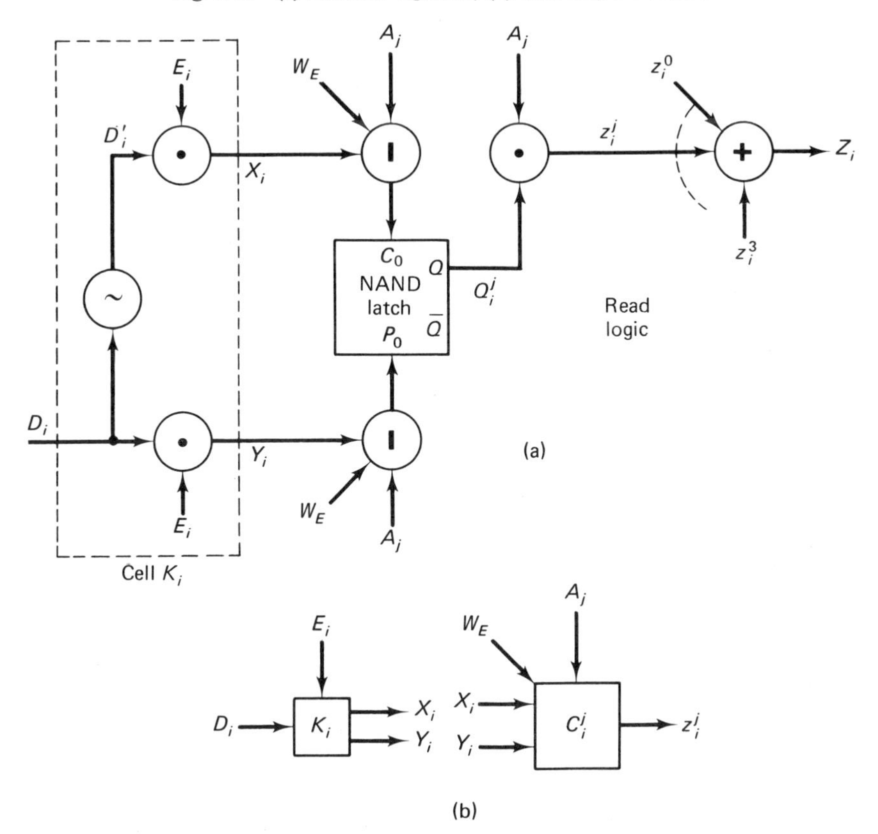

Fig. 8-56 (a) Basic memory cell C_i^j of a RAM; (b) symbols.

W_E—WRITE ENABLE input
E_i—BIT i ENABLE input
D_i—ith DATA bit
A_j—jth ADDRESS bit.

To *read* the contents Q_i^j of cell C_i^j, set $W_E = 0$ and $A_j = 1$. $W_E = 0$ forces $P_0 = C_0 = 1$, resulting in the passive condition to the latch. Only one A_j is 1 at any time; i.e., the A_j represent the decoded (1-out-of-4) address. Hence $z_i^j = Q_i^j$ and $z_i^k = 0$, for $k \neq j$. Therefore, the ith-bit output Z_i is Q_i^j.

To *write*, set $W_E = 1$ and $E_i = 1$, for all i, and set $A_j = 1$, if the word D_0, D_1, D_2, D_3 is to be stored in $Q_0^j, Q_1^j, Q_2^j, Q_3^j$. Then $P_0 = D_i'$, $C_0 = D_i$, and Q_i^j becomes 1 if D_i is 1; 0 if D_i is 0. Hence the word is properly stored.

A block diagram for a 4 × 4 RAM is shown in Fig. 8-57. The columns correspond to registers.

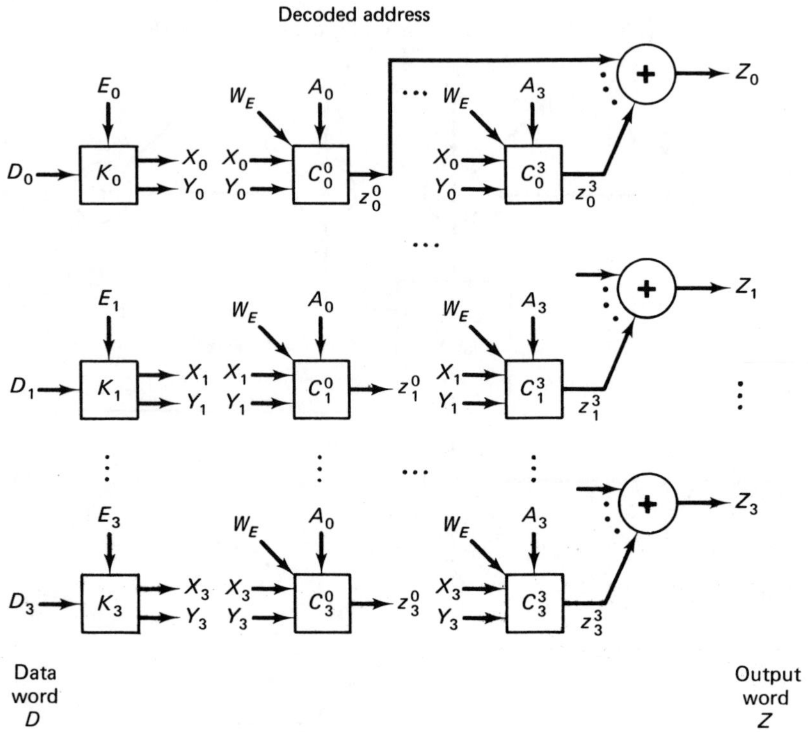

Fig. 8-57 RAM block diagram.

CAMs

Often it is desirable to find out whether a particular data word D belongs to a given list L of words. For example, the words in the list may represent items presently available in stock. If a customer asks for item D, it would require a relatively long time to compare D with all the items in the list sequentially, for example, by a computer program. A *content addressable memory* (*CAM*) performs this comparison in parallel, resulting in very fast operation.

The basic RAM cell of Fig. 8-56 and the cellular network of Fig. 8-57 can be easily modified to obtain CAM capabilities. We now explain how this is done. In Fig. 8-58 we add an extra CELL MATCH output m_i^j to the cell

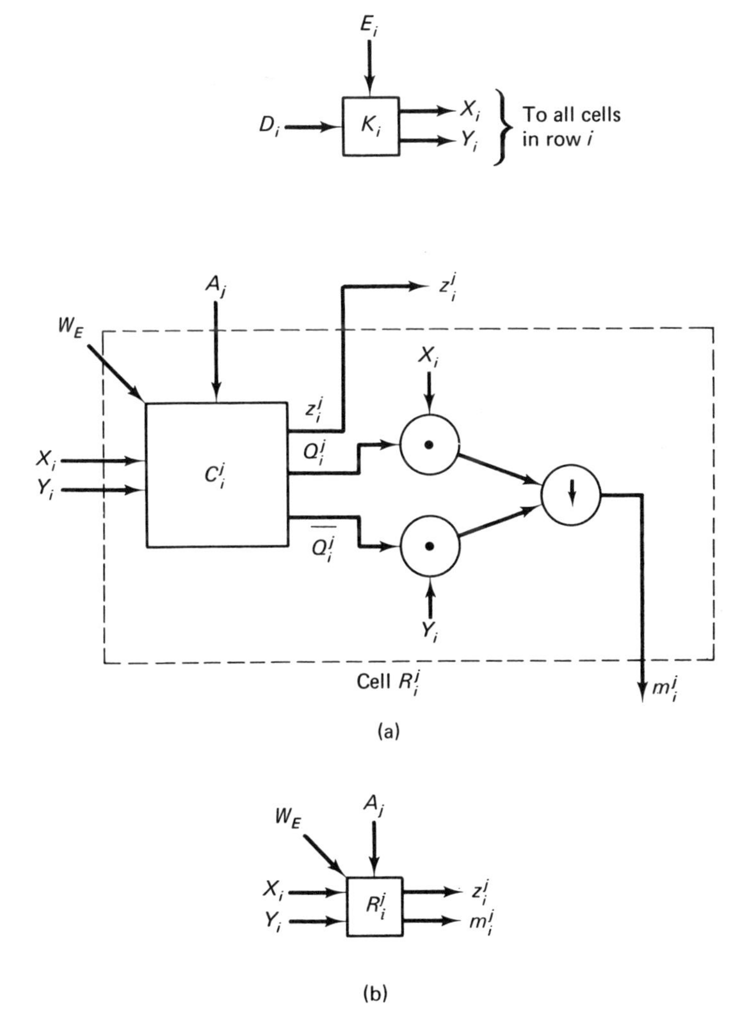

Fig. 8-58 (a) RAM/CAM cell R_i^j with auxiliary row-cell K_i; (b) RAM/CAM symbol.

C_i^j of Fig. 8-56. One verifies that

$$\begin{aligned} m_i^j &= (X_i Q_i^j + Y_i \overline{Q_i^j})' \\ &= (E_i D_i' Q_i^j + E_i D_i (Q_i^j)')' \\ &= (E_i (D_i \oplus Q_i^j))' \\ &= E_i' + D_i \ominus Q_i^j. \end{aligned}$$

For CAM operation, set all $E_i = 1$ (bit i enabled). Then $m_i^j = D_i \ominus Q_i^j$. Thus the CELL MATCH output m_i^j is 1 iff $D_i = Q_i^j$. Note that, for CAM operation, the WRITE ENABLE input W_E is set to 0. This causes $P_0 = C_0 = 1$, ensuring that the latch contents are not affected (see Fig. 8-56).

A block diagram of the entire CAM is shown in Fig. 8-59. It is clear that the output M_j,

$$M_j \triangleq m_0^j m_1^j m_2^j m_3^j,$$

is 1 iff the data word D matches the word in column register j. Finally, the output $M \triangleq M_0 + M_1 + M_2 + M_3$ is 1 iff the data word D matches at least one of the words stored in the memory.

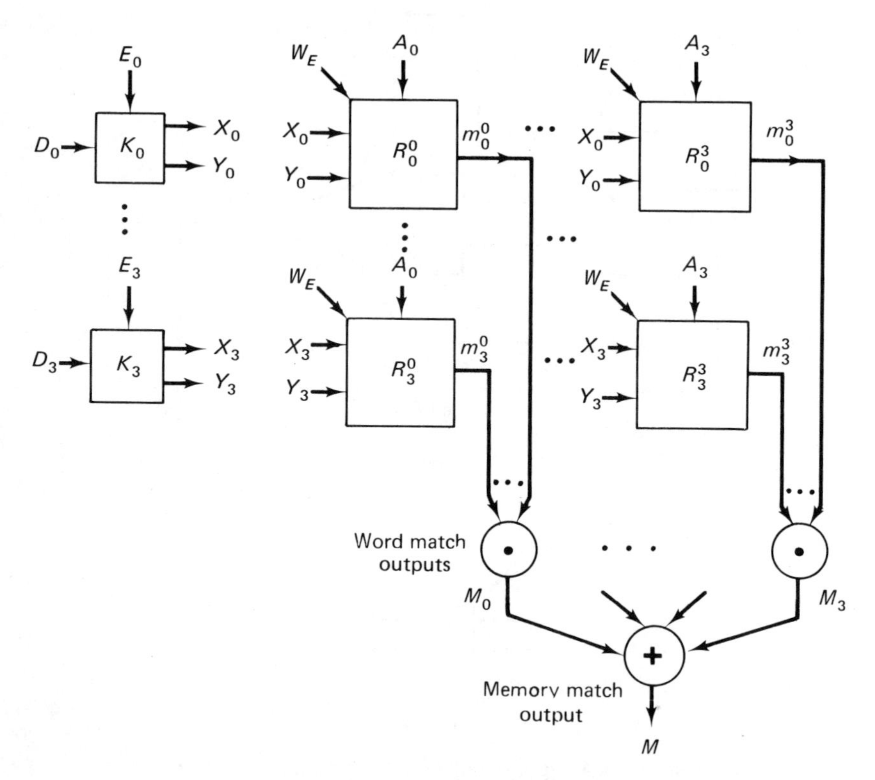

Fig. 8-59 CAM block diagram.

Another way of searching the list is provided with the aid of the BIT ENABLE inputs E_i. Suppose that we want to ask the following question: Given an item D, are there any items in the list that are "similar" to D, in the sense that they agree with D in some key bit positions, but are not necessarily identical to D? For example, the color of an item may be of no importance to the customer. In that case, one need not match the bits that represent the color.

This type of matching is easily implemented. If $E_i = 0$, then $m_i^j = 1$ for all j. Hence all the cells in row i behave as if a match existed for the ith bit.

A 4×4 RAM/CAM very similar to the one we have just described is available as an IC "Schottky Bipolar" package 3104 [INT]. It is a 24-pin package, compatible with appropriate TTL packages. Typical times for the various operations are:

Write time: From the arrival of the correct data until the time the information is properly stored: 40 ns.

Read time: From the arrival of the correct address until the arrival of the correct output: 14–30 ns.

Match time: From the arrival of the correct data word until the arrival of the correct match output: 15–30 ns.

IC Memory Modules

A large variety of memories are presently available as IC packages. In Table 8-1 we list some RAM packages which are similar to the RAM described earlier in this chapter.

Table 8-1 Some TTL and Static MOS RAMs

Type	Catalog Number	Reference	Number of Words	Word Length	Maximum Access Time (ns)
Schottky TTL	3101A	[INT]	16	4	35
TTL	5501	[MON]	16	4	50
Schottky TTL	3106A	[INT]	256	1	60
MOS (static)	1101	[TI1]	256	1	750
MOS (static)	2102–2	[INT]	1024	1	650

In all the RAMs of Table 8-1, the MAR decoder shown in Fig. 8-55 forms part of the IC package. The inputs consist of the k-bit address ($2^k = n =$ number of words), the data word D, and the WRITE ENABLE bit W_E (see Fig. 8-56). All the RAMs listed also have a CHIP SELECT (CS) input bit, the purpose of which will soon become clear. The only RAM output is usually the output word Z (see Fig. 8-57). In some 1-bit-word RAMs (e.g., the 1101), both the output bit and its complement are provided.

Larger RAMs are easily formed from available RAM packages. Figure 8-60 shows the formation of a 1,024-word × 2-bit RAM by means of eight 256 × 1 RAMs. According to the output of the 2-to-4 decoder, one of the

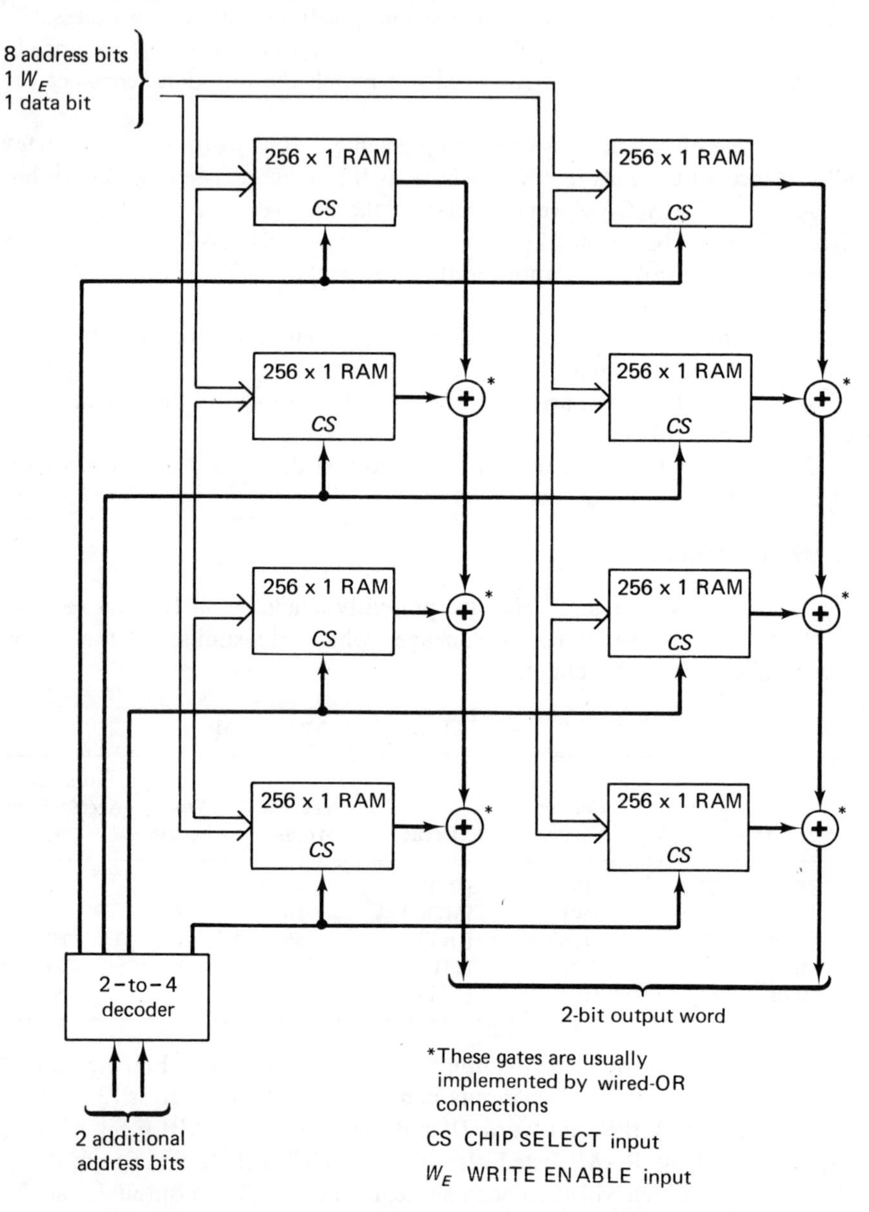

Fig. 8-60 Formation of 1024 × 2 RAM by means of eight 256 × 1 RAMs.

four rows of the array shown in Fig. 8-60 is selected, by means of the CHIP SELECT (CS) inputs. The other rows are disabled. The row outputs are interconnected as shown in Fig. 8-60. The RAM packages of Table 8-1 and similar RAMs provide wired-OR connecting facilities. In static MOS RAMs the "three-state" principle (see Appendix D) is used to provide this facility.

In Table 8-2 we list some examples of available static ROM packages. Their design follows the principle discussed in Section 2.8.

Table 8-2 Some TTL and Static MOS ROMs

Type	Catalog Number	Reference	Number of Words	Word Length	Maximum Access Time	Special Features
TTL	7488A	[TI2]	32	8	45 ns	
TTL	74187	[TI2]	256	4	60 ns	
Schottky TTL	3304	[INT]	{512 or 1024}	{8 or 4}	65 ns	
MOS (static)	1302	[INT]	256	8	1 μs	
MOS (static)	TMS2600	[TI1]	{256 or 512}	{8 or 4}	1 μs	
MOS (static)	1702A	[INT]	256	8	1 μs	Erasable and electrically reprogrammable

As in the case of RAMs, ROM packages may be combined to form larger ROMs. For this purpose, ROM packages also have CHIP SELECT inputs and provide wired-OR connecting facilities. ROM packages with alternative organizations (e.g., 3304) are equipped with an ORGANIZATION SELECT input, by means of which the desired organization is selected.

The 1702A ROM is equipped with a transparent quartz lid, which allows the user to expose the chip to ultraviolet light to erase the bit pattern. The chip can then be electrically reprogrammed. Two minutes are required for reprogramming the chip. For further details, see [INT].

Dynamic Memories

One very important aspect of MOS technologies is their capability of utilizing capacitive storage as temporary memory. Memories based on this principle are referred to as *dynamic*. The advantage of a dynamic memory cell is low power dissipation, low cost, and small cell size. Its disadvantage is the necessity for a periodic refresh operation, in order to retain its content.

As an example of a dynamic memory we describe the 2105-1 Dynamic MOS RAM [INT]. This is a 1,024-word × 1-bit RAM, with a maximum access time of 80 ns. Read, write, and chip-select operations are performed in a similar way as for static RAMs. The 2105-1 RAM has to be refreshed every 10 μs by applying a pulse (minimal width: 50 ns) to the REFRESH input. Such a refresh pulse causes each temporary memory cell to be read and its content to be rewritten into it. Most dynamic RAMs, however, require more complicated refresh procedures, since they have to be refreshed row by row.

PROBLEMS

1. Analyze the network of Fig. P8-1.

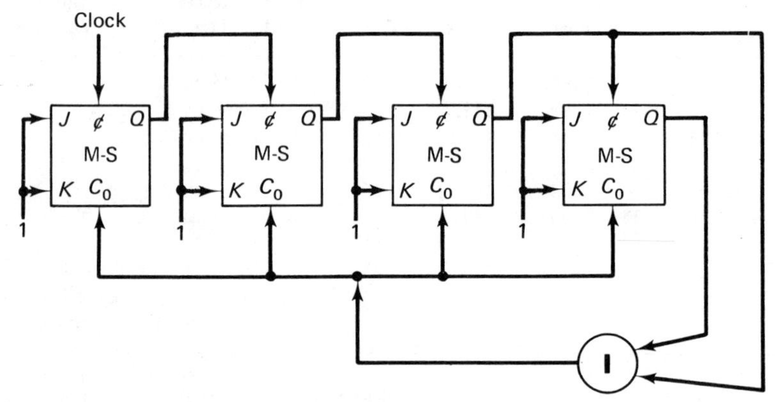

Figure P8-1

2. Analyze the synchronous-mode sequential network shown in Fig. P8-2.
(a) Find the excitation equations.
(b) Construct the excitation table.

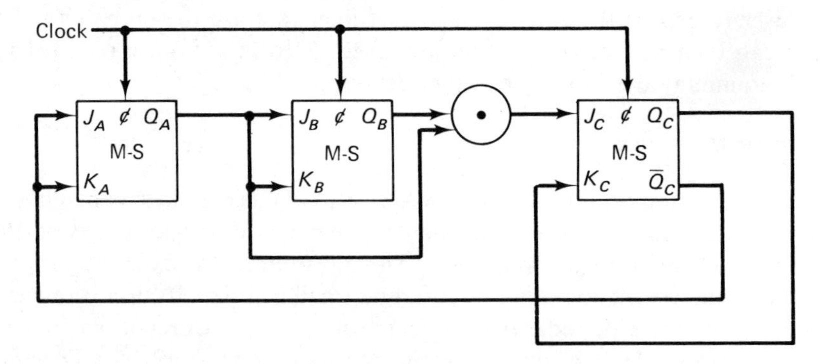

Figure P8-2

(c) Construct the transition table.
(d) Draw a state graph.
(e) What does the network do?

3. It is desired to produce a control sequence on three terminals x_1, x_2, and x_3 in such a way that $\perp x = \perp(x_1, x_2, x_3)$ goes through the successive values 4, 6, 7, 3, 1, 4, 6, 7, etc. In other words, the sequence 4, 6, 7, 3, 1 is to be repeated indefinitely. Design a sequential network with this behavior using *JK* master–slave flip-flops and gates. The network is to operate in synchronous mode under the control of clock ¢.

4. Repeat Problem 3, using (a) *RS* flip-flops; (b) *T* flip-flops; and (c) *D* flip-flops.

5. Design a synchronous modulo-6 counter using clocked master–slave *RS* flip-flops and AND gates, OR gates, and inverters.

6. Design a synchronous-mode modulo-3 up/down counter. The inputs are (1) a clock input ¢ and (2) an input x which does not change when ¢ $= 1$. There are to be two outputs z_1 and z_2. When $x = 0$, the network is to count the clock pulses modulo 3. When $x = 1$, the count is to be "backward"; i.e., go through the sequence 0, 2, 1, 0, 2, 1, 0 etc. The present count is to be given by $\perp(z_1, z_2)$. The network is to be implemented using *JK* master–slave clocked flip-flops and AND gates, OR gates, and inverters. Use minimal-gate sum-of-products forms for each function.

7. Repeat Problem 3 using the following components:
(1) One synchronous modulo-5 counter.
(2) One 1-out-of-8 decoder.
(3) A chip containing four NOR gates with fan-in of 3.
(4) A chip containing eight inverter gates.

8. You are given the following components:
(1) One modulo-8 binary synchronous counter, where the present count is given by $\perp(y_1, y_2, y_3)$, and a 1 on the CLEAR input (asynchronously) resets the counter to 0.
(2) One chip containing two 1-out-of-8 decoders, where if the enable bit e is 1, then $z_i = 1$ iff $i = \perp(s_1, s_2, s_3)$; and if $e = 0$, all outputs z_i are 0.
(3) One chip containing two *JK* master–slave flip-flops.
(4) One chip containing 8 inverters.
(5) One chip containing 4 OR gates with a fan-in of 2 each.
(6) Two chips containing 2 OR gates with a fan-in of 4 each.
(a) The execution of a certain computer instruction requires generating a sequence of control signals over five control lines numbered 0 through 4. The sequence is to be 02301144, meaning that after a 1-pulse appears at the GO input, line 0 is to have a 1, and the other lines are all 0's. After the first clock pulse, line 2 is to have a 1; after the second, line 3; etc. Design a network using the given packages (not all devices may be needed) to perform this sequencing.
(b) Repeat for the control sequence 023011441244.

9. [COE] Design a programmable pulse generator capable of producing a sequence of N pulses, where N may be preset to any value between 1 and 999. To prepare the pulse generator for starting an output sequence, the COUNT input is made 0. To start the output sequence, the COUNT input is set to 1. Use IC modules and a few gates, if necessary.

10. (a) Solve Problem 11 of Chapter 1 (p. 23) using the ideas of Section 8.5, i.e. treat each specification (a)–(d) as an iterative network.

(b) Reconsider Problem 11 of Chapter 1 with the modification that the signals $x_1, x_2, \ldots$ form a time sequence, i.e. represent the values of a binary input x at consecutive clock times. For each of the specifications (a)–(d) find a sequential network using clocked D flip-flops.

9 DESIGN OF DIGITAL NETWORKS

ABOUT THIS CHAPTER

So far we have discussed some systematic approaches to the design of small digital networks, both combinational and sequential. We have also introduced the reader to a large variety of readily available MSI and LSI IC packages. In this chapter we describe an IC-oriented approach to the design of medium-scale and large-scale digital networks. In such an approach, available IC packages are considered to be the basic building blocks.

No well-established theory exists for the design of large digital networks. Present techniques rely heavily on intuition and experience. For this reason the problems discussed here are challenging from an engineering viewpoint but do not use advanced mathematical methods.

We begin with the design of an automatic gain controller in Section 9.1. We present some background information about the problem and derive a precise description of the desired behavior. In the next step we decompose the network into smaller subnetworks. This very important step in the design process depends highly on intuition and experience. We then complete the design by implementing each subnetwork. Similar steps are carried out in the examples of Section 9.2, where we present three types of service request controllers. The solutions presented for our examples are far from being unique. However, we prefer to follow a particular approach in some detail, rather than to describe superficially a large number of alternatives.

Section 9.3 contains more general design ideas, namely a modern approach to the design of computation structures. The basic concept consists of separating the control functions of a system from its computing functions. We treat both synchronous-mode and asynchronous-mode control structures, and describe a novel approach to their implementation. Section 9.4 deals with the replacement of software algorithms by hardware. Although we only use two examples, the concepts can be easily extended.

This chapter makes use of most of the ideas presented earlier, with the exception of Chapter 3 and the starred parts of Chapter 4.

We should emphasize here that we have made no attempt to provide in this book an exhaustive listing of available IC modules. Rather, we have tried to develop an approach to modular design of digital networks, using a representative list of IC packages. New modules are appearing on the market almost daily, and the designer must keep up to date by following the current technical literature.

9.1. DIGITAL AUTOMATIC GAIN CONTROLLER

Our first example of digital network design is based on [FAR].

Background Information

The network N to be designed is to be a part of a larger system. The basic function of the system is to amplify an input signal. The amplified signal should always be at least as strong as a specified lower limit but should not exceed a specified upper limit. Thus the gain of the amplifier must be controlled automatically to ensure that the amplified signal falls in the desired range. The network N we are to design is to perform this gain control. The system is thus an amplifier with a "digital automatic gain controller."

A block diagram of the amplifier and digital network N to be designed is shown in Fig. 9-1. N is to control the attenuation of the amplifier. Each output z_i, $0 \leq i \leq 12$, corresponds to a certain attenuation. The outputs $z_4, z_5, \ldots, z_{12}$ correspond to 10 decibels (dB) each, whereas the first four outputs provide for a finer adjustment; i.e., outputs z_0, z_1, z_2, and z_3 correspond to 1, 2, 4, and 8 dB, respectively. Let the attenuation corresponding to z_i be w_i for $i = 0, \ldots, 12$. Then, if the output word of N is $z = z_0, \ldots, z_{12}$,

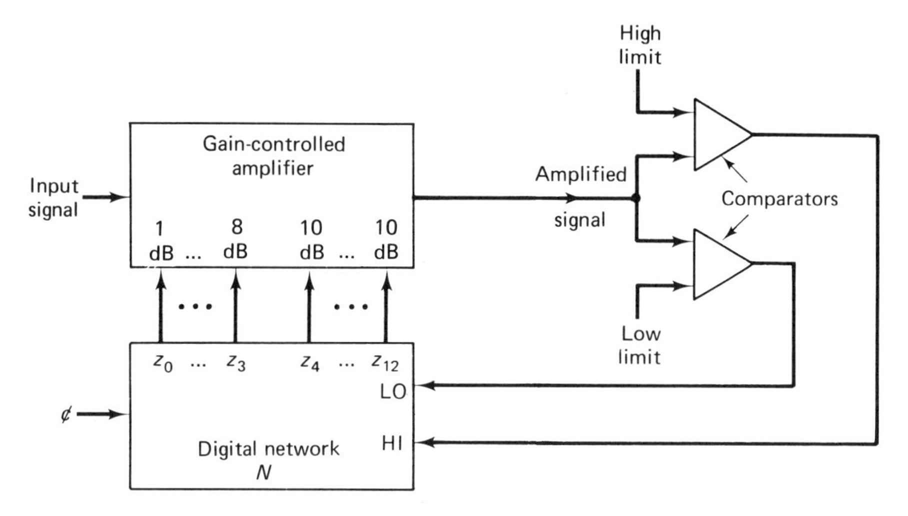

Fig. 9-1 Block diagram of a digital automatic gain controller (DAGC).

a total attenuation of $w = \sum_{i=0}^{12} w_i z_i$ decibels is inserted to reduce the amplifier gain. The amplified signal is compared with an upper limit and a lower limit. If the gain is too high, HI is set to 1, and if too low, LO becomes 1. If HI = 1, the control network is to increase the attenuation w by 1 dB, provided w is not yet maximal (99 dB), and if LO = 1, w is to be decreased by 1 dB, provided $w > 0$. The rate at which the comparisons and adjustments take place is under the control of a clock signal ¢.

We have now provided sufficient background information to describe the desired behavior of N.

Word Description of the Problem

The network N has a clock input ¢ and two control inputs HI and LO. It is to have 13 outputs $z_0, \ldots, z_{12}$, each of which corresponds to a weight w_i, where

$$w_i = 2^i \qquad \text{for } 0 \leq i \leq 3$$

and

$$w_i = 10 \qquad \text{for } 4 \leq i \leq 12.$$

The *overall output weight* w is defined by

$$w \triangleq \sum_{i=0}^{12} w_i z_i.$$

Note that in this context the logical values 0 and 1 of the z_i are interpreted as real values; i.e., the multiplication $w_i z_i$ and the summation $\sum$ is performed as in ordinary arithmetic.

Let $w(t)$ denote the value of w at the tth clock pulse. Then $w(t + 1)$ is to depend on $w(t)$, as well as LO(t) and HI(t), according to Table 9-1.

Table 9-1 Specifying the Network to Be Designed

LO(t)	HI(t)	Output Behavior
0	0	$w(t + 1) = w(t)$
1	0	$\begin{cases} w(t+1) = w(t) - 1, \text{ if } w(t) > 0 \\ w(t+1) = w(t), \text{ otherwise} \end{cases}$
0	1	$\begin{cases} w(t+1) = w(t) + 1, \text{ if } w(t) < 99 \\ w(t+1) = w(t), \text{ otherwise} \end{cases}$
1	1	This input can never occur

Decomposition Step

In this and similar design problems, one of the essential steps is the decomposition of the specified network into a number of simpler parts. If at all possible, this decomposition should be such that some of the parts become easily implementable by means of available MSI or LSI ICs. This decomposition step depends heavily on the experience and ingenuity of the designer.

In our present example a reasonable decision is to center the design around a BCD up/down counter. This decision leads to a decomposition, as shown in Fig. 9-2. It is now rather easy to specify the various subnetworks precisely and to arrive at their final implementation.

The required counter is obtained by cascading two BCD up/down counters 74190, as described in Sections 8.4 and 8.6. Let its output be $y_0^1, y_1^1, y_2^1, y_3^1, y_0^2, y_1^2, y_2^2, y_3^2$. This output is the *BCD* representation of the integer

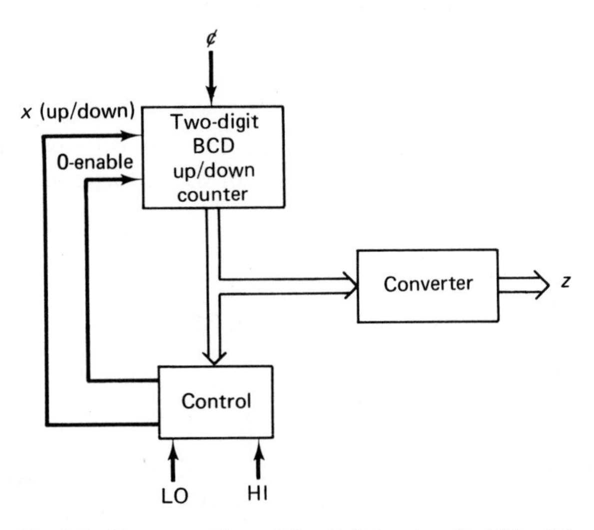

Fig. 9-2 Decomposition of the digital network of Fig. 9-1.

$$Y = 10Y^1 + Y^2,$$

where $Y^i = \perp(y_0^i, \ldots, y_3^i)$, $i \in \{1, 2\}$.

The CONVERTER is a combinational network that converts the counter output into z. This CONVERTER output z is to satisfy the condition $w = \sum_{i=0}^{12} w_i z_i = Y$, for $0 \leq Y \leq 99$.

The CONTROL is another combinational network. Its performance specification, which follows from Table 9-1, is given in Table 9-2. A dash (–) entry is to be interpreted as "irrelevant."

Table 9-2 Specifying the CONTROL Network of Fig. 9-2

LO	HI	Y	0-ENABLE	x (up/down)
0	0	–	1	–
1	0	$Y > 0$	0	1
1	0	$Y = 0$	1	–
0	1	$Y < 99$	0	0
0	1	$Y = 99$	1	–
1	1	–	–	–

Design of CONVERTER Subnetwork

We now turn to the detailed design of the CONVERTER. First, we conveniently choose $z_0 = y_3^2$, $z_1 = y_2^2$, $z_2 = y_1^2$, and $z_3 = y_0^2$. There remains the problem of converting the input $y^1 = y_0^1, y_1^1, y_2^1, y_3^1$ into the output $z_4, \ldots, z_{12}$ such that

$$Y^1 = \perp y^1 = \sum_{i=4}^{12} z_i \triangleq \sum_z \qquad \text{for } 0 \leq Y^1 \leq 9.$$

Note: Since there is no risk of ambiguity, we drop the superscript 1 from Y^1, y^1, and y_i^1 in the following discussion.

In a straightforward approach to this conversion problem one would first set up a table of combinations, such as Table 9-3 (don't-care input com-

Table 9-3 Table of Combinations Satisfying

$$Y \triangleq \perp y - \sum_{i=4}^{12} \hat{z}_i = \sum_z \quad \text{for } 0 \leq Y \leq 9$$

y_0	y_1	y_2	y_3	$\hat{z}_4$	$\hat{z}_5$	$\hat{z}_6$	$\hat{z}_7$	$\hat{z}_8$	$\hat{z}_9$	$\hat{z}_{10}$	$\hat{z}_{11}$	$\hat{z}_{12}$
0	0	0	0	0	0	0	0	0	0	0	0	0
0	0	0	1	1	0	0	0	0	0	0	0	0
0	0	1	0	1	1	0	0	0	0	0	0	0
0	0	1	1	1	1	1	0	0	0	0	0	0
0	1	0	0	1	1	1	1	0	0	0	0	0
...	...	...	...	...	...	...	...	...	...	...	...	...
1	0	0	0	1	1	1	1	1	1	1	1	0
1	0	0	1	1	1	1	1	1	1	1	1	1

binations are not shown). Note that the set of functions $\hat{z}_4, \ldots, \hat{z}_{12}$ of Table 9-3 represents just one possible solution to the problem. In fact, there are many solutions, since there are many ways of defining functions $z_4, \ldots, z_{12}$ so that $\sum_{i=4}^{12} z_i = \sum_z$ is correct for the given y. The particular choice of the $\hat{z}_i$ functions in Table 9-3 is quite natural but was made rather arbitrarily and without any attempt to minimize the network implementing the conversion. Of course, this table can be dealt with by the methods of Chapter 5. However, there is a considerably better choice for the z_i functions, as we describe below.

Notice in Table 9-3 that, as $\perp y$ goes through the values 0, 1, . . . , 8, 9, $\sum_z$ goes through the sequence (even, odd, . . . , even, odd). Suppose that we choose one of the z_i, say z_{12}, to account for this fluctuation in $\sum_z$. It is clear that we can set $z_{12} = y_3$, thus obtaining the correct parity for $\sum_z$ without using any gates at all.

Table 9-4

$\perp y$	$y_0\ y_1\ y_2$	$\sum_z - z_{12}$	$(\sum_z - z_{12})/2 = n$
0, 1	0 0 0	0	0
2, 3	0 0 1	2	1
4, 5	0 1 0	4	2
6, 7	0 1 1	6	3
8, 9	1 0 0	8	4

We can now consider a reduced problem, as shown in Table 9-4. If z_{12} is subtracted from $\sum_z$, then the values $2n$ and $2n + 1$ of $\perp y$, for $n = 0, 1, \ldots, 4$, are equivalent with respect to $\sum_z - z_{12}$, since in both cases we must have $\sum_z - z_{12} = 2n$. The value n is represented properly in the binary code by the bits y_0, y_1, and y_2. If we continued in the same fashion, we would let $z_{10} = z_{11} = y_2$. When n is even, the contribution of these two outputs to $\sum_z$ would be 0, and when n is odd, it would be 2. After removing z_{10} and z_{11} from $\sum_z - z_{12}$, we would have the sequence 0, 0, 4, 4, 8 to realize using the 6 outputs $z_4, \ldots, z_9$. This, however, is impossible since we require 8 z's to be 1 when $\perp y^1 = 8$ or 9, but only 6 are available. Therefore, this scheme must be modified. The scheme:

1. Choose 1 output (z_{12}) equal to y_3
2. Choose 2 outputs (z_{10}, z_{11}) equal to y_2
3. Choose 4 outputs ($z_6, \ldots, z_9$) equal to y_1

is shown in Table 9-5 for $\perp y = 0, 1, \ldots, 7$. The next step would be:

4. Choose 8 outputs equal to y_0.

Table 9-5

$\perp y$	y_0	y_1	y_2	y_3	z_4	z_5	z_6	z_7	z_8	z_9	z_{10}	z_{11}	z_{12}	
0	0	0	0	0	0	0	0	0	0	0	0	0	0	Regular scheme
1	0	0	0	1	0	0	0	0	0	0	0	0	1	
2	0	0	1	0	0	0	0	0	0	0	1	1	0	
3	0	0	1	1	0	0	0	0	0	0	1	1	1	
4	0	1	0	0	0	0	1	1	1	1	0	0	0	
5	0	1	0	1	0	0	1	1	1	1	0	0	1	
6	0	1	1	0	0	0	1	1	1	1	1	1	0	
7	0	1	1	1	0	0	1	1	1	1	1	1	1	
8	1	0	0	0	1	1	1	1	1	1	1	1	0	"End effects"
9	1	0	0	1	1	1	1	1	1	1	1	1	1	

Since there are only two outputs left, this cannot be done and we must give up for $\perp y = 8$ and 9 our nice scheme that uses no gates whatsoever. For 8 and 9 we can still use z_{12} to keep track of the parity. The remaining z_i are all forced to be 1 in order to produce the proper $\sum_z$ for 8 and 9. The complete scheme, including the "end effects," is shown in Table 9-5 and results in the following equations:

$$z_{12} = y_3$$

$$z_{10} = z_{11} = y_2 + y_0$$

$$z_6 = z_7 = z_8 = z_9 = y_1 + y_0$$

$$z_4 = z_5 = y_0.$$

Altogether, this conversion requires two 2-input OR gates.

Incidentally, the scheme we have just discovered suggests that the outputs $z_4, \ldots, z_{12}$ could be obtained from $\perp y$ without any logic if the code for $\perp y$ were (d_0, d_1, d_2, d_3), as shown in Table 9-6, rather than the binary or "the

Table 9-6 (2, 4, 2, 1) Weighted Decimal Code

	8	**4**	**2**	**1**	**2**	**4**	**2**	**1**
$\perp y$	y_0	y_1	y_2	y_3	d_0	d_1	d_2	d_3
0	0	0	0	0	0	0	0	0
1	0	0	0	1	0	0	0	1
2	0	0	1	0	0	0	1	0
3	0	0	1	1	0	0	1	1
4	0	1	0	0	0	1	0	0
5	0	1	0	1	0	1	0	1
6	0	1	1	0	0	1	1	0
7	0	1	1	1	0	1	1	1
8	1	0	0	0	1	1	1	0
9	1	0	0	1	1	1	1	1

(8, 4, 2, 1) weighted decimal code." The code of Table 9-6 is called "the (2, 4, 2, 1) weighted decimal code" because we can associate weights 2, 4, 2, 1 as shown in Table 9-6 and we verify that

$$\perp y = 2d_0 + 4d_1 + 2d_2 + 1d_3.$$

The conversion equations are obtained by inspection:

$$d_0 = y_0$$
$$d_1 = y_1 + y_0$$
$$d_2 = y_2 + y_0$$
$$d_3 = y_3.$$

The entire conversion process can be viewed now as shown in Fig. 9-3: a code converter [changing the (8, 4, 2, 1) code into the (2, 4, 2, 1) code] and wire-splitting to obtain the z_i.

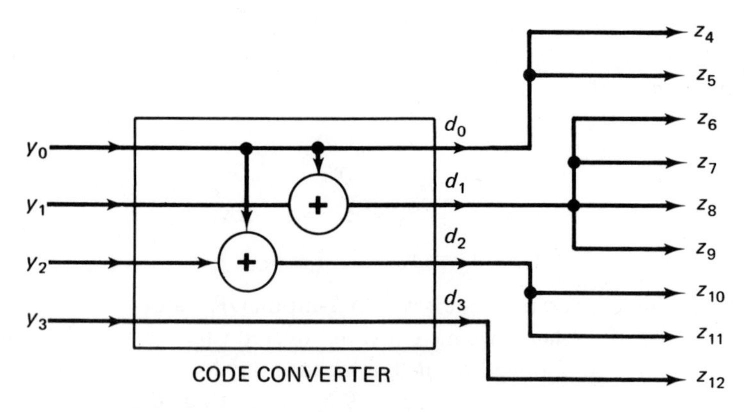

Fig. 9-3 A simple network satisfying $\perp y = \sum_z$.

Design of CONTROL Subnetwork

We now proceed to implement Table 9-2. For the 74190 counter, let UF denote the count of 0 (underflow), and let OF denote the count of 9 (overflow). These can be obtained as follows:

$$\text{OF} = y_0 \cdot \text{MM} \qquad \text{UF} = y_0' \cdot \text{MM}.$$

Table 9-2 leads to the following equations.

$$x = 1 \equiv \text{LO} = 1 \wedge \text{Y} > 0$$
$$\equiv \text{LO} = 1 \wedge (\text{Y}^1 \neq 0 \vee \text{Y}^2 \neq 0).$$

Thus

$$x = \text{LO} \cdot ((\text{UF}_1)' + (\text{UF}_2)').$$

$$\text{0-ENABLE} = 1 \equiv (\text{LO} = 0 \wedge \text{HI} = 0) \vee (\text{LO} = 1 \wedge \text{Y} = 0)$$
$$\vee\ (\text{HI} = 1 \wedge \text{Y} = 99)$$
$$\equiv (\text{LO} = 0 \wedge \text{HI} = 0) \vee (\text{LO} = 1 \wedge \text{Y}^1 = 0 \wedge \text{Y}^2 = 0)$$
$$\vee\ (\text{HI} = 1 \wedge \text{Y}^1 = 9 \wedge \text{Y}^2 = 9).$$

Therefore,

$$\text{0-ENABLE} = \text{LO}' \cdot \text{HI}' + \text{LO} \cdot \text{UF}_1 \cdot \text{UF}_2 + \text{HI} \cdot \text{OF}_1 \cdot \text{OF}_2.$$

A block diagram of the digital controller is shown in Fig. 9-4.

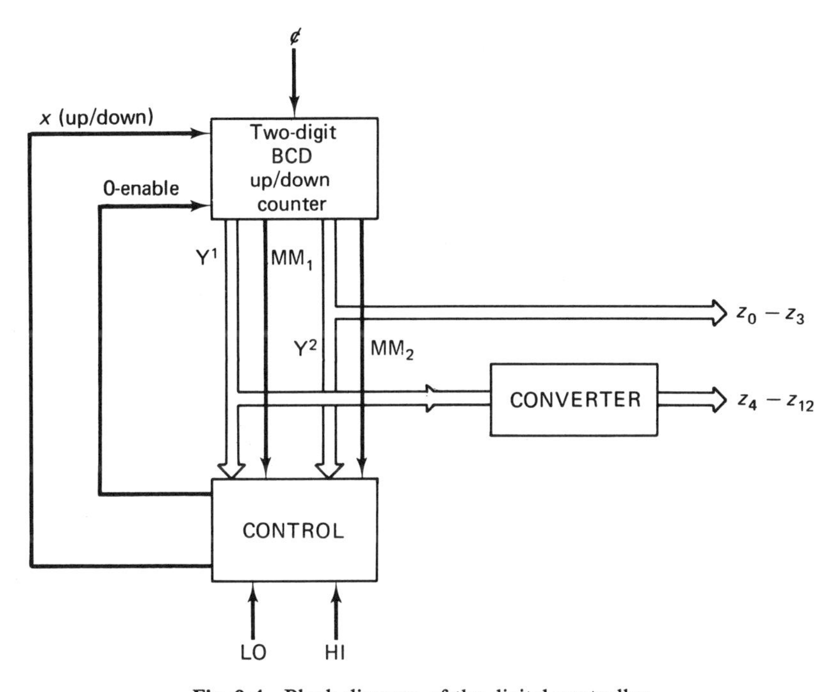

Fig. 9-4 Block diagram of the digital controller.

9.2. SERVICE REQUEST CONTROLLERS

Background

Consider a situation where a number of terminal stations (e.g., peripheral devices, dial telephones, etc.) are connected to a central processor (CP). Any terminal station may request service from the CP at any instant. Assume that the CP may only attend to one request at any given time. A service request controller is to detect incoming service requests and to pass them on one by one to the CP.

Some controllers of this type operate on the "first-come first-served" principle. They have to store all the incoming requests in order of their arrivals and then pass these requests on to the CP on a "first-in first-out" or "FIFO" basis. Other systems treat incoming service requests on a fixed priority basis, where each terminal station is allotted a predetermined priority. If a number of stations are competing for service, the station with highest priority is served first. Finally, many systems, especially various telephone switching systems, apply a random selection principle; i.e., the controller arbitrarily selects any station from all the stations requesting service.

In this section we illustrate design approaches according to the three selection principles mentioned.

Word Description (Priority Selection Principle)

Figure 9-5 shows the input–output connections of a typical service request controller (SRC). ST is the *start* input; a transition of ST from 0 to 1 is to

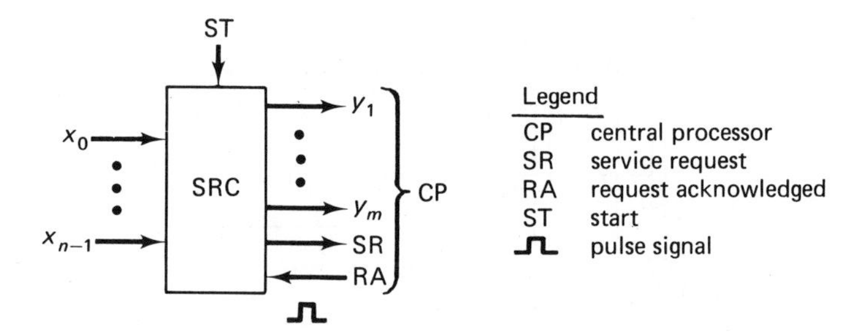

Fig. 9-5 Typical service request controller.

reset the controller to its starting state. The input $x_i = 1$ indicates that station *i* (e.g., a dial telephone station) requests service (e.g., wishes to originate a call). The controller is to inform the CP (central processor) of the existence of a service request by setting SR = 1, and to identify the station requesting service, say station *i*, by setting $\perp y = i$. We refer to y as the *address* of station *i*. It is assumed that $2^{m-1} < n \leq 2^m$, so that all the addresses can be represented. If two or more stations request service simultaneously, the controller selects the station with the lowest station number. After receiving the SR signal, the CP takes notice of the address of the calling station. The CP then returns a pulse signal to the SRC on its RA (request acknowledged) input and initiates a procedure to offer service to station *i* ($i = \perp y$). Immediately after receiving the RA pulse, the controller is to reset SR to 0. As soon as service is offered to station *i* (e.g., dial tone is connected to indicate that dialing may start), x_i will be reset to 0 by the equipment interfacing the stations and the controller. After alerting the CP (by setting SR to 1), the controller is to wait for both the appearance of the RA pulse and the resetting of x_i, before looking for further service requests. This arrangement ensures that only one service-offering procedure may be in progress at any time. After setting SR to 1, the controller is to freeze its outputs, until the RA pulse is received, in order to enable the CP to record the controller output without any disturbances. It is therefore essential that the SRC provides the correct station identification before or simultaneously with the setting of SR to 1, and that the SR and y outputs remain unchanged during the recording period, even if the station in question withdraws its service request in the meantime. The possibility of a premature withdrawal of a service request must, of course, be taken into account by the service-offering procedure.

Decomposition Step

The main task of the controller is to pick out one of the stations requesting service and to produce its address. Among the various stations requesting service, the one with the lowest station number (highest priority) is to be selected. An obvious choice for implementing this main task is the priority encoder described in Section 2.6.

We now turn to the application of an 8-line priority encoder to the design of an 8-line SRC. For the controller to perform properly, it must be able to "freeze" the inputs for a period long enough to reach a decision. This can be achieved by means of flip-flops. A corresponding block diagram, incorporating 8 D flip-flops and an 8-line priority encoder, is shown in Fig. 9-6. The x_i values will be sampled from time to time with the aid of the clock $¢_D$, and the sampled values will be stored in the D flip-flops. The clock $¢_D$ will be described later. Once a line requesting service has been selected by the SRC, the controller is to "watch" this line until the request disappears. This must be done somehow by the network A (yet to be designed) of Fig. 9-6. The "line watching" can be achieved with the aid of a data selector (Section 2.5). If the selector address is set to z_1, z_2, z_3 and if the inputs $x_0, x_1, \ldots, x_7$ are applied to its data inputs, the selector output will be equal to the input x_*, whose service request is being handled.

In Fig. 9-7 the network A of Fig. 9-6 is decomposed into a data selector and a control structure B. This control structure has four inputs (ST, RA, x_*, z_0) and two outputs (SR, $¢_D$).

Design of Network *B* of Figure 9-7

So far we have succeeded considerably in reducing the overall problem of designing a service request controller by suitably breaking up (decomposing) the original problem. There remains the task of designing network B of Fig. 9-7. A convenient decision at this stage is to design network B as a synchronous-mode sequential network. Recall that ST has been specified as a reset

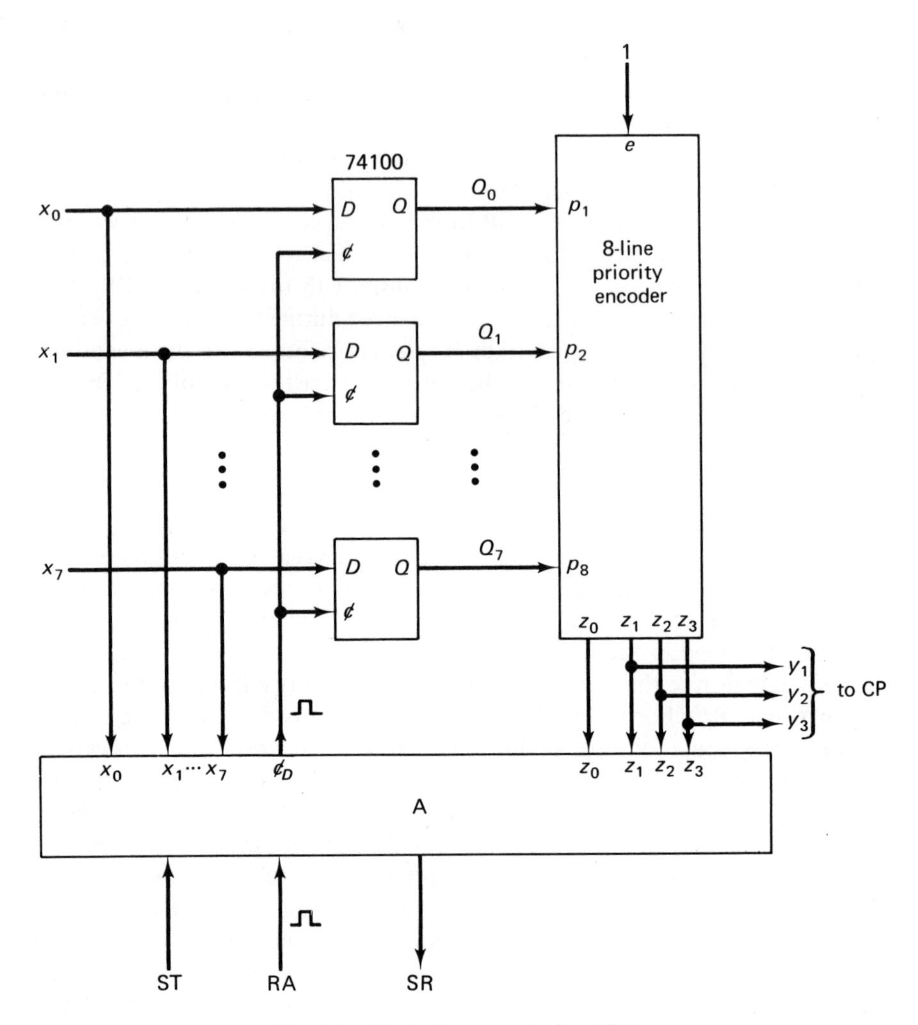

Fig. 9-6 Block diagram of 8-line SRC.

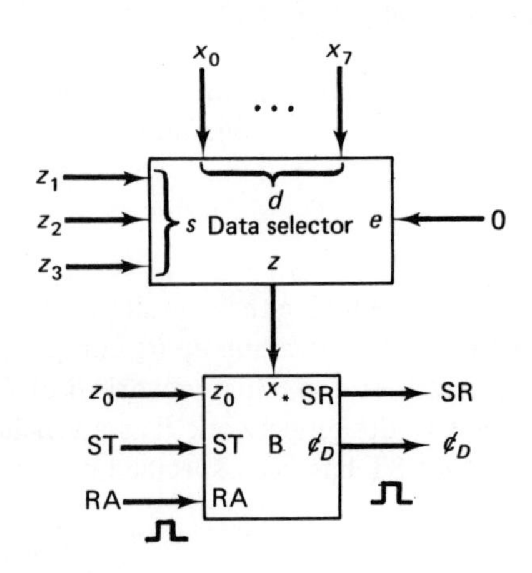

Fig. 9-7 Decomposition of network A of Fig. 9-6.

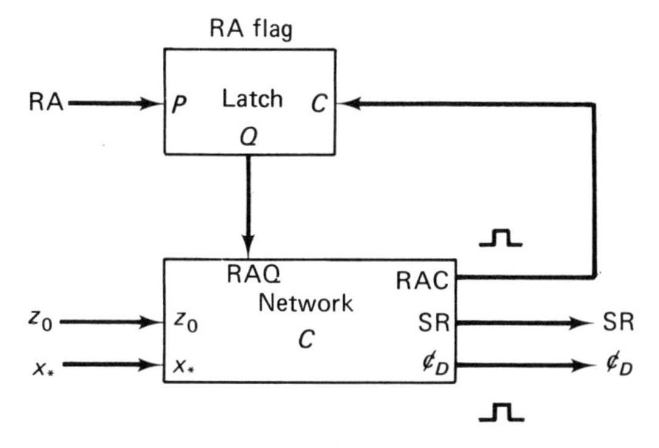

Fig. 9-8 Introduction of a flag on RA-line of network B (Fig. 9-7).

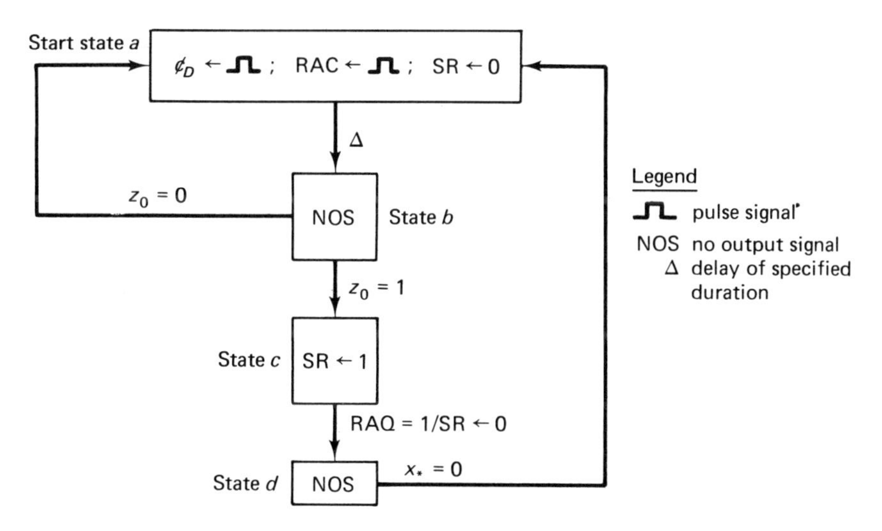

Fig. 9-9 Transition chart for network C (Fig. 9-8).

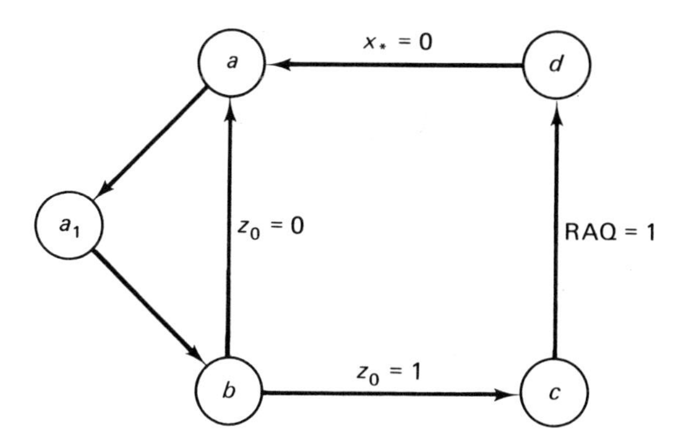

Fig. 9-10 State diagram for network C (Fig. 9-8).

input. It is convenient to delay the incorporation of such an input to the final stage of the design. A difficulty we now have to cope with is the fact that the RA input pulse will occur independently of the internal clock of network *B*. One method of overcoming this difficulty is the introduction of a latch ("*flag*") as shown in Fig. 9-8, where we have also omitted input ST.

The performance of network *C* may now be specified by a *transition chart* as shown in Fig. 9-9. This transition chart shows four states (*a–d*), each state represented by a rectangular box. Inside each box the output conditions for this particular state are indicated. For example, while in state *a*, the network is to provide a single pulse on output $¢_D$, a single pulse on output RAC, and a 0-level on output SR. A transition between states is to occur only if the associated condition is met (e.g., $z_0 = 1$) or a specified delay Δ has passed.

A brief description of the transition chart of Fig. 9-9 follows. The reader should also refer to Figs. 9-6, 9-7, and 9-8. While in the start state *a*, the network is to set up suitable initial conditions, i.e., to clear the RA flag and to set SR = 0. It initiates its operation by issuing a *sampling* pulse on $¢_D$. This sampling pulse will transfer the instantaneous status of the x_i inputs into the *D* latches. After a propagation delay Δ (180 ns for the encoder of Section 5.8), the 8-line priority encoder will provide the correct outputs. Accordingly, the network is to enter state *b* after Δ nanoseconds and to examine z_0. If $z_0 = 0$, no service request has been detected, and start state *a* is reentered. Alternatively ($z_0 = 1$), state *c* is entered, and the CP is informed of the existence of a service request by setting SR = 1. Note that the correct address has already been set up on outputs y_1, y_2, and y_3 (Fig. 9-6). While in state *c*, the network waits for the RA signal to arrive. As soon as RAQ becomes 1, SR is to be reset to 0, and state *d* is entered next. In this state the network waits for x_* (the input line whose address is $y = y_1, y_2, y_3$) to reset to 0. Thereupon, the network reenters state *a*.

Assume now that we choose a clock-pulse frequency of 10 MHz (megahertz), i.e., 10^7 pulses per second, or one clock pulse each 100 ns, and that this clock-pulse rate is satisfactory for all state transitions except the state *a* to state *b* transition, which requires a delay of $\Delta = 180$ ns. To overcome this difficulty, we introduce an auxiliary state a_1, as shown in the state diagram of Fig. 9-10. If speed is an important factor, this introduction of an additional state is preferable to slowing down all state transitions by choosing a lower clock-pulse frequency. The transition from *a* to *b* will now take place in two clock periods, i.e., in 200 ns, which is quite close to the minimum of 180 ns required by the priority encoder. The 10-MHz clock is called input ¢ in the following discussion.

In the state diagram of network *C* in Fig. 9-10 there are really 4 inputs: ¢, z_0, RAQ, and x_*. A shorthand version of the state table is shown in Fig. 9-11(a), where $s[p]$ means that *s* is the next state if $p = 1$.

s^n	s^{n+1}	
a	a_1	
a_1	b	
b	$a\,[z'_0]$;	$c\,[z_0]$
c	c[RAQ'];	d[RAQ]
d	$d\,[x_*]$;	$a\,[x'_*]$

(a)

	$Q_1^n\,Q_2^n\,Q_3^n$		
a	0 0 0	100	
a_1	1 0 0	101	
b	1 0 1	000 $[z'_0]$;	111 $[z_0]$
c	1 1 1	111 [RAQ'];	110 [RAQ]
d	1 1 0	110 $[x_*]$;	000 $[x'_*]$

(b)

	$Q_1^n\,Q_2^n\,Q_3^n$		$J_1 K_1 J_2 K_2 J_3 K_3$	
a	0 0 0	s_0	1 – 0 – 0 –	
a_1	1 0 0	s_4	– 0 0 – 1 –	
b	1 0 1	s_5	– 1 0 – – 1 $[z'_0]$;	– 0 1 – – 0 $[z_0]$
c	1 1 1	s_7	– 0 – 0 – 0 [RAQ'];	– 0 – 0 – 1 [RAQ]
d	1 1 0	s_6	– 0 – 0 0 – $[x_*]$;	– 1 – 1 0 – $[x'_*]$

(c)

Fig. 9-11 "Shorthand" tables for the state diagram of Fig. 9-10.

We require at least 3 variables, Q_1, Q_2, and Q_3, to represent the 5 states. The assignment shown in the transition table of Fig. 9-11(b) was arbitrarily chosen.

We choose to design the network using master–slave JK flip-flops. To illustrate a new approach, we use an address decoder as shown in Fig. 9-12. Then the excitation functions J_i, K_i will be obtained from the inputs and the decoded state variables $s_0, \ldots, s_7$, as shown in the excitation table of Fig. 9-11(c). One easily verifies that the following excitation equations satisfy Fig. 9-11(c):

$$J_1 = s_0 \qquad K_1 = s_5 z'_0 + s_6 x'_*$$

$$J_2 = s_5 z_0 \qquad K_2 = s_6 x'_*$$

$$J_3 = s_4 \qquad K_3 = s_5 z'_0 + s_7 \text{RAQ}.$$

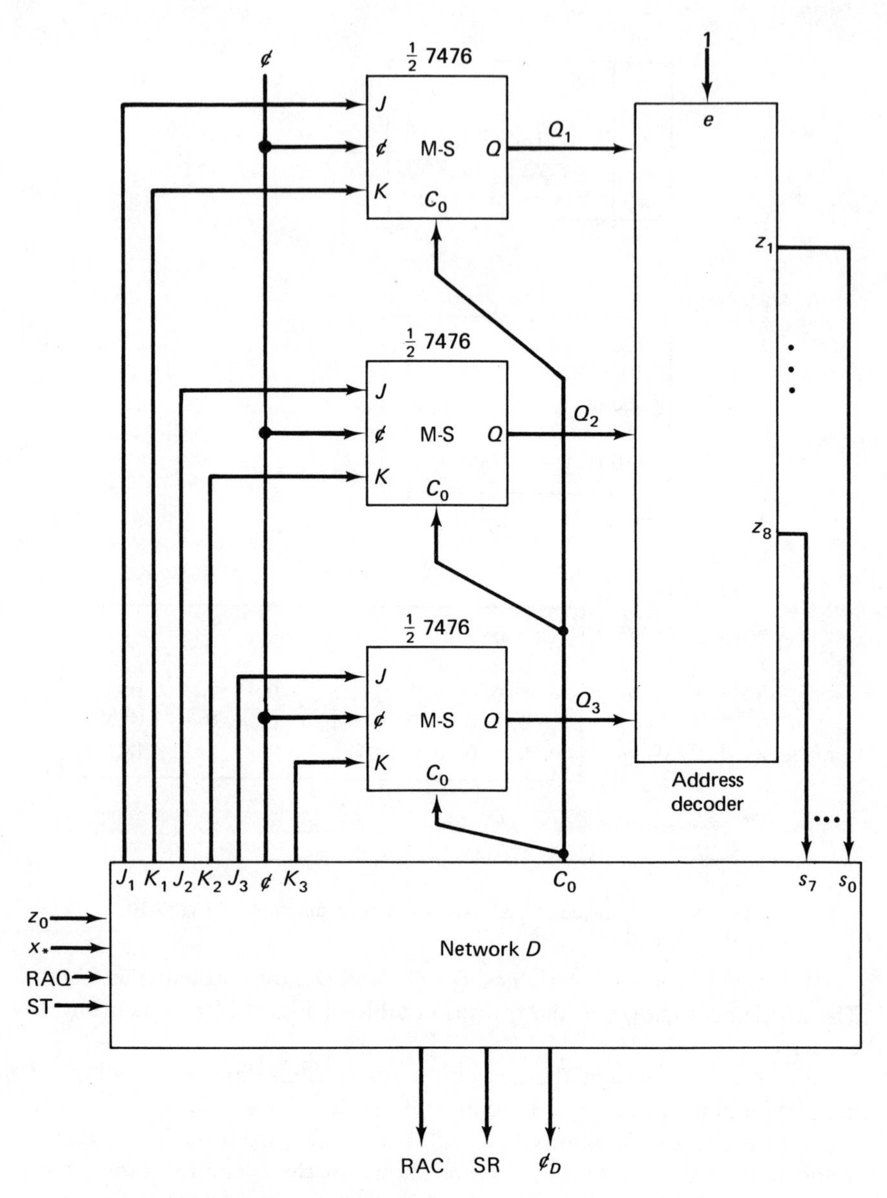

Fig. 9-12 Implementation of state diagram of Fig. 9-10.

The required outputs RAC, SR, and $¢_D$ may be obtained as follows:

$$¢_D = ¢s_0$$

$$\text{RAC} = ¢s_0$$

$$\text{SR} = s_7(\text{RAQ})'.$$

We still have to deal with the reset input ST. As specified, a transition from 0 to 1 of the ST input is to reset the controller to its starting state. An evident way to achieve this is to connect the ST input to a monostable multivibrator (74121) which will transform the 0–1 transition of ST into a pulse signal suitable for clearing all flip-flops. Thus the $\bar{Q}$ output of the multivibrator is to be connected to the C_0 inputs of the *JK* flip-flops.

The details of implementing network *D* of Fig. 9-12 are now straightforward and are left to the reader.

Service Request Scanner

A service request controller operating on the random selection principle is frequently referred to as a *service request scanner.* We now turn to the design of such a scanner. The main idea is the use of a binary modulo-*n* counter together with a data selector, as shown in Fig. 9-13.

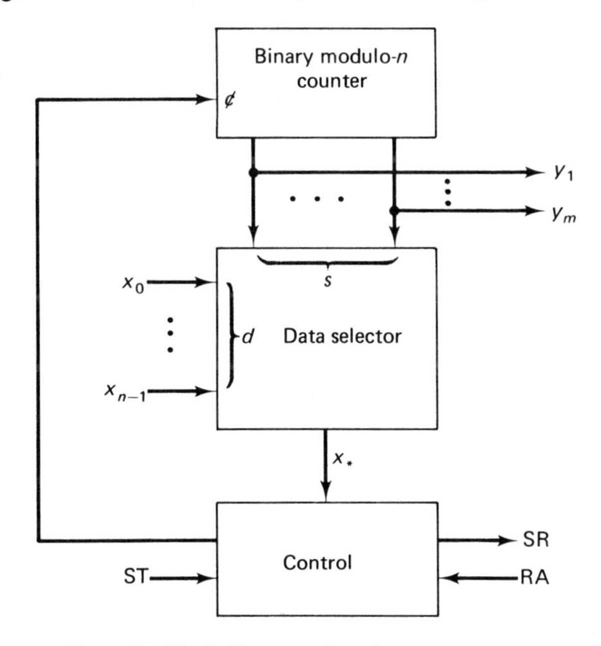

Fig. 9-13 Block diagram of service request scanner.

At any particular moment, the scanner examines the input line x_* selected by the output of the counter; i.e., $x_* = x_i$, where $i = \perp y$. If $x_* = 0$, the counter is incremented and the next input line is examined (where x_0 is "next" to x_{n-1}). If $x_* = 1$, the CONTROL network of Fig. 9-13 communicates with the CP, as required. After an RA pulse has been received and x_* has returned to 0, the counter is again incremented and the scanning continues.

The details of this design are left to the reader.

Queue Controller [PHI]

We conclude this section by discussing the design of a *queue controller*, i.e., a network indicating the station whose request was first. Such a queue controller is, of course, applicable to the design of an SRC operating on the "first-in first-out" principle.

We assume that the queue controller has n input lines $x_1, \ldots, x_n$ and n output lines $z_1, \ldots, z_n$. The controller sets z_i to 1 to indicate that station i is requesting service and that it is its turn to be served.

One easily verifies (see below) that the solution for $n > 2$ may be based on a solution for the case $n = 2$. Therefore, we consider the case $n = 2$ first.

We derive the state table of Fig. 9-14 as follows. There are two inputs x_1

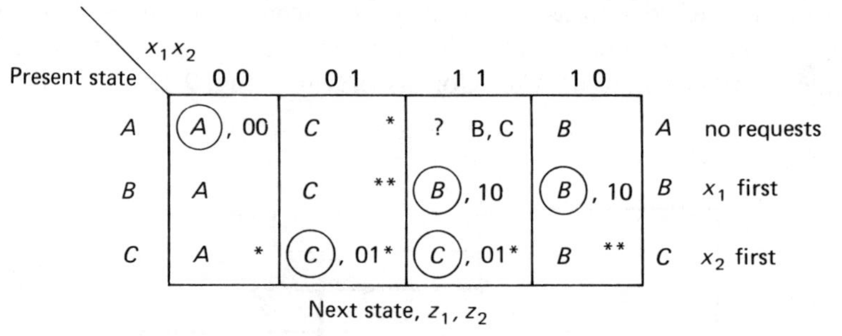

Fig. 9-14 State table for 2-input controller.

and x_2 and hence 4 input columns. When there are no requests, i.e., $x_1 = x_2 = 0$, there should be no output, i.e., $z_1 = z_2 = 0$, and the situation should not change unless the inputs change. Let this condition be represented by stable state A as shown in Fig. 9-14. When the input becomes $x_1 = 1, x_2 = 0$, z_1 is to be turned on and z_2 is to remain 0. Thus we need a stable state B in column 10 as shown for the total states $(1, 0; A)$ and $(1, 0; B)$. When x_2 requests service while the network is in state B, this should be ignored, since x_1 has already seized control. Thus the entry in total state $(1, 1; B)$ should be B. When x_1 becomes 0 again and $x_2 = 0$, the network should return to the waiting state A. These arguments give us the unstarred entries of Fig. 9-14. Because the network is symmetric in x_1 and x_2, by similar reasoning we find the entries marked * in Fig. 9-14, where state C is analogous to B except that x_2 seized the controller first.

Next note that the condition $(0, 1; B)$ can be reached only if the state is B with x_1 in control, and x_1 gives up its control while x_2 requests service. The network should then give the control to x_2, i.e., go to state C. Similarly, the entry in state $(1, 0; C)$ should be B. These two entries are marked ** in Fig. 9-14.

There remains the entry in $(1, 1; A)$ to be settled. This total state can be reached only if the network is in the "no-requests" state A and both lines

x_1 and x_2 request service simultaneously. It would not be correct to interpret this as a "no-request" situation and, on the other hand, it is impossible to handle both requests at the same time. One could give preference to one of the lines or, alternatively, one could permit a race to either B or C. We choose the latter solution for reasons that will become obvious soon.

If we ignore the outputs z_1 and z_2 temporarily and consider only the state behavior of Fig. 9-14, we easily recognize that this behavior is identical to that of the NAND latch [see Fig. 7-5(a), p. 197], if we identify the states as follows:

$$A\colon\quad \bar{Q} = 1,\, Q = 1$$

$$B\colon\quad \bar{Q} = 0,\, Q = 1$$

$$C\colon\quad \bar{Q} = 1,\, Q = 0,$$

and if we let $x_1 = C_0$, $x_2 = P_0$. Now consider the outputs z_1 and z_2. From the table of Fig. 9-15 it is clear that $z_1 = \bar{Q}'$ and $z_2 = Q'$. The entire network can now be implemented as shown in Fig. 9-16.

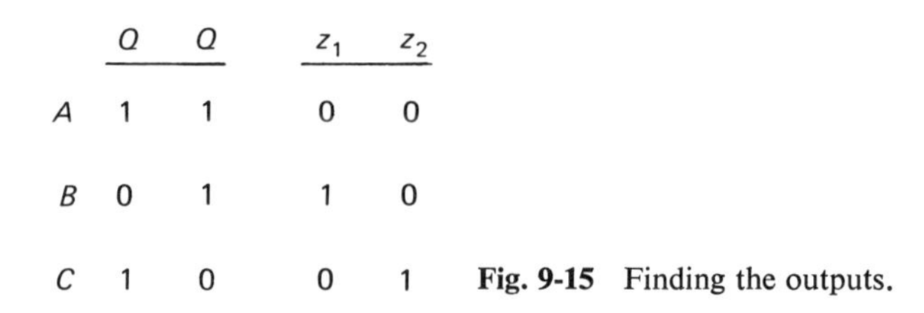

	Q	Q	z_1	z_2
A	1	1	0	0
B	0	1	1	0
C	1	0	0	1

Fig. 9-15 Finding the outputs.

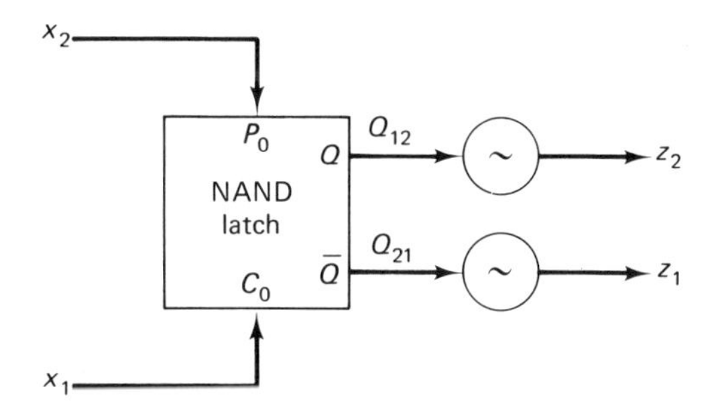

Fig. 9-16 Implementation of 2-input queue controller.

As explained in Section 7.1, NOR and NAND latches usually serve as binary storage cells. Then the output $\bar{Q}$ is always the complement of Q, ($\bar{Q} = Q'$), and the combination $Q = \bar{Q}$ is conventionally avoided by the designer. We have just demonstrated an application where the complete, unrestricted behavior of the NAND latch is efficiently utilized.

The idea just described can be extended to the case $n > 2$ as follows. Every input line x_i is made to compete with every other input line x_j $(i \neq j)$.

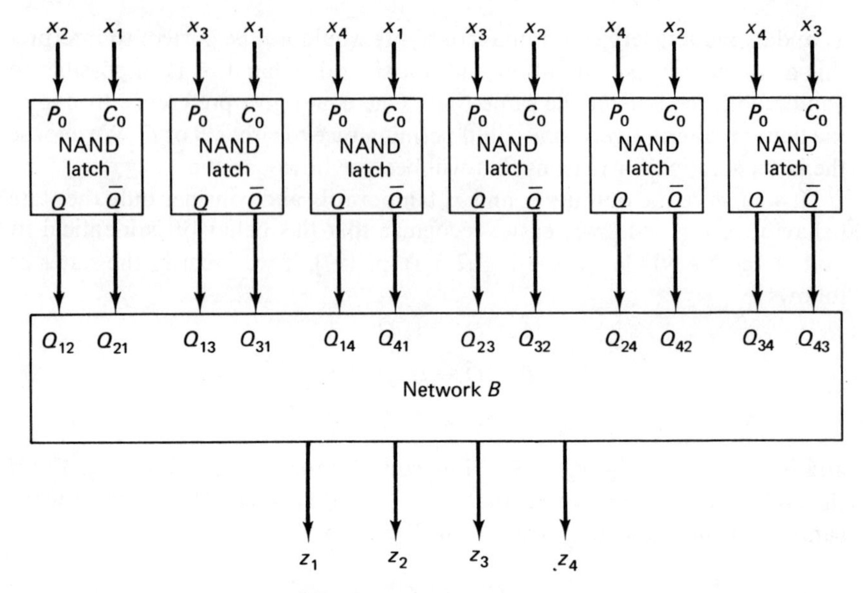

Fig. 9-17 Queue controller, $n = 4$.

For example, the arrangement for $n = 4$ is shown in Fig. 9-17. One verifies the following requirements for network B:

$$z_1 = 1 \equiv (Q_{21} = 0) \wedge (Q_{31} = 0) \wedge (Q_{41} = 0)$$
$$z_2 = 1 \equiv (Q_{12} = 0) \wedge (Q_{32} = 0) \wedge (Q_{42} = 0), \text{ etc.}$$

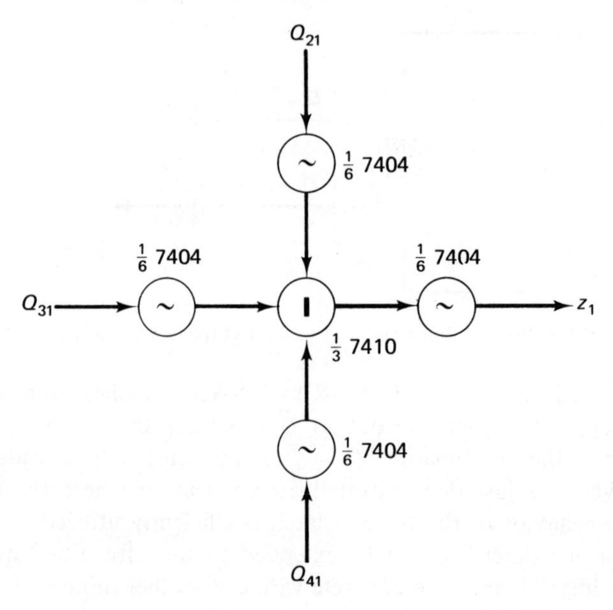

Fig. 9-18 Implementation of z_1-output of network B (Fig. 9-17).

The equivalence for z_1, for example, follows because $Q_{i1} = 0$ iff input 1 requests service before input i. Thus

$$z_1 = Q'_{21} \cdot Q'_{31} \cdot Q'_{41}$$
$$z_2 = Q'_{12} \cdot Q'_{32} \cdot Q'_{42}$$
$$z_3 = Q'_{13} \cdot Q'_{23} \cdot Q'_{43}$$
$$z_4 = Q'_{14} \cdot Q'_{24} \cdot Q'_{34}.$$

Note that we exclude the possibility of three or more simultaneous arrivals of requests.

An implementation of the equation for z_1, by means of the packages of Table 2-1, is shown in Fig. 9-18. The other equations are, of course, implemented in the same way.

9.3. SEQUENCE CONTROLLERS

Introduction

Many sequential networks may be considered as consisting of two parts: a *device structure* and a *control structure* [PA-DE]. The device structure consists of specific devices, such as adders, counters, code converters, etc. The control structure, also called *sequence controller* or *sequencer*, supervises the activities and sequencing of these devices.

A device either performs one specific task (e.g., an adder) or has multitask capabilities (e.g., an ALU). In this discussion we assume all devices to be of the single-task type. The extension of our study to multitask devices is straightforward.

An *asynchronous* device is given a *start* or *GO* command by the control structure to start its operation. Upon completion of its task the device returns a *completion* or *DONE* signal. A *clocked* device is started similarly but does not return a completion signal. The time it takes the device to perform its task (worst case) is known to the control structure and is either measured by means of an internal clock or, alternatively, by means of an external timer.

Both start commands and completion signals may be either of the pulse or level type. We assume that in both cases the $0 \longrightarrow 1$ transition represents the actual GO command or DONE signal, by convention.

In what follows we develop an approach to the design of digital networks using the notion of sequence controller. Although we present the ideas by means of examples, the methods are quite general and widely applicable.

Task Flow and State Transition Charts

The sequence in which given tasks are to be initiated by a sequence controller is conveniently specified by means of a *task flow chart*, an example of which is given in Fig. 9-19. In this chart TA*i* denotes the operation of performing task *i*. The nature of the task itself is of no interest to the sequencer except as noted below. The two-way branch symbols ("decision boxes") have the usual meaning. The p_j represent propositions (conditions) whose truth values determine the sequencing of the relevant tasks. BEGIN and END refer

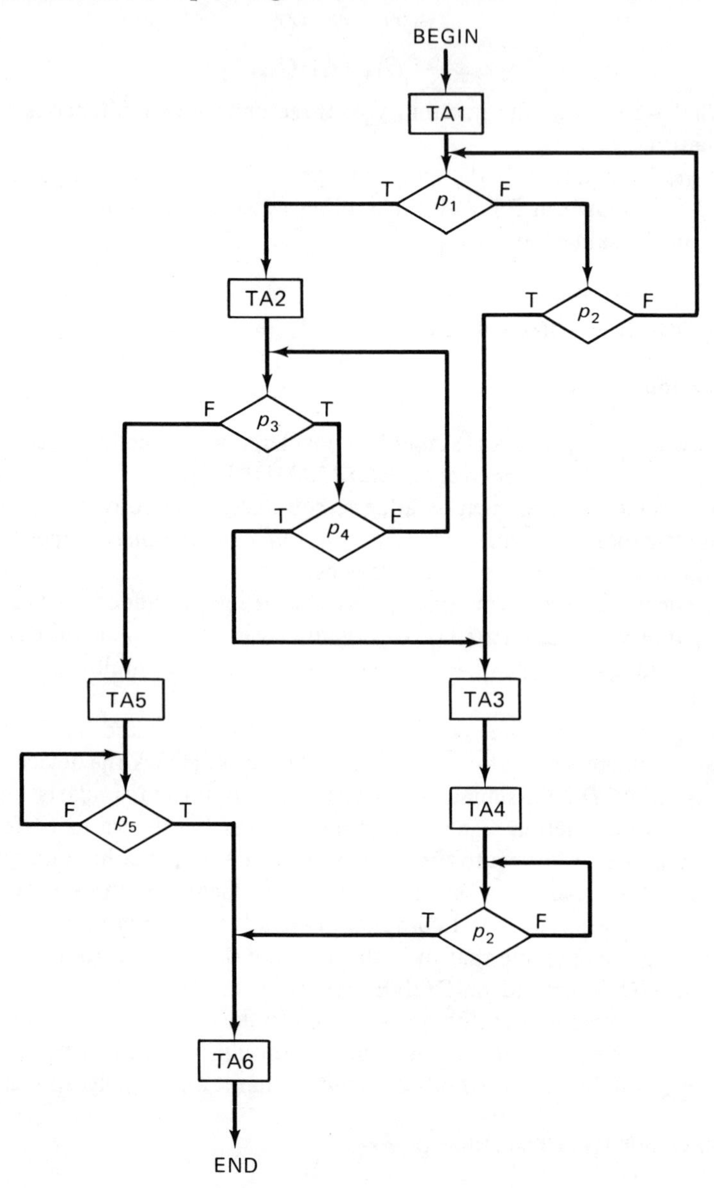

Fig. 9-19 Example of a task flow chart.

to the start command and completion signal (respectively) of the overall network.

With reference to Fig. 9-19, we now proceed to discuss the next design step. First, let us make the following specific assumptions. Tasks 1 and 3 are asynchronous, issuing pulse-type completion signals. All other tasks are to be timed internally by the sequence controller, which is to allot two time units to task 2 and a single time unit to tasks 4, 5, and 6. All GO commands, as well as the BEGIN and END signals, are level-type signals. The inputs and outputs of the sequence controller to be designed are shown in Fig. 9-20.

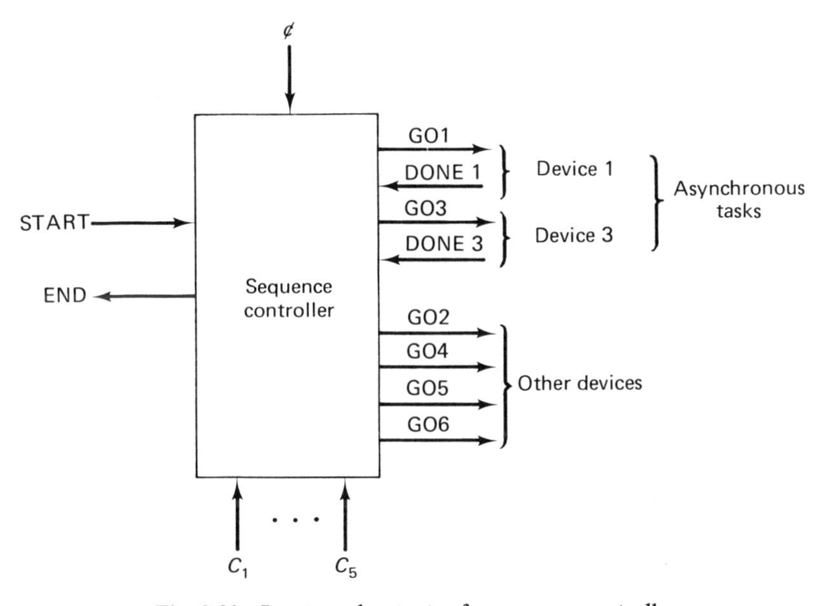

Fig. 9-20 Inputs and outputs of a sequence controller.

The C_i are the *condition inputs* corresponding to the p_i of Fig. 9-19; i.e., $C_i = 1 \equiv p_i$.

In Section 9.2 we have explained the purpose of introducing "flags" to interface clocked sequential networks with independent input pulses. The introduction of such flags in our example yields Fig. 9-21.

The next step is the preparation of a *transition chart* for the network A of Fig. 9-21. In our example we obtain Fig. 9-22. The reader will verify that the transition chart of Fig. 9-22 corresponds to the task sequencing chart of Fig. 9-19, the details specified in Fig. 9-21, and the additional assumptions that we have made. A few comments might be helpful. A $0 \longrightarrow 1$ transition of the START input is to reset the network to its proper initial conditions (i.e., state a, and all flags reset to 0). State d is an auxiliary state. Its purpose is to

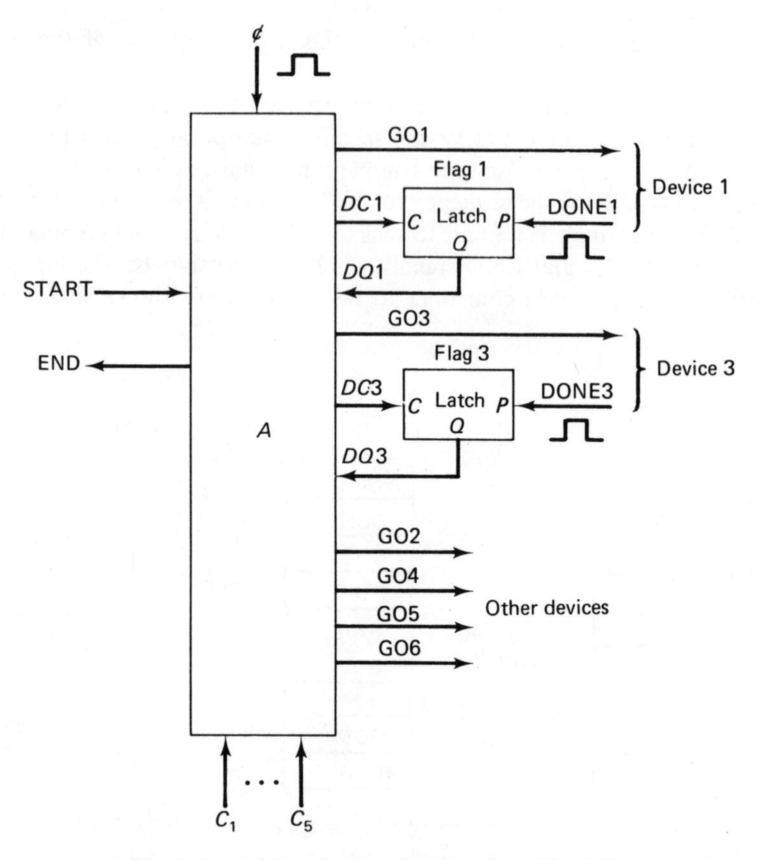

Fig. 9-21 Adding flags to the network of Fig. 9-20.

ensure that two time units will pass after the initiation of task 2, i.e., that task 2 will be completed before $p_3 \equiv C_3 = 1$ is tested.

Note that the output specified for a state is to persist only as long as the controller is in that state. For example, when state *a* is entered, GO1 becomes 1; and GO1 becomes 0 as soon as state *a* is left.

We shall return later to the implementation of transition charts. Presently we turn to another example of a task flow chart and the corresponding transition chart. Figure 9-23 is an example of a task flow chart which involves parallel processing. Upon completion of task 1, tasks 2 and 3 are to start concurrently. The *symbolic* "AND gate" (we use "$\wedge$" rather than "$\cdot$" for this purpose) indicates that only after the termination of both task 2 *and* task 3, condition p_1 is to be tested. Assuming that the first three tasks are asynchronous, setting $p_1 \equiv C_1 = 1$, and introducing flags as before, we obtain a network (B) having the inputs and outputs shown in Fig. 9-24. The corre-

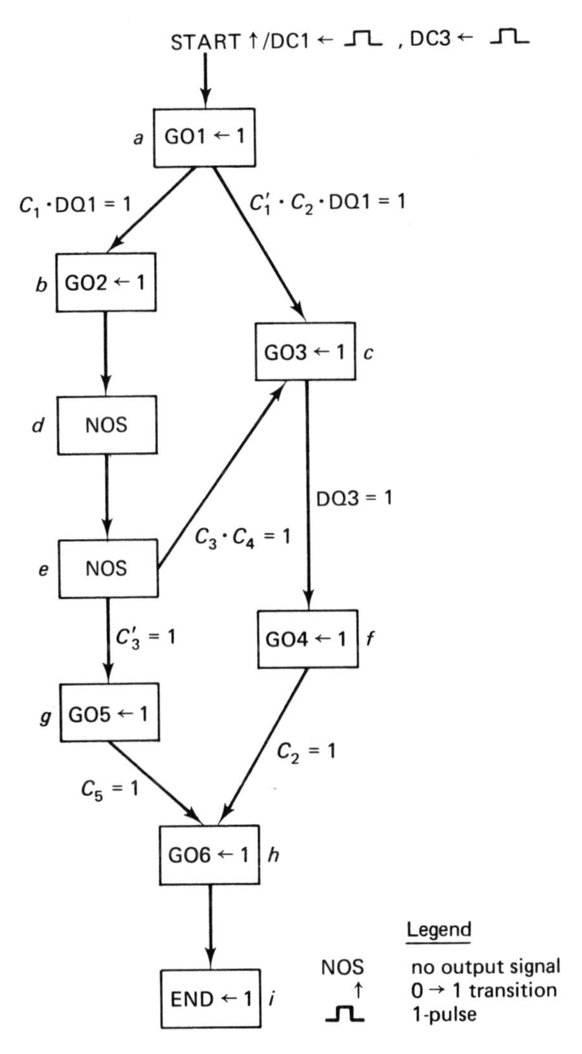

Fig. 9-22 Transition chart for network of Fig. 9-21.

sponding transition chart is easily derived (Fig. 9-25). Tasks 4 and 5 are clocked and require one time unit each.

MSI Implementation

We now turn to the implementation of transition charts. Any such chart with up to $2^4 = 16$ states is conveniently implemented by the network shown in Fig. 9-26, which makes use of standard MSI ICs as basic building blocks. The binary counter indicated is a modulo–16 counter similar to the decade

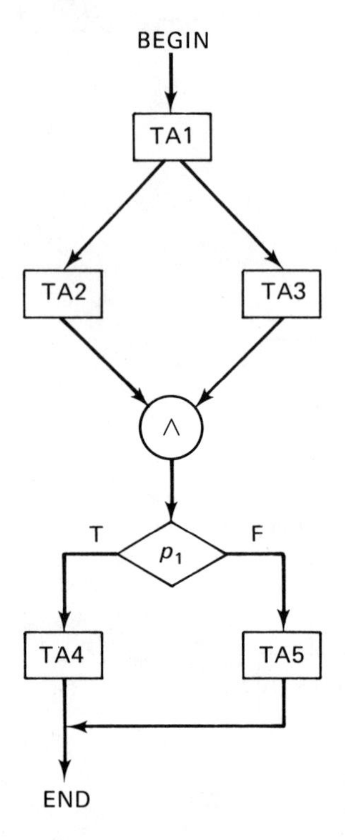

Fig. 9-23 Task flow chart involving parallel processing.

counter of Fig. 8-26. We show a single ENABLE input rather than the two inputs P and T of Fig. 8-26. N_1 and N_2 denote low-complexity combinational networks.

As an example, consider the transition chart of Fig. 9-22. Often the following first step may be very useful. Select a directed path of maximal length, starting at the initial state a. In our case there exists a unique path of length 7, namely $a \longrightarrow b \longrightarrow d \longrightarrow e \longrightarrow c \longrightarrow f \longrightarrow h \longrightarrow i$. We now assign to the kth state in this path $(0 \leq k \leq 7)$ the 4-bit word s, where $\perp s = k$. This yields the following partial assignment:

State	a	b	d	e	c	f	h	i
Assignment	0000	0001	0010	0011	0100	0101	0110	0111

We now show how to implement a transition chart consisting of a single directed path. We assume that the Q output of the binary counter (Fig. 9-26) indicates the assignment of the present state. The transitions that correspond to the path selected are to be achieved by incrementing the counter, provided the proper conditions are met. To ensure this, we set $P = y$ in N_2 and deter-

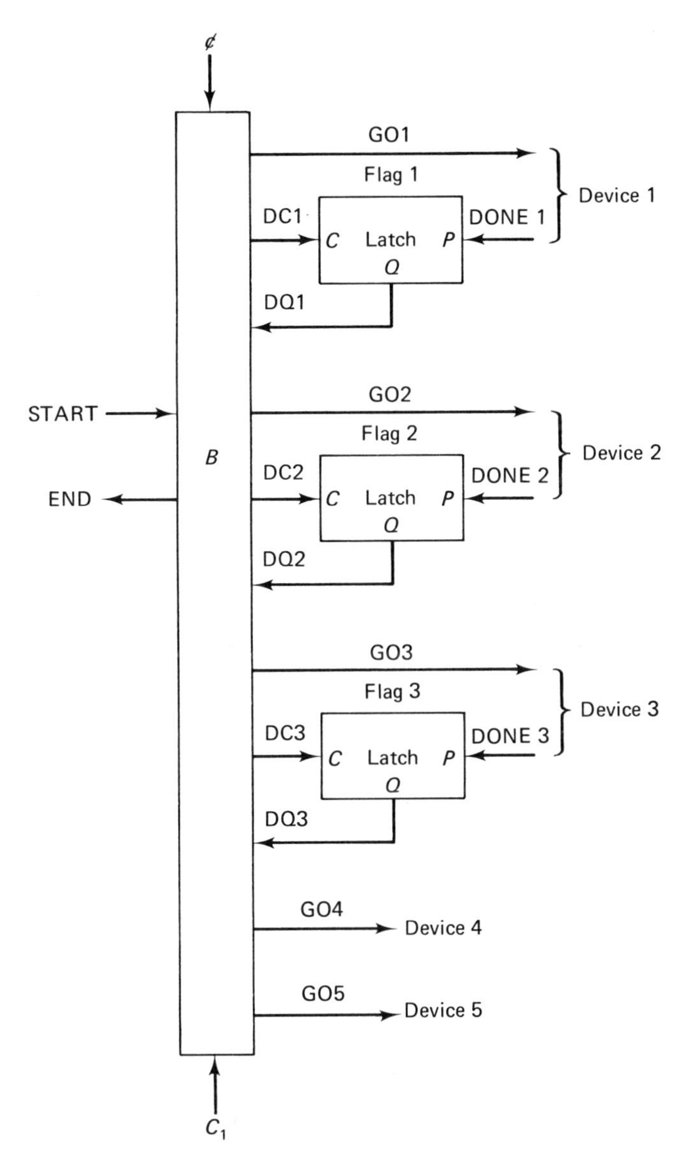

Fig. 9-24 Block diagram corresponding to Fig. 9-23.

mine N_1 accordingly, obtaining the following equations:

$$d_0 = C_1 \cdot DQ1 \qquad d_4 = DQ3$$
$$d_1 = 1 \qquad d_5 = C_2$$
$$d_2 = 1 \qquad d_6 = 1$$
$$d_3 = C_3 \cdot C_4 \qquad d_7 = \cdots = d_{15} = 0.$$

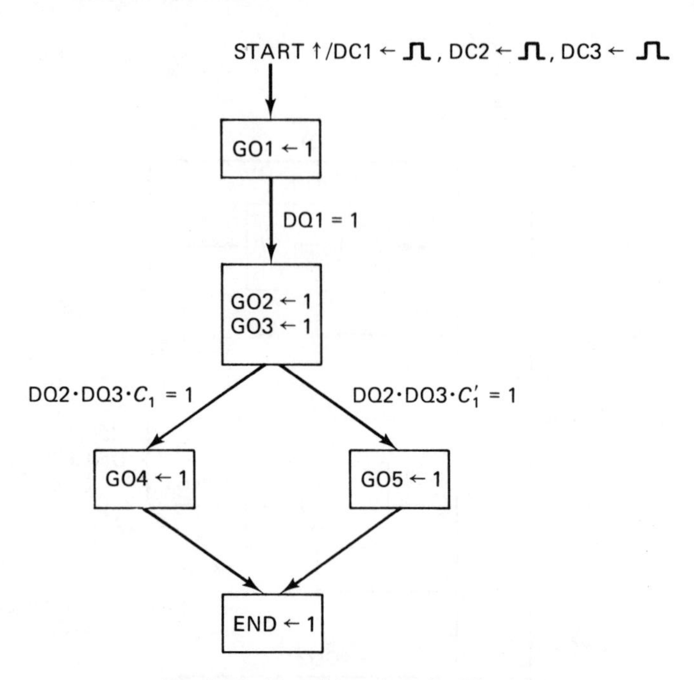

Fig. 9-25 Transition chart for Fig. 9-23.

To illustrate, assume that the present state is e; i.e., the Q output is 0011. The DATA SELECTOR sets $z = d_3$, since $\perp 0011 = 3$. It follows that $d_3 = C_3 \cdot C_4$ becomes the ENABLE input of the counter. If the condition $C_3 \cdot C_4 = 1$ is met, the counter will advance to 0100, i.e., to state c, as required in Fig. 9-22. Otherwise it will not advance; i.e., it will remain in state e.

From the discussion above, it is clear that any chart consisting of a single path can be implemented in this manner. Next, we have to implement the transitions of Fig. 9-22, not contained in the path selected. This is done by utilizing the load facilities of the counter. The only state not included in the path selected is state g. If we assign, say, 1100 to this state, we obtain the following listing of transitions not yet accounted for (i.e., not included in the path selected).

From		To		
State	**Assignment**	**State**	**Assignment**	**Condition**
a	0000	c	0100	$C_1' \cdot C_2 \cdot DQ1 = 1$
e	0011	g	1100	$C_3' = 1$
g	1100	h	0110	$C_5 = 1$

Note that the Q output of the counter is 0000 (i.e., the present state is a) iff the input x_0 to the network N_2 (Fig. 9-26) equals 1. Similarly, $x_3 = 1$ iff the

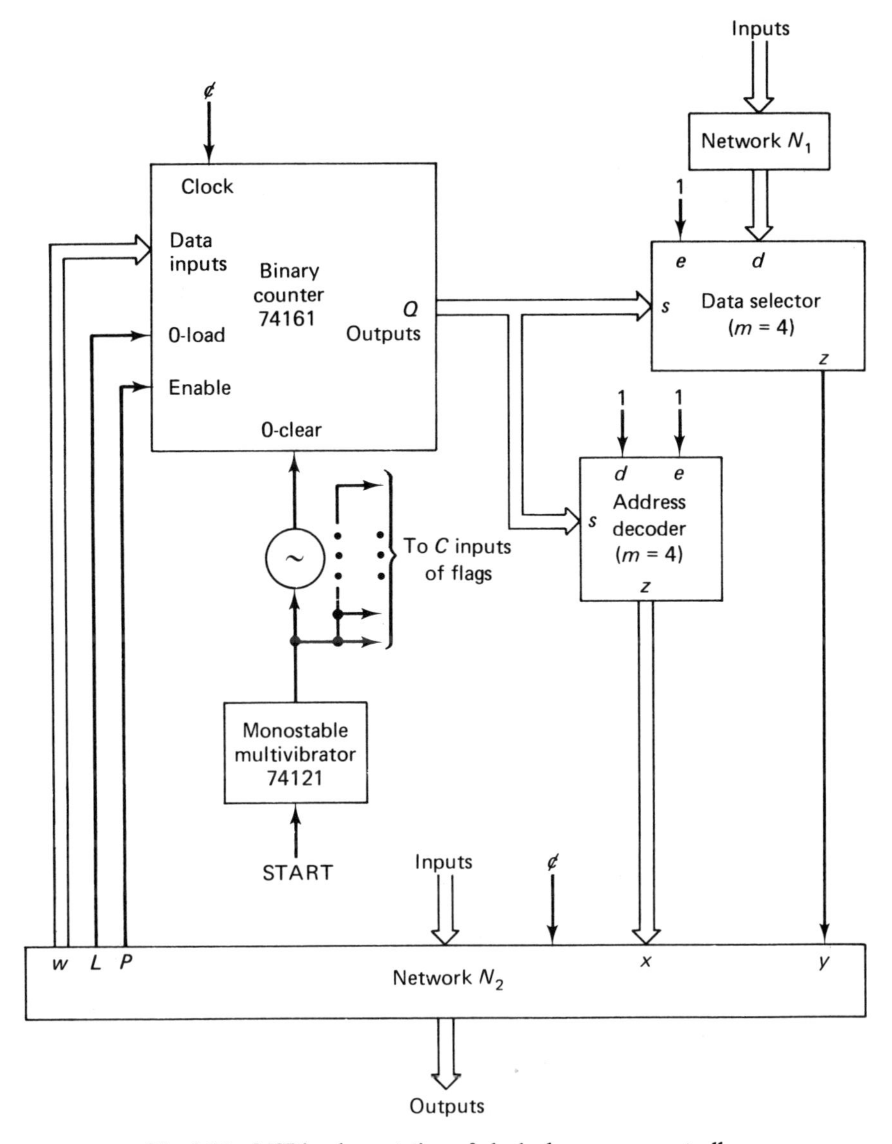

Fig. 9-26 MSI implementation of clocked sequence controller.

present state is e (Q output 0011), and $x_{12} = 1$ iff the present state is g (Q output 1100). To enforce the transitions specified in the listing above, we set up the following counter loading conditions by means of the network N_2 of Fig. 9-26:

$$L' = x_0 \cdot C'_1 \cdot C_2 \cdot DQ1 + x_3 \cdot C'_3 + x_{12} \cdot C_5$$

$$w_1 = x_3, \quad w_2 = 1, \quad w_3 = x_{12}, \quad w_4 = 0.$$

These equations assume that the outputs w_1, w_2, w_3, and w_4 of network N_2 are connected to the DATA inputs D, C, B, and A of the counter, respectively. L' rather than L appears in the first equation, since the L output of N_2 is the 0-LOAD input of the counter, and the right-hand side of this equation specifies the conditions under which loading is to take place.

The outputs of network N_2 provide the overall outputs of the sequence controller. The output equations are easily set up, by reference to Fig. 9-22, Fig. 9-23, and the state assignments decided upon. For example, if the controller is in state a, the required output conditions are GO1 $= 1$. This can be achieved by setting GO1 $= x_0$. Similarly, we obtain GO2 $= x_1$, GO3 $= x_4$, GO4 $= x_5$, GO5 $= x_{12}$, GO6 $= x_6$, and END $= x_7$.

Asynchronous Sequence Controllers

If all the tasks controlled by a sequence controller are either asynchronous or timed externally, the controller can be designed without an internal clock. To illustrate, we refer again to Fig. 9-23, but assume this time that all five tasks are asynchronous. Since the controller is now to operate *asynchronously*, i.e., without an internal clock, our strategy to provide interfacing flags has to be modified accordingly. In the case of tasks 1, 4, and 5, the controller may respond directly upon receipt of a completion signal. On the other hand, if,

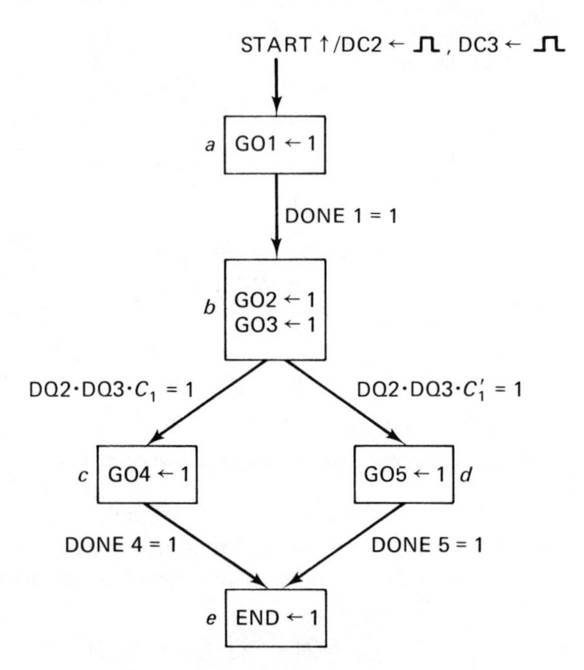

Fig. 9-27 Asynchronous transition chart.

say, task 2 is completed before task 3, the controller has to memorize this fact. Hence flags are required for tasks 2 and 3. This leads to the obvious modification of Fig. 9-24 and to the transition chart shown in Fig. 9-27. An *asynchronous control structure* suitable for implementing the chart of Fig. 9-27 is shown in Fig. 9-28.

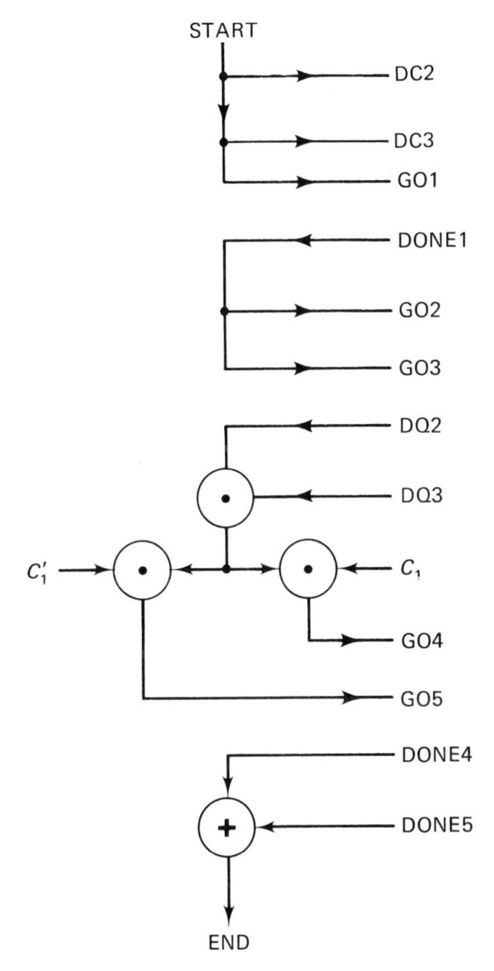

Fig. 9-28 Implementation of chart of Figure 9-27.

The simple method of implementation shown in Fig. 9-28 can easily be extended to similar task flow charts, where all the tasks are asynchronous. In fact, even if some tasks are clocked, a modified version of this approach can be used. What is needed is a completion signal from each of the clocked devices. Such a signal can be obtained by starting a synchronous counter together with the device. The counter can be preset as required and made to count down to zero. Upon reaching zero the counter should stop and issue a completion signal. We suggest that the reader implement the task flow charts of Figs. 9-19 and 9-23 (with tasks as on pp. 312 and 316) according to this approach.

9.4. IMPLEMENTATION OF NUMERICAL ALGORITHMS [PEA]

The present-day availability of powerful MSI and LSI packages makes it both feasible and attractive to replace slow software implementations of numerical algorithms used in early-generation computers by high-speed hardware configurations. Such algorithm-implementing networks are either completely combinational, such as the BCD-to-binary and binary-to-BCD converters of Section 2.7, or contain some sequential part, as a compromise between the requirements of high speed on the one hand and low package count on the other.

In this section we discuss some examples of such algorithm-implementing networks.

Four-Bit Multiplier

As an example of a numerical algorithm to be hardware-implemented, consider the multiplication of the two 4-bit words $x = x_0, x_1, x_2, x_3$ and $y = y_0, y_1, y_2, y_3$, yielding the 8-bit result $z = z_0, z_1, \ldots, z_7$, where $\perp z = (\perp x) \times (\perp y)$. For example, let $x = 1101$ and $y = 1011$. The multiplication can be carried out as indicated in Fig. 9-29. This algorithm can be implemented by means of 4-bit parallel adders (specified in Section 2.9), as shown

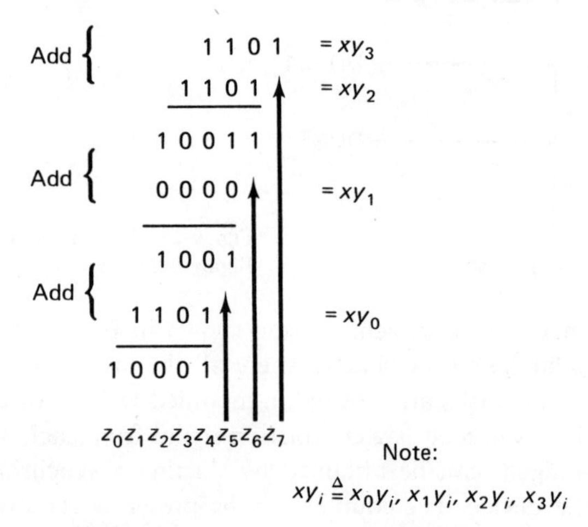

Fig. 9-29 Example of binary multiplication.

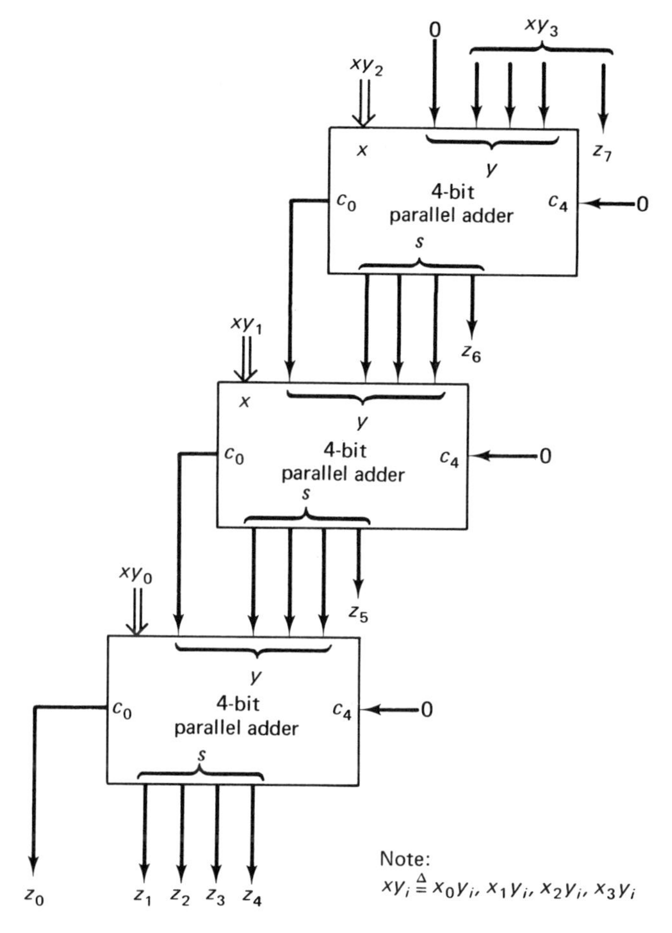

Fig. 9-30 Implementation of a 4-bit multiplier using parallel adders.

in Fig. 9-30. Note that this diagram completely specifies the algorithm. In addition to the 4-bit parallel adders shown in Fig. 9-30, the arrangement requires 16 two-input AND gates, to generate all $x_i y_j$ combinations. Alternatively, four 4-bit ALUs (74181) can be used together with 3 inverters and 4 two-input AND gates, as shown in Fig. 9-31 [MOT3]. (See Table 2-5)

Binary-to-BCD Conversion Revisited

Combinational binary-to-BCD converters have been discussed in Section 2.7. Figure 2-14 shows such a converter, applicable to any 7-bit input word, consisting of 5 BDMs. A similar converter for 12-bit input words requires a total of 18 BDMs.

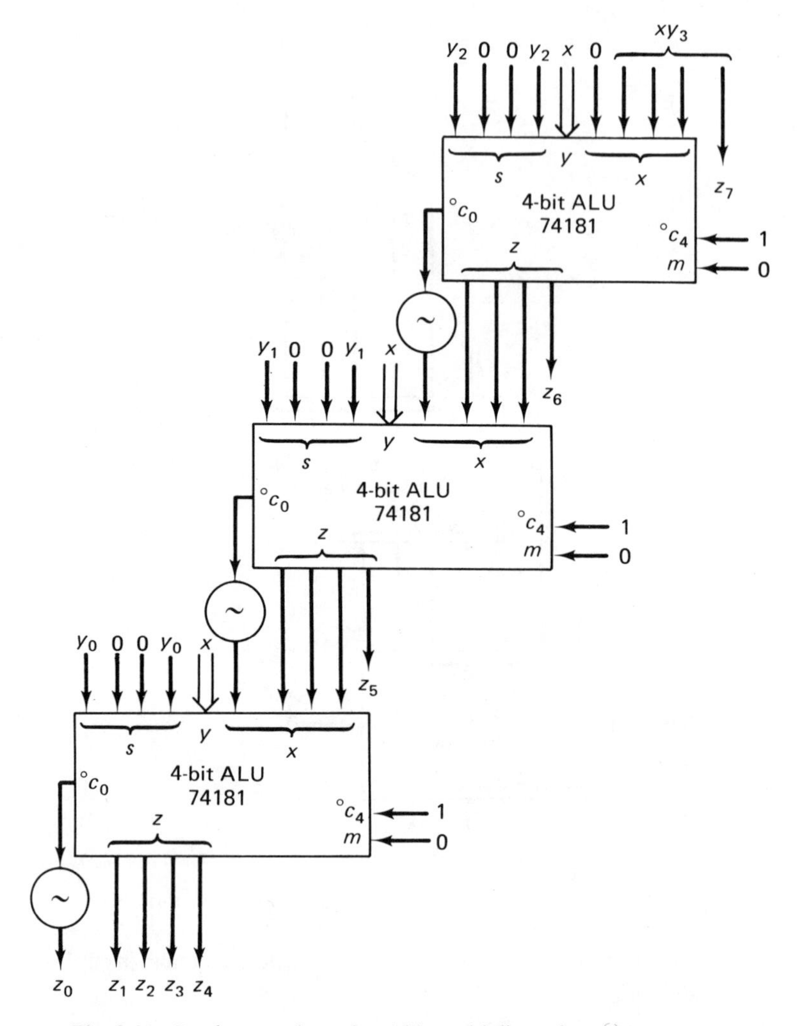

Fig. 9-31 Implementation of a 4-bit multiplier using ALUs.

As the length of the binary input word increases, the amount of hardware required for a purely combinational converter becomes excessive, and some form of sequential implementation has to be considered.

To illustrate the replacement of a combinational converter by a sequential one, we refer to Fig. 9-32, which shows a sequential network performing the same computation as the converter of Fig. 2-14. The network of Fig. 9-32 requires only two BDMs, but now the two registers R_1 and R_2 and a sequential CONTROL network have to be included. It follows that the replacement of the combinational converter by a sequential one will reduce the overall package count, only if the binary input word is long enough.

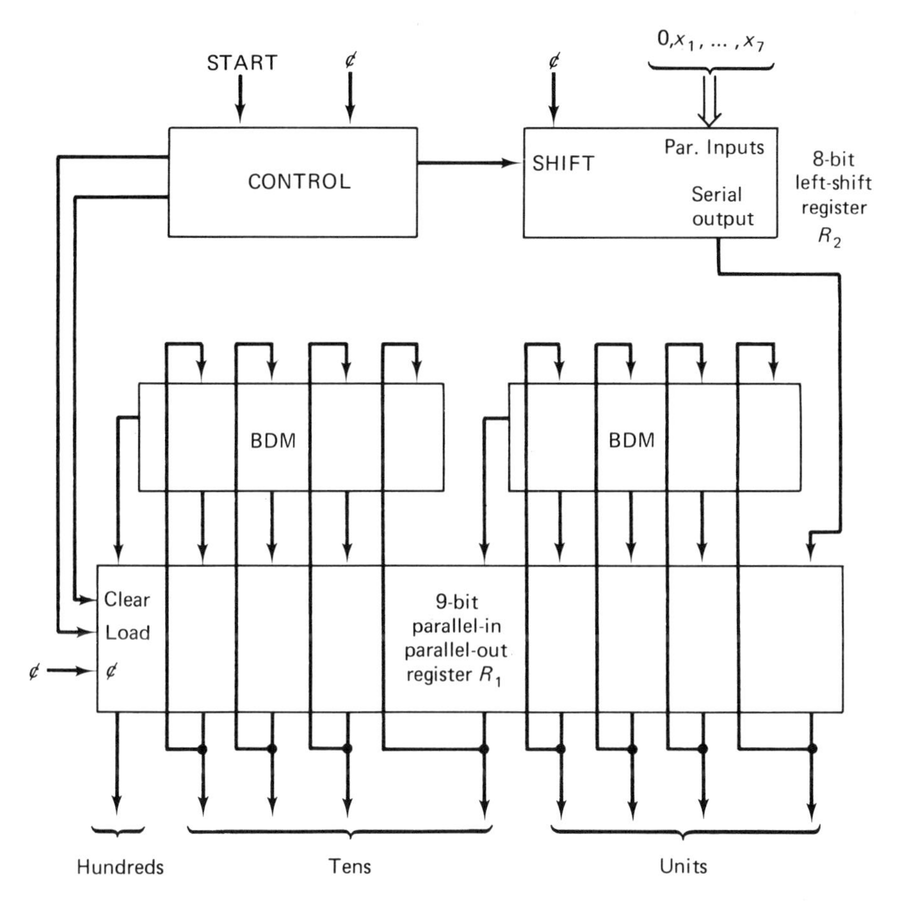

Fig. 9-32 Sequential binary-to-BCD converter.

The functioning of the sequential converter of Fig. 9-32 is easily explained. Let N_t be the integer whose BCD representation appears as the output of the converter at the tth clock pulse. Assume that x_t is the SERIAL OUTPUT of R_2 at that time. Assuming further that the LOAD input of R_1 is enabled, we shall have (see Section 2.7)

$$N_{t+1} = 2N_t + x_t \qquad 0 \leq t \leq 7, \tag{1}$$

where $x_0 \triangleq 0$ and N_0 is set to 0 (i.e., R_1 is cleared at time $t = 0$) by the CONTROL unit. Provided that the necessary CLEAR, LOAD, and SHIFT commands are issued by the CONTROL unit, the network of Fig. 9-32 will execute operation (1) for $t = 0, 1, \ldots, 7$. It follows that the final output will be the same as the output of the combinational converter of Fig. 14 of Chapter 2. The CONTROL unit is easily designed by means of the general principles discussed in Section 9.3. Also, the detailed design of registers R_1

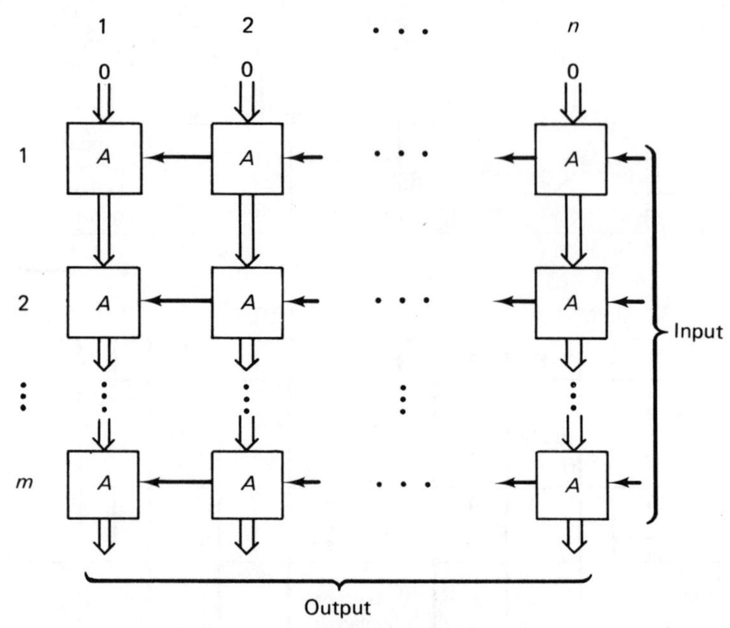

Fig. 9-33 Two-dimensional cellular array.

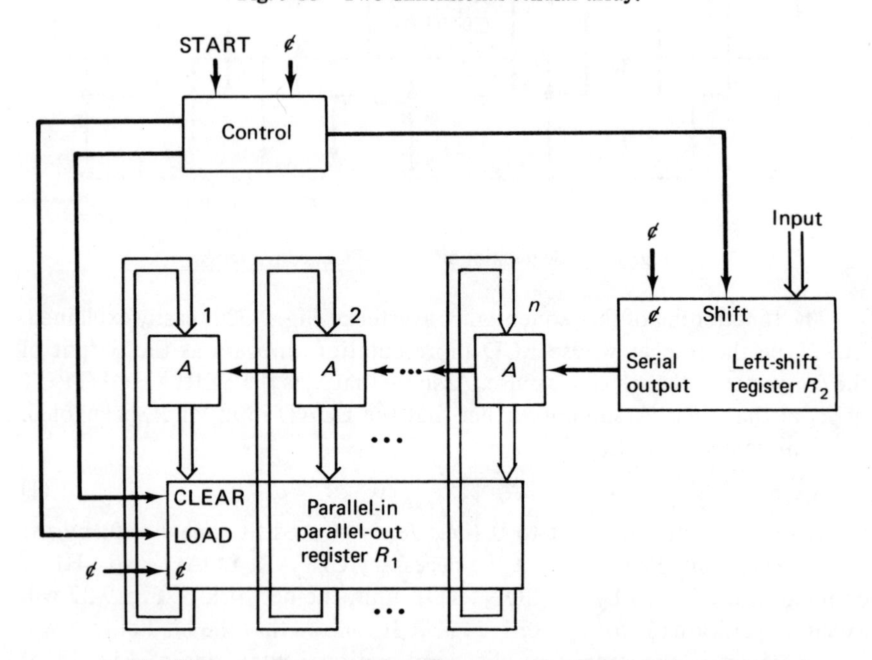

Fig. 9-34 *n*-cell sequential replacement of cellular $m \times n$ array (Fig. 9-33).

and R_2, by means of the MSI packages mentioned in Section 8.2, is straightforward.

Refer now to the cellular array of Fig. 9-33 for an explanation of the principle on which the replacement of Fig. 2-14 by Fig. 9-32 is based. This cellular array, consisting of $m \times n$ identical cells A, can be replaced by a sequential network of n cells, as shown in Fig. 9-34. By suitably activating this network m times, a performance equivalent to that of the $m \times n$ array (Fig. 9-33) is achieved. The register R_2 performs a parallel-to-serial conversion of the input.

The network of Fig. 9-34 corresponds to a single row of the $m \times n$ array of Fig. 9-33. Alternatively, an $m \times n$ cellular array may be replaced by an m-cell sequential network which corresponds to a single column of the $m \times n$ array.

PROBLEMS

1. *Time-Division Multiplexing* A simplified version of a 16-line *time-division multiplexing* (TDM) system is shown in Fig. P9-1. We shall say that x_i is *through-connected* iff x_i is effectively connected to the output z_i via the data transmission channel DTC; i.e., we have x_i = DTC = z_i.

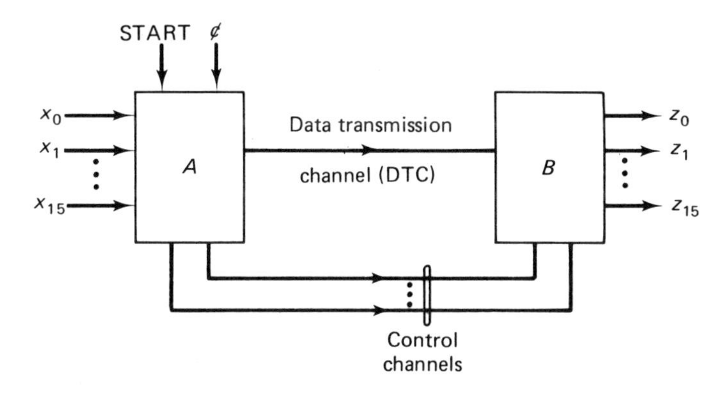

Figure P9-1

The TDM system of Fig. P9-1 is to perform as follows. As long as START = 0, x_0 is to be through-connected. A $0 \longrightarrow 1$ transition of START will initiate a count of the clock pulse ¢. At the nth clock pulse, x_j is to be through-connected, where $j \equiv n \pmod{16}$.

(a) Assume that 4 control channels (see Fig. P9-1) are available. Design the networks A and B using only IC packages referred to in this text.
(b) Modify your design to meet the assumption that only two control channels are available.
(c) Modify the design of part (b) to obtain a 20-line TDM system.

2. *Digital Clocks* The *COS/MOS Application Notes* [RCA] discuss the design of battery-operated digital clocks and watches. Crystal oscillators serve as main timing elements. Since their size and cost decrease with increasing frequency, oscillator frequencies of $2^{13} = 8{,}192$ Hz (pulses per second) or higher are generally used. Frequency division is then applied to provide pulses every second, minute, and hour. These pulses are processed to drive numerical displays of the 7-segment type (see Section 5.8) as illustrated in Fig. P9-2.

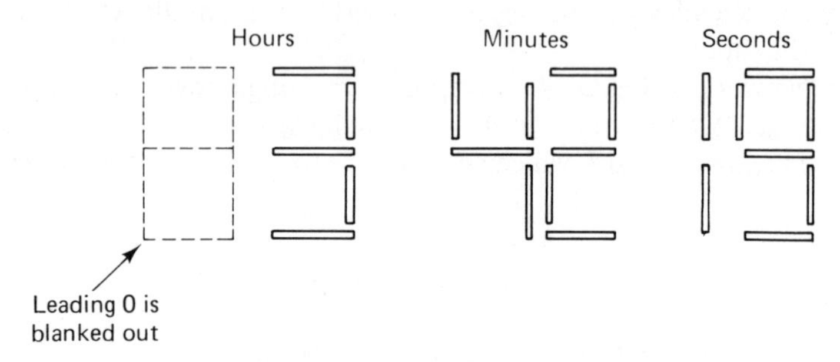

Figure P9-2

The hours range is usually 1–12. Three manual switches, HOUR RESET, MINUTE RESET, and SECOND RESET, provide facilities to reset the corresponding readings to zero. By means of the manual switches HOUR UPDATE and MINUTE UPDATE the corresponding readings can be advanced at the rate of one step per second.

Digital clock designs utilize BCD-TO-SEVEN-SEGMENT decoders similar to those discussed in Section 5.8. In the *nonmultiplexing scheme*, each of the six 7-segment displays is driven by a separate decoder. In the *multiplexing scheme* a single decoder is shared by all six displays. In this scheme the correct information is supplied to one display at a time, while the eye retains the previous state of the other five displays. The multiplexing scheme is particularly applicable if "light-emitting diode" (LED) displays are used.

(a) Design a digital network that has a 1-Hz clock signal as input and provides the outputs necessary to operate the six numerical 7-segment displays of a digital clock. You may use any of the IC packages mentioned in the text, as well as 5 BCD-TO-SEVEN-SEGMENT decoders of the type defined in Section 5.8.

(b) Extend the design of part (a) by incorporating the manual RESET and UPDATE switches mentioned earlier. Furthermore, assume that the timing element available is an 8,192-Hz oscillator rather than a 1-Hz clock.

(c) Modify the design of part (b) by incorporating a suitable multiplexing scheme. The basic frequency of the multiplexing network should be 1,024 Hz.

3. *Arithmetic-Logic Unit* Design a 16-bit arithmetic-logic unit (ALU) in accordance with the following specification. The ALU has a 16-bit *data* input $x = x_1, \ldots, x_{16}$, a 4-bit *instruction control* input $s = s_1, s_2, s_3, s_4$, and a

1-bit *start* input GO. The unit contains a 16-bit register A (*accumulator*), the state $A \triangleq A_1, \ldots, A_{16}$ of which is available as an external output. Additional outputs are the *carry* bit c and the *completion* signal DONE.

By means of the instruction control input s, the ALU can be set to perform the following 16 instructions.

$\perp s$	Instruction	$\perp s$	Instruction
0	NO-OP	8	OR
1	CLEAR	9	XOR
2	SET TO 1	10	ADD
3	COUNT UP	11	SUBTRACT
4	COUNT DOWN	12	SHIFT LEFT
5	LOAD	13	SHIFT RIGHT
6	COMPLEMENT	14	ROTATE LEFT
7	AND	15	ROTATE RIGHT

A detailed explanation of the instructions follows.

NO-OP: No change takes place.

CLEAR: The accumulator is cleared; i.e., all the A_i are set to 0.

SET TO 1: All the A_i are set to 1.

COUNT UP: $\perp A$ is increased by 1 modulo 2^{16}.

COUNT DOWN: $\perp A$ is decreased by 1 modulo 2^{16}.

LOAD: $A \leftarrow x$; i.e., the word x is loaded into A.

COMPLEMENT: Each A_i is replaced by A_i'.

AND: $A \leftarrow A \cdot x$, where $A \square x \triangleq A_1 \square x_1, \ldots, A_{16} \square x_{16}$ for any Boolean operation $\square$.

OR: $A \leftarrow A + x$.

XOR: $A \leftarrow A \oplus x$.

ADD: $\perp(c, A) \leftarrow \perp A$ PLUS $\perp x$, where c, A denotes the 17-bit word $c, A_1, \ldots, A_{16}$ and "$\leftarrow$" assumes the meaning "obtains the value of."

SUBTRACT: $\perp(c, A) \leftarrow \perp A$ PLUS $(2^{16} - \perp x)$.

SHIFT LEFT: $A \leftarrow A_2, \ldots, A_{16}, 0$.

SHIFT RIGHT: $A \leftarrow 0, A_1, A_2, \ldots, A_{15}$.

ROTATE LEFT: $A \leftarrow A_2, \ldots, A_{16}, A_1$.

ROTATE RIGHT: $A \leftarrow A_{16}, A_1, \ldots, A_{15}$.

The execution of any instruction is initiated by the rising edge of a pulse signal appearing on the GO input. The ALU indicates completion by a pulse signal on its DONE output.

For the design of this ALU you may use any of the IC packages mentioned in this text. Furthermore, you may assume the availability of a suitable clock.

10 ERRORS AND HAZARDS

ABOUT THIS CHAPTER The first two sections of this chapter are concerned with errors (failures) that may occur in digital networks. In the first section we discuss methods by which a single error (i.e., failure of a single output bit) may be detected. The second section deals with more sophisticated methods by which a single error may be corrected; i.e., the correct output may be reconstructed.

Starred Section 10.3 is a continuation of the material on hazards discussed earlier in Sections 3.5 and 4.5. It is based on [EIC]. Section 10.4 relates the theory of hazards to our analysis methods of Chapters 6 and 7.

10.1. ERROR DETECTION IN DIGITAL NETWORKS [FR-ME, SE-HS-BE]

Frequently, the designer of digital networks has to take into account the fact that some of the components of a digital network may fail. In some cases it is advisable to design a network in such a way that failures, at least of a certain type, will be automatically detected and their presence indicated by a corresponding output signal. This signal is sometimes utilized to replace the faulty unit automatically by a stand-by unit.

In this section we describe some commonly used methods of detecting errors.

Networks with Coded Outputs

Consider a digital network with an n-bit output word z. In general, the output word may assume any one of the 2^n values in $\{0, 1\}^n$. Some ability to detect errors is obtained if the network is designed in such a way that z will only assume values from a set Z, where Z is a proper subset of $\{0, 1\}^n$. The set Z is referred to as the (output) *code set*, or simply *code*. Suppose that an error occurs in z and the output is $\hat{z}$ instead of z. Suppose further that we can check whether or not the output is in Z. Then, if $\hat{z} \notin Z$, we have detected an error. See the block diagram of Fig. 10-1. Of course, the FAIL signal is correct only if the error detector is itself free of errors. Otherwise, we may have a "false alarm."

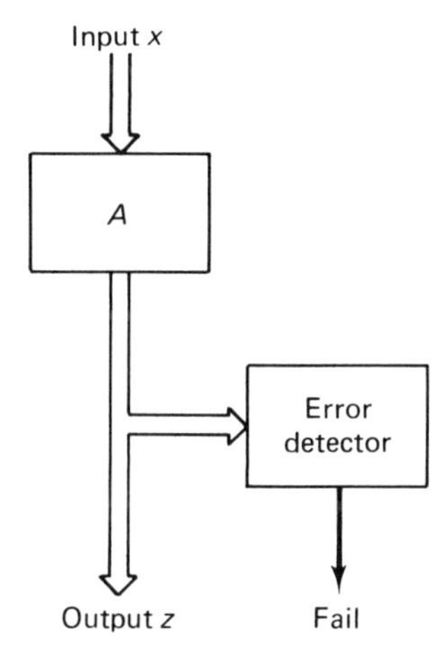

Fig. 10-1 Error detection for coded-output networks.

A simple example of this idea is provided by any network whose output is represented in the BCD code. If z is a BCD representation of a decimal digit, then the output code set is $Z = \{z \in \{0, 1\}^4 | \perp z \leq 9\}$. In this case we have FAIL $= 1 \equiv \perp z > 9$. One easily verifies that FAIL $= z_1 z_2 + z_1 z_3$. Thus the error-detector network of Fig. 10-1 is easily implemented.

The BCD example shows that many errors may go undetected because the resulting word $\hat{z}$ is still a valid code word. For instance, if $\perp z = 0$ and any bit becomes wrong, this cannot be detected in such a scheme. Thus the BCD code is not very good from the point of view of error detection, and other codes should be considered.

We say that a *single error* (*multiple error*) occurs if exactly one (more than one) bit of z is incorrect. Multiple errors are often not taken into account, for the following reasons. Suppose that the probability of a single error is p and that the various errors are caused by independent events. In that case the probability of a double error is p^2, of a triple error p^3, etc. Usually p is very small, and therefore p^2, p^3, etc., may be small enough to be negligible. For example, if $p = 10^{-3}$ ("one-in-a-thousand"), then $p^2 = 10^{-6}$ ("one-in-a-million"). Even if multiple errors occur relatively often, one may still decide to ignore them, because the cost of detecting them may be prohibitive. In Fig. 10-1 the complexity of the error-detector network should be much smaller, in general, than the complexity of the network A; otherwise, the cost of the error detector may be too large. In most error-detecting schemes the cost of detecting multiple errors exceeds significantly the cost of detecting single errors.

Unless stated otherwise, we assume that only single errors are to be detected. Let $z \in Z$ be the correct output and $\hat{z}$ the actual incorrect output with a single error. It is clear that we must have $\hat{z} \notin Z$ in order to detect the error. Thus a necessary and sufficient condition that a code Z be *single-error-detecting* is that if $z \in Z$, then no word adjacent to z should be in Z. Alternatively, we can describe this condition using the notion of distance. For any two n-bit words x and y, let the (*Hamming*) *distance* between x and y be defined as

$$d(x, y) \triangleq w(x_1 \oplus y_1, x_2 \oplus y_2, \ldots, x_n \oplus y_n),$$

where $w(z_1, z_2, \ldots, z_n) = \sum_{i=1}^{n} z_i$ is the *weight* of z, i.e., the number of 1's in z. Thus $d(x, y)$ is simply a nonnegative integer k corresponding to the number of components in which x and y differ. In terms of the distance concept, a code Z is single-error-detecting iff $d(x, y) \geq 2$, for any two different code words $x, y \in Z$.

Two simple single-error-detecting codes are:

$$\textit{Odd-parity code:} \quad Z_o \triangleq \{z \mid z_1 \oplus z_2 \oplus \cdots \oplus z_n = 1\}$$

$$\textit{Even-parity code:} \quad Z_e \triangleq \{z \mid z_1 \oplus z_2 \oplus \cdots \oplus z_n = 0\}.$$

Thus the odd-parity (even-parity) code consists of all the n-tuples with an odd (even) number of 1's. Also often used are the *k/n (k-out-of-n) codes*,

$$Z_{k/n} \triangleq \{z \in \{0, 1\}^n \mid w(z) = k\},$$

where $w(z)$ is the weight of z. They are sometimes referred to as *constant-weight codes*. These codes are also single-error-detecting, since any single error causes a change in the weight.

Some of these ideas are now illustrated in the example below.

Example 1

The network A of Fig. 10-1 is a BCD-to-decimal decoder (7442), as defined in Section 2.7. The output set Z forms a 1/10 code. An error detector for this code would require quite a few gates. We may compromise, in view of the obvious observation that Z is a subset of the odd-parity code for $n = 10$, and design an error detector that checks for odd parity. The TTL package 74180 [SIG, TI2] (Fig. 10-2), which we will now analyze, is con-

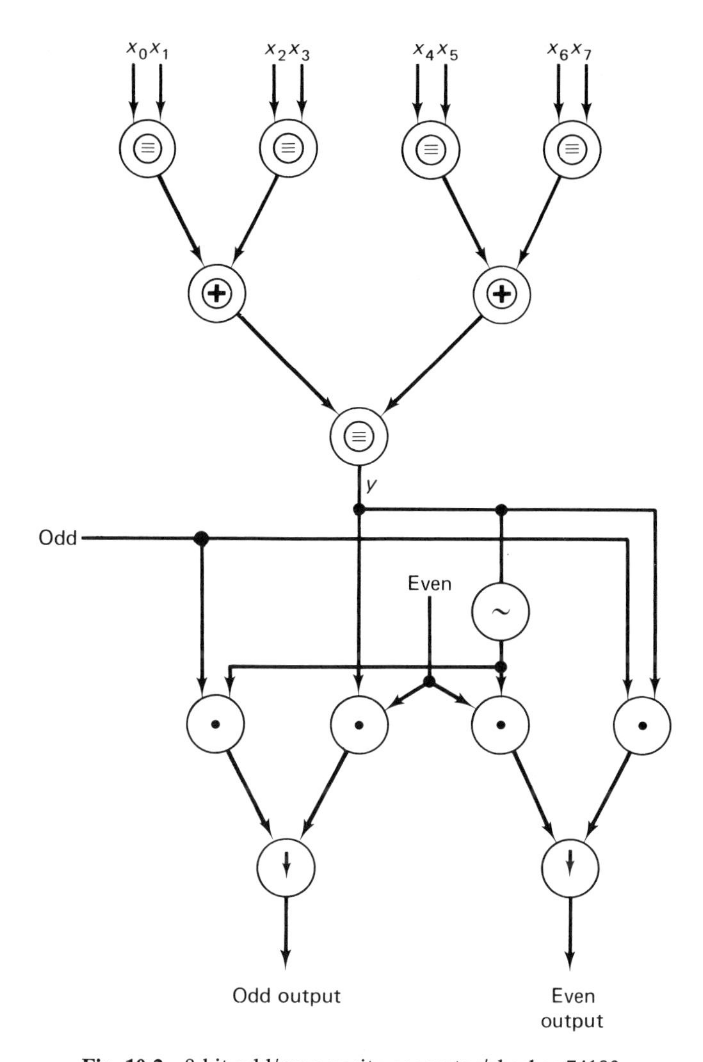

Fig. 10-2 8-bit odd/even parity generator/checker 74180.

venient for this purpose. One easily verifies that $y = (x_0 \oplus x_1 \oplus \cdots \oplus x_7)'$. Furthermore,

$$\text{ODD OUTPUT} = (y \cdot \text{EVEN} + y' \cdot \text{ODD})'$$

$$\text{EVEN OUTPUT} = (y \cdot \text{ODD} + y' \cdot \text{EVEN})'.$$

If we set EVEN = 1 and ODD = 0, we have

$$\text{EVEN OUTPUT} = y = (x_0 \oplus \cdots \oplus x_7)'.$$

Thus EVEN OUTPUT is 1 iff the parity of $x = x_0, \ldots, x_7$ is even. Similarly, when EVEN = 1 and ODD = 0, we find ODD OUTPUT = 1 iff the parity of x is odd. The ODD and EVEN OUTPUT logic is redundant if we view the module as an 8-bit parity checker only. However, these outputs are used for the parity generator function and for other applications of the module as will be described. If x belongs to any arbitrary code and EVEN $= 1$, ODD $= 0$, we find that the ODD OUTPUT is 0 if the parity of x is even, and it is 1 if the parity of x is odd. Thus the parity of the word $(x_0, \ldots, x_7,$ ODD OUTPUT) is always even. The addition of such a parity bit converts an arbitrary code into an even-parity code.

The 74180 package can be easily extended to provide a 10-bit odd-parity checker as illustrated in Fig. 10-3. The FAIL signal will appear whenever the

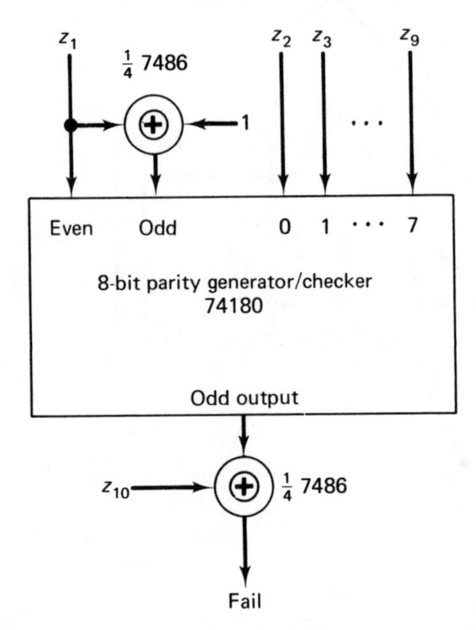

Fig. 10-3 10-bit odd-parity checker.

word $z = z_1, \ldots, z_{10}$ has even weight. If a single z_i output of the checked network is in error, then either all the z_i become 0, or two of the z_i become 1. Both cases will be indicated by a FAIL signal. The error detector of Fig. 10-3 is not intended to provide for cases where more than one of the z_i's fail. The appearance of a FAIL signal may, of course, also be caused by a failure in the checker network.

Digital Data Transmission

Error checking is especially important in connection with digital data transmission systems. The basic principle is illustrated in Fig. 10-4. The purpose of the system is to transmit binary *message words* x of fixed length, say n, over some distance. Thus normally $z = x$. However, the transmission

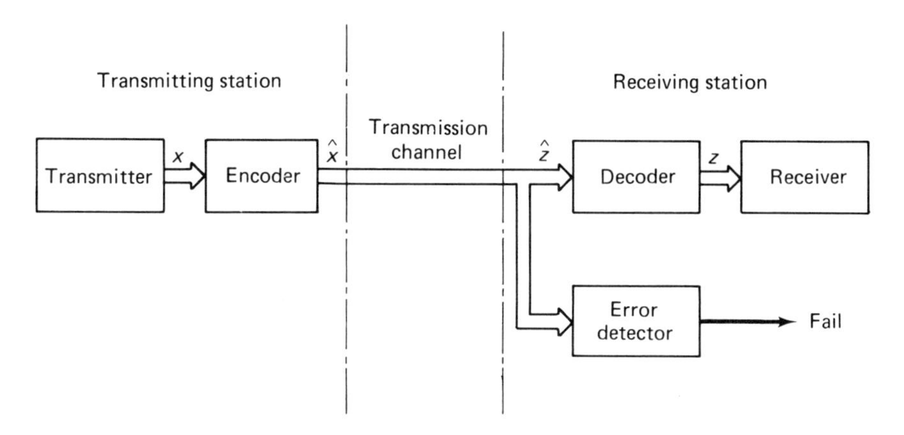

Fig. 10-4 Digital data transmission with error checking.

channel is liable to cause errors in the transmission. The encoder and error detector serve the purpose of error checking. The transmitted words $\hat{x}$, which are longer than the original message words x, form some code set $\hat{X}$. Normally $\hat{z} = \hat{x}$. The error detector checks whether $\hat{z} \in \hat{X}$. If this is not the case, a transmission error has been discovered, provided that the encoder and error detector are failure-free. The decoder normally regenerates the original message word $z = x$ from $\hat{z} = \hat{x}$.

Example 2

Design a system as in Fig. 10-4, which is capable of transmitting arbitrary 9-bit words and incorporates even-parity checking.

In this example $n = 9$, and $\hat{x} = x, p$, where p is a parity check bit such that $\hat{x}$ is an even-parity code word. Thus

$$x_1 \oplus \cdots \oplus x_9 \oplus p = 0 \qquad \text{and} \qquad p = x_1 \oplus \cdots \oplus x_9.$$

The encoder outputs x as is, and in addition generates the check bit p. This can be done by the arrangement shown in Fig. 10-5. The error detector is identical to that of Fig. 10-3. The decoder simply omits the last bit of $\hat{z}$ to produce z.

Example 3

Design a system as in Fig. 10-4, where all admissible words x form a 1/10 code and the words $\hat{x}$ form a 2/5 code. Thus 5 binary transmission lines are used to transmit 10 possible message words.

A 2/5 code is shown in Table 10-1. This particular correspondence between the 1/10 code and the 2/5 code is shown conveniently in Fig. 10-6, which also specifies the required behavior of the encoder and the decoder. Namely,

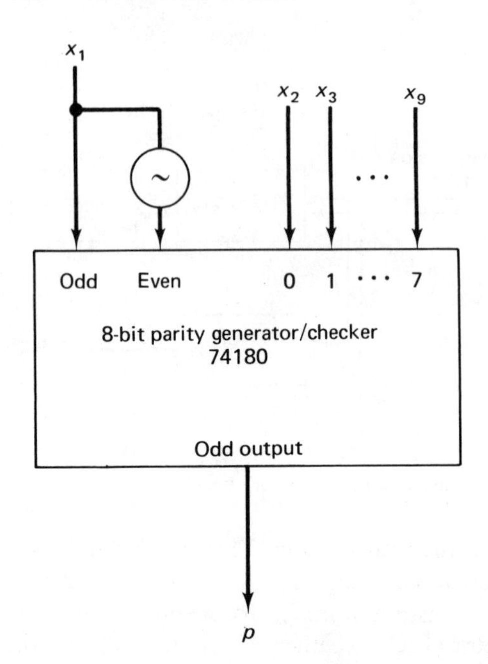

Fig. 10-5 9-bit odd-parity generator.

Table 10-1 2/5 Code

Digit	1/10 Code x_1	x_2	···	x_9	x_{10}	2/5 Code $\hat{x}_1$	$\hat{x}_2$	$\hat{x}_3$	$\hat{x}_4$	$\hat{x}_5$
0	1	0	···	0	0	1	1	0	0	0
1	0	1	···	0	0	0	1	1	0	0
2						0	0	1	1	0
3						0	0	0	1	1
4			·			1	0	0	0	1
5			·			1	0	1	0	0
6			·			0	0	1	0	1
7						0	1	0	0	1
8	0	0	···	1	0	0	1	0	1	0
9	0	0	···	0	1	1	0	0	1	0

an edge labeled k between vertices i and j indicates that

$$x_k = 1 \equiv \hat{x}_i = 1 \wedge \hat{x}_j = 1,$$

and similarly,

$$z_k = 1 \equiv \hat{z}_i = 1 \wedge \hat{z}_j = 1.$$

Table 10-1 or Fig. 10-6 yield the following equations for the encoder:

$$\hat{x}_1 = x_1 + x_5 + x_6 + x_{10} \qquad \hat{x}_2 = x_1 + x_2 + x_8 + x_9$$

$$\hat{x}_3 = x_2 + x_3 + x_6 + x_7 \qquad \hat{x}_4 = x_3 + x_4 + x_9 + x_{10}$$

$$\hat{x}_5 = x_4 + x_5 + x_7 + x_8.$$

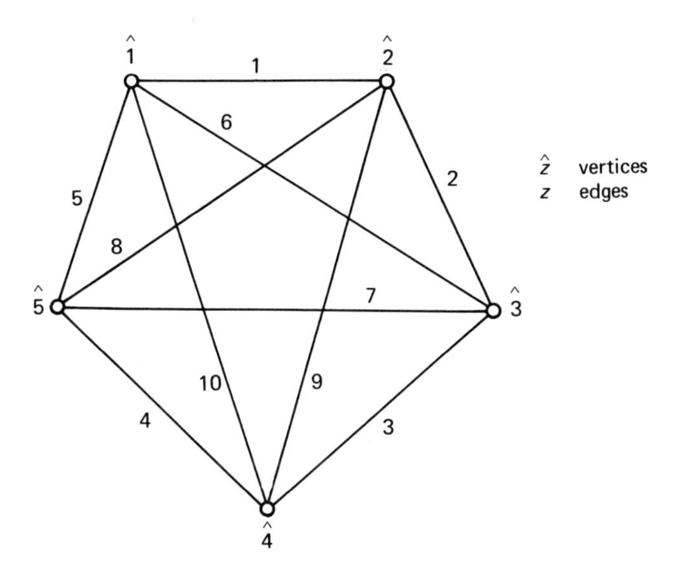

Fig. 10-6 Graph showing relationship between the 1/10 code and the 2/5 code.

The following equations specify the decoder:

$$z_1 = \hat{z}_1\hat{z}_2, \quad z_2 = \hat{z}_2\hat{z}_3, \quad z_3 = \hat{z}_3\hat{z}_4, \quad z_4 = \hat{z}_4\hat{z}_5, \quad z_5 = \hat{z}_5\hat{z}_1$$

$$z_6 = \hat{z}_1\hat{z}_3, \quad z_7 = \hat{z}_3\hat{z}_5, \quad z_8 = \hat{z}_5\hat{z}_2, \quad z_9 = \hat{z}_2\hat{z}_4, \quad z_{10} = \hat{z}_4\hat{z}_1.$$

One method of designing the error detector is discussed in Section 4.7 (see Fig. 4-15 and Table 4-10). An alternative, and more efficient implementation of the error detector may be obtained if the words z, rather than the words $\hat{z}$ (see Fig. 10-4), are taken as the inputs to the error detector [SCH]. Let $Z_a = \{z_1, z_2, z_3, z_4, z_5\}$ and $Z_b = \{z_6, z_7, z_8, z_9, z_{10}\}$. Now refer to Fig. 10-6 or to the decoder equations to verify that $w(\hat{z}) > 2$ implies $z_i = 1$ for some $z_i \in Z_a$ as well as $z_j = 1$ for some $z_j \in Z_b$. [It is sufficient to check this when $w(\hat{z}) = 3$.] This leads to the implementation of the error detector shown in Fig. 10-7. Indeed, if $w(\hat{z}) < 2$, then all the z_i are 0; thus $y_1 = y_2 = 0$ and FAIL $= 1$. If $w(\hat{z}) > 2$, then $y_1 = y_2 = 1$; whence FAIL $= 1$. On the other hand, if $w(\hat{z}) = 2$, then $w(z) = 1$. Thus $y_1 \neq y_2$ and FAIL $= 0$.

Error Checking of Arbitrary Combinational Networks

An arbitrary network can be error-checked by means of duplication, as indicated in Fig. 10-8. Normally, A and $\bar{A}$ are identical and $z = \bar{z}$. The comparator is specified by $z \neq \bar{z} \equiv$ FAIL $= 1$. A 4-bit comparator may be implemented for example by means of the 4-bit magnitude comparator package 7485 of Section 2.4, or the ALU of Section 2.9, as explained there.

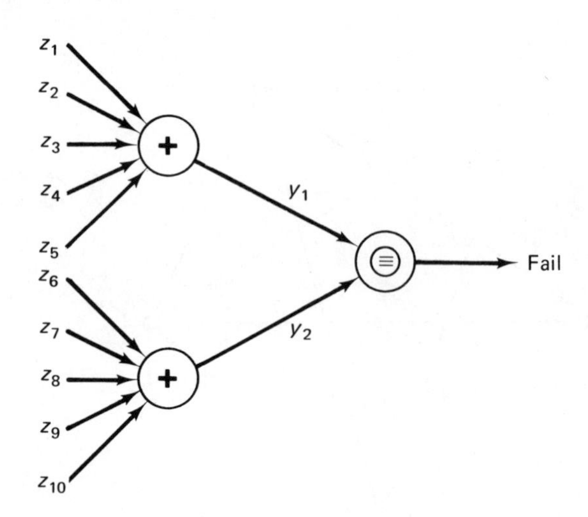

Fig. 10-7 Implementation of error detector for Example 4.

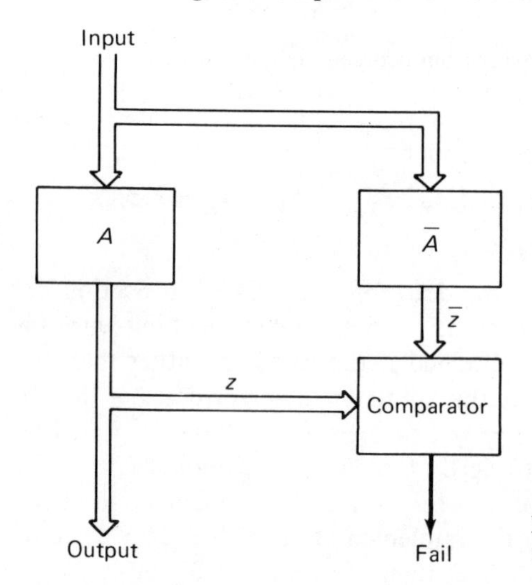

Fig. 10-8 Error-checking by duplication.

Two checking schemes for multi-output combinational networks which are less expensive than the duplication scheme are shown in Fig. 10-9. Referring to Fig. 10-9(a), we denote by z the actual output for the input word x, and by $\bar{z}$ the expected, error-free output. For $z = z_1, \ldots, z_n$, we set $p(z) \triangleq z_1 \oplus \cdots \oplus z_n$. The output of the parity predictor in Fig. 10-9(a) is $p(\bar{z})$. Hopefully, the network required to compute only the parity of the output word is considerably simpler than the network A computing the entire output word. The checking network is specified by

$$\text{FAIL} = 1 \equiv p(z) \neq p(\bar{z}) \equiv p(z) \oplus p(\bar{z}) = 1.$$

Thus

$$\text{FAIL} = z_1 \oplus \cdots \oplus z_n \oplus p(\bar{z}).$$

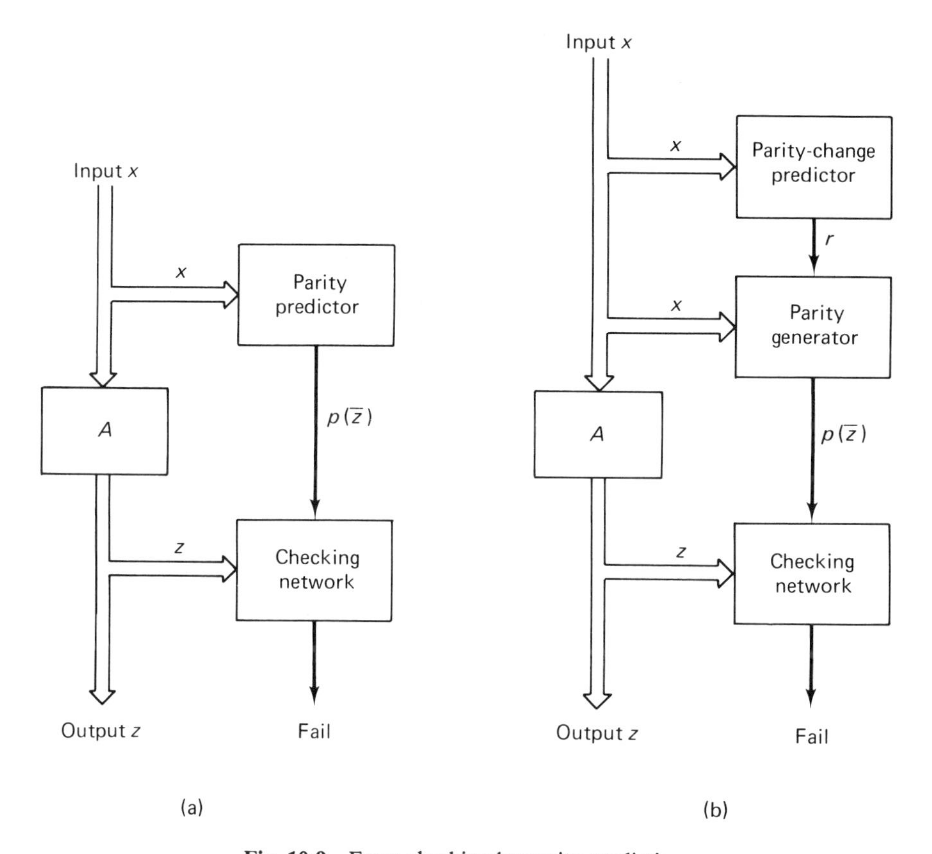

Fig. 10-9 Error-checking by parity prediction.

In Fig. 10-9(b), $p(\bar{z})$ is obtained in two steps; this is sometimes more convenient than the method of Fig. 10-9(a). The parity-change predictor of Fig. 10-9(b) produces a signal r that is equal to $p(x) \oplus p(\bar{z})$. This can be done without actually realizing $p(\bar{z})$, as we explain in Example 4 below. Next, the parity generator produces the output $p(x) \oplus r = p(\bar{z})$. The checking network is the same as in Fig. 10-9(a).

Note that for single-output combinational networks, the schemes of Figs. 10-9(a) and 10-8 coincide.

We now illustrate the use of the scheme of Fig. 10-9(b).

Example 4

Let the network A of Fig. 10-9(b) correspond to the combinational part of a BCD counter. Thus, for $\perp x < 9$, $\perp \bar{z} = 1$ PLUS $\perp x$, and for $\perp x = 9$, $\perp \bar{z} = 0$. The corresponding table of combinations, together with the resulting values of r, are shown in Table 10-2. Applying the map method, we obtain for r the map of Fig. 10-10, which immediately yields

$$r = x'_4 + x'_2 x_3.$$

Table 10-2 Table of Combinations for Example 4

$\perp x$	x	$\perp \bar{z}$	$\bar{z}$	r
0	0000	1	0001	1
1	0001	2	0010	0
2	0010	3	0011	1
3	0011	4	0100	1
4	0100	5	0101	1
5	0101	6	0110	0
6	0110	7	0111	1
7	0111	8	1000	0
8	1000	9	1001	1
9	1001	0	0000	0

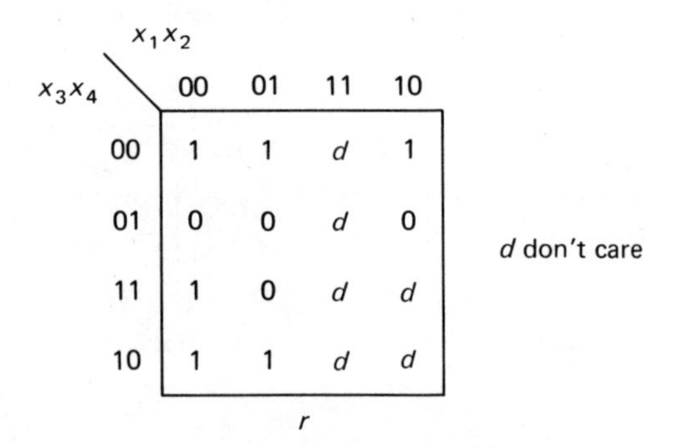

Fig. 10-10 Map for r of Table 10-2.

The parity generator that is specified by

$$p(\bar{z}) = r \oplus p(x) = r \oplus x_1 \oplus x_2 \oplus x_3 \oplus x_4$$

can be implemented by the package 74180 of Fig. 10-2. Another package 74180 is required to implement the checking network, specified by

$$\text{FAIL} = z_1 \oplus z_2 \oplus z_3 \oplus z_4 \oplus p(\bar{z}).$$

Error Checking of Counters

Some of the ideas on error checking developed so far also apply to sequential networks. Thus the scheme of Fig. 10-1 is applicable to output-coded sequential networks, and the duplication scheme of Fig. 10-8 offers an easy method for error-checking arbitrary sequential networks. We now discuss an error-checking technique that is particularly suitable for counting networks. Error checking of counters is of special interest in digital computers, since a failure of the program counter, for example, would cause a complete breakdown of the computer.

The principle of parity prediction (Fig. 10-9) is easily adapted to the error checking of counters. Figure 10-11 shows a suitably modified version of Fig. 10-9(b). In Fig. 10-11, y denotes the present output (state) of the

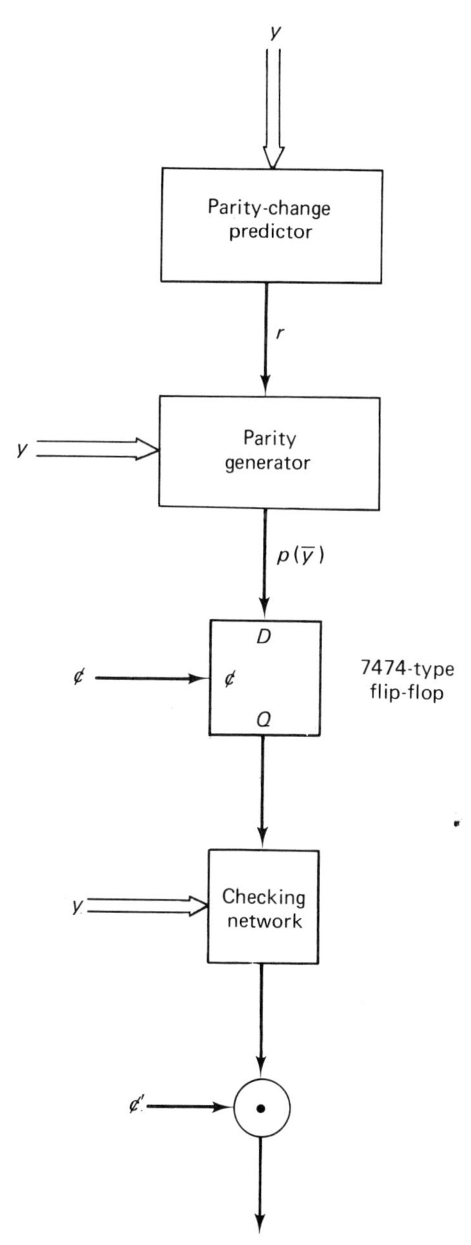

Fig. 10-11 Error-checking of counter by parity prediction.

counter and $\bar{y}$ the expected next output. The predicted parity of the next output, $p(\bar{y})$, is obtained similarly as in Fig. 10-9(b). $p(\bar{y})$ is stored in a 7474-type (positive-edge sensitive D type) flip-flop, while the clock rises from 0 to 1. The checking network compares $p(y)$, the actual parity of the counter output, with Q, the memorized predicted parity. We assume the propagation delays and clock pulse width to be such that the checking network will produce the correct output not later than the 1–0 transition of the clock signal $\not{c}$. This assumption leads to the inclusion of the AND gate at the bottom of Fig. 10-11.

10.2. ERROR CORRECTION [HAM]

In Fig. 10-4 we have shown a digital data transmission system with error-checking facilities. This principle may be expanded as shown in Fig. 10-12, to provide for *error correcting* rather than error checking. If an error occurs in the system of Fig. 10-12, the receiving station decoder uses information obtained from the error detector in order to "guess" the transmitted word $\hat{x}$, from which it then deduces the original message word x.

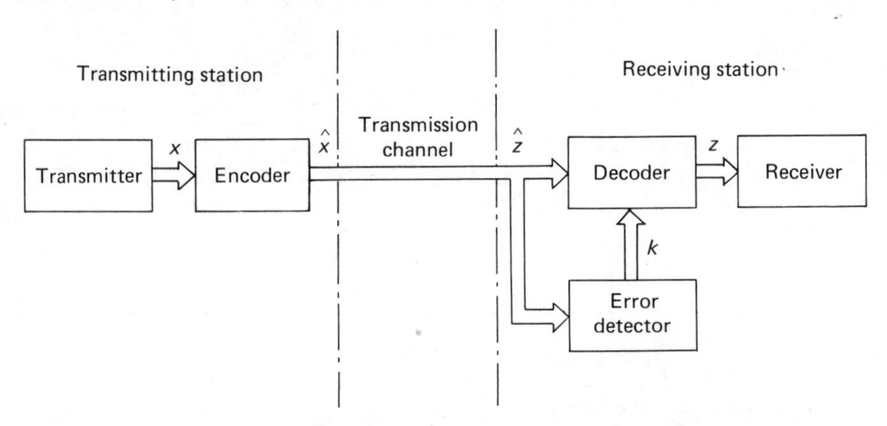

Fig. 10-12 Digital data transmission with error correcting.

Example 5

We consider a digital data transmission system for $n = 1$; i.e., the possible message words are simply 0 and 1. Let 000 be the code word for 0 and 111 the code word for 1. Thus $\hat{X} = \{000, 111\}$. Assume now that the word $\hat{z} = 101$ was received. Since $\hat{z} \notin \hat{X}$, an error has obviously occurred. If we are to guess the transmitted word $\hat{x}$, a reasonable choice would be 111, rather than 000, since 101 differs from 111 in a single bit, whereas 101 differs from 000 in two bits. Similarly, if 001 was received, it is more probable that 000, rather than 111, was transmitted.

From this simple example we can deduce the necessary and sufficient condition for a code X to be single-error-correcting. This condition is

$$d(x^i, x^j) \geq 3 \qquad \text{for all } x^i, x^j \in X.$$

Clearly, if this condition is satisfied and a single error occurs yielding x^*, there can be only one code word x adjacent to x^*. Thus we can always determine uniquely the transmitted word x, under the single error assumption. Conversely, if there exist two code words x^i and x^j such that $d(x^i, x^j) < 3$, one easily verifies that there are single errors that cannot be corrected.

More generally, the receiving station decoder of an error-correcting system applies the "minimal-distance" principle. That is, if $\hat{z}$ is the word received, it selects from $\hat{X}$ the word $\hat{x}$ such that $d(\hat{x}, \hat{z})$ is minimal and assumes $\hat{x}$ to be the word transmitted.

An elegant and well-known method for designing error-correcting codes is due to Hamming. An example of a single-error-correcting Hamming code for $n = 4$ follows.

Example 6

We describe a digital transmission system with single error correction (Fig. 10-12) for $n = 4$. All 16 four-bit words are admissible message words. The code set $\hat{X}$ is defined by means of the following matrix:

$$A \triangleq \begin{Vmatrix} 0 & 0 & 1 \\ 0 & 1 & 0 \\ 0 & 1 & 1 \\ 1 & 0 & 0 \\ 1 & 0 & 1 \\ 1 & 1 & 0 \\ 1 & 1 & 1 \end{Vmatrix}$$

If A_i denotes the ith row of A, then $\perp A_i = i$, for $1 \leq i \leq 7$. We now define $\hat{X}$ as the set of all 7-bit words $\hat{x}$ such that $\hat{x}A = 0$, where 0 denotes an all-zero vector. As usual, the expression $\hat{x}A$ refers to the multiplication of the row vector $\hat{x}$ by A, except that modulo-2 addition ($\oplus$) replaces ordinary addition. The condition $\hat{x}A = 0$ is thus equivalent to the following system of equations:

$$\begin{aligned} \hat{x}_4 \oplus \hat{x}_5 \oplus \hat{x}_6 \oplus \hat{x}_7 &= 0 \\ \hat{x}_2 \oplus \hat{x}_3 \oplus \hat{x}_6 \oplus \hat{x}_7 &= 0 \\ \hat{x}_1 \oplus \hat{x}_3 \oplus \hat{x}_5 \oplus \hat{x}_7 &= 0. \end{aligned} \tag{1}$$

This system of equations can be solved as follows. Set the values of $\hat{x}_3$, $\hat{x}_5$, $\hat{x}_6$, and $\hat{x}_7$ arbitrarily (i.e., either 0 or 1), and determine $\hat{x}_1$, $\hat{x}_2$, and $\hat{x}_4$, accordingly. Thus $\hat{x}_4 = \hat{x}_5 \oplus \hat{x}_6 \oplus \hat{x}_7$, etc. We now identify the message word

$x = x_1, x_2, x_3, x_4$ with the word $\hat{x}_3, \hat{x}_5, \hat{x}_6, \hat{x}_7$. This identification, in conjunction with (1), yields the following encoder specification:

$$\begin{aligned}
\hat{x}_3 &= x_1 \\
\hat{x}_5 &= x_2 \\
\hat{x}_6 &= x_3 \\
\hat{x}_7 &= x_4 \\
\hat{x}_1 &= x_1 \oplus x_2 \oplus x_4 \\
\hat{x}_2 &= x_1 \oplus x_3 \oplus x_4 \\
\hat{x}_4 &= x_2 \oplus x_3 \oplus x_4.
\end{aligned} \tag{2}$$

The error detector is designed to produce the word $k = k_1, k_2, k_3$ specified by $k = \hat{z}A$; i.e.,

$$\begin{aligned}
k_1 &= \hat{z}_4 \oplus \hat{z}_5 \oplus \hat{z}_6 \oplus \hat{z}_7 \\
k_2 &= \hat{z}_2 \oplus \hat{z}_3 \oplus \hat{z}_6 \oplus \hat{z}_7 \\
k_3 &= \hat{z}_1 \oplus \hat{z}_3 \oplus \hat{z}_5 \oplus \hat{z}_7.
\end{aligned} \tag{3}$$

It follows from the definition of $\hat{X}$ that $\hat{z} \in \hat{X}$ iff $\perp k = 0$. If $\perp k > 0$, an error has occurred. We shall show that in this case there exists a unique code word $z^* \in \hat{X}$ which has Hamming distance 1 from $\hat{z}$. Thus the "minimal-distance" principle leads to the guess that z^* was transmitted.

To verify our preceding assertion, we assume that $\perp k > 0$, and define $\hat{k}$ as the 7-bit word specified by

$$\hat{k}_i = \begin{cases} 1 & \text{if } i = \perp k \\ 0 & \text{if } i \neq \perp k. \end{cases}$$

It follows that $\hat{k}A = A_{\perp k} = k$ by construction of A. We now set $z^* \triangleq \hat{z} \oplus \hat{k}$ (bit-by-bit XOR). Then

$$z^*A = (\hat{z} \oplus \hat{k})A = \hat{z}A \oplus \hat{k}A = k \oplus k = 0.$$

Thus $z^* \in \hat{X}$. Since $w(\hat{k}) = 1$, we have

$$d(\hat{z}, z^*) = w(\hat{z} \oplus z^*) = w(\hat{z} \oplus \hat{z} \oplus \hat{k}) = w(\hat{k}) = 1.$$

Furthermore, one verifies that z^* is the only 7-bit word satisfying the conditions $z^* \in \hat{X}$ and $d(\hat{z}, z^*) = 1$.

The receiving station decoder produces z^* and selects from z^* the four bits that correspond to the original 4-bit message word x. Its output thus becomes:

$$\begin{aligned}
z &= z_3^*, z_5^*, z_6^*, z_7^* \\
&= \hat{z}_3 \oplus \hat{k}_3, \hat{z}_5 \oplus \hat{k}_5, \hat{z}_6 \oplus \hat{k}_6, \hat{z}_7 \oplus \hat{k}_7.
\end{aligned} \tag{4}$$

IC XOR-gate packages (7486) may be used to implement (2), (3), and (4). $\hat{k}$ is easily produced from k by means of an address decoder.

*10.3. THEORY OF LOGIC HAZARDS

Introduction

In Section 3.5 we have considered the poset $\langle L_0, \leq \rangle$, where $L_0 = \{0, \frac{1}{2}, 1\}$ and $0 < \frac{1}{2} < 1$, and the corresponding lattice $\langle L_0, +, \cdot \rangle$, where for $x,y \in L_0$, we define

$$x + y \triangleq x \lceil y \tag{5}$$

and

$$x \cdot y \triangleq x \lfloor y. \tag{6}$$

Recall also that the unary operation $*$ is defined as follows:

$$x^* = 1 - x \qquad \text{for } x \in L_0. \tag{7}$$

The system $\langle L_0, +, \cdot, *, 0, 1 \rangle$ is a distributive lattice with universal bounds. This lattice is not complemented and therefore not Boolean. However, the $*$ is a complement-like operation satisfying, for all $x,y \in L_0$,

$$(x + y)^* = x^* \cdot y^* \tag{8}$$

$$(x \cdot y)^* = x^* + y^* \tag{9}$$

$$(x^*)^* = x. \tag{10}$$

In Section 4.5, the operations $+$, $\cdot$, and $*$ were extended to functions in L_n, where

$$L_n \triangleq \{f \mid f : L_0^n \longrightarrow L_0\},$$

in the natural way, paralleling the extension of operations in B_0 to B_n. The system $\langle L_n, +, \cdot, *, ||0||, ||1|| \rangle$ is also a distributive lattice with universal bounds and satisfies (8)–(10).

To each Boolean expression E in $\mathcal{E}_n$ we assign a function $||E||$ in L_n (Definition 8, Chapter 4), which is an extension of the function $|E|$ in the sense that for every $x \in B_0^n$, $||E||(x) = |E|(x)$. We associate $|E|(x)$ with the steady-state behavior and $||E||(x)$ with transient behavior.

Logic Hazards [EIC]

We now examine conditions that can result in erroneous outputs when the inputs change. In general, such conditions are called *hazards*. More precisely, we have the following definitions:

Definition 1

For $y \in L_0^n$ and $z \in B_0^n$, we say that z *is in the neighborhood of* y iff $\mathrm{ABS}(z_i - y_i) \leq \frac{1}{2}$, for every $i \in \{1, \ldots, n\}$, where $\mathrm{ABS}(x)$ denotes the absolute value of x. Let the set of all z in the neighborhood of y be $N(y)$.

Notice that if $\mathrm{ABS}(z_i - y_i) = 0$, then either $z_i = y_i = 0$ or $z_i = y_i = 1$; i.e., z_i and y_i agree. The condition $\mathrm{ABS}(z_i - y_i) = \frac{1}{2}$ occurs iff $y_i = \frac{1}{2}$; i.e., y_i is changing. Thus z is in the neighborhood of y iff for each i either z_i agrees with y_i or y_i is changing.

Definition 2

Let $E \in \mathcal{E}_n$ and let $h \in L_0^n - B_0^n$. We say that h is a (logic) *0-hazard* of E iff

(a) $|E|(z) = 0$ for every z in the neighborhood of h.
(b) $||E||(h) = \frac{1}{2}$.

We say that h is a *1-hazard* if (a) is replaced by (a′) $|E|(z) = 1$ and (b) remains the same. A logic hazard h is called *static* iff exactly one coordinate of h equals $\frac{1}{2}$ (only one variable is changing).

These concepts are best illustrated by an example. Let $E = x_1x_2 + x_1x_3 + x_1'x_4$ and let $h = (\frac{1}{2}, 1, \frac{1}{2}, 1)$. To evaluate $||E||(h)$, set $x_1 = x_3 = \frac{1}{2}$ and $x_2 = x_4 = 1$. Then (recalling Definition 8, Chapter 4)

$$||E||(h) = \tfrac{1}{2}\cdot 1 + \tfrac{1}{2}\cdot\tfrac{1}{2} + (\tfrac{1}{2})^*\cdot 1 = \tfrac{1}{2}.$$

Consider now any z in the neighborhood of h. Since $x_2 = x_4 = 1$, we must have $z_2 = z_4 = 1$. On the other hand, since $x_1 = x_3 = \frac{1}{2}$, these two variables are changing. Consequently, z_1 and z_3 could have any value in $\{0, 1\}$. Therefore, the neighborhood $N(h)$ of h consists of four 4-tuples:

$$N(h) = \{(0, 1, 0, 1), (0, 1, 1, 1), (1, 1, 0, 1), (1, 1, 1, 1)\}.$$

To see whether h is a 1-hazard of E we must verify that $|E|(z) = 1$ for each z in $N(h)$. Now $x_2 = x_4 = 1$ implies $|E| = x_11 + x_1x_3 + x_1'1 = 1$. Therefore, both conditions of Definition 2 are verified and h is indeed a 1-hazard, obviously not a static one.

The application of these concepts to the analysis of transient conditions is the following. Suppose that $z = (0, 1, 0, 1)$ is an input combination. Then $|E|(z) = 1$. Suppose further that $y = (1, 1, 1, 1)$. Here also $|E|(y) = 1$. Under ideal conditions, if we have a network corresponding to E, and the input changes from z to y, the output should remain at 1. However, when this input change takes place, the first and third coordinate become indeterminate ($\frac{1}{2}$). Thus the transient input is properly represented by $h = (\frac{1}{2}, 1, \frac{1}{2}, 1)$. The fact that $||E||(h) = \frac{1}{2}$ is a warning that a spurious 0-pulse may occur during this change. In fact, consider the network of Fig. 10-13. Before the input change, we have $|E| = 1$ because the output of gate G_2 is 1. After the change, $|E|$ is 1 because the outputs of gates G_1 and G_3 are 1. If the output

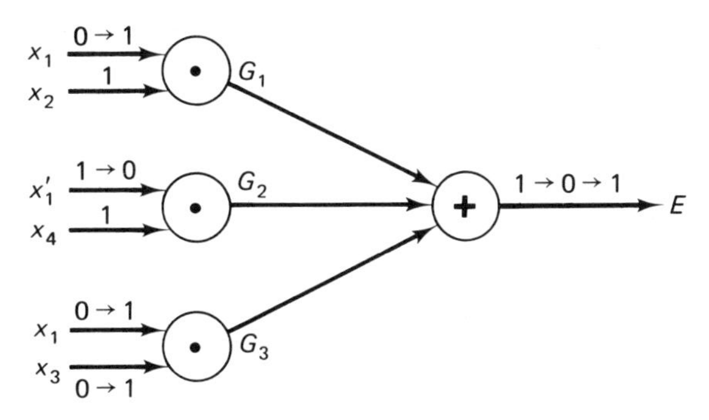

Fig. 10-13 Illustrating a 1-hazard.

of G_2 becomes 0 before either the output of G_1 or that of G_3 becomes 1, $|E|$ will temporarily become 0.

In general, if an expression E has a 0-hazard (respectively, 1-hazard), then a spurious 1-pulse (respectively, 0-pulse) may occur on the output of the corresponding network.

Hazards in SPL Expressions

In this section we will show that the problem of detecting hazards in any Boolean expression E can be reduced to that of detecting hazards in an equivalent "SPL" expression F. First we define the notion of "hazard equivalence" or H-equivalence.

Definition 3

Two Boolean expressions E and F in $\mathcal{E}_n$ are *H-equivalent*, $E \underset{H}{\sim} F$, iff $||E|| = ||F||$.

Definition 4

An *SPL expression* is a Boolean expression that is a sum of products of literals, where the products may contain both a variable x_i and its complement x_i'.

Proposition 1

Let $E \in \mathcal{E}_n$ be any Boolean expression. There exists an SPL expression F that is H-equivalent to E.

Proof

We proceed by induction on the number r of operators in E. If $r = 0$, then $E \in \{0, 1, x_1, \ldots, x_n\}$. Thus E is already SPL. Now assume that the proposition is true for E_1 and E_2, each with r or less operators. Thus there exist SPL expressions F_1 and F_2, where $F_1 \underset{H}{\sim} E_1$, $F_2 \underset{H}{\sim} E_2$; i.e., $||F_1|| = ||E_1||$ and $||F_2|| = ||E_2||$.

(a) If $E = E_1 + E_2$, let $F = F_1 + F_2$. Clearly,

$$||E|| = ||E_1 + E_2|| = ||E_1|| + ||E_2|| = ||F_1|| + ||F_2||$$
$$= ||F_1 + F_2|| = ||F||.$$

Thus $E \underset{H}{\sim} F$, and F is obviously SPL.

(b) If $E = E_1E_2$, let $F_1 = P_1 + \cdots + P_i$ and $F_2 = Q_1 + \cdots + Q_j$, where $P_1, \ldots, P_i, Q_1, \ldots, Q_j$ are products of literals. Let $F = P_1Q_1 + \cdots + P_1Q_j + \cdots + P_iQ_1 + \cdots + P_iQ_j$. Now $||E|| = ||E_1E_2|| = ||E_1|| \cdot ||E_2|| = ||F_1|| \cdot ||F_2|| = ||P_1 + \cdots + P_i|| \cdot ||Q_1 + \cdots + Q_j|| = (||P_1|| + \cdots + ||P_i||) \cdot (||Q_1|| + \cdots + ||Q_j||)$. By the distributive law, the last expression is equal to $||P_1|| \cdot ||Q_1|| + \cdots + ||P_1|| \cdot ||Q_j|| + \cdots + ||P_i|| \cdot ||Q_1|| + \cdots + ||P_i|| \cdot ||Q_j|| = ||F||$. Again $E \underset{H}{\sim} F$, where F is SPL.

(c) If $E = E'_1$, let F be F'_1 converted to an SPL form by using De Morgan's laws and the distributive laws. The verification that $||F|| = ||E||$ is straightforward. □

Example 7

$E = (x_1 + x_2)(x_1x'_2)'$. To obtain an SPL expression F, we have

$$E \underset{H}{\sim} (x_1 + x_2)(x'_1 + x_2) \underset{H}{\sim} x_1x'_1 + x_1x_2 + x_2x'_1 + x_2x_2 = \bar{F}.$$

The SPL expression $\bar{F}$ can be simplified by using any laws valid in the lattice L_n. Thus $\bar{F} \underset{H}{\sim} x_1x'_1 + x_1x_2 + x'_1x_2 + x_2 \underset{H}{\sim} x_1x'_1 + x_2 = F$. Note, however, that L_n is not a Boolean lattice and $xx' = 0$ is not necessarily true. Thus $x_1x'_1$ must remain as is.

Detection of Static Hazards

The next result characterizes static hazards, i.e., those hazards that can result in an indeterminate output while only a single input is indeterminate. This result permits us to test whether a given SPL expression is free of static hazards.

Theorem 1

Let $E \underset{H}{\sim} F \triangleq x_nE_1 + x'_nE_2 + x_nx'_nE_3 + E_4$, where $E \in \mathcal{E}_n$ and $E_i \in \mathcal{E}_{n-1}$ for $i = 1, \ldots, 4$. Then

(a) The n-tuple $h \triangleq (\hat{h}, \frac{1}{2})$, where $\hat{h} \in B_0^{n-1}$, is a static 1-hazard of E iff $|E_1|(\hat{h}) = |E_2|(\hat{h}) = 1$ and $|E_4|(\hat{h}) = 0$.

(b) The n-tuple h is a static 0-hazard of E iff $|E_1|(\hat{h}) = |E_2|(\hat{h}) = |E_4|(\hat{h}) = 0$ and $|E_3|(\hat{h}) = 1$.

Proof

(a) If h is a 1-hazard of E, we must have $||E||(h) = \frac{1}{2}$ or equivalently $||F||(h) = \frac{1}{2}$. Now $|E_4|(\hat{h}) = 1 \equiv ||E_4||(\hat{h}) = 1$ since $\hat{h} \in B_0^{n-1}$. Thus $|E_4|(\hat{h}) = 1$ implies $||F||(h) = 1$. Therefore, we must have $|E_4|(\hat{h}) = 0$.

Next, the neighborhood of h consists of $(\hat{h}, 0)$ and $(\hat{h}, 1)$. We find that

$$|E|(\hat{h}, 0) = 1 \quad \text{implies } |E_2 + E_4|(\hat{h}) = 1$$
$$|E|(\hat{h}, 1) = 1 \quad \text{implies } |E_1 + E_4|(\hat{h}) = 1.$$

Altogether, if h is a 1-hazard, we must have

$$|E_4|(\hat{h}) = 0 \qquad \text{and} \qquad |E_2|(\hat{h}) = |E_1|(\hat{h}) = 1.$$

Conversely, it is easy to verify that these conditions are sufficient for h to be a 1-hazard.

(b) The proof is very similar and is left as an exercise. □

Detection of Arbitrary Logic Hazards

Theorem 2 will provide a method of detecting whether a given SPL expression has a 1-hazard. We require some preliminary results.

Lemma 1

Let $E \in \mathcal{E}_n$, $x \in B_0^n$, and $y \in L_0^n$. If x is in the neighborhood of y, then

$$d \triangleq \text{ABS } (||E||(x) - ||E||(y)) \leq \tfrac{1}{2}.$$

Proof

Because of Proposition 1 we can assume that E is an SPL expression. Consider first the case when $d = 1$ because $||E||(x) = 1$ and $||E||(y) = 0$. At least one product P of E must be 1; i.e., $||P||(x) = 1$. Thus all the literals of P must be assigned the value 1. Because x is in the neighborhood of y, the literals that appear in P must be assigned in y either the value of 1 or $\frac{1}{2}$. Thus $||P||(y)$ is a product of some 1's and some $\frac{1}{2}$'s. In any case, $||P||(y) \geq \frac{1}{2}$, implying $||E||(y) \geq \frac{1}{2}$, which contradicts our assumption. Hence this case is impossible. The argument is similar when $d = 1$ because $||E||(x) = 0$ and $||E||(y) = 1$, and this case is also impossible. □

We now introduce the following notation. Define

$$w^0 = w' \qquad \text{if } w \in B_n$$
$$w^0 = w^* \qquad \text{if } w \in L_n$$
$$\left.\begin{array}{r} w^1 = w \\ w^{1/2} = 1 \end{array}\right\} \text{for } w \in B_n \text{ or } w \in L_n.$$

Lemma 2

Let $q \in B_n$ be a product function and let Q be a product expression for q (i.e., $q = |Q|$). If $||Q||(h) = 1$ for some $h \in L_0^n$, then $|x_1|^{h_1} \cdots |x_n|^{h_n} \leq q$.

Proof

One verifies that there is a unique n-tuple k in L_0^n such that $q = |x_1|^{k_1} \cdots |x_n|^{k_n}$. Since $||Q||(h) = 1$, we have $h_1^{k_1} \cdots h_n^{k_n} = 1$. Hence $h_i^{k_i} = 1$ for

$i = 1, \ldots, n$. Now $h_i^{k_i} = 1$ implies $k_i = \frac{1}{2}$ or $k_i = h_i = 0$ or $k_i = h_i = 1$. In all cases we have $|x_i|^{h_i} \leq |x_i|^{k_i}$, and the claim follows. □

With the aid of these two lemmas, we now give a characterization of 1-hazards in SPL expressions.

Theorem 2

Let E be an SPL expression. Then E has no 1-hazards iff every prime implicant p of $|E|$ is represented by some term P (i.e., $p = |P|$) of E.

Proof

Assume that every prime implicant of $|E|$ is represented by some term of E. Let $h \in L_0^n - B_0^n$ and assume that $|E|(x) = 1$ for every x in the neighborhood of h. Now consider the Boolean product function $p = |x_1|^{h_1} \cdots |x_n|^{h_n}$, and let P be the corresponding Boolean product expression for p. One verifies that $||P||(h) = 1$ and that $p(x) = 1$ iff x is in the neighborhood of h. Thus $p(x) = 1$ implies $|E|(x) = 1$; i.e., $p \leq |E|$. Let q be the prime implicant of $|E|$ such that $p \leq q \leq |E|$. By assumption, E contains a term Q such that $|Q| = q$. It follows that $||Q|| \leq ||E||$. Furthermore, $p \leq q$ implies $L(Q) \subseteq L(P)$; hence $||P|| \leq ||Q||$. Together, $||P|| \leq ||Q|| \leq ||E||$, and $||P||(h) = 1$ implies $||E||(h) = 1$. Hence h is not a 1-hazard of $||E||$.

Conversely, let E be an SPL expression and let p be a prime implicant of $|E|$ which is not represented by any term of E. Let h be the unique n-tuple in L_0^n such that $p = |x_1|^{h_1} \cdots |x_n|^{h_n}$. We will show that h is a 1-hazard of E.

Indeed, let $y \in B_0^n$ be in the neighborhood of h; i.e., $\text{ABS}(y_i - h_i) \leq \frac{1}{2}$, for $i = 1, \ldots, n$. If $h_i \neq \frac{1}{2}$, we must have $y_i = h_i$, implying $y_i^{h_i} = 1$. If $h_i = \frac{1}{2}$, then $y_i^{h_i} = 1$, by definition. Thus $p(y) = y_1^{h_1} \cdots y_n^{h_n} = 1$ and $|E|(y) = 1$, since p is an implicant of $|E|$.

Next, consider $||E||(h)$. Assume that for some term Q of $|E|$, $||Q||(h) = 1$. By Lemma 2, $p = |x_1|^{h_1} \cdots |x_n|^{h_n} \leq |Q|$. Since p is a prime implicant of $|E|$ and $|Q| \leq |E|$, we must have $p = |Q|$. This, however, contradicts our assumption that p is not represented by any term of E. Consequently, $||Q||(h) \neq 1$, for every term Q of E, implying that $||E||(h) \neq 1$. On the other hand, $|E|(y) = 1$ for every $y \in B_0^n$ in the neighborhood of h. By Lemma 1, $\text{ABS}(||E||(y) - ||E||(h)) \leq \frac{1}{2}$. But $||E||(y) = |E|(y) = 1$. Hence $||E||(h)$ cannot be 0. This leaves $||E||(h) = \frac{1}{2}$, completing the proof that h is a 1-hazard of E. □

To detect whether a given expression E has a 0-hazard, we may check whether E' has a 1-hazard. This is justified by the following.

Proposition 2

A Boolean expression E has a 0-hazard iff E' has a 1-hazard.

Proof

The verification is straightforward. □

Recall that an SP expression does not allow both x_i and x_i' to appear in one product expression. In case a given expression E has an H-equivalent SP expression, the following result is applicable.

Theorem 3

If E is an SP expression, then E has no 0-hazards.

Proof

Suppose that $h = h_1, \ldots, h_n \in L_0^n$ is a 0-hazard of E. Then $||E||(h) = \frac{1}{2}$, implying at least one product P of E satisfies $||P||(h) = \frac{1}{2}$. Let $k \in L_0^n$ be the unique n-tuple such that $|x_1|^{k_1} \cdots |x_n|^{k_n} = p = |P|$. Then $h_i^{k_i} \geq \frac{1}{2}$. We now define $z \in B_0^n$ as follows:

$$\text{if } k_i = 1, \text{ let } z_i = 1$$
$$\text{if } k_i = 0, \text{ let } z_i = 0$$
$$\text{if } k_i = \tfrac{1}{2}, h_i \neq \tfrac{1}{2}, \text{ let } z_i = h_i$$
$$\text{if } k_i = h_i = \tfrac{1}{2}, \text{ let } z_i = 0.$$

It is easy to verify that $|P|(z) = 1$, implying $|E|(z) = 1$. We now claim that z is in the neighborhood of h, contradicting the assumption that h is a 0-hazard of E.

To prove the claim, note that if $k_i = 1$, the condition $h_i^{k_i} \geq \frac{1}{2}$ implies $h_i \geq \frac{1}{2}$. Since we have chosen $z_i = 1$, we have $\text{ABS}(z_i - h_i) \leq \frac{1}{2}$. Next, if $k_i = 0$, $h_i^0 \geq \frac{1}{2}$ implies $h_i^* \geq \frac{1}{2}$ or, equivalently, $h_i \leq \frac{1}{2}$. Since we have chosen $z_i = 0$ here, we have again $\text{ABS}(z_i - h_i) \leq \frac{1}{2}$. If $k_i = \frac{1}{2}$, then, by definition, either $\text{ABS}(z_i - h_i) = 0$ or $\text{ABS}(z_i - h_i) = \frac{1}{2}$. This proves the claim and concludes the proof of the theorem. □

In conclusion, any Boolean function f can be implemented without any logic hazards, namely by the network corresponding to its complete sum or, dually, to its complete product.

*10.4. APPLICATIONS OF HAZARD THEORY

To demonstrate an application of the preceding theory, we refer to Fig. 10-14. Assume that the present input is $x = 0111$. It follows that $f(x) = 1$; i.e., the NAND-latch inputs are in the passive state. Assume further that $z = 0$; i.e., the latch is in the cleared state. If x_1 changes from 0 to 1, the new input becomes $\bar{x} = 1111$. Note that $f(\bar{x}) = f(x) = 1$. Thus we expect the NAND-latch inputs to continue in the passive state. However, a hazard analysis shows us that $h = (\frac{1}{2}, 1, 1, 1)$ is a 1-hazard. Indeed, let F be the Boolean expression representing the four-gate combinational network in Fig. 10-14. Then

$$||F|| = [(x_1 \cdot x_2)^* \cdot (x_1 \cdot x_3)^* \cdot (x_1^* \cdot x_4)^*]^*$$

and

$$||F||(h) = [(\tfrac{1}{2} \cdot 1)^* \cdot (\tfrac{1}{2} \cdot 1)^* \cdot (\tfrac{1}{2} \cdot 1)^*]^* = \tfrac{1}{2}.$$

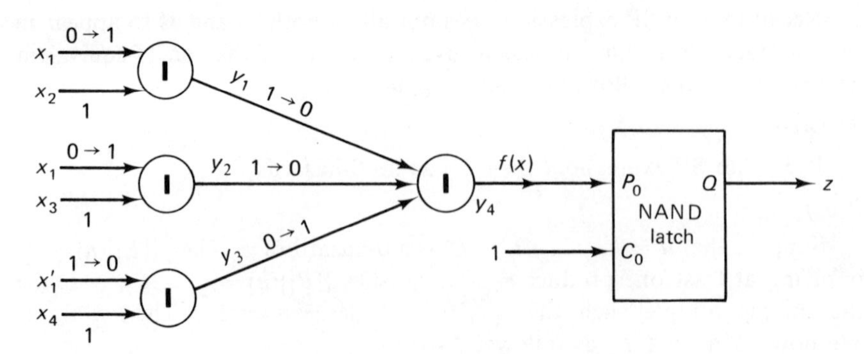

Fig. 10-14 A logic hazard.

Thus the input change from $x = 0111$ to $\bar{x} = 1111$ might cause a 0-pulse at the P_0 input of the NAND latch. Such a 0-pulse would preset the NAND latch and change the output z to 1, although we expected the latch to remain cleared.

If logic 1-hazards are to be prevented, we may apply Theorem 2. This means that we have to find the complete sum F_C for $f = |F|$ and then replace the combinational network of Fig. 10-14 by a network whose Boolean expression is H-equivalent to F_C. First, we find an SP expression E for f:

$$E = x_1 \cdot x_2 + x_1 \cdot x_3 + x_1' \cdot x_4.$$

Next, we may apply the iterated consensus method to find the complete sum

$$F_C = x_1 \cdot x_2 + x_1 \cdot x_3 + x_1' \cdot x_4 + x_2 \cdot x_4 + x_3 \cdot x_4.$$

F_C may be implemented by the network of Fig. 10-15. One easily verifies that the Boolean expression representing the network of Fig. 10-15 is H-equivalent to F_C. It follows that the network of Fig. 10-15 is free of 1-hazards; i.e., 0-pulses due to transient phenomena will not occur.

We now relate this hazard theory to the models of sequential networks presented in Chapter 6. First let us analyze the NAND-gate network producing $f(x)$ of Fig. 10-14, treating it by means of the GSW model.

Let $y = y_1, y_2, y_3, y_4 = 1101$ be the initial gate state which is stable for the input $\hat{x} = 0111$. When $\hat{x}$ changes to $x = 1111$ we find that the first 3 gates become unstable. The final outcome for this transition is the stable state $\bar{y} = 0011$. One verifies that in the relation diagram for R_x one of the paths is:

$$\underline{1}\,\underline{1}\,\underline{0}\,1,\quad \underline{1}\,\underline{1}\,1\,\underline{1},\quad \underline{1}\,\underline{1}\,1\,0,\quad \underline{1}\,0\,1\,\underline{0},\quad 0\,0\,1\,\underline{0},\quad 0\,0\,1\,1.$$

Observe that in this path y_4 changes from 1 to 0 and back to 1. This corresponds precisely to the 0-pulse predicted by the hazard theory! The relation diagram also shows that in some paths no such 0-pulse appears.

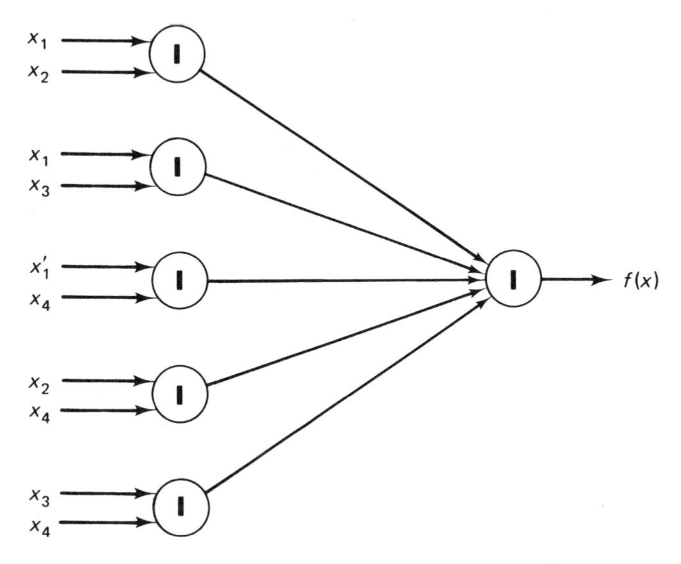

Fig. 10-15 A 1-hazard free implementation of the combinational network of Fig. 10-14.

Thus the relation diagram provides considerably more information about the network than does the hazard theory. In particular, one can find from R_x which delay distributions actually result in hazard pulses. Also, various other more complex transient phenomena in the outputs can be detected, for example, changes like $0 \longrightarrow 1 \longrightarrow 0 \longrightarrow 1 \longrightarrow 0$.

Alternatively, to analyze the network of Fig. 10-14, we can treat it as one 6-gate network, without partitioning it into a *combinational* part and a *memory* part. The GSW analysis of this 6-gate network shows that it has a critical race.

In summary, we can use three different ways of analyzing networks like that in Fig. 10-14.

1. Consider the overall network as one gate network (6 NAND gates in our example) and analyze it by the methods of Chapter 6, i.e. associate a state variable with each gate output.
2. Decompose the network into a number of smaller subnetworks connected in cascade, and analyze each subnetwork as in 1. In this approach, one has to consider the effect of transient output phenomena in one subnetwork on the behavior of the subsequent subnetworks. As we have seen in our example, a transient output condition in a subnetwork may cause a critical race in the overall network.

   Note that the decomposition of a network need not be restricted to the case of a combinational part driving a sequential part. (See for example, our analysis of the edge-sensitive D flip-flop, pp. 219–23. Although we have not emphasized this point there, one must verify that the overall network will not malfunction as a result of transient conditions on the outputs y_2 and y_3 of the network N_1 of Fig. 7-34.)

3. Decompose the network into a loop-free combinational part driving a sequential part. Analyze the sequential part as in 1, and apply the hazard theory of Section 10.3 to the combinational part. If hazards are detected, one must then study their effect on the sequential part. Note that the existence of a hazard does not necessarily imply a malfunction of the network.

We have stressed method 1 in Chapters 6 and 7, because it corresponds to our basic model of sequential network behavior. If method 2 is applicable it will be more manageable computationally, without loss of information. When applicable, method 3 provides a faster analysis of the combinational part. However, some information is lost.

For a comparison of these methods with the conventional approach see [BR-YO2].

PROBLEMS

1. Let A be the network of Problem 5-11. Design the corresponding parity predictor and checking network, in accordance with Fig. 10-9(a). Use IC packages only.

2. Repeat Problem 1 for the DBM specified in Section 2.7.

3. Design an error-checking network in accordance with Fig. 10-11 for the 74190 up/down counter of Fig. 8-41. Assume the inputs $G, L, D_1, \ldots, D_4$ to be suitably fixed and disregard the outputs RC and MM. Use IC packages only.

4. Consider a digital data transmission system for the message words 00, 01, and 10. Design a code set suitable for single-error correcting. Prove that the minimal length of the code words is 5 bits. Design the corresponding encoding and error-correcting networks. You may use gates and IC modules.

5. For the Boolean expression

$$E = (x_1' + x_2 \cdot x_3') \cdot (x_2 + x_4') \cdot (x_1 + x_2')$$

(a) Find all static hazards.
(b) Find all logic hazards.
(c) Find a hazard-free equivalent expression.

A *PROPOSITIONAL CALCULUS

In this appendix we provide a very brief introduction to the calculus of propositions. We use some of the notation of propositional calculus as a precise and concise way of describing gate and contact networks (e.g., Chapters 1 and 2). We also use it as basic mathematical notation (e.g., Chapter 3 and Appendix B). Finally, the propositional calculus can be viewed as a Boolean algebra (Section 4.6) and so is of interest to us also from this point of view.

Consider statements such as the following: (a) $2 + 2 = 4$; (b) $1 \geq 3$; (c) 7 is a negative integer; (d) Rome is the capital of Italy; and (e) The statement (e) is false. Statements (a) and (d) are true, whereas statements (b) and (c) are false. We shall say that a true statement has the *truth value* T (TRUE), and a false statement has the truth value F (FALSE). Statement (e) is self-contradictory and consequently has no definite truth value. A *proposition* is a statement that has a definite truth value (either T or F). Henceforth we restrict our considerations to propositions.

One can construct *compound propositions* from given propositions by means of the *logical connectives*. For instance, if p and q are propositions, the compound proposition $(p \wedge q)$ is defined to be true iff (if and only if) both p *and* q are true. Generally, the truth value of a compound proposition is uniquely determined by the truth values of its constituent propositions. Such a relationship of truth values is conveniently specified by means of a *truth table*. Table A-1 gives the truth tables of some frequently used logical connectives: $\sim$ (NOT or *negation*), $\wedge$ (AND), $\vee$ (OR), $\rightarrow$ (*arrow* or *conditional*), $\leftrightarrow$ (*double arrow* or *biconditional*), XOR (exclusive OR), NAND (negated AND), and NOR (negated OR).

Table A-1 Truth Tables for Some Logical Connectives

p	$\sim p$
F	T
T	F

p	q	$(p \wedge q)$	$(p \vee q)$	$(p \longrightarrow q)$	$(p \longleftrightarrow q)$	$(p \text{ XOR } q)$	$(p \text{ NAND } q)$	$(p \text{ NOR } q)$
F	F	F	F	T	T	F	T	T
F	T	F	T	T	F	T	T	F
T	F	F	T	F	F	T	T	F
T	T	T	T	T	T	F	F	F

For example, let $p \triangleq 2 + 2 = 4$, $q \triangleq 1 \geq 3$, and $r \triangleq 2 > 1$, where $\triangleq$ means "is defined as." Then $(p \longleftrightarrow q)$ has the truth value F, since p is true and q is false, whereas $((p \longleftrightarrow q) \text{ NAND } r)$ has the truth value T.

We now wish to consider such strings of symbols as "$((p \longleftrightarrow q) \text{ NAND } r)$" independently of any assignment of truth values to the *propositional letters* p, q, and r. We introduce the following formal definition of *formula* (*in* $p_1, \ldots, p_n$).

Definition 1

Let $p_1, \ldots, p_n$ denote n distinct symbols (*proposition letters*), all different from T and F. Then

(a) F, T, $p_1, \ldots, p_n$ are formulas.

(b) If f and g are formulas, then so are $\sim f$, $(f \wedge g)$, $(f \vee g)$, $(f \longrightarrow g)$, $(f \longleftrightarrow g)$, $(f \text{ XOR } g)$, $(f \text{ NAND } g)$, and $(f \text{ NOR } g)$.

(c) The only formulas are those given by (a) and (b).

If one substitutes specific propositions for the proposition letters of a formula, or alternatively assigns truth values (T or F) to all the proposition letters in a formula, the truth value of the formula can be evaluated by means of Table A-1. A *truth table* for a given formula f is a listing of all possible truth-value combinations for the proposition letters of f, together with the corresponding truth values for f. The following table illustrates the evaluation of the truth table for the formula $f \triangleq ((p \wedge \sim q) \vee \text{F})$.

p	q	$\sim q$	$(p \wedge \sim q)$	$((p \wedge \sim q) \vee \text{F})$
F	F	T	F	F
F	T	F	F	F
T	F	T	T	T
T	T	F	F	F

If the formula f contains n distinct proposition letters, then the truth table for f consists of 2^n rows.

Two formulas are *equivalent* (written $f \equiv g$) iff they have identical truth tables. One verifies easily, for example, that

$$(p \land q) \equiv \sim(\sim p \lor \sim q).$$

A formula f is a *tautology* iff in every row of its truth table the entry for f is T. An immediate consequence is the following.

Proposition 1

Let f and g be formulas. f and g are equivalent iff $f \leftrightarrow g$ is a tautology.

Let f and g be formulas over the same propositional letters. f *implies* g, written $f \Rightarrow g$ iff for every combination of truth values for which f is true, g is also true. The following two propositions are easily verified:

Proposition 2

$f \Rightarrow g$ iff $f \rightarrow g \equiv \mathrm{T}$

Proposition 3

Let f, g, and h be formulas over the same propositional letters. Then:
(a) $f \equiv g$ iff $f \Rightarrow g$ and $g \Rightarrow f$.
(b) If $f \Rightarrow g$ and $g \Rightarrow h$, then $f \Rightarrow h$.
(c) $f \land g \Rightarrow f$.
(d) $f \Rightarrow f \lor g$.
(e) If $f \Rightarrow g$, then $f \Rightarrow f \land g$.
(f) If $f \Rightarrow g$, then $f \lor g \Rightarrow g$.

Table A-2 lists basic properties of the logical connectives $\land$, $\lor$, and $\sim$. They can be easily verified by constructing truth tables. For convenience, in such formulas as $((f \lor g) \lor h)$, we omit the outer parentheses and write $(f \lor g) \lor h$. Also the $\sim$ operator has precedence over the other operators. Note that the properties in Table A-2 are listed in "dual" pairs in the sense

Table A-2 Properties of the Logical Connectives $\land$, $\lor$, and $\sim$

P1. $f \lor f \equiv f$	P1′. $f \land f \equiv f$	(idempotent laws)
P2. $f \lor g \equiv g \lor f$	P2′. $f \land g \equiv g \land f$	(commutative laws)
P3. $f \lor (g \lor h)$ $\equiv (f \lor g) \lor h$	P3′. $f \land (g \land h)$ $\equiv (f \land g) \land h$	(associative laws)
P4. $f \lor (f \land g) \equiv f$	P4′. $f \land (f \lor g) \equiv f$	(absorption laws)
P5. $f \lor \mathrm{F} \equiv f$	P5′. $f \land \mathrm{T} \equiv f$	(laws for F and T)
P6. $f \lor \mathrm{T} \equiv \mathrm{T}$	P6′. $f \land \mathrm{F} \equiv \mathrm{F}$	
P7. $f \lor \sim f \equiv \mathrm{T}$	P7′. $f \land \sim f \equiv \mathrm{F}$	(laws for negation)
P8. $\sim \sim f \equiv f$		
P9. $f \lor (g \land h)$ $\equiv (f \lor g) \land (f \lor h)$	P9′. $f \land (g \lor h)$ $\equiv (f \land g) \lor (f \land h)$	(distributive laws)
P10. $\sim(f \lor g) \equiv \sim f \land \sim g$	P10′. $\sim(f \land g) \equiv \sim f \lor \sim g$	(De Morgan's laws)

that property Pn' can be obtained from property Pn by interchanging $\vee$ with $\wedge$ and F with T.

In Table A-3 we list various conversion rules that enable us to replace one logical connective by means of others. In view of the conversion rules 1–5 of Table A-3, all the connectives of Table A-1 can be expressed by $\wedge$, $\vee$, and $\sim$. In view of rule 6, $\wedge$ and $\sim$ suffice. Similarly, by rule 7, $\vee$ and $\sim$ are sufficient to express all the other connectives. By rules 8 and 9, $\sim$ and $\wedge$ can be replaced by NAND. Thus NAND alone suffices to replace all other connectives, as also does NOR, in view of rules 8 and 10.

Table A-3 Conversion Rules for Logical Connectives

1. $f \longrightarrow g \equiv \sim f \vee g$
2. $f \longleftrightarrow g \equiv (f \wedge g) \vee (\sim f \wedge \sim g) \equiv (f \vee \sim g) \wedge (\sim f \vee g)$
3. $f \text{ XOR } g \equiv (f \wedge \sim g) \vee (\sim f \wedge g) \equiv (f \vee g) \wedge (\sim f \vee \sim g)$
4. $f \text{ NAND } g \equiv \sim (f \wedge g) \equiv \sim f \vee \sim g$
5. $f \text{ NOR } g \equiv \sim (f \vee g) \equiv \sim f \wedge \sim g$
6. $f \vee g \equiv \sim (\sim f \wedge \sim g)$
7. $f \wedge g \equiv \sim (\sim f \vee \sim g)$
8. $\sim f \equiv f \text{ XOR T} \equiv f \text{ NAND T} \equiv f \text{ NAND } f \equiv f \text{ NOR F} \equiv f \text{ NOR } f$
9. $f \wedge g \equiv \sim (f \text{ NAND } g)$
10. $f \vee g \equiv \sim (f \text{ NOR } g)$

B SETS

B.1. NOTATION

The following list gives our terminology and notation pertaining to sets.

Symbol	Meaning
$x \in X$	*x is an element of X*
$x \notin X$	*x is not an element of X*
$\{z \in X \mid z \text{ has property } P\}$	the set of all elements of X that have property P
$X = Y$	the sets X and Y are *equal*
$X \subseteq Y$ or $Y \supseteq X$	*X is a subset of Y*
$X \subset Y$ or $Y \supset X$	*X is a proper subset of Y*
$\varnothing$	the *empty set*
$X \cap Y$	the *intersection* of X and Y
$X \cup Y$	the *union* of X and Y
$X - Y$	the *difference* of X and Y, $X - Y \triangleq \{z \mid z \in X \text{ and } z \notin Y\}$
$X \triangle Y \triangleq (X - Y) \cup (Y - X)$	the *symmetric difference* of X and Y
I	thc *universal set*
$\bar{X} \triangleq I - X$	the *complement* of X in I
$X \cap Y = \varnothing$	X and Y are *disjoint*
$P(X)$	the *power set* of X; i.e., the set of all subsets of X

*B.2. INTRODUCTION TO SET THEORY

The notion of a set is one of the primitive notions in mathematics. However, there is a much stronger reason for including a brief discussion of sets here. We show in Chapter 3 that each Boolean algebra can be represented by a certain algebra of sets. This characterization provides a very convenient model for Boolean algebra.

A *set* is a collection of elements. For example, the set that has the integers 1, 2, and 3 as elements is written $\{1, 2, 3\}$. "$x \in X$" stands for "x is an element of X" and "$x \notin X$" for "$\sim(x \in X)$." Thus the proposition $2 \in \{1, 2, 3\}$ is true whereas $5 \in \{1, 2, 3\} \equiv \mathrm{F}$. For any integer x, the following equivalence holds:

$$x \in \{1, 2, 3\} \equiv (x = 1) \vee (x = 2) \vee (x = 3).$$

More generally, the set having the objects $a_1, \ldots, a_n$ as its elements is written $\{a_1, \ldots, a_n\}$, and we have the following equivalence for every x:

$$x \in \{a_1, \ldots, a_n\} \equiv (x = a_1) \vee \cdots \vee (x = a_n). \tag{1}$$

The statement that equivalence (1) holds for every x needs some clarification. Such a statement is to be understood as follows: Every substitution for x that yields a proposition on either side will also yield a proposition on the other side, and both propositions will have the same truth value.

The method of describing a set by listing all its elements can only be applied to *finite* sets, i.e., sets having a finite number of different elements. Infinite sets can be described in various ways. First, we may introduce special symbols to denote particular infinite sets. For example, we shall denote by $\mathbf{Z}$ the set of all integers. Thus the equivalence

$$x \in \mathbf{Z} \equiv x \text{ is an integer}$$

holds for all x. Second, we may select from a given (finite or infinite) set X all the elements z possessing some property P. The corresponding set is written

$$\{z \in X \mid z \text{ has property } P\}$$

or, alternatively,

$$\{z \mid (z \in X) \wedge (z \text{ has property } P)\}.$$

The following equivalence thus holds for all x:

$$x \in \{z \in X \mid z \text{ has property } P\} \equiv (x \in X) \wedge (x \text{ has property } P). \tag{2}$$

For example, the set $\mathbf{N}$ of all nonnegative integers can be defined as follows:

$$\mathbf{N} \triangleq \{z \in \mathbf{Z} \mid z \geq 0\}.$$

Third, we shall introduce various constructions, which will enable us to generate new sets from given ones.

Although some of these constructions will use set descriptions of the form $\{z \mid z \text{ has property } P\}$ to define the "set of all objects having property P," we do not admit *arbitrary* set descriptions of this form. Such a general description rule leads, for example, to the famous *Russell paradox*, which deals with the "set" $T \triangleq \{X \mid X \notin X\}$. One easily verifies that the statement $T \in T$ is self-contradictory, just as statement (e) in Appendix A was.

We consider two sets X and Y to be *equal*, written $X = Y$, iff they contain the same elements. Thus

$$X = Y \text{ iff } z \in X \equiv z \in Y \qquad \text{for all } z. \tag{3}$$

X is a *subset* of Y (or X is *included* in Y), written $X \subseteq Y$ or $Y \supseteq X$, iff every element of X is also an element of Y. Hence

$$X \subseteq Y \text{ iff } z \in X \Rightarrow z \in Y \qquad \text{for all } z. \tag{4}$$

X is a *proper* subset of Y, $X \subset Y$, iff $X \subseteq Y$ and $X \neq Y$.

The *empty* set $\varnothing$ may be defined by

$$z \in \varnothing \equiv \text{F} \qquad \text{for all } z; \tag{5}$$

i.e., it is *the* set that has no elements. [Note that, by (3), all such sets are equal.]

If X and Y are given sets, their *intersection* $X \cap Y$, their *union* $X \cup Y$, their *difference* (or *relative complement*) $X - Y$, and their *symmetric difference* $X \triangle Y$ are defined by the following equivalences, which hold for all z:

$$z \in X \cap Y \equiv (z \in X) \wedge (z \in Y) \tag{6}$$

$$z \in X \cup Y \equiv (z \in X) \vee (z \in Y) \tag{7}$$

$$z \in X - Y \equiv (z \in X) \wedge (z \notin Y) \tag{8}$$

$$z \in X \triangle Y \equiv (z \in X) \text{ XOR } (z \in Y). \tag{9}$$

If $X \subseteq I$, the *complement* $\bar{X}$ of X *in* I is the set $I - X$. If $X \cap Y = \varnothing$, X and Y are said to be *disjoint*. Given a set X, its *power set* $P(X)$ is defined to be the set of all subsets of X. Thus

$$S \in P(X) \equiv S \subseteq X \qquad \text{for all } S. \tag{10}$$

For example, $P(\{1, 2, 3\}) = \{\varnothing, \{1\}, \{2\}, \{3\}, \{1,2\}, \{1,3\}, \{2,3\}, \{1,2,3\}\}$.

The following proposition lists various properties of the inclusion relation $\subseteq$.

Proposition 1

Let X, Y, and Z be sets. Then

(a) $X \subseteq X$ (reflexivity)
(b) $X \subseteq Y$ and $Y \subseteq X$ implies $X = Y$ (antisymmetry)
(c) $X \subseteq Y$ and $Y \subseteq Z$ implies $X \subseteq Z$ (transitivity)
(d) $X \cap Y \subseteq X$
(e) $X \cap Y \subseteq Y$
(f) $Z \subseteq X$ and $Z \subseteq Y$ implies $Z \subseteq X \cap Y$
(g) $X \cup Y \supseteq X$

(h) $X \cup Y \supseteq Y$
(i) $Z \supseteq X$ and $Z \supseteq Y$ implies $Z \supseteq X \cup Y$
(j) The following statements are all equivalent:
(1) $X \subseteq Y$ (2) $X \cap Y = X$ (3) $X \cup Y = Y$ (4) $X - Y = \varnothing$
(k) $X - Y \subseteq X$

Proposition 1 is easily proved. To illustrate, we prove (b) and (e).

Proof of (b)

Assume that $X \subseteq Y$ and $Y \subseteq X$. By (4), $z \in X \Rightarrow z \in Y$ and $z \in Y \Rightarrow z \in X$ for all z. By Proposition 3(a) of Appendix A, $z \in X \equiv z \in Y$ for all z. Hence $X = Y$, by (3). ▯

Proof of (e)

By (6), $z \in X \cap Y \equiv (z \in X) \wedge (z \in Y)$ for all z. By P2′ (Table 2 of Appendix A), $(z \in X) \wedge (z \in Y) \equiv (z \in Y) \wedge (z \in X)$ for all z. By Proposition 3(c) of Appendix A, $(z \in Y) \wedge (z \in X) \Rightarrow z \in Y$ for all z. Hence, by Proposition 3(a) and (b) of Appendix A, $z \in X \cap Y \Rightarrow z \in Y$ for all z. Thus $X \cap Y \subseteq Y$, by (4). ▯

Algebraic properties of the operators $\cap$, $\cup$, and $^{-}$ are listed in Table B-1, which is arranged along the same pattern as Table A-2. We assume X, Y, and Z to be subsets of the (*universal*) set I. $\bar{X}$ denotes the complement of X in I.

Table B-1 Properties of $\cap$, $\cup$, and $^{-}$

S1. $X \cup X = X$	S1′. $X \cap X = X$	(idempotent laws)
S2. $X \cup Y = Y \cup X$	S2′. $X \cap Y = Y \cap X$	(commutative laws)
S3. $X \cup (Y \cup Z) = (X \cup Y) \cup Z$	S3′. $X \cap (Y \cap Z) = (X \cap Y) \cap Z$	(associative laws)
S4. $X \cup (X \cap Y) = X$	S4′. $X \cap (X \cup Y) = X$	(absorption laws)
S5. $X \cup \varnothing = X$	S5′. $X \cap I = X$	(laws for $\varnothing$ and I)
S6. $X \cup I = I$	S6′. $X \cap \varnothing = \varnothing$	
S7. $X \cup \bar{X} = I$	S7′. $X \cap \bar{X} = \varnothing$	(laws for complements)
S8. $\bar{\bar{X}} = X$		
S9. $X \cup (Y \cap Z) = (X \cup Y) \cap (X \cup Z)$	S9′. $X \cap (Y \cup Z) = (X \cap Y) \cup (X \cap Z)$	(distributive laws)
S10. $\overline{X \cup Y} = \bar{X} \cap \bar{Y}$	S10′. $\overline{X \cap Y} = \bar{X} \cup \bar{Y}$	(De Morgan's laws)

Comparing Table B-1 with Table A-2, we see that every law of Table B-1 can be obtained from the corresponding law of Table A-2, by replacing the symbols f, g, h, F, T, $\wedge$, $\vee$, $\sim$, and $\equiv$ by the symbols X, Y, Z, $\varnothing$, I, $\cap$, $\cup$, $^{-}$, and $=$, respectively.

Since $X \subseteq I$, laws S5′ and S6 follow immediately, by Proposition 1(j). The other laws of Table B-1 are also easily derived by means of Table A-2, and the equivalences (3)–(8) of this section. We give two examples.

Proof of S4

By (7) and (6), we have for all z:

$$z \in X \cup (X \cap Y) \equiv z \in X \vee z \in (X \cap Y)$$
$$\equiv z \in X \vee (z \in X \wedge z \in Y).$$

By P4 of Table 2, Appendix A,

$$z \in X \vee (z \in X \wedge z \in Y) \equiv z \in X.$$

Thus $z \in X \cup (X \cap Y) \equiv z \in X$ for all z. Hence $X \cup (X \cap Y) = X$, by (3). ▯

Proof of S10′

The following equivalences hold for all z:

$z \in \overline{X \cap Y}$

$\equiv z \in I - (X \cap Y) \equiv z \in I \wedge \sim(z \in X \cap Y)$ [by (8)]

$\equiv z \in I \wedge \sim(z \in X \wedge z \in Y)$ [by (6)]

$\equiv (z \in I \wedge \sim(z \in X)) \vee (z \in I \wedge \sim(z \in Y))$ (by P10′ and P9′)

$\equiv z \in I - X \vee z \in I - Y$ [by (8)]

$\equiv z \in \bar{X} \vee z \in \bar{Y} \equiv z \in \bar{X} \cup \bar{Y}.$ [by (7)]

Hence $\overline{X \cap Y} = \bar{X} \cup \bar{Y}$, by (3). ▯

At this point, let us briefly review the proof techniques we have used so far in this section. Our proofs are precise in the sense that they do not rely on our intuition, as far as sets are concerned, but use only the equivalences (3)–(8), which we may consider as axioms or postulates. However, in addition to the set-theoretic equivalences (3)–(8) and the results from propositional calculus that we have quoted explicitly in our proofs, we have also used various rules based on intuitively evident logical reasoning. For example, we have invoked in the proof of S4 the equivalence P4 of Table A-2, to justify the equivalence

$$z \in X \vee (z \in X \wedge z \in Y) \equiv z \in X,$$

without formulating and referring to the corresponding substitution rule. Although we do not intend to set up the rather complex formalism required to replace these "intuitively obvious" rules, it is important that the reader is at least aware of the existence of such an underlying structure of rules of inference, and that he uses these rules correctly. For further discussion of this point, the reader is referred to suitable textbooks on mathematical logic, e.g. [CH-LE].

The next proposition summarizes properties of the symmetric difference operator Δ.

Proposition 2

Let X, Y, and Z be sets. Then

(a) $(X \triangle Y) \triangle Z = X \triangle (Y \triangle Z)$
(b) $X \triangle Y = Y \triangle X$
(c) $X \triangle \varnothing = X$
(d) $X \triangle X = \varnothing$
(e) $X \triangle Y = (X - Y) \cup (Y - X)$
(f) $X \cup Y = X \triangle Y \triangle (X \cap Y)$
(g) $X \cap (Y \triangle Z) = (X \cap Y) \triangle (X \cap Z)$

To prove the statements of Proposition 2, one first shows that the corresponding equivalences hold for the logical connective XOR (see also Table 1-9). The proof then proceeds similarly to the proofs for S4 and S10′ above.

In order to illustrate some of the properties of sets, it is often convenient to use a diagram called the *Venn diagram*. The universal set I is represented by the set of points inside a square, and subsets of I by sets of points contained in circles within the square. Figure B-1 illustrates the union, intersection, complement, and symmetric difference operations.

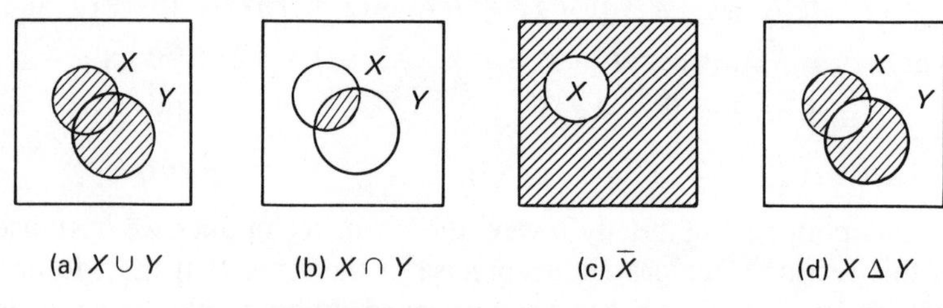

Fig. B-1 The basic operations.

The Venn diagram can also be used to illustrate various set identities. For example, the first of De Morgan's laws (S10) is demonstrated by the Venn diagram of Fig. B-2.

Table B-1 is representative of Boolean algebras discussed in Chapter 3.

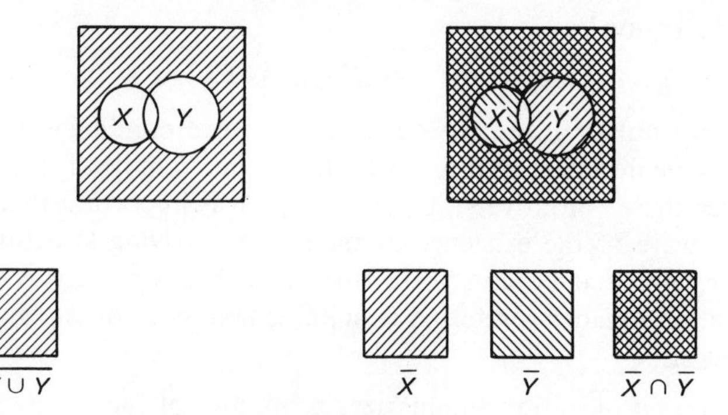

Fig. B-2 Illustrating $\overline{X \cup Y} = \bar{X} \cap \bar{Y}$.

C *FUNCTIONS AND RELATIONS

C.1. FUNCTIONS

Any mapping f of a set X into a set Y is called a *function from X into Y*, written $f: X \longrightarrow Y$. For every element $x \in X$, there exists a unique element y such that $f(x) = y$. X is called the *domain* of f and Y its *codomain*. Let

$$Y' \triangleq \{y \in Y \mid y = f(x) \text{ for some } x \in X\}.$$

Then Y' is the *image* of f. In case $Y = Y'$, we say that f is *surjective* (or that f is a function from X *onto* Y, or that f is *onto*). f is *injective* (or *one-to-one*) when $x_1 \neq x_2$ in X implies $f(x_1) \neq f(x_2)$. f is *bijective* (or one-to-one and onto) iff it is both injective and surjective. We write $f: X \cong Y$ to state that the function $f: X \longrightarrow Y$ is bijective. For example, if $\mathbf{Z}$ is the set of all integers, let $g(z) = z + 1$ for every $z \in \mathbf{Z}$. Then $g: \mathbf{Z} \cong \mathbf{Z}$.

It is sometimes convenient to describe a function by its *assignment rule*. For example, the function g above can be described by

$$g: z \longrightarrow z + 1 \qquad z \in \mathbf{Z}.$$

By an *n-tuple*, $n \geq 2$, we mean an ordered sequence of n elements, not necessarily all different. The n-tuple with a_i as its ith element ($i = 1, \ldots, n$) is written $(a_1, \ldots, a_n)$ or $\langle a_1, \ldots, a_n \rangle$. Two tuples $(a_1, \ldots, a_n)$ and $(b_1, \ldots, b_m)$ are *equal* iff $m = n$ and $a_i = b_i$ for all $i \in \{1, \ldots, n\}$. An *ordered pair* is a 2-tuple.

The *cartesian product* $X \times Y$ of two sets X and Y is defined to be the set of all ordered pairs (x, y), where $x \in X$ and $y \in Y$. In symbols,

$$X \times Y \triangleq \{(x, y) \mid x \in X \wedge y \in Y\}.$$

The cartesian product of n sets $X_1, \ldots, X_n$ is defined similarly. The nth *cartesian power* of X is $X^n \triangleq X \times \cdots \times X$ (n times); i.e., X^n is the set of all n-tuples with elements in X.

A *unary operation* on X is a function from X into X. A *binary operation* on X is a function from $X \times X$ into X. If $\Box$ denotes an arbitrary binary operation on X, we usually write $x \Box y$ for $\Box((x, y))$. Thus "$2 + 3 = 5$" stands for "$+((2, 3)) = 5$."

An *algebraic system* is a set X together with one or more operations on X which satisfy specified postulates. Selected elements of X can also participate in the description of some algebraic systems. For example, a *semigroup* is an ordered pair $\langle X, \Box \rangle$, where X is a set and $\Box$ is a binary operation on X which is associative. Thus $\langle \mathbf{Z}, + \rangle$ is a semigroup. A *monoid* is a 3-tuple $\langle X, \Box, e \rangle$, where $\langle X, \Box \rangle$ is a semigroup and e is an element of X that is a *unit* for $\Box$; i.e., $e \Box x = x \Box e = x$ for every $x \in X$. Thus $\langle \mathbf{Z}, +, 0 \rangle$ is a monoid. Several other algebraic systems are introduced in Chapter 3.

If $f: X \longrightarrow Y$ is a function and y is an element of Y, we define $f^{-1}(y)$ to be

$$f^{-1}(y) \triangleq \{x \in X \mid f(x) = y\}.$$

C.2. RELATIONS

Fundamental Notions

A *binary relation R between* sets X and Y is any subset of $X \times Y$. If $(x, y) \in R$, we also write xRy and say that x is related by R to y. A binary relation *on* X is a binary relation between X and X. In accordance with this approach we view, for example, the proposition "$3 < 5$" to stand for "$(3, 5) \in <$" and identify the relation $<$ with the set of all ordered pairs of numbers (x, y) such that x is less than y.

The reader can easily verify that a binary relation R between X and Y is a function from X into Y iff:

(a) For every $x \in X$ there exists a $y \in Y$ such that xRy.
(b) xRy_1 and xRy_2 implies $y_1 = y_2$.

Let R be a binary relation on X; i.e., $R \subseteq X \times X$.

1. R is *reflexive* iff xRx for all $x \in X$.
2. R is *symmetric* iff xRy implies yRx for all $x, y \in X$.
3. R is *antisymmetric* iff xRy and yRx implies $x = y$ for all $x, y \in X$.
4. R is *transitive* iff xRy and yRz implies xRz for all $x, y, z \in X$.

A binary relation on X which is reflexive, antisymmetric, and transitive is called a *partial order* of X. By Proposition 1(a), and (b), and (c), Appendix B, $\subseteq$ is a partial order of the power set $P(X)$ for any set X. Also $\leq$ and $\geq$ are both partial orders of $\mathbf{Z}$.

A binary relation on X which is reflexive, symmetric, and transitive is called an *equivalence* (relation) on X. The equality relation I_X on X defined by

$$I_X \triangleq \{(x, x) \mid x \in X\}$$

is obviously an equivalence on X.

Given an equivalence relation E on a set X, and given any $x \in X$, let $[x] \triangleq \{y \in X \mid xEy\}$. Then $[x]$ is called the E-equivalence class of x. One easily verifies that $[x] = [y]$ iff xEy. Also if $[x] \neq [y]$, then $[x]$ and $[y]$ are disjoint. Let $X/E \triangleq \{[x] \mid x \in X\}$. X/E is called the *quotient set* of X by E.

A set π is a partition of a set X iff the elements of π are subsets of X, and for every $x \in X$ there exists exactly one $S \in \pi$ such that $x \in S$. Clearly X/E is a partition of X. Conversely, let π be a partition of X. Define

$$E \triangleq \{(x, y) \in X\cdot \mid x \in S \wedge y \in S \text{ for some } S \in \pi\}.$$

Then E is an equivalence on X and $\pi = X/E$.

The *converse* of a binary relation $R \subseteq X \times Y$ is the binary relation $R^{-1} \subseteq Y \times X$ defined by

$$R^{-1} \triangleq \{(y, x) \mid xRy\}.$$

The converse of the relation $\subseteq$ is $\supseteq$ and $(<)^{-1} = >$. Clearly, $(R^{-1})^{-1} = R$. Assume now that $R \subseteq X^2$. The reader can verify that R^{-1} is reflexive (respectively, antisymmetric, transitive) iff R is. Furthermore, R is symmetric iff $R = R^{-1}$.

Composition and Closure Operations

Let R_1 and R_2 be binary relations. Their *composition*, written $R_1 \circ R_2$ or $R_1 R_2$, is defined whenever $R_1 \subseteq X \times Y$ and $R_2 \subseteq Y \times Z$. It is the binary relation between X and Z defined by

$$R_1 R_2 \triangleq \{(x, z) \in X \times Z \mid xR_1y \wedge yR_2z \text{ for some } y \in Y\}.$$

For example, consider the following binary relations R_1 and R_2 on $\{1, 2, 3, 4\}$:

$$R_1 \triangleq \{(1, 2), (2, 3), (3, 4), (4, 2)\} \quad R_2 \triangleq \{(2, 4), (3, 1), (3, 2)\}.$$

They are represented, in an evident way, in Fig. C-1(a) and (b). Their composition $R_1 R_2$ is shown in Fig. C-1(c) and $R_2 R_1$ in Fig. C-1(d).

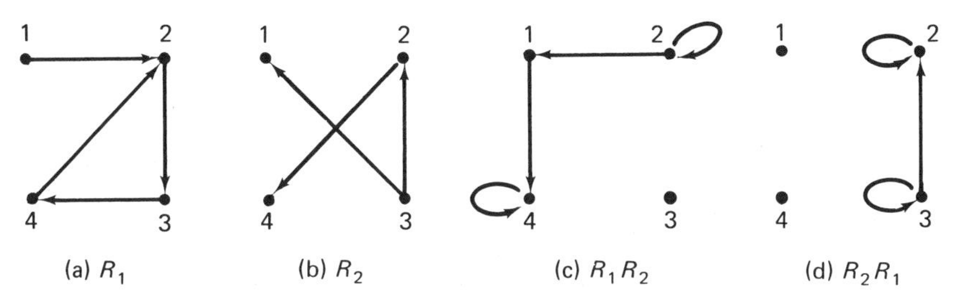

(a) R_1 (b) R_2 (c) $R_1 R_2$ (d) $R_2 R_1$

Fig. C-1 Composition of binary relations.

One easily verifies that the composition of three relations R_1, R_2, R_3, whenever defined, is associative; i.e., $(R_1R_2)R_3 = R_1(R_2R_3)$. Furthermore, $(R_1R_2)^{-1} = R_2^{-1}R_1^{-1}$.

For any binary relation R on X, let $R^1 \triangleq R$, and $R^{n+1} \triangleq RR^n$, for $n \geq 1$. Evidently, R is transitive iff $R^2 \subseteq R$. If R is both reflexive and transitive, then $R^2 = R$. The *transitive closure* R^+ of R is defined to be

$$R^+ \triangleq \{(x, y) \mid xR^hy \text{ for some positive integer } h\}.$$

R^+ is the smallest transitive relation that includes R; i.e., we have:

Proposition 1

Let R be a relation on the set X. Then R^+ is the unique relation on X satisfying the following conditions:
(a) $R \subseteq R^+$.
(b) R^+ is transitive.
(c) If $T \subseteq X^2$, $T \supseteq R$, and $T^2 \subseteq T$ (i.e., T is transitive), then $T \supseteq R^+$.
The proof is left to the reader.

Figure C-2 shows R_1^+ and R_2^+ for the relations R_1 and R_2 of Fig. C-1. They can be constructed by noting that xR^+y iff in the diagram of R there exists a directed path from x to y.

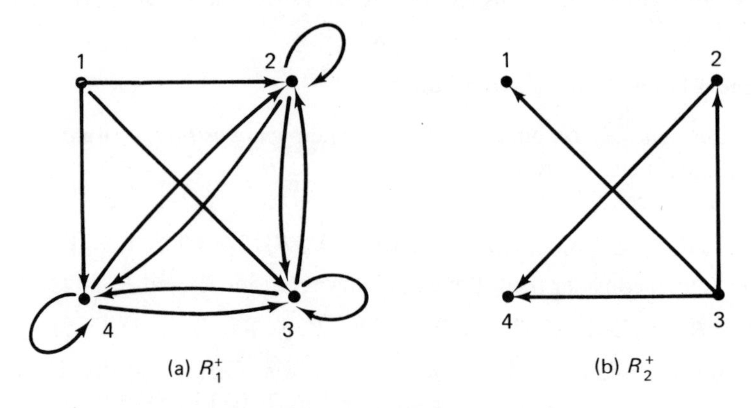

Fig. C-2 Transitive closures of R_1 and R_2 (Fig. C-1).

If X is a finite set, we denote by $\#X$ the number of elements in X. For finite sets X, R^+ can be obtained from R as shown in the following proposition.

Proposition 2

Let X be a finite set and R a binary relation on X. Then
(a) $R^+ = R \cup R^2 \cup \cdots \cup R^n$, where $n \triangleq \#X$.
(b) If R is reflexive, and $n > 1$, then $R^+ = R^{n-1}$.

Proof

Let $T \triangleq R \cup R^2 \cup \cdots \cup R^n$, and suppose that xR^+y, i.e., xR^hy, for some positive integer h. Assume h to be the least positive integer such that xR^hy. Then there exist $x_0, x_1, \ldots, x_h$ in X such that $x_0 = x$, $x_h = y$, and

$$x_0Rx_1, x_1Rx_2, \ldots, x_{h-1}Rx_h. \qquad (*)$$

The sequence $x_1, \ldots, x_{h-1}$ cannot contain any repetition. For if $x_i = x_j$, with $j > i$, then the chain

$$x_0Rx_1, \ldots, x_{i-1}Rx_i, x_jRx_{j+1}, \ldots, x_{h-1}Rx_h$$

would show that h was not minimal. Now suppose that the sequence $x_0, x_1, \ldots, x_{h-1}$ contains a repetition. This is possible only if $x_0 = x_j$ for some j, $0 < j \leq h - 1$. But then the sequence $x_jRx_{j+1}, \ldots, x_{h-1}Rx_h$ contradicts the fact that h is minimal. A similar argument shows that the sequence $x_1, \ldots, x_h$ contains no repetitions. Since $\#X = n$, we must have $h \leq n$. Hence $(x, y) \in T$. Consequently, $R^+ \subseteq T$. But clearly, $T \subseteq R^+$. Thus $R^+ = T$, proving part (a).

Assume now that R is reflexive, and that xR^hy, where $h > 1$ is minimal. If $h = n$, we must have $x_0 = x_h$, since the sequence

$$x_0, x_1, \ldots, x_h,$$

which is of length $h + 1 = n + 1$, can have no other repetition. But then x_0Rx_h, contradicting the minimality assumption about h. Thus

$$R^+ = R \cup R^2 \cup \cdots \cup R^{n-1}.$$

Since R is reflexive, the chain (*) can always be extended by x_hRx_h. Thus $R^h \subseteq R^{h+1}$ for every $h \geq 1$. It follows that $R^+ = R^{n-1}$. $\square$

The *reflexive* closure of a relation R on X is the relation $R \cup I_X$. It is the smallest reflexive relation that includes R.

The *reflexive-and-transitive* closure of $R \subseteq X^2$ is the relation $R^* = R^+ \cup I_X$. One easily sees that $R^* = (R \cup I_X)^+$. R^* is the smallest reflexive and transitive relation that includes R. We leave it as an exercise to the reader to verify that

$$R^+ = (R^+)^+ = RR^* = R^*R$$

and

$$R^* = (R^*)^* = (R^*)^2.$$

D MODELS OF STATIC MOS GATES

We have been treating gates as basic black boxes, specified by their input–output behavior, without going into details of their composition. In this appendix we introduce more fundamental building blocks, corresponding essentially to resistors and transistors and show how gates and gate networks may be composed by means of these building blocks. In particular, we explain how static MOS and COS/MOS gates are actually constructed from simpler elements (which we call T-elements and $\bar{\text{T}}$-elements).

D.1. Contact-Resistor Networks

As a preliminary step we discuss contact-resistor networks. Consider the network of Fig. D-1, consisting of a two-terminal contact network N, in

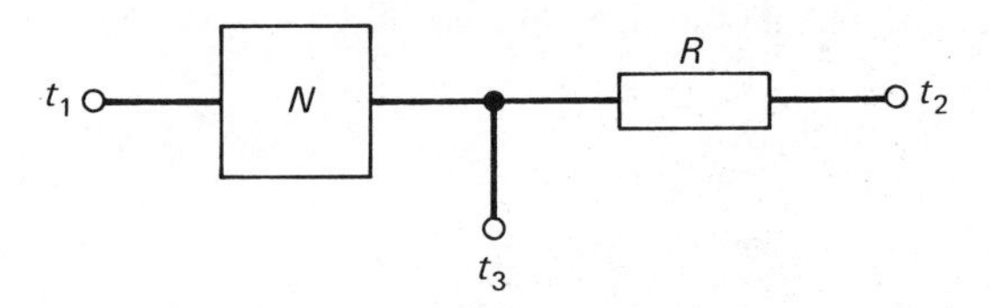

Fig. D-1 The basic contact-resistor network.

series with a resistor R. Assume that binary signals (0 or 1), represented by either the positive or the negative logic convention, are connected to the terminals t_1 and t_2. Electric circuit considerations enable us to determine the binary signal appearing at t_3 as follows. Let f be the output function of N and $\hat{t}_i$ the binary signal at terminal t_i.

Then

$$f = 1 \text{ implies } \hat{t}_3 = \hat{t}_1$$

and (1)

$$f = 0 \text{ implies } \hat{t}_3 = \hat{t}_2.$$

We may consider Fig. D-1 as a mathematical model, the behavior of which is specified by (1).

Let us now consider the contact-resistor network of Fig. D-2. We denote

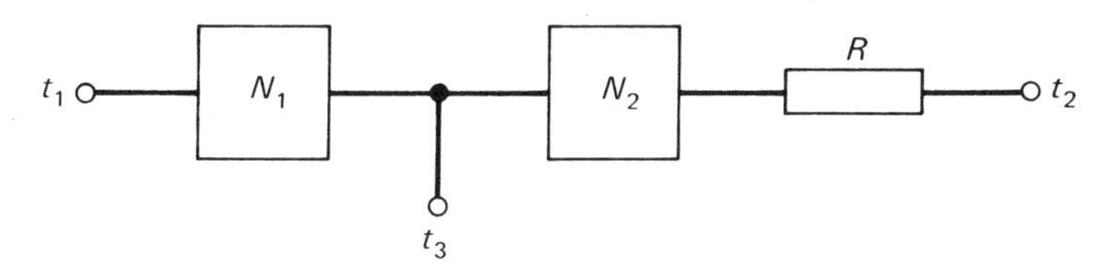

Fig. D-2 Network with a third state.

by f_i the output function of N_i, $i \in \{1, 2\}$. If, for some input combination, $f_1 = 1$, then $\hat{t}_3 = \hat{t}_1$. If $f_1 = 0$ and $f_2 = 1$, then $\hat{t}_3 = \hat{t}_2$. If, however, $f_1 = f_2 = 0$, then the signal $\hat{t}_3$ is neither 0 nor 1. We shall refer to this situation as the *third state*. If $f_1 + f_2 = 1$ for every input combination, then the case $f_1 = f_2 = 0$ cannot occur; i.e., the possibility of t_3 being in the third state is eliminated. This is the case, for example, if $f_1 = x_1' + x_2'$ and $f_2 = x_1 + x_2$. Hence, in the contact-resistor network of Fig. D-3, both outputs z_1 and z_2 are well specified. Indeed, we have $z_1 = 0 \equiv (x_1' + x_2' = 1)$; hence $z_1 = (x_1' + x_2')' = x_1 \cdot x_2$, and $z_2 = [(x_1' + x_2') \cdot (x_1 + x_2)]' = x_1 \cdot x_2 + x_1' \cdot x_2'$.

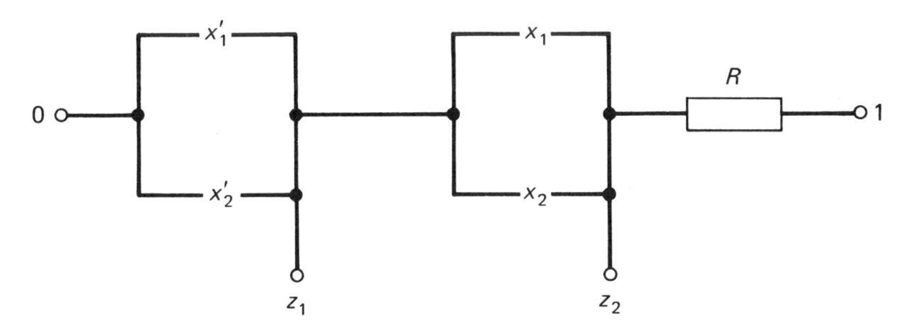

Fig. D-3 Example of a two-output contact-resistor network.

D.2. T-ELEMENT

We now introduce an abstract switching element, shown symbolically in Fig. D-4(a). (A similar symbol is introduced in [GSC] to discuss resistor-transistor logic.) The input x is a binary variable represented by voltage levels. We shall refer to this element as a *T-element*. Under suitable conditions the

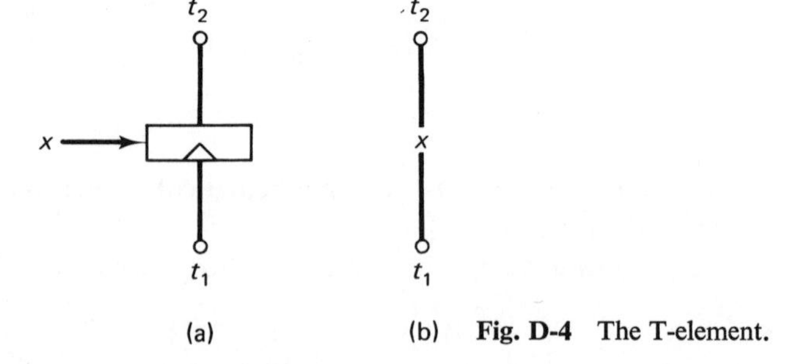

Fig. D-4 The T-element.

T-element of Fig. D-4(a) will perform in the same way as the x contact shown in Fig. D-4(b). By means of Fig. D-5 we specify conditions that are sufficient for this to be the case. Namely, (a) all the T-elements form a two-terminal series–parallel network M, connected as shown in Fig. D-5(a), and (b) all the T-elements of M are oriented as indicated in Fig. D-5(b).

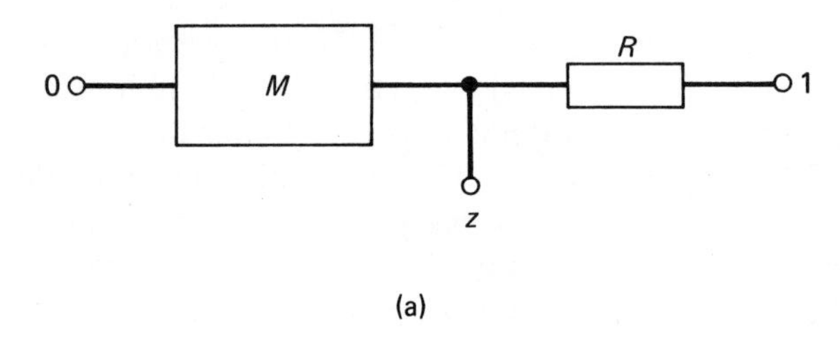

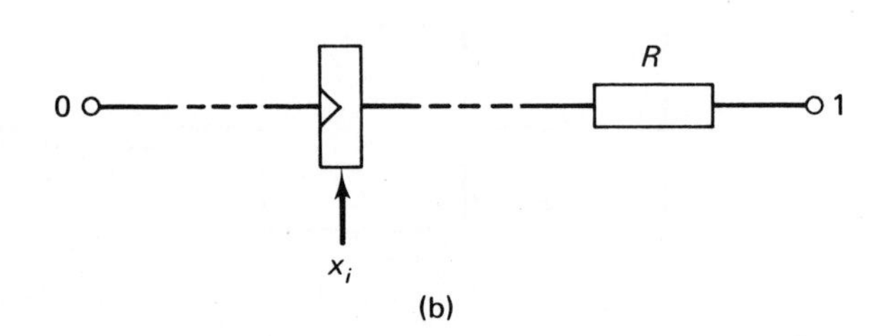

Fig. D-5 Valid connections of T-elements.

T-elements may be implemented, for example, by N-channel MOS transistors, and the signals 0 and 1 then correspond to ground and a positive voltage, respectively. Although our model is capable of explaining the fundamental logical properties of static MOS gates, it is an idealized model. In particular, what we represent by a short circuit and an open circuit in the T-element of Fig. D-4(a) is more closely modeled by a low resistance and a high resistance, respectively. Also the resistor R of Fig. D-5 is usually replaced by an MOS transistor. However, since we are not concerned with the actual electrical properties, our simplifying assumptions are justified.

D.3. STATIC MOS GATES

Figure D-6 shows examples of networks that satisfy conditions (a) and (b) above. If each T-element is replaced by the corresponding contact, con-

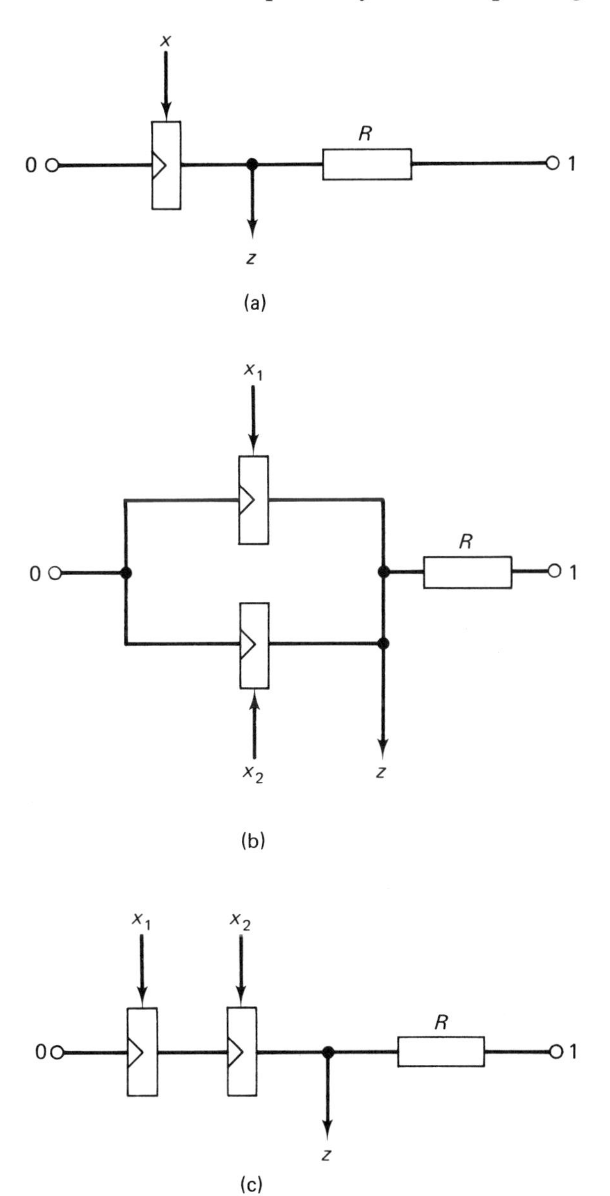

Fig. D-6 MOS realization of basic gates: (a) inverter; (b) NOR; (c) NAND.

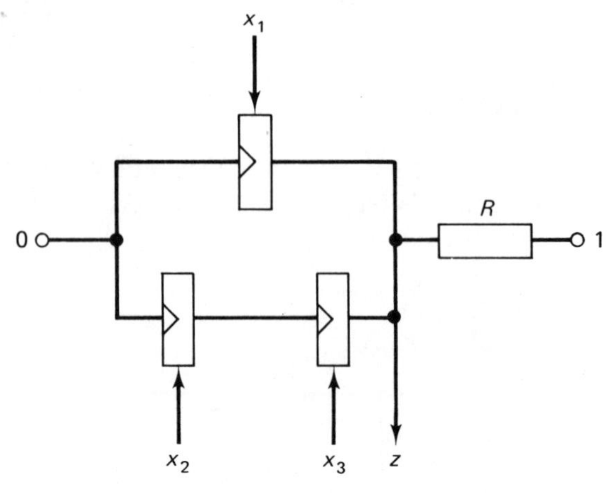

Fig. D-7 Network for $x'_1 \cdot (x'_2 + x'_3)$.

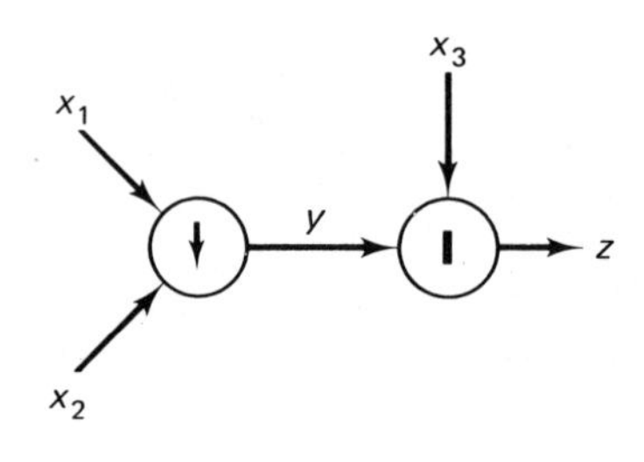

(a)

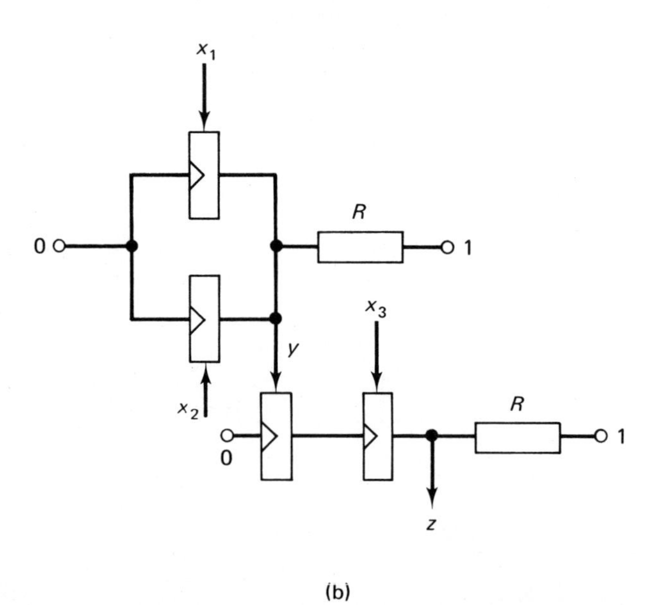

(b)

Fig. D-8 Interconnections of gates.

tact-resistor networks of the type shown in Fig. D-1 are obtained, and rule (1) becomes applicable. It follows that Fig. D-6(a) represents an inverter ($z = x'$), Fig. D-6(b) a NOR gate ($z = x_1 \downarrow x_2$), and Fig. D-6(c) a NAND gate ($z = x_1 | x_2$). For the network of Fig. D-7, which may be analyzed similarly, we obtain

$$z = 0 \equiv x_1 + x_2 \cdot x_3 = 1$$

or

$$z = (x_1 + x_2 \cdot x_3)' = x_1' \cdot (x_2' + x_3').$$

Gates of the type shown in Fig. D-6 may be interconnected as usual. Thus the gate network of Fig. D-8(a) may be implemented as shown in Fig. D-8(b).

In Fig. D-9 we indicate somewhat more general conditions under which T-elements can be replaced by their corresponding contacts. Here, M_1 and M_2 are two-terminal, series–parallel networks of T-elements, and all the T-elements are to be oriented as indicated in Fig. D-5(b).

Figure D-10 shows a T-element network that is equivalent to the contact-resistor network of Fig. D-3. The network of Fig. D-10 corresponds to

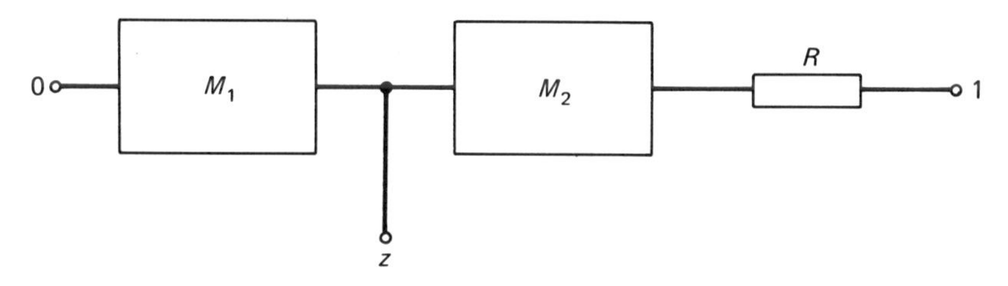

Fig. D-9 More general network of T-elements.

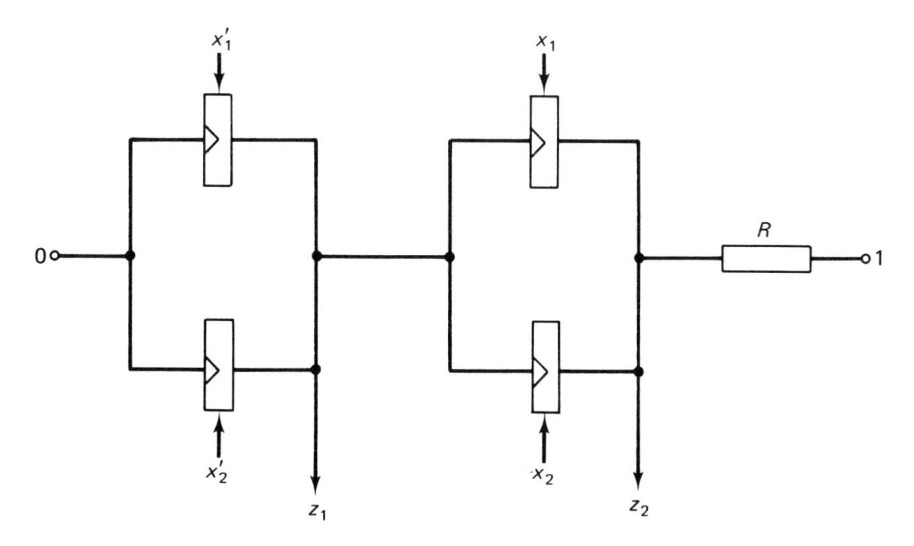

Fig. D-10 Two-output T-element network.

Fig. D-9, as far as the output z_1 is concerned, and to Fig. D-5 as far as z_2 is concerned.

Figure D-11 illustrates a useful application of the third state discussed earlier. In Fig. D-11, z_1 will be in the third state if $e = 0$, and z_2 will be in this state if $e = 1$. Hence the outputs z_1 and z_2 may be tied together (i.e., wired-OR connected). The resulting output will equal z_1 if $e = 1$, and it will equal z_2 if $e = 0$. This idea is easily extended to the wired-OR connection of more than two networks, provided that always exactly one of the networks is enabled and the outputs of all networks that are not enabled are in the third state.

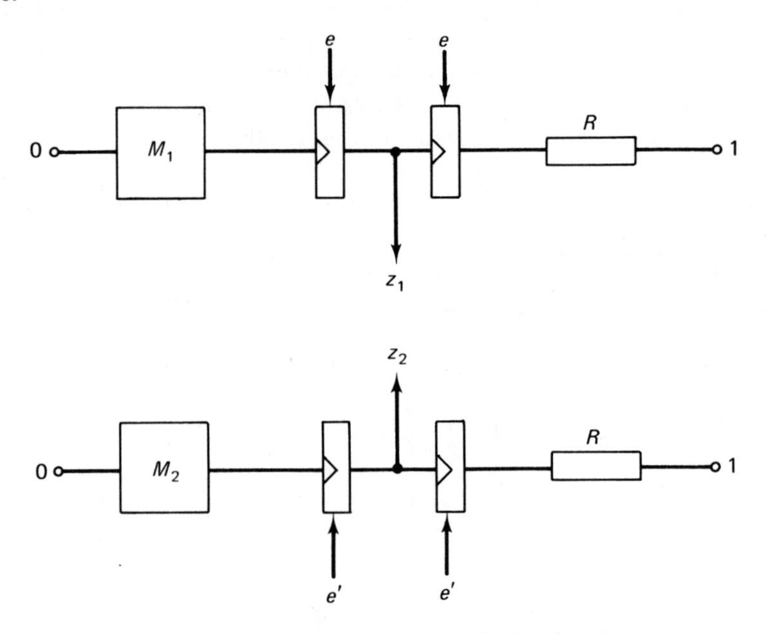

Fig. D-11 Principle of the "wired-OR" connection.

D.4. COS/MOS GATES

For the purpose of modeling COS/MOS gates (see Section 2.2), we introduce another switching element, the $\bar{T}$-element, which may be considered the "dual" of the T-element. We use the symbol of Fig. D-12(a) to denote a $\bar{T}$-element, which performs in the same way as the x' contact of Fig. D-12(b) provided that suitable conditions are met.

In what follows we shall be concerned with networks having the configuration of Fig. D-13(a). Here, M_1 and M_2 are series–parallel networks of T-elements and $\bar{T}$-elements, respectively. All the T-elements of M_1 are connected as shown in Fig. D-13(b), and all the $\bar{T}$-elements of M_2 are connected as shown in Fig. D-13(c). In such networks the T-elements and the $\bar{T}$-elements may be replaced by their corresponding contacts. $\bar{T}$-elements may be imple-

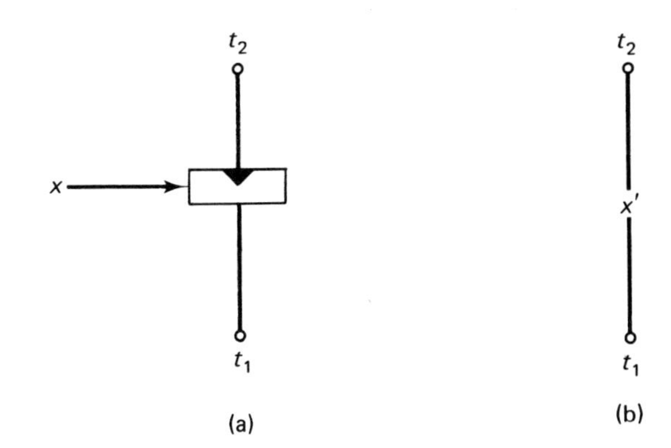

Fig. D-12 The $\bar{\text{T}}$-element.

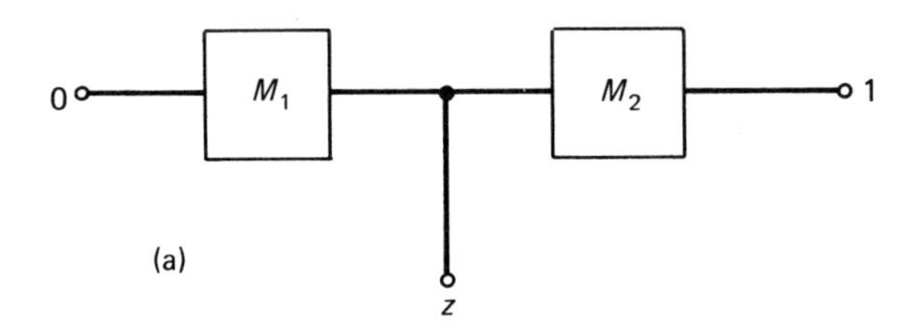

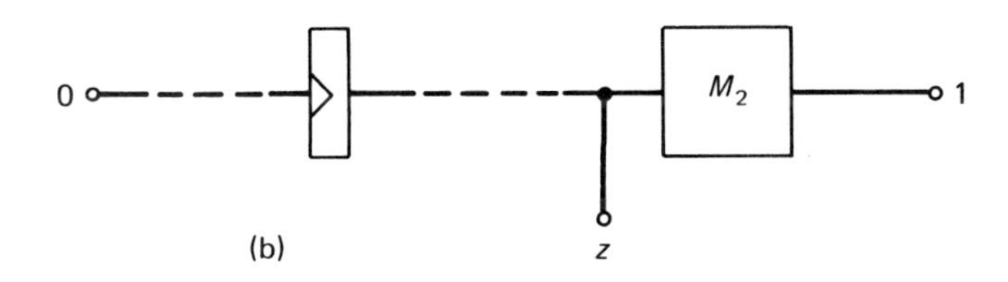

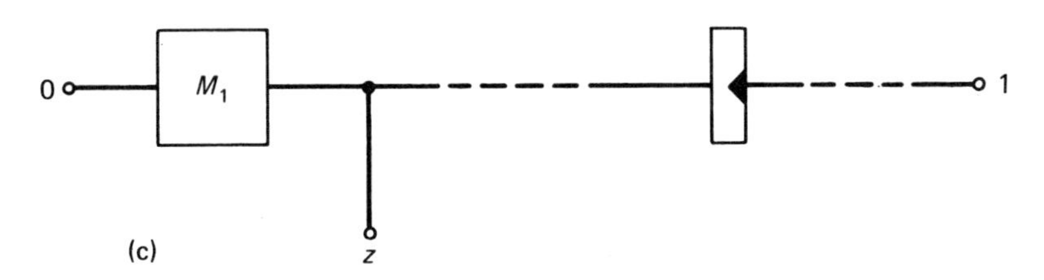

Figure D-13

mented by *P*-channel MOS transistors provided that the positive logic convention is adopted. COS/MOS gates have the configuration shown in Fig. D-13. They avoid the necessity of resistors, yielding higher speed and lower power dissipation.

Let f_i be the output function of the contact network equivalent to M_i

($i = 1, 2$) in Fig. D-13(a). If $f_1 = f_2 = 0$, z is neither 0 nor 1; i.e., z is in the third state. If $f_1 = f_2 = 1$, the output terminal is connected to both 0 and 1, a situation that has to be avoided. Hence we must always have $f_1 \cdot f_2 = 0$. If the third output state is also to be avoided, we must also have $f_1 + f_2 = 1$. The two conditions $f_1 \cdot f_2 = 0$ and $f_1 + f_2 = 1$ are equivalent to $f_2 = f'_1$. Thus the condition $f_2 = f'_1$ is necessary and sufficient for the output z of Fig. D-13(a) to be always uniquely specified.

Figure D-14 shows three COS/MOS gates having the configuration of Fig. D-13 and satisfying the condition $f_2 = f'_1$. Figure D-14(a) shows a COS/MOS inverter, Fig. D-14(b) a COS/MOS NOR gate, and Fig. D-14(c) a COS/MOS NAND gate.

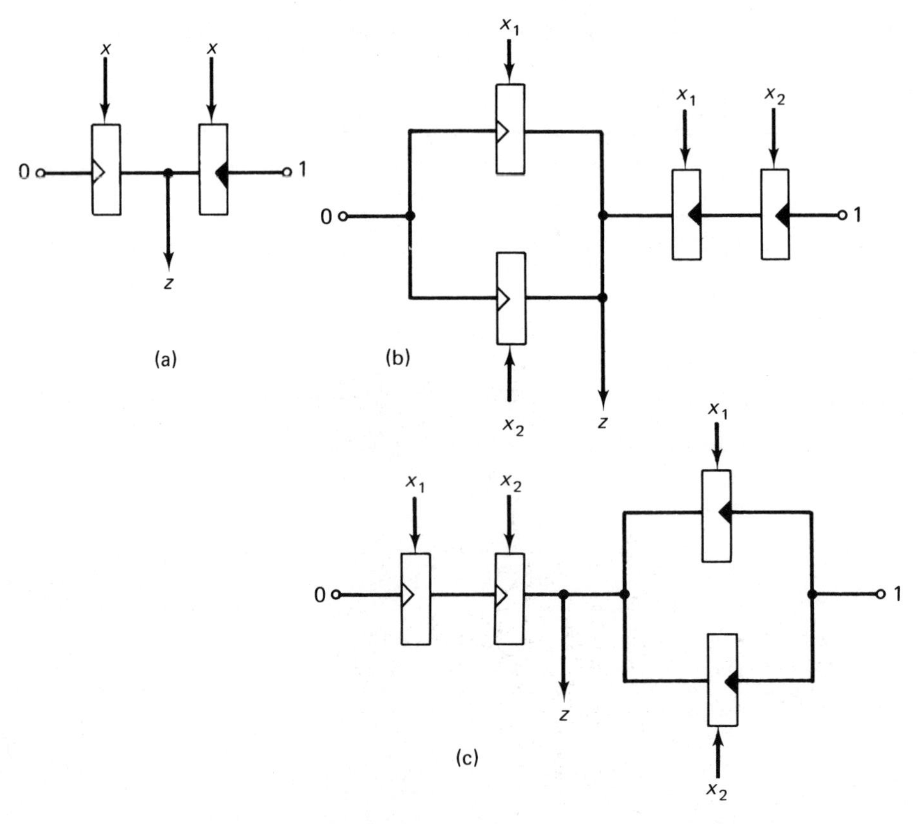

Fig. D-14 COS/MOS gates: (a) inverter; (b) NOR; (c) NAND.

D.5. MOS LATCHES AND FLIP-FLOPS

The MOS networks discussed so far may be used to design latches and flip-flops. For example, the NOR latch of Fig. 7-1(a) (p. 195) may be implemented by means of the MOS NOR gate of Fig. D-6(b). The resulting MOS NOR latch is shown in Fig. D-15.

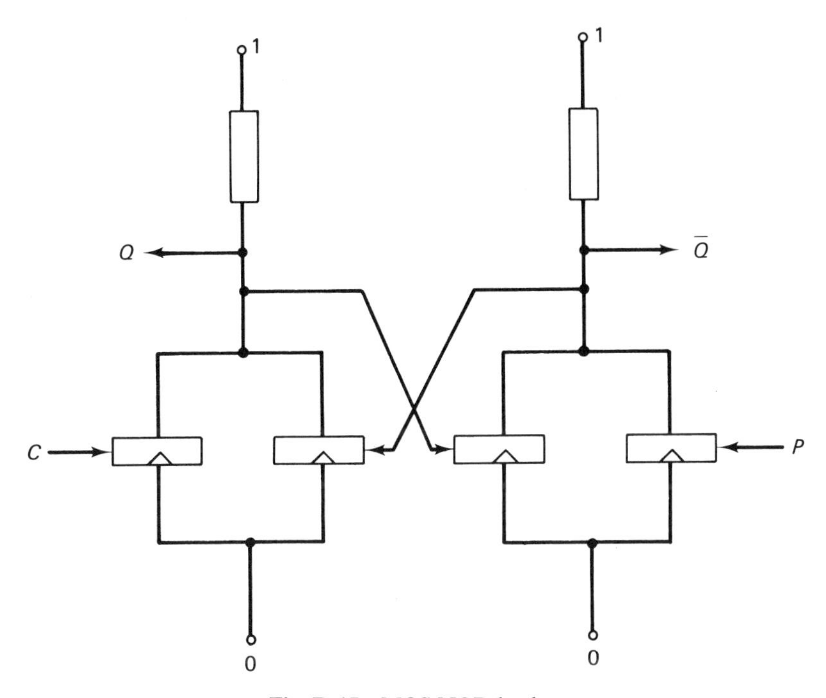

Fig. D-15 MOS NOR latch.

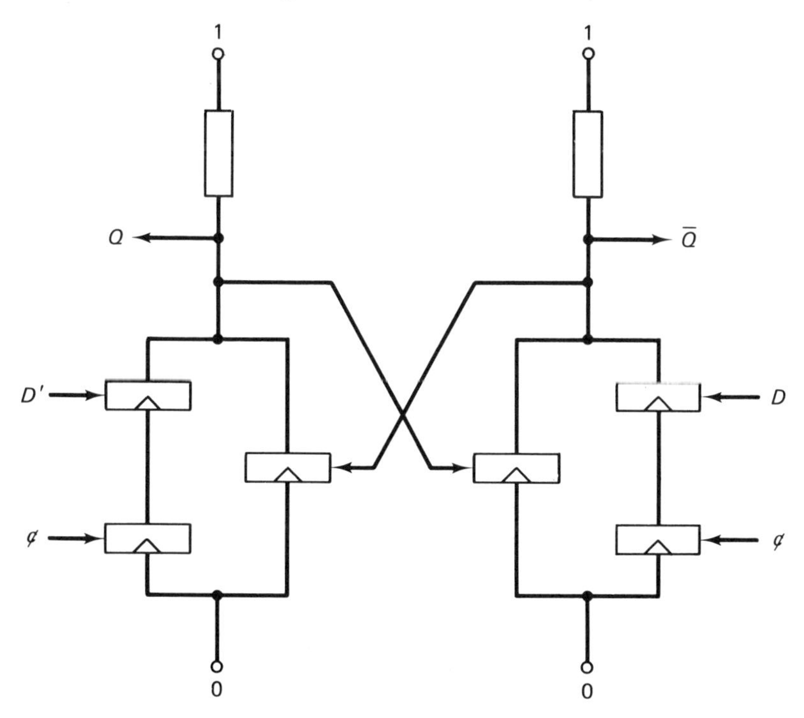

Fig. D-16 D flip-flop.

Figure D-16 shows an MOS network that is easily shown to be functionally equivalent to the basic D flip-flop of Fig. 7-11(a) (p. 203).

Other MOS latches and flip-flops are easily designed in a similar way. For further reading on MOS gate networks, including the important area of "dynamic" MOS logic, we suggest [AMS, CA-MI].

REFERENCES

[AMS] American Micro-Systems Staff, *MOS Integrated Circuits.* New York: Van Nostrand Reinhold Company, 1972.

[BEO] BEOUGHER, L. C., "A Method for High Speed BCD-to-Binary Conversion." *Computer Design*, Vol. 12, March 1973, pp. 53–59.

[BRE] BREUER, M. A., "Logic Synthesis" *in* M. A. Breuer, ed., *Design Automation of Digital Systems.* Englewood Cliffs, N.J.: Prentice-Hall, Inc., 1972, pp. 21–100.

[BR-YO1] BRZOZOWSKI, J. A., and M. YOELI, "Models for Analysis of Races in Sequential Networks," in A. Blikle, ed., *Mathematical Foundations of Computer Science.* (Lecture Notes in Computer Science, Vol. 28.) Berlin: Springer-Verlag, 1975, pp. 26–32.

[BR-YO2] BRZOZOWSKI, J. A., and M. YOELI, "A Practical Approach to Asynchronous Gate Networks," Proc. IEE, Vol. 123, June 1976, pp. 495–498.

[CA-MI] CARR, W. N., and J. P. MIZE, *MOS/LSI Design and Application.* New York: McGraw-Hill Book Company, 1972.

[CH-LE] CHANG, C.-L., and C.-T. LEE, *Symbolic Logic and Mechanical Theorem Proving.* New York: Academic Press, 1973.

[COE] COERS, G., "Preset Generator Produces Desired Number of Pulses." *Electronics*, July 3, 1972, p. 88.

[EIC] EICHELBERGER, E. B., "Hazard Detection in Combinational and Sequential Switching Circuits." *IBM Journal*, Vol. 9, March 1965, pp. 90–99.

[FAR] FARLEY, M. F., "Digital Approach Provides Precise, Programmable AGC." *Electronics*, August 30, 1971, pp. 52–56.

[FR-ME] FRIEDMAN, A. D., and P. R. MENON, *Fault Detection in Digital Circuits.* Englewood Cliffs, N.J.: Prentice-Hall, Inc., 1971.

[GAR] Garrett, L. S., "Integrated Circuit Digital Logic Families." *IEEE Spectrum*, Vol. 7, October 1970, pp. 46–58; November 1970, pp. 63–72; December 1970, pp. 30–42.

[GOL] Golomb, S. W., *Shift Register Sequences.* San Francisco: Holden-Day, Inc., 1967.

[GSC] Gschwind, H. W., *Design of Digital Computers.* New York: Springer-Verlag, 1967.

[HAM] Hamming, R. W., "Error-Detecting and Error-Correcting Codes." *Bell System Tech. Journal*, Vol. 29, 1950, pp. 147–160.

[HAR] Harrison, M. A., *Introduction to Switching and Automata Theory.* New York: McGraw-Hill Book Company, 1965.

[HUN] Huntington, E. V., "Sets of Independent Postulates for the Algebra of Logic." *Trans. Amer. Math. Soc.*, Vol. 5, 1904, pp. 288–309.

[INT] Intel Staff, *Intel Data Catalog.* Santa Clara, Ca. Intel Corporation, 1973.

[KAU] Kautz, W. H., "The Necessity of Closed Circuit Loops in Minimal Combinational Circuits." *IEEE Trans. Computers*, Vol. C-19, 1970, pp. 162–166.

[MCC] McCluskey, E. J., *Introduction to the Theory of Switching Circuits.* New York: McGraw-Hill Book Company, 1965.

[MO-MI] Morris, R. L., and J. R. Miller, eds., *Designing with TTL Integrated Circuits.* New York: McGraw-Hill Book Company, 1971.

[MON] Monolithic Memories Staff, "64 Bit TTL Random Access Memory," Sunnyvale, Ca., 1970.

[MOT1] Motorola Application Note, "BCD-to-Binary/Binary-to-BCD Number Converter MC-4001P." Phoenix, Arizona: Motorola Semiconductor Products, Inc., August 1969.

[MOT2] Motorola Application Note AN-537, "The MC4023, an MTTL 4-Bit Universal Counter." Phoenix, Arizona: Motorola Semiconductor Products, Inc.

[MOT3] Motorola Staff, *MECL 10,000.* Phoenix, Arizona: Motorola, Inc., 1971.

[MUR] Muroga, S., *Threshold Logic and Its Applications.* New York: John Wiley & Sons, Inc., 1971.

[MU-IB] Muroga, S., and T. Ibaraki, "Design of Optimal Switching Networks by Integer Programming." *IEEE Trans. Computers*, Vol. C-21, June 1972, pp. 573–582.

[NIL] Nilsson, N. J., *Learning Machines: Foundations of Trainable Pattern-Classifying Systems.* New York: McGraw-Hill Book Company, 1965.

[PA-DE] Patil, S. S., and J. B. Dennis, "The Description and Realization of Digital Systems." *Revue Française d'Automatique, Informatique et de Recherche Operationelle*, February 1973, pp. 55–69.

[PEA] PEATMAN, J. B., *The Design of Digital Systems*. New York: McGraw-Hill Book Company, 1972.

[PHI] Philips Application Book, *Control System Design for 60-Series Norbits*. Eindhoven, Netherlands: Philips, 1968.

[QUI1] QUINE, W. V., "On Cores and Prime Implicants of Truth Functions." *Amer. Math. Monthly*, Vol. 66, 1959, pp. 755–760.

[QUI2] QUINE, W. V., "A Way to Simplify Truth Functions." *Amer. Math. Monthly*, Vol. 62, 1955, pp. 627–631.

[RCA] RCA Staff, *COS/MOS Digital Integrated Circuits*. Sommerville, N.J.: RCA Corporation, 1972.

[RE-CO] RENWICK, W., and A. J. COLE, *Digital Storage Systems*, 2nd ed. London: Chapman & Hall Ltd., 1971.

[SCH] SCHNORF, A., "Das Elektronische Bausteinsystem." *Albiswerk-Berichte*, Vol. 18, May 1966, pp. 36–42.

[SE-HS-BE] SELLERS, F. F., M. Y. HSIAO, and L. W. BEARNSON, *Error Detecting Logic for Digital Computers*. New York: McGraw-Hill Book Company, 1968.

[SIG] Signetics Staff, *Digital Bipolar Circuits*. Sunnyvale, Ca: Signetics Corporation, 1972.

[TI1] Texas Instruments Staff, *The Integrated Circuits Catalog*. Dallas, Texas: Texas Instruments, Inc.

[TI2] Texas Instruments Staff, *The TTL Data Book for Design Engineers*. Dallas, Texas: Texas Instruments, Inc., 1973. See also: *Supplement to the TTL Data Book for Design Engineers*, 1974.

[YOE] YOELI, M., "Counting with Nonlinear Binary Feedback Shift Registers." *IEEE Trans. Electronic Computers*, Vol. EC-21, 1963, pp. 357–361.

[YO-RI] YOELI, M., and S. RINON, "Application of Ternary Algebra to the Study of Static Hazards." *J. Assoc. Comp. Mach.*, Vol. 11, January 1964, pp. 84–97.

[illegible] The Design of [illegible] [illegible] New York: McGraw-Hill [illegible]

[illegible]

[illegible] Vol. 6, 1958, pp. [illegible]

[illegible] Vol. 12, [illegible] pp. 627-631.

[illegible]

[illegible] Prentice-Hall [illegible]

[illegible]

[illegible]

[illegible]

[illegible]

[illegible]

[illegible]

[illegible]

INDEX

B

C

D

E

F

G

M

N

T